Atlantis Advances in Quaternary Science

Volume 2

Series editor

Colm O'Cofaigh, Department of Geography, Durham University, Durham, UK

The aim of the Atlantis book series 'Advances in Quaternary Science' is to bring together texts in the broad field of Quaternary Science that highlight recent research advances on aspects of glaciation and sea level change, the development and application of Quaternary geochronological methods, records of climate change from marine and terrestrial settings, geomorphology and landscape evolution and regionally-focused reviews of Quaternary environmental change. The series comprises monographs and edited volumes that require extensive illustration and substantial space, and which provide state of the art thematic and regional reviews on Quaternary related topics often focusing on processes and associated responses within the fields of geology, geomorphology, glaciology, geochronology and palaeo-biology. In the last two decades technological developments in dating methods, remote sensing and techniques for the analysis and interpretation of sedimentary and climatic archives have resulted in significant advances of climate and ocean change across a range of time-scales from annual to millennial. Publications in the Atlantis book series 'Advances in Quaternary Science' capture these developments and show how they have increased understanding of Pleistocene to Holocene climate, cryosphere and ocean change across a range of spatial and temporal scales.

More information about this series at http://www.springer.com/series/15358

Helgi Björnsson

The Glaciers of Iceland

A Historical, Cultural and Scientific Overview

Julian Meldon D'Arcy, University of Iceland:
English translation

Helgi Björnsson
School of Engineering and Natural Science,
    Institute of Earth Sciences
University of Iceland
Reykjavík
Iceland

Atlantis Advances in Quaternary Science
ISBN 978-94-6239-206-9   ISBN 978-94-6239-207-6   (eBook)
DOI 10.2991/978-94-6239-207-6

Library of Congress Control Number: 2016947792

Printed on acid-free paper

The publication of this book was supported by Pálmi Jónsson's Nature Conservation Fund, Náttúruverndarsjóður Pálma Jónssonar.

NÁTTÚRUVERNDARSJÓÐUR
PÁLMA JÓNSSONAR
STOFNANDA HAGKAUPS

The translation of the book was supported by the Icelandic Literature Center.

MIÐSTÖÐ ÍSLENSKRA BÓKMENNTA
ICELANDIC LITERATURE CENTER

I dedicate this book to my wife, Þóra Ellen Þórhallsdóttir, and my children Valgerður, Þórhallur, Ásdís, Björn and Svanhildur

Foreword

Glaciers are an essential part of the Icelandic identity. They hover in their beauty and magnificence on the horizons far and wide around the country, whatever their names: Snæfellsjökull, Öræfajökull, Eyjafjallajökull. These white giants play such an important role in creating Icelanders' self-image wherever they are found, as they majestically reign over the landscape, extending over 10% of Iceland. All true Icelanders are fascinated by glaciers from childhood, and many of those who live in the area of our capital on the southwest coast begin the day by looking out across the bay towards the glacier Snæfellsjökull, the old beacon for seamen on perilous fishing grounds. For those in the west of the country it is our own mystic mountain, our Fujiyama.

The quest for knowledge of these white giants, which serve as guides to where we are located in life, commenced here earlier than elsewhere in the world, for Iceland is truly a land of glaciers. Scientists and daring pioneers, both Icelanders and other foreign nationals, were determined to learn all about these glaciers at whatever the cost—for conditions were so harsh that their work was all but impossible in the early days. Accounts of what they achieved and endured sound incredible: their packhorses inch their way forward over slippery ice at the edge of a precipice, before tumbling into crevasses, where they wait patiently until they are hoisted to safety—or until they expire.

In an account of one of the first crossings of the great Vatnajökull glacier in the first years of the twentieth century, two Scottish pioneers were the first to record their impressions of the indefinable beauty of the white frozen wilderness, as they lay in a blizzard, imprisoned in their tent, reading Cervantes' *Don Quixote* and other masterpieces of world literature.

We owe a huge debt of gratitude to the scientists who have explored our glaciers, both past and present. They have brought to us vital knowledge about these white giants which will provide essential testimony in the great forthcoming trial concerning the potentially catastrophic case of global warming.

Glaciologist Helgi Björnsson is certainly a first among equals in the science of observing the life of our glaciers. His book, the fruit of four decades' work, is a unique testament to the history of Iceland's glaciers over countless centuries, and to their destiny. *The Glaciers of Iceland* is now finally available in English translation at a crucial time for the future of the world. It will be an important reference book on the beauty and magnificence of these phenomena of nature, which will not remain intact for ever, as we had so optimistically believed only a few years ago.

The glaciers of the world and of Iceland are no longer symbols of permanence and eternity—instead they remind us that everything in the world is transitory, even glaciers themselves—where beauty reigns alone, as our great writer Laxness expressed it in *World Light*. Glaciers are most certainly beacons of light in the world, and it is our profound wish that we may continue to live in their light, and not in the shadows.

Vigdís Finnbogadóttir
Former President of Iceland

Preface

Wherever one looks in Iceland, its landscape bears witness to the impact of ice and fire. The terrain appears to be moulded either by a glacier that covered Iceland 18,000 years ago, or by lava which had flowed after the glacier had thawed. Long after glaciers had disappeared, extant landforms indicated their previous existence, for they have chiselled bedrock, scooped out corries, shattered cliff faces, and left behind massive and jagged sculptures and sharp mountain pinnacles. Glaciers have gouged deep and narrow fjords far out into Iceland's continental shelf, and hollowed out valley floors and troughs that are now full of lakes. Taking a closer look at the landscape, more refined pieces of evidence can be seen: striated rocks, serrated crags, polished whalebacks and erratic boulders. Glacial rivers have harrowed out ravines, often during catastrophic floods, and discharged sediments over outwash plains. In many places there are visible signs of volcanic eruptions beneath glaciers. Palagonite ridges rise above volcanic fissures and precipitous table mountains, some of the most magnificent in the country, tower high above their surroundings, bearing witness to the thickness of an ice-age glacier. Glacial moraines illustrate the power of previously advancing outlet glaciers, and both fresh-water lakes and the ocean have sediment strata which have been borne and dispersed there by glacial rivers. Iceland's flora still reveals signs of the vegetation which had been destroyed in a glacial age.

Glaciers now cover only about 10 % of Iceland and are retreating rapidly. Ancient glacial plains and valleys have become the country's most fertile agricultural areas. But there are signs of life left in the glaciers, nonetheless, they still sometimes expand and even surge forward, responding quickly to changes in the climate. The greatest rivers of the country flow from them into power stations, groundwater systems and the ocean, and they also provide the greatest storage facility for our fresh water. Glacial rivers continue to need bridging or containing with defensive levees. Huge outburst floods (jökulhlaups) still rush from proglacial lakes in geothermal areas and from subglacial volcanic eruptions.

By researching present-day glaciers we can discover the basic laws of their formation and behaviour and their relationship to climate change. Here, as in other geophysical sciences, the key to the future can be found in the past. Questions may be asked about glaciers as to when and how they originated, how large were they during the settlement of Iceland, and how did they thrive while the nation endured a long-term cold period? A knowledge of present-day glaciers is no less a key to the future. Will glaciers be able to grow again and advance in the coming years, or will they shrink and retreat so much that glacial rivers will run dry? What effect will that have on hydroelectric power stations, groundwater systems, and supplies of drinking water? What would the hitherto hidden mountains and valleys look like, should the glaciers that cover them disappear?

The aim of this book is to record a history of the knowledge and understanding of the origins, habitats and behaviour of glaciers and how we evaluate their role in nature. The first part traces this history from the first settlement of Iceland in the ninth century right up until modern science has revealed the island's hidden, subglacial terrain. It also reveals how research into remnants of ancient glaciers has made mankind realise how Earth's climate is in a constant state of fluctuation. The second part contains a detailed study of all of Iceland's major glaciers as they now are in the beginning of the twenty-first century.

In writing this book I have used a wide variety of historical and scientific sources, from the Sagas of the Icelanders to recent academic research, from pencil drawings to computer-generated and satellite images. I have tried to produce a text that, while avoiding an overuse of scientific discourse, can nonetheless present precise and valid explanations of glaciological phenomena and data in a lucid manner accessible to the general reader and geoscientists alike.

Glaciers are now rapidly receding all over the world and the surface of the ocean is rising and threatening our coastlines and Earth's hydrologic cycles, while global warming is stimulating increasingly volatile climate changes. The questions and answers relating to glaciers are thus of vital relevance to all of mankind for the foreseeable future.

Reykjavík, Iceland   Helgi Björnsson

Translator's Note

In keeping with the spirit of Helgi Björnsson's multidisciplinary approach to the scientific, cultural and historical content of *The Glaciers of Iceland*, I have attempted to provide an English translation which, while retaining formal elements of glaciological and scholarly discourse, will hopefully remain accessible, informative, interesting and enjoyable to the general reader. All translations of Björnsson's Icelandic sources are mine, unless otherwise stated in the reference sections. Sources in English are, of course, quoted from the original texts.

I have followed the common practice of translating all Icelandic personal and place-names in their original Icelandic spelling in the nominative case, except in instances where a place-name (on maps) is specifically declined, e.g. the river Jökulsá á Fjöllum. Although I sometimes attempt to indicate the kind of geographical entity certain place-names imply (hill, spur, tongue, bog, etc.), many common and frequently used suffixes are not continually repeated in English, and it is hoped the reader will quickly grasp the meaning of most of them. These include, most importantly, the following: *-jökull* = glacier, outlet glacier; *-fjörður* = fjord; *-flói* = bay; *-fjall*, *-fell* = mountain (*-fjöll*, pl. mountains); *-heiði* = highland moor, mountain; *-á*, *-fljót*, *-kvísl* = river; *-vatn, -lón* = lake, reservoir; *-dalur* = valley, dale; *-hraun* = lava field; *-sandur* = outwash or gravel plain; *-öræfi* = wilderness. The Icelandic *jökulhlaup* (for pro- or subglacial outburst flood) has been internationally accepted as the scientific term for this phenomenon.

A rough guide to the pronunciation of the consonants and vowels special to the Icelandic alphabet, as compared to general, standard RP English pronunciation, is as follows: The non-diacritic vowels a, e, i, o and u, are very similar to those in English (bat, bed, bid, bog, bun), the Icelandic medial 'r' is often trilled, and the double 'll' is usually pronounced 'tl' as in kettle or the Welsh 'll' in Llangollen; Þ þ = unvoiced 'th' as in: Beth, bath, path (always at beginning of words); ð is a voiced 'th' as in: then, this, that (always in middle or end of words); Á á = 'ou' as in bound, found, round; É é = 'ye' as in y̲et, y̲ellow, y̲esterday; Í í = 'ee'/'ea' as in seen, keen, lean, mean; Ó ó = 'o'/'oe' as in go, no, foe, doe; Ú ú = 'oo' as in moon, boon, doom; Ö ö = 'ur' as in burn, turn, urn; Æ æ = 'i' as in bite, kite, trite.

I would like to thank Helgi Björnsson for his boundless patience and assistance with geophysical and glaciological terminology, and to express my gratitude to the Pálmi Jónsson Nature Preservation Fund and the Icelandic Literature Centre for grants towards this translation.

Julian Meldon D'Arcy
University of Iceland

Acknowledgements

This book is the result of more than 40 years of work and research on Icelandic glaciers, which could not have been completed without the help and support of so many of my colleagues, both past and present, at the Science Institute of the University of Iceland. I wish to express my sincere thanks to them for their help and collaboration over this long period of time. I am also grateful to many institutes for both financial support and field assistance, including The National Power Company of Iceland, The Road and Coastal Administration of Iceland, The Division of Signal Processing of the University of Iceland (headed by Prof. Sigfús Björnsson), The Icelandic Meteorological Office, The Iceland Glaciological Society, The National Energy Authority of Iceland, and the Reykjavík Geothermal Agency. I would also like to express my gratitude to several research funds for financial grants: the Icelandic Science Fund, the University of Iceland Research Fund, the Research Fund of the European Union, the Research Fund of Kvísker, the Icelandic Parliament Financial Committee, and the Research Fund of Eggert V. Briem.

I am especially grateful to many colleagues for reading over chapters of the Icelandic version and for providing valuable information: Ólafur Grímur Björnsson, Jökull Sævarsson, Eiríkur Þormóðsson, Bragi Þ. Ólafsson, Örn Hrafnkelsson, Björn S. Stefánsson, Sveinbjörn Björnsson, Leó Kristjánsson and Þóra Ellen Þórhallsdóttir. I also wish to express my gratitude to many of my closest colleagues at the Science Institute for long-lasting collaboration in field work and data processing: Finnur Pálsson, Sverrir Guðmundsson, Eyjólfur Magnússon, and Magnús T. Guðmundsson.

Many photographers have allowed me the use of their photographs. Sven Þ. Sigurðsson gave me permission to use photographs from his father's collection (Sigurður Þórarinsson). The National and University of Iceland Library, the National Museum of Iceland, the Árni Magnússon Institute, and The Royal Library in Copenhagen have all allowed me the use of photographs from their collections. Hagþenkir supported the writing of the Icelandic version, which was published by Opna. I am grateful to the publishers Sigurður Svavarsson and Guðrún

Magnúsdóttir and to their layout and graphic designers Líba Ásgeirsdóttir and Björn Valdimarsson.

Finally, I would like to thank the Pálmi Jónsson Conservation Fund and the Icelandic Literature Center for helping fund the translation of this book. I am also extremely grateful to Prof. Julian Meldon D'Arcy of the University of Iceland both for translating the text into English and for his continuous encouragement.

Contents

Abbreviations

AMI	The Árni Magnússon Institute for Icelandic Studies, Reykjavík
CLIMAP	Climate: Long range Investigation, Mapping, and Prediction (project)
COHMAP	Cooperative Holocene Mapping Project
DCPEM	Department of Civil Protection and Emergency Management (Almannavarnir)
DGI	Danish Geodetic Institute
DGS	Danish General Staff
EPICA	European Project for Ice Coring in Antarctica
GPS	Global Positioning System
HB	Helgi Björnsson (photographs)
HSD	Hydrological Service Division of NEAI
ICAA	Icelandic Civil Aviation Administration (Flugmálastjórn)
IESUI	Institute of Earth Sciences of the University of Iceland (Jarðvísindastofnun Háskóla Íslands)
IGS	Iceland Glaciological Society (Jöklarannsóknafélag Íslands)
IMO	Iceland Meteorological Office (Veðurstofa Íslands)
IPCC	Intergovernmental Panel on Climate Change
IRCA	Iceland Road and Coastal Administration (Vegagerð Ríkisins)
ISRC	Icelandic State Research Council
ITA	Iceland Touring Association (Ferðafélag Íslands)
LM	Loftmyndir (Aerial photographs)
MODIS	Moderate Resolution Imaging Spectroradiometer (NASA)
NEA	National Energy Authority (of Iceland) (Orkustofnun)
NLSI	National Land Survey of Iceland (Landmælingar Íslands)
NMI	National Museum of Iceland (Þjóðminjasafn)
NPCI	National Power Company of Iceland (Landsvirkjun)
NULI	National and University Library of Iceland (Lands- og Háskólabókasafns Íslands)

RLC	Royal Library of Copenhagen
SCSI	Soil Conservation Service of Iceland (Landgræðsla Ríkisins)
SIUI	Science Institute of the University of Iceland
SPOT	French: Satellite Pour l'Observation de la Terre. (Satellite for observation of Earth)
USAMS	US Army Map Service

Part I
The Origins and History of Glaciers and Glaciology

Chapter 1
Origins and Nature of Glaciers

'Out of whose womb came the ice? And the hoary frost of
heaven, who hath gendered it?' Job 38:29.

Abstract A knowledge of the nature of glaciers and ice sheets is required to
understand their important role in the Earth's hydrologic cycle. The glaciers of
today provide an insight into both past and future glaciers and climate changes. In
this chapter, the basic concepts and terminology of glaciers and glaciology are
presented as an introduction to the accounts of Icelandic glaciers in the later sections of the book. This includes an explanation of how ice masses are formed from
snow, which falls and settles in an accumulation zone, hardens into firn snow and
consequently glacial ice, and then flows downward and across the equilibrium line
into an ablation zone, where it melts. The mechanics of how ice masses move are
outlined, as well as the occurrence of surges and the formation of crevasses, ogives,
and moulins. The effects of subglacial, geothermal and volcanic activity on glaciers
are also explained, particularly the formation of ice cauldrons and ice-dammed
lakes, along with the reasons for the phenomenon of outburst floods, or jökulhlaups,
from glaciers. The landforms resulting from glacial movements and fluvial processes are also defined, e.g. moraines, canyons, and gravel outwash plains.

1.1 The Frozen World and Its Imprints

Glaciers cover polar landmasses and the highest mountains of all the continents of
the world, with the exception of Australia, and blanket about a tenth of the Earth's
land surface. Their influence extends even further than this, however, for glacial
rivers flow through land to the sea, ice shelves break up in the seas around the
world's polar regions, and icebergs and ice floes are borne by ocean currents into
shipping lanes. Glaciers were previously much larger than they are now and almost
every part of the world's land mass has been covered by glacial ice at some time in
its history. The evidence glaciers leave behind bear witness to this (Figs. 1.1, 1.2
and 1.3). They have chiselled at the toughest bedrock, gnawed out mountains and
basins, broken up cliff faces, and carved out pointed summits, leaving behind sharp,

© Atlantis Press and the author(s) 2017

H. Björnsson, *The Glaciers of Iceland*, Atlantis Advances
in Quaternary Science 2, DOI 10.2991/978-94-6239-207-6_1

Fig. 1.1 The sharp edges of Veðurárdalsfjöll mountains bear witness to the erosives powers of glaciers. Breiðamerkurjökull descends south along the mountains towards Jökulsárlón lagoon on the outwash plain Breiðamerkursandur (see Sect. 8.5). In the centre of the photograph, a glacier in the wide valley of Svöludalur has pushed up a curvilinear moraine, Veðurárdalsrönd. The North Atlantic can be seen in the far distance. HB, September 1994

Fig. 1.2 Hvalfjörður, a 35-km-long fjord abraded by an ice-age glacier. View down Botnsdalur valley over Botnsvogur inlet, Þyrilsnes peninsula, Hvammsvík bay and Reyniháls ridge. Furthest away is Mt. Akrafjall. HB, September 1997

Fig. 1.3 The glacially eroded valley Fossadalur on the left of Lambatungnajökull, a present day outlet of the eastern parts of Vatnajökull. In the *top left* corner can be seen the nunataks (mountain peaks protruding through a glacier's surface) in Hofellsjökull (Sect. 8.6). HB

craggy peaks, and huge, jagged rock sculptures. They have gouged out deep and narrow fjords far out onto continental shelves, eroded broad valley basins and large trenches now filled by lakes, and levelled out and polished uneven ground in the lowlands. Erratic boulders are scattered far and wide, and if one looks closer, more refined traces can be found: glacially striated rocks, grooves in boulders, smooth roches moutonnées (Fig. 1.4). In glacial forefields, terminal moraines stand alone as silent witnesses to the former might of the glaciers. Further afield, glacial rivers have gouged out ravines and their contents created gravel plains and sediment strata in both lakes and the sea. This is the manner in which glaciers have left behind traces that tell us the history of climate change, even long after the glaciers themselves have disappeared.

1.2 Understanding Ice in the Global Environment

Glaciers and ice sheets are an important link in the Earth's hydrologic cycle (Fig. 1.5). Their existence and size is determined by climate, the transport of moisture and warmth around the globe through atmospheric and oceanic currents, the locations and levels of seas, and Earth's crustal movements. Ice masses adjust

Fig. 1.4 Scraped and grooved roches moutonnées just to the north of Gullfoss waterfall (see Chap. 6). The striation marks show the direction of the flow of the ice-age glacier. The grinding tools are still strewn all over the bedrock. Vegetation protected the roches moutonnées for thousands of years, but since Iceland has been settled, wind erosion has removed all flora and the topsoil. HB, July 1997

rapidly to changes in climate, but they may also influence the climate themselves. Glaciers contain the largest reservoirs of fresh water on Earth, storing water as ice during cold periods over timescales of hundreds to thousands of years. In the short term, they gather snow during winters and supply meltwater through the summers, providing water to plants, animals and humans when other sources may be scant during dry times.

Knowledge of the origins, existence and nature of glaciers is thus important for a full comprehension of the global environment. Such an understanding comes to a great extent from research into modern-day glaciers, for the nature of glaciers is the same now as it was in the past, and as it will be, for as long they exist: the present is a key to the past and the future. This applies to how ice masses expanded and once shaped continents, as well as to what their responses might be to climate change in the future. Moreover, the glaciers of today preserve a long history of climate, and evidence of changes in the composition of the Earth's atmosphere can be found in both the chemical composition of their layers of ice and in their frozen air bubbles.

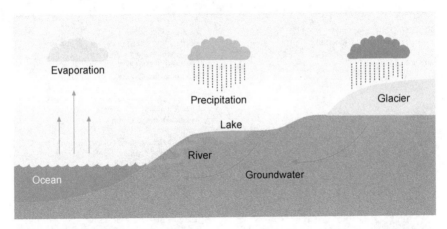

Fig. 1.5 Glaciers in the Earth's hydrologic cycle. Earth has been called both the Blue Planet, because 70 % of its surface is covered with ocean, and the Frozen Planet, because about 10 % of the land masses are covered with ice. It can even be said that we still live in an ice age, for the triple point of water, whereby it can coexist in all three of its phases, water vapour, liquid water, and frozen ice, is 0 °C. Ice masses on Earth are important for their influence on the climate and its fluctuations, the flow of meltwater to rivers, and Earth's sea levels. Water remains longest in its glacial forms during its global cycle

Glaciers are all similar in nature even though they may have dissimilar shapes, sizes and appearances (Fig. 1.6), and can be found all over the world, from the polar regions to the equator. They are all created from snow, but their development and behaviour can vary according to the amount of snow they accumulate, how cold and hard the ice is, how they discharge their meltwater, and if they creep forward over hard bedrock or soft basal sediments. Glaciers expand or recede in accordance with alterations to their mass balance, but their responses to climate change can vary from one glacier to another and their dynamics may be affected by

MAIN GROUPS	SUBCATEGORIES
Fig. 1.6a Ice sheets Move in all directions from their centres and over land, which is already mostly covered by ice; the surface does not usually reflect the subglacial topography. Glacier	**Continental glaciers, inland ice** The largest ice sheets like those existing today in Greenland and Antarctica. **Ice caps** Smaller and thinner than continental glaciers, e.g. the main ice masses in Iceland (Ch. 3) such as Vatnajökull, Hofsjökull, Mýrdalsjökull and Eiríksjökull. Also common on islands in the Arctic Ocean and in the southern part of Alaska.

Fig. 1.6 Glaciers may be divided into three main categories according to their geomorphological shape and size

Fig. 1.6b Outlet glaciers
Move in one direction, determined by
the landscape. Can be branches of
an ice sheet.

Piedmont glaciers.
Glaciers that spread out like a fan once
they reach lowland plains, e.g.
Skeiðarárjökull (Ch. 8) and Múlajökull (Ch. 6)
and Malaspina in Alaska.

Valley glaciers. Glaciers in valleys, e.g.
Svínafellsjökull in Öræfi and Fláajökull in
Austur-Skaftafell County.

Valley-head glaciers. Glaciers limited
to the head of a valley, e.g. Bægisárjökull
(Ch. 7).

Cirques (corries). Glaciers in rounded
valley hollows, e.g. Barkárdalsjökull.

Hanging glaciers. Glaciers in
hanging valleys, e.g. above the main valley
on Tröllaskagi peninsula (Ch. 7).

Ice aprons. Ice carapaces on mountain
sides, e.g. in the highest reaches of
Tröllaskagi (Ch. 7).

Mixed glaciers. Glacial tracts in the
highlands broken up by nunataks, e.g.
to the north of Öræfajökull towards
Mávabyggðir and Esjufjöll (Ch. 8.5).

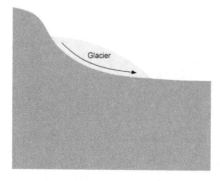

Fig. 1.6c Ice shelves.
The part of a glacier that floats
on the sea or a lake and calves
at its precipitous margins.

The largest ice shelves of Antarctica,
e.g. the Ross Ice Shelf and the
Filchner-Ronne Ice Shelf.

The ice cover on Lake Grímsvötn and the
blue snout of Breiðamerkurjökull, which
breaks up (calves) into the Jökulsárlón
lagoon (see Ch. 8).

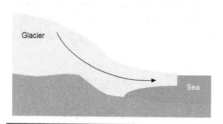

Fig. 1.6 (continued)

processes that are not directly related to climate, e.g. glaciers that surge on land or
calve icebergs into the sea. A detailed description of the complex nature of glaciers
will not be addressed here, but an overview of basic glaciology will follow below.
For further details see Sharp (1991), Paterson (1994), Knight (1999), Hambrey and
Alean (2004), Hooke (2005), Cuffey and Paterson (2010), Benn and Evans (2010),
van der Veen (2013).

1.3 Introduction to Glaciers

A glacier is a body of ice that has been made from snow, collected year by year above the snowline (Fig. 1.7), and which becomes buried deeper and deeper below fresh layers of snow, gradually being transformed into ice that flows forward because it does not have the strength to sustain its own weight (Fig. 1.8). The ice then moves downstream until far below the snowline, where ablation eventually eradicates all the ice that has arrived from higher reaches. The glacial surface area above the snowline is called the accumulation zone, and the lower part of the glacier, which receives ice flows from above it, is called the ablation zone. All snow which falls on this area melts away. The ablation zone only exists because of the mass transport of ice from the accumulation zone. The boundary between these two zones is indeed a kind of snowline, but here it is called the equilibrium line, because at that elevation the glacier's mass balance is in equilibrium: ablation equals accumulation. Increased accumulation over many years will enlarge the accumulation zone and instigate a greater downward transportation of ice, which in turn will lower the equilibrium line and induce glacial tongues to advance. Conversely, a diminished accumulation zone will lead to a reduction in the transportation and expansion of ice, and the glacier would then have to recede in order to restrain itself for a more frugal operation.

In this way, glaciers can be seen as maintaining production, storage and distribution systems, and they grow or recede in accordance with their glacial economy. The equilibrium line is an economic index line. The altitude of the equilibrium line fluctuates with the climate and indicates the glacier's economic health: the lower

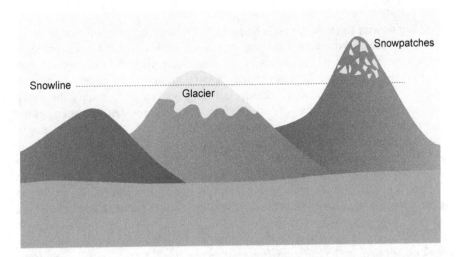

Fig. 1.7 Glaciers are formed above the snowline and can advance far below this height. Although pointed mountain peaks can be much higher than the snowline, only snow patches collect on them instead of massive glaciers due to their steep flanks

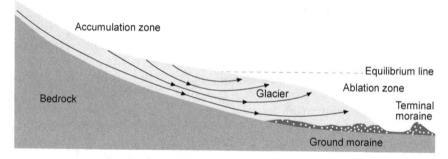

Fig. 1.8 Longitudinal section of a glacier from its head to its snout. In the accumulation zone, previous snowfalls are buried annually under new layers of snow. All the snow buried in this zone is carried down the glacier along the flow lines depicted until it resurfaces in the ablation zone. On the boundary between the accumulation and ablation zones, melting maintains a mass balance, income equalling expenditure. This is the equilibrium line of a glacier. Snow that survives the summer on a glacier is called firn snow, and the firn zone reaches as far down as the firn line. On glaciers where meltwater refreezes and superimposed ice survives the summer, the equilibrium line is lower than the firn line

the line, the more a glacier will expand; the higher the line, the more a glacier will retreat. Thus, if there is a balance between production, distribution and consumption, the surface of a glacier may remain unchanged, even though it is actually in continual motion.

1.4 Formation of a Glacier

It all begins with snow. Moist air is borne from the ocean up onto land and flows up mountainsides or above cold, low-lying air masses and cools at dew point, invisible water vapour solidifying into drops of water around particles of dust and salt that float in the atmosphere, the supercooled vapour forming an ice nucleus. Drops of water and ice crystals float in clouds, colliding and expanding until the rising current of air can no longer support them and they begin to fall to earth as either rain or snow. The higher the air currents reach, the colder it becomes and the greater the precipitation. The size and shape of the snow crystals is determined by the moisture and temperature of the air. Hard and brittle shards of ice fall in high degrees of frost, but when it is warmer the nuclei stick together in flakes. In calm weather the snowflakes remain polygonal, the snow loose and full of air, but in stormy weather the sharp points break off and the snow is packed together into hard snow.

Once snow has fallen to earth, the crystals that have been formed at high altitude seek equilibrium in their new environment. While it still retains considerable frost, the snow is transformed slowly and can remain as soft snow, but when it becomes warmer it settles more rapidly and becomes coarse-grained snow. Molecules are in

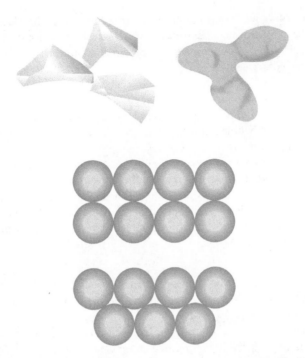

Fig. 1.9 Glaciers are made from snowflakes that are transformed into ice. The sharp points of grains of snow are blunted so that they become spherical in shape. Molecules are in constant motion, both within and on the surface of the grains. Water vapour also occupies the air around them. When the grains of snow have all become spherical and equal in size they are compacted in such a way that each horizontal row is diagonally juxtaposed with the next so that the centre of each sphere is directly beneath the surface contact point of the two spheres in the row directly above it. At this juncture, 40 % of the compacted snow is comprised of air cavities and its bulk density is 550 kg/m³

constant motion, both on the surface and within the grains, and water vapour also occupies the air spaces in between them. Through sublimation, molecules are released from sharp points and borne in water vapour to concave hollows on the surface of the grains, settling there and turning into ice once more. The movement of molecules is thus faster the nearer the temperature is to freezing point. The sharp points are gradually blunted so that the grains of snow become spherical in shape and then gradually equal in size because the vapour pressure is higher above smaller grains than larger ones. Vapour is thus forced from the smaller to the larger grains of snow, obliterating the smaller ones and increasing the size of the larger ones. The snow grains are then compacted so tightly that 40 % of the volume of packed snow is comprised of air cavities and its mass density has become 550 kg/m³ (Fig. 1.9). In glaciers in Iceland, snow reaches this stage by the end of each summer, but this process takes many years on the cold polar glaciers and does not take place until the snow has been buried dozens of metres below fresh snowfalls. Snow that survives longer than one summer on a glacier is called firn snow.

The grains of snow are then compressed together at their interfaces, the more rapidly as the pressure increases from new layers of snow accumulating on the glacier. The air in the snow is gradually constricted and finally stops circulating, becoming trapped in bubbles. At this point, glacial ice is said to have been formed (density ca. 800–850 kg/m^3). In glaciers in Iceland, this takes about 5–6 years and at a depth of 20–30 m, though it needs a depth of 60–100 m after 100 years on the coldest glaciers on Earth. With increased weight and pressure when further downstream, the air bubbles in the glacial ice are compressed even more and its mass density increases (up to 900 kg/m^3).

The transformation of snow into glacial ice at a temperature below freezing has been described above. If the sharp points of the snow grains melt, this can speed up the conversion. Meltwater seeps down into the snow and where it still retains frost it will refreeze and form icicles that spread through the layer of snow, but when it meets water-resistant crusts of snowdrift compacted by strong winds, meltwater will freeze in thin, horizontal layers.

Ice Crystals Grow in Size Downglacier

The glacier's snout contains the oldest ice and the largest ice crystals, for it has travelled the furthest along the glacier, from the upper parts of the accumulation zone. Ice crystals grow larger in time as they move down the glacier, the more speedily the higher the temperature of the ice. The largest ice crystals are as big as a man's head in glacial snouts in Iceland, but can reach up to a metre in diameter at the base of the Antarctica, for indeed the ice there is a thousand times older than in Iceland. The outlines of large crystals can be seen with the naked eye on the surface of glacial snouts, for furrows are formed at their edges during melting, both on sunny as well as rainy days. Also visible within the ice are oblong bubbles of air that have been flattened out under the weight of overlying ice and the tensions of the glacier's directional movements. These bubbles, which may resemble jewels, are full of water vapour and preserve atmospheric contents from ancient times.

The stress caused by the glacier's movement has for a long time been centred on the ice crystals, the structures of which are being continuously transformed. The greatest tension builds up where the crystals' basal planes are directly transverse to the direction of the ice flow, and so agitated molecules abandon them and these crystals gradually disappear. When their planes are parallel to the direction of the ice flow, on the other hand, they give way and slide forwards like a pack of cards; there is little tension and the crystals grow on their journey down the glacier because they collect molecules that have fled from stress elsewhere.

Mass Balance Measurements

The mass balance of a glacier is measured in the spring and autumn (Fig. 1.10). In spring, pits are dug or holes bored through the layer of winter snow at many different locations spread equally over the glacier. By measuring the snow's thickness and density, the layer's water equivalent can be estimated. Snow accumulation in the mountains can be many times greater than precipitation alone, because snow is windborne, drifts into hollows, collects on the lee sides of cirques and basins, and falls from cliffs in avalanches. In the autumn, the firn snow on the accumulation zone is assessed the same way, but the summer melt in the ablation zone is gauged from calibrated sticks and wires that have been lowered down boreholes drilled into the glacial ice in the spring. At the end of the glacial year, in the autumn, the annual mass balance is then calculated at each location and for the glacier as a whole by averaging the measurements from the various locations.

The mass balance over a period of decades may also be calculated by comparing maps of glacial surfaces. One surface map is subtracted from the other, and the difference shows the mass balance between the mapping dates. It is important that the map covers the entire glacier because localised changes in elevation can be caused by ice being transported from one location to another within the glacier. In Iceland the accumulation zones are rather larger than the ablation ones, typically around 60–70 % of the glacier's surface.

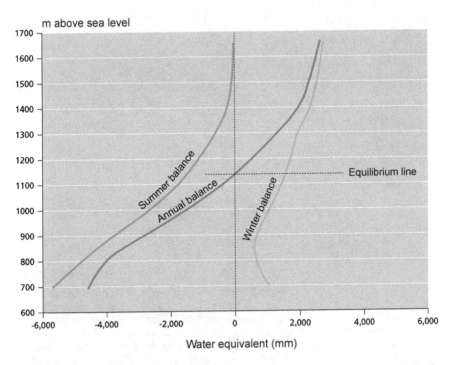

Fig. 1.10 Mass balance of Tungnaárjökull 1991–1992. Mass balance measurements in the summer, winter, and end of the glacial year in autumn at various elevations on Tungnaárjökull (Chap. 8). (Measurements from Science Institute of the University of Iceland)

1.5   Elements of the Glacial Transport System

The glacier acts as a conveyor belt that delivers ice from the accumulation zone down to the ablation zone, where meltwater flows into glacial rivers. The velocity of a glacier's ice flow normally increases down the incline of the accumulation zone, because once it has reached the equilibrium line it is bearing the bulk of the entire ice mass that has accumulated above it. Its speed is reduced in the ablation zone, however, for further down the incline it reaches, the more ablation increases, and so there is less and less ice to be transported as it disappears from the glacier in the form of meltwater (Fig. 1.11).

The ice, which has been transported with increasing velocity down the accumulation zone, confronts resistance when it meets the retarding ice flow in the

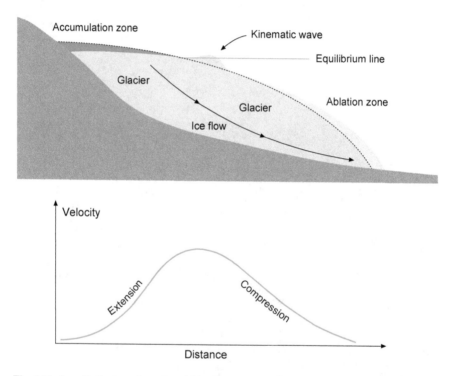

Fig. 1.11 Longitudinal section of a glacier from the accumulation zone to the ablation zone. A longitudinal section of a glacier showing how an ice flow's velocity increases downstream from the accumulation zone towards the equilibrium line, and then gradually decreases the further down the ablation zone it reaches. On the boundary between the accumulation and ablation zones, a bulging wave of ice rises onto the surface of the glacier as ice descends from the accumulation zone into the ablation zone, and this wave moves on the glacier's surface at about fourfold the speed of the glacier's main flow. This is reminiscent of a worm that moves onwards by forming a loop in a section of its body that is then distended through its entire length, although it can only advance forward itself a short distance of the loop at a time

ablation zone, and so to progress further down the glacier it must throw up a bulge of ice (called a kinematic wave) at the equilibrium line, which then continues downstream over the surface of the ablation zone much faster than the average speed of the bulk of the glacier. Hence an increased mass balance of the accumulation zone would be indicated by a more rapid response at the glacier's terminus than could be deduced from the average velocity of the whole glacier. The ice wave on its surface thus transports this information faster than the bulky glacier itself.

A kinematic wave tends to be deflated by increased deformation of the ice at its crest, however, and may never reach the glacier's snout. On flat, slow-moving glaciers, ice waves are usually eliminated and only manage to reach as far as the snout during long-term climate changes that last for decades, or even centuries. However, the waves are more likely to reach as far as the snouts of steep, fast-moving glaciers, which may thus respond quickly (within a few years) to changes in mass balance. Small, steep and fast-moving glaciers thus reveal more short-lived fluctuations in weather conditions, while large glaciers are tardier and only reveal long-term changes in the mass balance and climate.

If an ice flow from the accumulation zone replaces exactly the amount that is eliminated in the ablation zone, the surface of the glacier remains unchanged; the glacier is in equilibrium and moves at a so-called balance velocity. By comparing the surface speed and the calculated balance velocity, it can be determined if the glacier is moving fast enough to transport the ice it has accumulated. If it is not, a build-up of ice might result in a glacial surge (see Sect. 1.10).

Details of glacier movements. Glaciers advance slowly under their own weight (Fig. 1.12). The thicker and steeper the bulk of ice is, the greater the consequent internal stress and deformation of the ice. The most frictional resistance is at the base and margins, so it is the centre of a glacier that moves forwards most rapidly. A glacier's speed is reduced when thrust upwards at a bedrock ridge, but accelerates again on the lee side because of an increase in downward slope. Advancing over a ridge head, the ice is stretched, and the larger the crest, the more taut the ice becomes.

Unless it is frozen solid to the bedrock, the entire mass of ice can also slide over a slippery bed of either hard rock or soft basal sediments. The base becomes slick as water spreads in a thin film beneath it, and the greater the basal water pressure, the more the glacier's overburden on its base is reduced. There is most water present in glaciers in the summer, and so they usually move faster then than during winter.

A glacier, when thrust by its flow onto a bump on its bed, moves forwards when replacing ice, which melts in front of the obstacle. This is because when meeting an obstruction, the glacial pressure increases, lowering its melting point below freezing point at its base, and so the ice melts. The meltwater drains away, flowing over the obstacle and freezing on the lee side of it, where the pressure is now reduced. When water refreezes there, it releases a latent heat of fusion that is conducted back through the solid obstacle to where the ice first melted. This cycle of water and heat maintains the melting process. The glacier slides more rapidly over obstacles less than a few cm long. The basal slide is thus slower the larger the hindrance is, as this lengthens the route of thermal conduction, which delivers heat for melting.

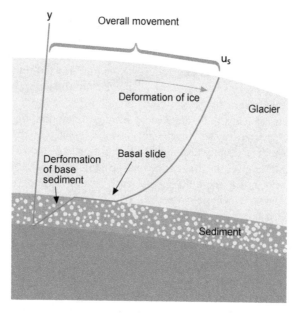

Fig. 1.12 Longitudinal section of a glacier from its surface to its base. Glaciers advance by sliding along a slippery base and through the deformation of ice and sediments beneath them. If a glacier is frozen to its bed it does not slide at all. Deformation really begins once the ice has become more than some 50 m thick. Ice crystals are made of parallel layers that give way under a heavy weight so that they are displaced forward like a pack of cards. The speed of ice deformation increases in relation to the glacier's thickness (four squared) and surface gradient (three squared). The rigidity of the ice is dependent on its temperature, increasing a hundredfold if it is lowered by about 20 °C

Crevasses Produced by Glacier Movements

Crevasses are cracks in the surface of a glacier caused by tension pulling the ice apart (Figs. 1.13 and 1.14). Transverse crevasses open up above an uneven bedrock base, while icefalls cascade from cliff edges. Marginal crevasses open at an angle of 45° upstream on a glacier where ice meets resistance from the mountainside, the glacier increasing speed towards the centre, furthest away from this resistance. In the accumulation zone, transverse crevasses can split open right across a glacier's directional flow because the downstream crevasse wall is travelling faster than the upstream one; the ice flow's velocity increases further down the glacier, the glacier expands and the crevasses grow wider. In valley glaciers, even moderate transverse crevasses can stretch from one side of the valley to the other. In the ablation zone, crevasses turn upstream because the speed of the ice flow has been reduced further down the glacier and the body of ice has become more compressed.

Longitudinal crevasses are formed when a glacier spreads out below a bottleneck, and splayed radial crevasses fan out like the outstretched fingers of a hand, heading downstream at the snout of piedmont glaciers, though they can also form on glacial knolls. There are also other forms of crevasses. When a glacier frees itself from a mountainside rock face, a gap a few metres wide, or randkluft (bergschrund), can be created. When glaciers calve icebergs into the sea, crevasses are formed right across the glacial tongue, where its ice shelf begins to float and is subject to tidal waters.

In Iceland, crevasses are not usually more than 20–30 m deep. Ice is compressed by its own weight further down the crevasse and they gradually close. Once crevasses have been borne with the ice flow down the glacier, they begin to deform and revolve, like rafts bumping into each other on a river. When confronted with obstacles on the glacial bed, crevasses are compacted and thin lines of blue ice may be the only witnesses to their previous existence. In polar regions, where the ice is colder and more rigid, crevasses can be 100 m deep, even up to 500 m deep in Antarctica.

Ogives Indicate Ice Flows

Ogives are curvilinear bands of light and dark ice at regular intervals on the surface of a glacier below an icefall where the glacier has moved very rapidly, splintering in the process, but being compressed together again further downstream (Fig. 1.15). The ice mass descending an icefall during the summer forms a dark band because of the melting and rapid compression of ice and snow and the sand that blows onto it, but in winter, with cleaner snow and slower motion, the bands are pale. The shape of the ogives reveals how the velocity of a glacier is greatest at its centre and reduced at its margins. The summer bands are rather broader than the winter ones as the glacier moves more rapidly in the summer. The ogives are gradually eliminated when the hollows between them are filled with snow. Dark summer layers melt quicker than the paler winter ones because of the difference in the reflection of solar radiation, so that waves can form on the surface; white winter bands form ridges, summer layers, hollows. This undulation on the glacier's surface is gradually eliminated when the hollows between crests are filled in with snow. The curved ogives Sveinn Pálsson, a pioneer in glacial observations, viewed from up on Öræfajökull in 1794 were indeed an indication for him that glaciers move (see Sects. 4.6 and 8.4). Ogives can also be found in Kvíárjökull and Morsárjökull (Sect. 8.4).

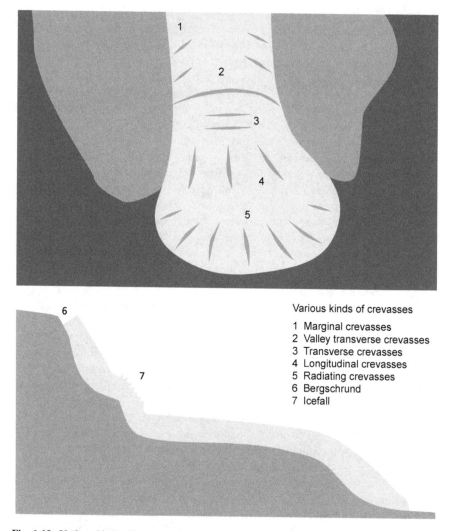

Fig. 1.13 Various kinds of crevasses; view from above and along a longitudinal section

1.6 Thermal Characteristics of Glaciers

The thermal regime of glaciers significantly controls their dynamics, hydrology and erosional capacity. The transformation of ice to water is subject to changes in temperature. The deformation of glacial ice varies according to how cold and hard the ice is (Fig. 1.16).

On the coldest glacial areas on Earth, Antarctica and the polar regions of Canada and northern Greenland, there is no melting because the air temperature remains below freezing all year round. At around a depth of 15 m, the temperature of the

Fig. 1.14 The crevasse-riddled Vatnajökull ice cap has covered land for more than 2000 years. This land could see the light of day within the next two to three centuries (Chap. 8). HB, October 1994

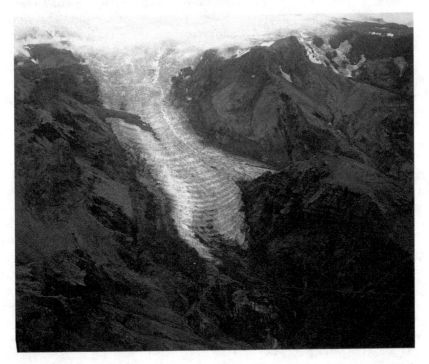

Fig. 1.15 Ogives, with their darker summer and paler winter layers, below the icefall from Hábunga on Mýrdalsjökull (Chap. 5). HB, July 1999

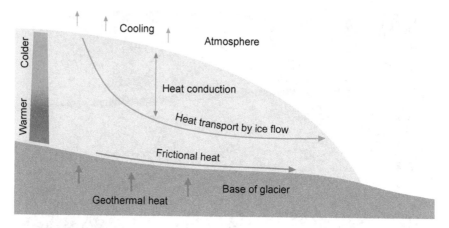

Fig. 1.16 Heat fluxes through glaciers. The temperature of a glacier is determined by surface cooling, thermal conduction from its base, and heat transference within the ice flow as well as frictional heat from the glacier's movements. A glacier functions as an insulator and can be thick enough to prevent frost from permeating to its base

glacier is equal to the mean annual temperature of the air above it because frost can permeate the ice to such a depth within a year through thermal conduction. With the passing of time, frost has permeated most glaciers to some depth, even down to the bed. Shallow glaciers in polar regions are likely to be frozen to their beds, whereas thick glaciers may have temperate bases (i.e. at melting point). The almost 4000-m-thickness of the Antarctica glacier is sufficient to insulate its base from frost over large areas. Its base temperature is thus around freezing point, even though its surface is the coldest place on Earth with an average temperature of −50 to −60 °C. Indeed, there are hundreds of lakes, full of water in its liquid form beneath the Antarctica ice sheet, the best known of them being Lake Vostok, which is all of 500 m deep and 250 km long (14,000 km^2).

Surface melting takes place further away from the polar regions, where winds blow warm air over the glaciers and the sun rises higher in the sky. Meltwater from the surface percolates down into cold snow and freezes, but in doing so it releases a latent heat of fusion that transmits warmth into adjacent layers of snow. A few days of melting every summer is sufficient to eradicate all the winter frost in the accumulation zone. In the ablation zone, meltwater percolates through a thin layer of winter snow to the surface of the old glacial ice and refreezes there to form a layer of superimposed ice. A slow working conduction bears the released latent heat down into the underlying glacial ice. In northerly sub-polar regions, the summers are so short, however, that even though frost may disappear from the firn zones, it is still maintained in the ice of the snouts, where cold winter temperatures can reach to a depth of 10 m in the glacial ice, as there is only a single, year-old layer of snow covering it, just 2–4 m thick. Frost is eradicated in this layer of snow as described above. But in the ice below, the conduction of latent heat released by the freezing of meltwater on its surface does not last long enough in the short summertime to raise the temperature to melting point for the entire layer of frozen ice in the ablation zone. Thus, in sub-polar regions glaciers may be temperate in the accumulation zone, whereas frost can still be maintained in the ablation zone. In Iceland, on the other hand, frost is eliminated from the entire ablation zone during the summer. All frost disappears from glaciers and they are at freezing point except for the topmost layer of snow during winter.

1.7   Energy Fluxes and Glacial Melting

Glaciers are created and maintained by a cold climate. As soon as a glacier has claimed its territory, its immediate environment becomes favourable for its continuing growth and existence. Once formed, it begins to influence its environment so as to sustain its future existence. When a glacier has covered mountains that had previously attracted precipitation, its surface will be colder than the mountain's original one and so snowfalls will increase. The ice cap towers at a great height and can now attract precipitation itself from the air currents flowing up to and over it, growing colder in the process. The energy for melting is also reduced because the snow reflects a substantial amount of solar radiation like a mirror. On the other hand, when a warmer climate causes a reduction of snow and ice cover on land and sea, an increased amount of solar radiation is absorbed by the Earth's surfaces; this accelerates warming and thus enhances melting, which activates still more absorption of radiation. A cyclical chain reaction is thus instigated.

A glacier's survival is thus threatened by the heat advection of warm air masses as well as radiation energy, for both solar and atmospheric radiation transmit heat to glaciers (Figs. 1.17 and 1.18). Only a portion of the sun's radiation permeates a glacier, because snow reflects a large amount of it like a mirror, seven to nine tenths of it from clean snow, and half of it from surface ice, as sunburned travellers on glaciers are only too well aware. The dirt that collects on a glacier during the summer reduces this reflection and thus the radiation that penetrates a glacier's surface can be fairly evenly distributed throughout the summer and well into August, even though solar radiation has been diminished. This radiation can penetrate to a depth of 1 m in snow, but to around a depth of 10 m in clear ice.

Atmospheric radiation reaches a glacier from various gases in the atmosphere (especially carbon dioxide) and moisture in the clouds. This kind of radiation is not reflected from the surface but hardly penetrates a glacier either, so that only a very thin surface layer of the glacier absorbs all this energy. Moreover, a glacier also exudes terrestrial radiation into the atmosphere, and the temperatures of the atmosphere and the glacial surface consequently determine which component of radiation is the stronger. A glacier usually suffers a deficit in this exchange of radiation energy with the atmosphere because air grows colder the greater its altitude above the Earth. When the night skies are clear and the atmospheric and terrestrial radiations dominate, everything rigidifies rapidly on the glacier. The air, however, can occasionally be so warm at the height of summer, and cloud coverage

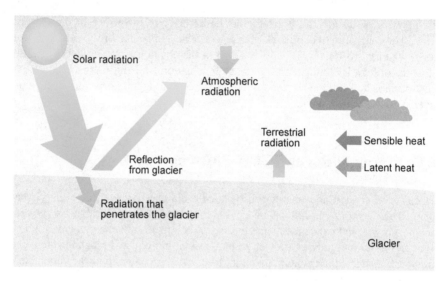

Fig. 1.17 Energy flows on the surface of a glacier. A glacier receives heat from solar and atmospheric radiation, warm air and the condensation of moisture. Snow reflects a great deal of solar radiation, but a small amount permeates the glacier's surface. Atmospheric radiation reaches a glacier from different kinds of air molecules in the atmosphere and water in the clouds. On the other hand, a glacier also exudes terrestrial radiation into the Earth's atmosphere. Rainfall may bring some heat to a glacier. Heat is utilised to eradicate frost in a glacier, and once that has been done it proceeds to melt snow and ice

so thick, that atmospheric radiation will be greater than radiation emitted from the glacier; indeed, the glacier cannot become warmer than freezing point.

When warm air is borne over a glacier, a heat flux exudes from it, its strength depending on its temperature and speed. The more uneven a glacier is, the more an air current will be disturbed, enabling warm turbulent eddies to reach its surface. In calm weather there is hardly any melting because colder and denser air lies just above the surface, preventing warmer and lighter air from reaching down to the glacier. Latent heat is released when moist air cools so that vapour condenses on a glacier's surface. The strength of this flow depends on the humidity, velocity and turbulence of the air. The air in Iceland is seldom so dry that vapour rises from a melting glacier, but on sunny, frosty days there can be sublimation, i.e. the direct transition from ice to vapour. Finally, a glacier also receives heat from rainfall, but this contains little energy compared to other sources. If the amalgamated heat is enough to eliminate frost on a glacier's surface, it then proceeds to melt snow and ice.

There is hardly any frictional heat produced within glaciers because they move so slowly, but where there is most friction, at the base of the glacier, this can cause some melting, especially during surges. Geothermal heat from the Earth can maintain the base of the thickest polar glaciers at freezing point, but will melt only

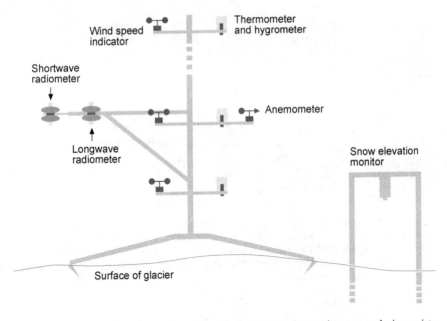

Fig. 1.18 Weather observations on a glacier. Automated weather stations on a glacier register various meteorological data in order to calculate the energy fluxes that cause summer melting on glaciers. They measure temperature, humidity, wind speed, wind direction, and all forms of radiation (shortwave solar radiation, shortwave reflection from the surface of the glacier, longwave radiation from the atmosphere and the surface of the glacier). Snow levels are measured sonically to record quantity of melting. Measuring all the forms of heat exchange is a complicated task and thus often simplified to finding a regular, empirical relationship between melting and various meteorological data, e.g. temperature, cloud cover, wind speed and humidity. Summer temperatures are an important factor in evaluating melting because of the high correlation between air temperature and the radiation and heat borne by winds over the glacier

about 1 cm a year at the base of temperate glaciers. In geothermal areas found under ice caps in Iceland, the melting of ice can be a few metres per year (Sect. 8.3).

1.8   The Plumbing System of a Glacier

Glaciers contain complex drainage systems. Meltwater leaks from the surface through the snow down to the underlying glacial ice. Some water seeps through very narrow veins on the edges of crystals and trickles down to the base of the glacier. Most of the water, however, accumulates in channels on the glacier's surface and pours into crevasses and very deep drains that are called moulins or sinkholes (Figs. 1.19 and 1.20). Moulins are formed where there is a tension expanding cracks in the ice and thawing water manages to trickle down through the ensuing chinks to the glacial base. The frictional heat from running water melts ice

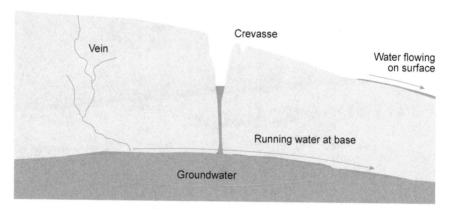

Fig. 1.19 Runoff meltwater channels from the surface of a glacier to its base. Microscopic, continuous veins of water at the edges of ice crystals are believed to be interconnected all the way to the base of a glacier. Soluble material that is transmitted from the ice to water during constant re-crystallization is flushed through them. Further downstream in glaciers, the ice can become purer in its composition than distilled water

Fig. 1.20 Moulins on Sólheimajökull (Chap. 5). On warm summer days, a loud murmur of running water can be heard from afar as meltwater cascades down moulins. They can thus easily be seen and heard, but in the autumn travellers must be very wary of them once snow has drifted over their openings. Small depressions covered in snow should then be avoided

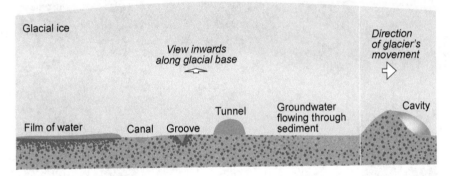

Fig. 1.21 Water channels at the base of a glacier. Cross section of glacial base. Water can flow as a thin film or run in runnels in the bedrock, or channels in the loose sediment, or in ice tunnels that are incised into the glacier. Water can also accumulate in cavities behind obstacles on the glacial bed. When water emerges from under a glacial snout, it has previously converged into a few large channels so that only a few main rivers flow from glaciers. Groundwater also flows beneath the glacier

walls and moulins are continuously enlarged during the summer, their circular openings even growing up to a few metres in diameter. During winter, when melting ceases, the pressure of the glacier's weight can constrict the moulins, only for thawing water to be channelled into and widen them once again when melting resumes the following spring. In the end, the moulins are transported by the ice flow beyond the crevassed area and fail to open up again when the spring meltwaters begin to run. New moulins, however, are constantly being created in the crevassed zone of an extending ice flow in similar areas as before.

Surface meltwater enters the glacier base from widely-spaced sources, but on its way along the glacier's bed, it collects in a few main waterways, which become outlets for larger and fast-flowing glacial rivers emerging from beneath the glacial margins (Fig. 1.21). These channels, full of water, criss-cross under the glacier like tree branches into its main trunk, which heads downstream. This is because the bigger channels with a high discharge have lower water pressure and thus tend to collect water from smaller neighbouring channels, which contain water subject to higher pressure. Thus, large channels are enlarged at the expense of smaller ones. When the water current increases, the ice tunnels expand because of the melting of their walls due to frictional heat, and thereby their water pressure is reduced. The resulting meltwater displaces only nine tenths of the volume previously filled by the ice that has melted.

The combined influence of the glacier's basal terrain's incline and the pressure of the water determine the speed and direction in which the water flows. There are channels full of water beneath a thick glacier, its overburden exerting great pressure on these conduits along its base. The overburden weight is typically greatest below the centre of the glacier, where it is thickest, and it thus directs the water to its sides, even up slopes, until the basal water channels have risen tenfold steeper than the glacier's surface; otherwise it runs along the sides of mountain ridges. The

Fig. 1.22 An ice tunnel formed by frictional heat from a subglacial waterway in Hrútárjökull in the Öræfi district (Chap. 8). Helgi Björnsson, farmer at Kvísker, stands in the tunnel's opening in the centre of the photograph. Hálfdán Björnsson 1965

accumulation zone of valley glaciers is concave so that water seeks the centre of the glacier, but in convex ablation zones the water tends to be directed out of each side of it and in these instances glacial rivers often emerge at both sides of its snout. A third drainage channel can still be in the centre of the glacier, nonetheless, if the valley sides are so steep that water flows along the valley's basin. Many glacial rivulets can sometimes emerge from half-full ice tunnels beneath the glacier's margins because the weight of the ice is not sufficient to press it down into the water and the only pressure on the water is thus atmospheric (Fig. 1.22).

Waterways at the base of the glacier become enlarged during the summer because the melting of the ice tunnel walls has the better over the pressure of the ice above trying to throttle them. Thus meltwater from the surface has a continually easier passage through the glacier to a river at its snout, its maximum flow being constant earlier in the day. When thawing finishes in the autumn, the weight of the ice regains the upper hand and the waterways are narrowed during the winter until they begin to expand again with the following spring's meltwater.

Water pressure can increase for a short while, nonetheless, due to water being suddenly borne into a channel during thawing, an extremely heavy rainfall, or a jökulhlaup, so that the channel cannot expand by melting rapidly enough to cope with the excess water. Such increased pressure is usually short-lived, however, because, given time, the tunnels can expand quickly with melting, thus reducing the pressure once again. With increased water pressure, the effective weight of the ice on the glacier's base is lessened, frictional resistance is reduced, and basal slide increases. The effects are tied to specific waterways because the water flows in demarcated channels. Where a glacier slides over uneven ground, increased water pressure can lift the glacier up off its bed, as if a hydraulic jack has been employed, so that a water-filled vault is created on the obstruction's lee side. The glacier then slides so quickly past the hindrance that it does not sink down onto the glacial base behind it.

Usually, there are only a few water-filled cavities, but it can so happen that they are formed one after another along the whole glacial bed and are connected by narrow ice tunnels that lie transverse to the direction of the glacier's movement. Such a water flow system is completely different to what has been described above, for now water pressure increases with an increase in flow. Melting ice from frictional heat does not suffice to ease the water pressure because the glacial slab settles instantaneously onto a thin film of water. Wherever water accumulates, its pressure increases so that it spreads over the glacial base, being driven from one place to another. The water is spread thinly beneath the glacier under such high pressure, that the glacier's basal slide increases vastly and it can surge forward as a whole.

1.9   Glacial Lakes and Outburst Floods (Jökulhlaups)

An outburst flood from a glacial lake is so common in Iceland, that the Icelandic word for them, jökulhlaup, has become internationally accepted as a glaciological term (Fig. 1.23). Jökulhlaups are sudden, fast, outburst floods from lakes in which rain and meltwater has accumulated instead of draining away in glacial rivers. They have often proved hazardous to populated regions, and over the centuries they have shaped and created landscapes, especially outwash plains and canyons (Sects. 8.2 and 8.8.6).

Jökulhlaups can originate in a subglacial lake, or in marginal lake, which is formed when a glacier dams a gully or tributary valley (Fig. 1.24). Water in the lakes rises until its pressure forces a way under the ice-dam, opening up channels along the glacier's base. At first these channels are very small, but they begin to expand because ice walls melt from frictional heat. The overburden of ice does not provide a sufficient hindrance to counteract the rate of melting, even if the water pressure were to be reduced as water flows out of the lake. An unstoppable surge of water then bursts from beneath the glacier into glacial rivers, its flow increasing much more rapidly than in currents caused by ordinary meltwater or rainfall. A typical jökulhlaup increases in volume exponentially over time until the weight of the ice finally succeeds in throttling the tunnels, thus bringing the flood to a sudden end, even before the lake has been emptied, for by then the expansion caused by melting ice is not sufficient to resist the influx of the overlying ice. The ice tunnels have expanded and their strength is reduced and the frictional warmth in the water has an increasingly further distance to reach the ice walls of the tunnel. At the end of a jökulhlaup, the drainage outlet is dammed once more and water begins to accumulate until the next outburst. There are about 15 marginal glacier lakes in

Fig. 1.23 A jökulhlaup on Skeiðarársandur outwash plain 1972 (see Chap. 8). HB

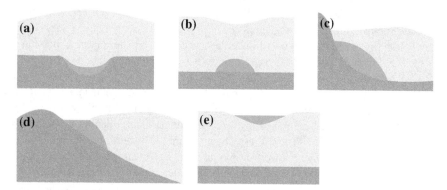

Fig. 1.24 Cross sections showing different kinds of glacial lakes. **a** Lake in a hollow in base bedrock. There are many such lakes under the Antarctic ice sheet. **b** Water cupolas under ice cauldrons in subglacial geothermal areas. Example (see Chap. 8): Skaftárkatlar (10 km northwest of Grímsvötn) and cauldrons in Mýrdalsjökull and Kverkfjöll mountains. **c** Subglacial lake on mountainside in geothermal area. Example: Grímsvötn in northwest Vatnajökull. **d** Dammed glacial lake (e.g. Grænalón by Skeiðarárjökull). **e** Lake on surface of glacier (e.g. in ice cauldrons)

Iceland, but their jökulhlaups are relatively small (more details in chapters on individual glaciers of Iceland). Floods from ice-dammed lakes at the margins of ice caps at the end of the last glacial period were much larger.

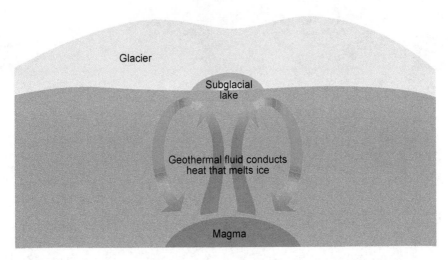

Fig. 1.25 When a glacier covers a shallow magma chamber, a geothermal area and subglacial lake can form

In some places the ice-dam is so large and dense that water cannot find a way under it and the water level rises in the lake until it flows in a steady stream over a col of bedrock or sediments. There are examples of such earthen dams giving way under the weight of water, or from ice and rocks crashing down onto them, the lake emptying very quickly. Such a jökulhlaup occurred when a rock face fell onto Steinsholtsjökull in 1967 (see Sect. 5.2).

Glacial lakes can also form in geothermal areas beneath glaciers. Ice melts at the base of the glacier and the meltwater is trapped beneath depressions consequently formed on the surface of the glacier (Fig. 1.25). Ice flows constantly inwards toward the centre beneath the depression and melts, forming water cupolas on the glacial bed. Slowly but surely, the ice cauldron becomes shallower, water pressure increases at the base of the lake, water rills reach further and further beneath the damming ice, and eventually the water forces its way out from beneath it, and so a jökulhlaup begins. The waterways grow bigger in a meltwater flood as the water cupola is reduced and the cauldron becomes deeper on the surface of the glacier. There are cauldrons like this on both Vatnajökull and Mýrdalsjökull, and they will be discussed in the relevant chapters (Chaps. 5 and 8).

Finally, a lake can form in the depressions on a glacier's surface but will seldom be very large because its water quickly reaches the base of a glacier through crevasses and moulins. Such lakes have been seen in cauldrons in Mýrdalsjökull and Vatnajökull, and also in depressions in ablation zones, e.g. on the Esjufjöll mountains (Sects. 5.1 and 8.5).

With an increasingly warmer atmosphere, more and more lakes have been spotted during the summer on the surface of the Greenland ice cap. If they remain unemptied, they freeze during the winter and amalgamate with the glacial ice. On

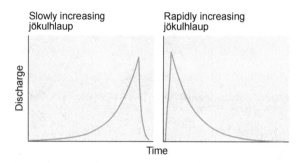

Fig. 1.26 Typical flow patterns in jökulhlaups. **a** Slowly increasing floodwaters. Most outburst floods from Grímsvötn are like this (Chap. 8). **b** Rapidly accelerating floodwaters. Examples of this include Katla (Chap. 5), many Skaftá jökulhlaups, and sudden Grímsvötn floods, e.g. in the autumn of 1996

the other hand, when water drains down to the base of the ice, this can accelerate a glacier's basal sliding.

A jökulhlaup usually begins before water pressure is sufficient to raise the ice-dam to flotation. There are a few examples, however, of water levels in a glacial lake rising until the ice-dam begins to float, and instead of its outburst being restricted to one channel, a broad stream of water floods from the lake, the layer of water spreading out beneath the entire glacier and advancing as a wave with so much pressure that the glacier is lifted to allow the water to make its way down to the snout. So great is the pressure of the water that it can surge up onto the surface of the glacier through crevasses and moulins, and there are even examples of it fracturing the ice bulk as it does so. Thus the flood can reach a higher discharge much more rapidly than when melting opens up a passage through an ice tunnel at the glacier's base. Such jökulhlaups increase almost linearly in time (Fig. 1.26). The broad wave of water can then break up into individual channels that bear the water speedily from beneath the glacier.

During volcanic eruptions beneath a glacier, magma melts the ice and jökulhlaups begin. The most meltwater comes when the eruption continues for a long time under a thick glacier. When a depression has formed on the glacier's surface (a supraglacial cauldron), water can accumulate in a cupola after which any eruption will be inside this subglacial lake. A typical jökulhlaup does not cause a glacial surge, but with volcanic activity beneath a glacier it can have rapid basal slides in certain locations. On steep, glacially-clad volcanoes, very dangerous mudslides and lahars can occur (e.g. on Öræfajökull, see Sect. 8.4).

Massive Jökulhlaups in Iceland
A jökulhlaup can have a tremendous impact on land formation by gouging out ravines and transporting sludge and icebergs over the outwash plains, destroying cultivated land and damaging roads and bridges. The destructive,

eroding power of a jökulhlaup can be gigantic and form canyons like Jökulsárgljúfur, Markarfljótsgljúfur, Hvítárgljúfur and Hafrahvammagljúfur (see Chaps. 5, 6 and 8). They also spread out over the lowlands and indeed jökulhlaups have played a part in forming Skeiðarársandur, which is currently the largest and most active glacial outwash plain on Earth (Sect. 8.2). Jökulhlaups threaten farmsteads, communities and hydro-electric power plants that use glacial waters, and they have even caused tsunamis in coastal regions. Mud in a powerful flow of water increases its destructive capabilities and if the flow is saturated with sediment it runs as a slow floating liquid that can also turn into mudslides (mud flows, debris flows and lahars). Jökulhlaups from subglacial lakes in Iceland can transport and disperse 10,000,000 tonnes of sediment and, in conjunction with a volcanic eruption, as much as 100,000,000 tonnes.

The most tremendous jökulhlaups in Iceland have probably emerged from the large, ice-filled calderas of Bárðarbunga and the Kverkfjöll mountains to the north of Vatnajökull (Sect. 8.8). The jökulhlaup that gouged out Jökulsárgljúfur canyon probably came from there, a flood of 400,000–1,000,000 m^3/s. There have been major jökulhlaups from Öræfajökull since the beginning of historical times (Sect. 8.4). The first contemporary description of such a flood is from the eruption of Öræfajökull in 1362. It ended within twenty-four hours and may have reached a maximum flow of 100,000 m^3/s. The meltwater was created in a 500-m-deep caldera at a height of 2000 m and emerged from beneath an outlet glacier at a height of about 100 m. A large number of farmsteads were swept away in this catastrophe and the subsequent ash fall caused even more damage. Similar jökulhlaups emerged from Eyjafjallajökull in 1612 and 1821–1823 and during the Hekla eruptions of 1845 and 1947 (Sects. 5.2 and 5.5). Knowledge of the nature and origins of jökulhlaups are necessary in order to give timely flood warnings. Seismographic data give advance warnings of volcanic eruptions, and maps of glacial beds are useful in assessing which areas will be most at risk. The largest jökulhlaups that now occur on Earth have a maximum flow rate of 1,000,000 m^3/s. Research on the interplay between glaciers and volcanoes can also increase our understanding of the effects of volcanic activity on other planets, e.g. on Mars.

1.10   Glacial Surges

Surges are temporal instabilities in a glacier's movement. As has been described earlier, glaciers transport ice in a gentle but steady stream from the accumulation zone down to the snout, although this progress is so slow that it is invisible to the naked eye. There are glaciers, however, that do not move quickly enough to

Fig. 1.27 Síðujökull surges forward, its steep, crevassed snout breaking up (Sect. 8.9). HB, March 1995

transport all the volume of snow that has piled up in the accumulation zone, while at the same time the glacial snout is receding and thinning out, so that a glacier becomes steeper with each passing year; for some as yet unclear reason, their basal slides are insufficient. Thus the pressure on the walls of the ice tunnels bearing basal water continually grows to the point where they eventually collapse and water is widely dispersed through numerous connected subglacial vaults beneath the glacier instead of flowing forwards in a few ice tunnels. The sediment bed beneath a glacier can also become waterlogged and lose its solidity, so that a glacier can slide on what is effectively sludge. Water under high pressure lubricates the base of the glacier and lifts it, the velocity of the glacial slide increasing considerably, forming deep tension-fracture crevasses; indeed, a glacial surge is first noticed when crevasses become visible high up on a glacier. The chain of crevasses becomes larger all the way down the glacier until it finally reaches the snout, where the glacier's margin begins to break up (Figs. 1.27 and 1.28). Brownish-coloured water now trickles forwards from all of a glacier's margins in a myriad of small channels and not from a few waterways as before. The speed of the glacial slide can increase a hundredfold, the ice moving as far in one day as it had previously done in six months or a year. The snout then surges forward and the accumulation zone descends for dozens of metres while the glacial tongue rises and advances hundreds of metres, even a few kilometres, within a few months. This great upheaval increases the muddy sediments in the glacial rivers. All this time, water-filled cavities are retained behind obstacles on the glacier's bed. A fast ice slide holds in check the

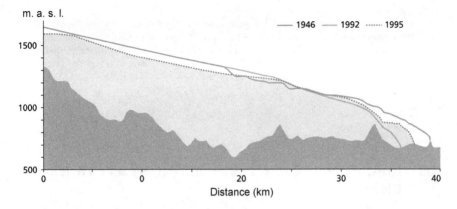

Fig. 1.28 Elevation changes in Tungnaárjökull during surges (Chap. 8). Tungnaárjökull surged in 1946, then retreated about 4 km until 1992 and grew thicker above an elevation of 1100 m. In 1995 it surged forward about a kilometre, but its surface above an altitude of 1100 m subsided

high-pressure water and the cavities so that the glacier can move forward rapidly over them. But when a glacier begins to spread out and the basal slide gradually slows down, water manages to distend the vaults and is dispersed into ice tunnels with low water pressure which discharge the water from beneath the glacier very speedily, in what could even be called a flood. This is how a surge ends. Rivers find their old fast-flowing channels. The high water pressure that had originally put a glacier into fast motion, then ultimately brings it to a halt by kicking the skateboards from under its feet, so to speak.

This is how masses of snow can be rapidly transported by surges from the accumulation to ablation zones of glaciers instead of being only borne onwards in a steady ice flow. After such a surge, surface melting increases because a snout has reached a lower elevation than previously and its surface has expanded and fractured. Surges thus cause long-term fluctuations in water runoff and sediment transport from glaciers. The glacier then retreats over the next few years, even decades, until a surge may commence once more.

A Glacier 'On the Move'

'It sometimes happens that large outlet glaciers all of sudden begin moving to an unusual extent, without there actually being any jökulhlaup as such, but its movements are of a similar nature and are probably caused by a huge volume of water in the glacier or sometimes by volcanic outbursts from far away. The glacier is then said to be "on the move." When this happens, the snout bulges outwards tremendously, tearing itself apart, turning itself inside out, water gushing out of the fissures; the pressure of the water and the glacier pushes more and more ice down to the end of the snout so that it grows in height and is in constant motion, pushing up earth, moraines, glacial till and boulders into piles and ridges; the glacier often presses so hard into the earth adjacent to it that it pushes it up in many folds, rolling up whole swathes of grass within them. After some time, the glacier reaches an equilibrium and comes to a halt, the water drains away, and all the jumble of ice that has piled up at its snout's end melts, and everything returns to normal, except that the glacial moraines have been driven much further forward than they would otherwise have reached under normal circumstances.' (Thoroddsen 1907–1911, Vol. 2 (1911), pp. 12–13).

1.11   Erosive and Creative Powers of Glaciers and Their Rivers

Glaciers have shaped some of the most spectacular landscapes on Earth. They can even be likened to a coarse rock crusher with a powerful conveyor belt and water sluicing system. At the base of a glacier there are large boulders, rocks and stones with sharp edges, gravel, sand, silt, clay and pulverised stones, which are all mixed together. A glacier pushes these digging tools forwards under a tremendous weight (Fig. 1.29). They carve out tracks, plough grooves, and scrape furrows into the underlying earth and rock. The digging machines crush and grind, while sediment layers beneath the glacier creep forwards, scraping, scratching, and polishing. The thrusting power of glaciers is colossal (Fig. 1.30). Glacial water also simultaneously rinses away ensuing dirt and debris so that the digging machines can continue to carve into the pristine bed, and channels meander all over the base so that water continually succeeds in flushing away materials from new places. The capacity of glacial gouging depends on the speed of the basal slide, the hardness of the digging tools, the power with which they are applied, and the durability of the underlying bedrock. Most sediments can be produced and transported by glaciers in surges. Glaciers prefer to burrow down deeply in the accumulation zone and erosion is usually most beneath the equilibrium line, where the glaciers are typically thickest and move most rapidly. By the time the snout is reached, both the speed and transporting capacity of a glacier are reduced and glacial till can settle at the base, especially in hollows. In active glaciers, the till reaches right to the end of the snout and is pushed up along with earth into a moraine at the sides and front of the terminus. If the glaciers lose power and retreat in a warming climate, they leave behind ground moraine, scattering erratic boulders and rocks in the wake of their flight back to the highest mountains. Kettle holes are created where stagnant ice and icebergs have melted (Fig. 1.31). This is how poorly-vegetated, gravelly screes and muddy sediment have come into existence in the forefields of glaciers that have been retreating for most of the 20th century (Fig. 1.32).

Braided gravel channels (eskers) meander through land where water once ran through them in ice tunnels at the base of a glacier. Water erosion can be considerable in large rivers beneath glaciers, especially when excavating water flows so rapidly (more than 15 m/s) that its surface tension cannot hold it together and, in a

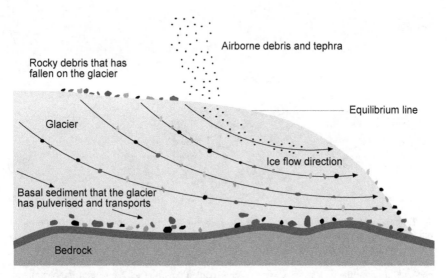

Fig. 1.29 Sediment transportation routes along a glacier. Most of the material that a glacier transports has been torn up from its bed and moved along its base. The glacier also bears ash, dust and dissolved matter that has fallen onto its surface from the atmosphere or from mudslides and avalanches, or has been shattered and removed from mountainsides eroded by frost weathering. In the accumulation zone, everything that settles on the surface is englaciated in the ice flow and re-emerges in the ablation zone, after which it remains on the surface of the glacier. The further down the glacier that debris is enveloped by ice, the higher up in the ablation zone it will resurface. This moraine overcoat can be so great at its snout, especially if landslides have fallen on the glacier, that the ice itself is no longer visible and a moraine-covered glacier is thus produced. It looks like a rocky moraine, but under it the ice advances; its overcoat can be such an effective insulator that the glacier can be preserved for centuries even when other glaciers in the vicinity have disappeared. In other instances an intermingled mixture of rocks and ice may form so-called rock glaciers, many of which can be found between Skagafjörður and Eyjafjörður and they have hardly retreated in the last half a century (Chap. 7)

process called cavitation, vacuum cavities are formed in the currents. These cavities then suddenly collapse and very sudden changes in pressure fracture rock. This happens in catastrophic floods such as those that are thought to have formed the Jökulsárgljúfur canyon north of Vatnajökull (Sect. 8.8.6). Potholes are common where moulins have reached a glacier's base. There can be considerable chemical erosion at the base of a glacier because glacial water continuously releases carbon dioxide out of air bubbles that are trapped in the ice when it is formed.

Glacial rivers transport eroding materials from beneath the glacier down into the valley and disperse them over the outwash plains. The sediment load fluctuates with the flow of water, from night to day, from summer to winter. When these rivers are greatly swollen, the trundling of boulders in them is such that loud thuds can be heard and riverbanks be seen to shudder. Glacial surges and jökulhlaups bear the greatest sediment loads and have the swiftest effect on forming the landscape. As the rivers reach further away from the glaciers, they become more diluted and dispersed, reducing the flow of water and its ability to transport sediments. The

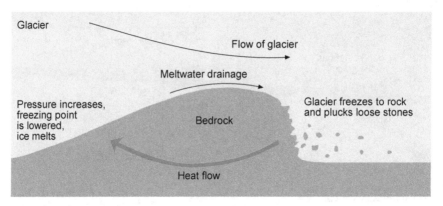

Fig. 1.30 Formation of a roche moutonnée (or whaleback). When ice that is at the melting point collides with a bedrock bump, it melts because under increased pressure the melting point of ice is lowered: its own actual temperature does not change, but it is now above melting point. The produced meltwater drains away. The heat required for melting ice upstream of the bump is conducted through the rock from where the ice is cooled down on the lee side of the obstruction and frozen to the rock. Pressure is less behind than in front of the obstruction and so ice freezes to the rock base there. Frost fractures the rock and the ice flow tears away clumps of it. Due to fluctuations in water pressure behind the bump, the glacier can be suddenly uplifted, as if by hydraulic jack, ripping up pieces of rock. Therefore, the lee side of an obstruction is rough and craggy, but its top and sides are polished by the basal slide, lubricated by meltwater. This is how a roche moutonnée is formed. On beds where ripples are only a few centimetres long, the meltwater may flow straight over the hindrance and freezes on the lee side releasing a latent heat of fusion which is then conducted back through the obstacle to the place where the ice first melted. If such a heat-conducting cycle is established, the glacier will not freeze to its bed anywhere

Fig. 1.31 Two young girls exploring a kettle hole in Skeiðarársandur from a jökulhlaup of 1895 or 1902 (see Chap. 8). Kettle holes are circular hollows in gravel plains created by melting blocks of ice that have been stranded there after a meltwater flood. HB, September 2005

Fig. 1.32 The many different traces left by glaciers. The smooth gabbro rock of Geitafell on the margin of Hoffellssandur (see Sect. 8.6). HB, July 1999

largest rocks thus quickly sink to the river bed, water spreads out over an ever-widening area, and gradually a more and more fine-grained sediment settles in layers on the bottom. The most minute, pulverised, erosive material is borne the furthest, and the rivers maintain a milky colouring all the way to the sea, even during wintertime.

What began with the story of a snowflake falling from the sky, now ends with the revelation of its ultimate destiny: as part of one of nature's most powerful tools in eroding and shaping the landscapes of Earth.

References

Benn, D., & Evans, D. (2010). *Glaciers and glaciation* (2nd ed.). London: Hodder Education.

Cuffey, K., & Paterson, S. (2010). *The physics of glaciers* (4th ed.). Amsterdam: Academic Press.

Hambrey, M., & Alean, J. (2004). *Glaciers* (2nd ed.). Cambridge, UK: Cambridge University Press.

Hooke, R. (2005). *Principles of glacier mechanics*. Cambridge, UK: Cambridge University Press.

Knight, P. (1999). *Glaciers*. Cheltenham: Stanley Thornes.

Paterson, W. (1994). *The physics of glaciers* (3rd ed.). London: Pergamon.

Sharp, R. (1991). *Living ice: Understanding glaciers and glaciation*. Cambridge, UK: Cambridge University Press.

Thoroddsen, Þ. (1907–1911) *Lýsing Íslands*, 2 vols. Copenhagen: Hið íslenzka bókmenntafélag.

Van der Veen, C. (2013). *Fundamentals of glacier dynamics* (2nd ed.) CRC Press.

Chapter 2
Reading the Landscape

'[A glacier] is an endless scroll, a stream of time, upon whose stainless ground is engraven the succession of events, whose dates far transcend the living memory of man.' (Forbes 1843, p. 22).

Abstract In this chapter the history of European knowledge of glaciers and their influence on landscape and climate is traced, from Ignatz Venetz-Sitten and Jean de Charpentier's glacial hypothesis, to Louis Agassiz's and Karl Schimper's theory of the Ice Age, and the gradual acceptance of the latter by the end of the 19th century. Further details of climate history were later revealed through fossilised flora and fauna in sediment layers (strata) and various dating methods. By the late 20th century, temperature records of several glacial and interstadial periods had been derived from oxygen isotopes in fossilised shells in deep-sea sediment strata and glacial ice cores from Greenland and Antarctica. Theories on the astronomical causes of climate fluctuations are outlined, e.g. those of James Croll and Milutin Milankovitch on the influence of Earth's orbital pattern and axial precession and tilt on solar radiation penetrating the Earth's atmosphere. A summary of glacial epochs from 130,000 years ago until the present time is then related, with reference to specific periods, e.g. late glacial, the Holocene and the Little Ice Age. The role of the North Atlantic and the transportation of tropical oceanic heat in climate fluctuations is also discussed. A brief history of glaciers in Iceland follows, from the settlement of the country until the present day. Finally, there is a discussion on the current climate of Earth and possible future changes.

2.1 The Long, Winding Road to Understanding Ice Ages

For thousands of years, the peoples of the northern reaches of Europe and America inhabited an environment shaped by ice-age glaciers, living among glacially polished rock formations and erratic boulders on ancient outwash plains and in glacially-gouged valleys between jagged crests of mountains (Figs. 2.1, 2.2, 2.3 and 2.4). But even though these traces of ancient glaciers were there for all to see, it

© Atlantis Press and the author(s) 2017
H. Björnsson, *The Glaciers of Iceland*, Atlantis Advances
in Quaternary Science 2, DOI 10.2991/978-94-6239-207-6_2

A GRETTISTAK.

Fig. 2.1 'Grettir's lift' or erratic boulder. Metcalfe (1861). NULI

Fig. 2.2 Ground moraine and an erratic boulder in the central interior of Iceland on the Sprengisandur highland route to the north of Lake Þórisvatn. HB, 2004

Fig. 2.3 Glacial terrain. The jagged peaks of Þverártindur (1554 m) with glacially eroded corries and steep outlet glaciers to the west of Kálfafellsdalur, inland from the Suðursveit district of Austur-Skaftafell County (Chap. 8). The pinnacles Karl ('man') and Kerling ('woman') are on the far right

was not until the 19th century that it finally dawned on scholars that a glacier had previously covered their territories (see Imbrie and Imbrie 1979, for a fundamental source of this chapter). Up until then, they had been convinced that lands and oceans had been virtually unchanged for as far back as their ancestors' memories and recorded sources reached. Accordingly, they believed fertile fields, forests, marshes, and steppes had always existed and that mountains and valleys were a permanent part of the landscape. No narratives or experiences had been preserved of their predecessors who, 12,000–14,000 years ago, had to fight for their very survival at the edges of ice-age glaciers in Europe and North America. Nordic mythology, however, did at least preserve some legends of an Earth and sky originating from a frozen giant, who had originally been created from the mists of thawing ice that had previously enveloped the universe. And in the end, the world would not be destroyed by fire, but by the freezing of lands and oceans at Ragnarök (Sturluson 1987, pp. 52–53).

In countries with glaciers, such as Iceland, Norway and Switzerland, farmers and travellers had long noticed that large boulders and stony screes in otherwise green and vegetated valley basins, and even on flat outwash plains, were similar to those they were familiar with far away up by the glaciers (Fig. 2.5). In Iceland these boulders were called 'Grettir's lifts' and in other countries 'erratics' because the rocks were unlike anything in their nearest environments and it was if they had somehow strayed there. Those who had often visited glaciers knew that their snouts

Fig. 2.4 Single-standing mountains are sometimes all that remain of high promontories between deep valleys carved out by ice-age glaciers. Kirkjufell (463 m a.s.l.) in Grundafjörður on the Snæfellsnes peninsula. HB, August 1998

could advance with tremendous power, shaping the land they traversed, and they thought it only natural to conclude that glaciers had previously borne onto their home fields and meadows the gigantic boulders that no human power could ever possibly remove. Such reflections had been preserved in oral traditions, but the first written observation that erratic boulders could reveal the size of ancient glaciers came in 1787 from the Swiss lawyer Bernard Kuhn (1762–1825). Eight years later, in 1795, a Scottish geologist, James Hutton (1726–1797), the founder of modern geology, stated that he believed a great glacier had covered the Alps in former times and had deposited granite rocks all over the Jura Mountains, which are themselves made of limestone. Hutton's (1795, 1959) observations were given scant attention, however, as his writings were stilted, a chore to read, and often obscure. Early in the 19th century, a German geologist, Leopold van Buch (1774–1853), put forward the theory that erratic boulders had come from Mont Blanc, the highest mountain of the Alps, which had suddenly surged upwards and scattered these rocks far and wide; erratic boulders in northern Europe had similarly arrived from Scandinavia. Neither craters from such a shower of rocks, nor scattered debris from such boulders shattering on impact, were anywhere to be seen, however. Most natural scientists thought it more likely that an enormous flood had transported these rocks, while at the same time creating kettle holes on mountain summits. It was well known how

Fig. 2.5 View from Steinasandur up to Brókarjökull, just visible at the head of Kálfafellsdalur (Chap. 8). It is hard to believe this apparently small outlet glacier had once extended right out onto the outwash gravel plain. HB, July 1999

rivers left behind large rocks, gravel and sand in their beds, and so until the second decade of the 19th century, it was generally considered that all the debris dispersed over the land, containing a mixture of large rocks, gravel and sand, had been deposited there by one or more catastrophic floods similar to those described in the tales of ancient peoples, such as the Egyptians. Moreover, as such a flood had been recorded in the Bible, it was thus an irrefutable fact. On the other hand, however, there were no comparable examples of more recent catastrophic floods, though there was evidence to the fact that in 1755 a huge flood wave (what today would be called a tsunami) had catapulted large boulders ashore at Lisbon during the great earthquake which took place just off the shores of Portugal.

Did icebergs transport erratic boulders? It was obvious that large boulders could not float and that, no matter how powerful, no current of water would be capable of moving granite rocks 1500 m high up onto the valley sides of the Jura mountains, on the border between France and Switzerland, or the hundreds of kilometres from Norway over the North Sea to the northern parts of England and Germany. The English geologist Charles Lyell (1797–1875) put forward the hypothesis in 1833 (*Principles of Geology*) that the boulders had been transported by floating icebergs, which had drifted from the polar regions when the sea level was higher than it was now (Lyell 1830–1833, 1865, 1990–1991). Originally frozen to the icebergs, the boulders finally broke off and sank when the icebergs melted. Such icebergs were familiar to seafarers in the northern seas and fossilized fish and

Fig. 2.6 Early thinkers of glacial landscaping. (From *left to right, top to bottom*) James Hutton (1726–1797), Scottish geologist. Jens Esmark (1763–1839), Norwegian geologist. Ignatz Venetz-Sitten (1788–1859), Swiss civil engineer. Jean de Charpentier (1786–1855), German-Swiss geologist. Louis Agassiz (1807–1873), Swiss scholar of natural sciences. Karl Friedrich Schimper (1803–1867), German natural scientist. William Buckland (1784–1856), English geologist. Georges Cuvier (1769–1832), French palaeontologist. Otto Andreas Lowson Mørch (1828–1878), Danish conchologist

shells in sediment strata in Europe revealed that the ocean had once covered land. It thus seemed more logical to assume that ice had travelled from northern regions by floating across the ocean rather than by crawling over land. Icebergs could also have been borne south in gigantic floods. The icebergs highest in the Jura Mountains, however, must presumably have originated from ice in lakes. Drifting ice not only carried large boulders, but also dispersed a haphazard mixture of clay, sand, stones and gravel, which was why this kind of debris was given the name 'drift' in English.

Did glaciers disperse erratics? This is how things stood until early in the 19th century (1815), when Jean-Pierre Perraudin, a guide in the southern part of the Swiss Alps, opined that one might conclude from the striation in the bedrock, which all slanted downwards, that glaciers had previously descended slowly along these valleys and bore with them rocks. He failed to arouse the interest of the German-Swiss geologist Jean de Charpentier (1786–1855) in his 'glacier hypothesis,' but he did succeed in gaining the attention of the civil engineer Ignatz Venetz-Sitten (1788–1859), who in 1821 published the conclusions to his own observations on landscape formation far from glaciers supporting the idea that glaciers in the Alps had previously extended far down into the valleys and moulded the topography there (Venetz-Sitten 1861). In 1824, the Danish-Norwegian geologist Esmark (1763–1839, 1824) came to the conclusion, completely independently from the Swiss scientists, that glaciers had formerly covered a large part of Norway and its offshore seabed and had transported erratic boulders and pushed up moraines (Fig. 2.6).

2.2   Venetz-Sitten, de Charpentier and the Glacial Hypothesis

In 1829, Venetz-Sitten (1861), in an instructive lecture for the Swiss Society of Natural Sciences, maintained that the lowlands to the north of the Alps, and indeed the whole of northern Europe, had previously been covered by a glacier. He described erratic boulders in the valleys, often high up on mountainsides, which were both striated and with sharp edges similar to rocks in the vicinity of active glaciers; he also gave an account of the curved ridges of gravel and rock that stretched right across vegetated valleys, similar to those in pastures high up near the Alpine glaciers. This now aroused the curiosity of Jean de Charpentier, the overseer of Swiss salt mines, who had become interested in glaciers in 1818 after a large number of people perished in a flood caused by a proglacial lake bursting its ice-dam. De Charpentier himself began examining erratic boulders and glacial moraines, and at a meeting of the Swiss Society of Natural Sciences in 1834, he systematically presented a variety of evidence supporting the idea of landscapes being formed by glaciers that had long-since disappeared. De Charpentier was a meticulous and highly-regarded scientist, and he presented facts that are now considered indisputable evidence of ancient glaciers, but his lecture was nonetheless greeted rather indifferently by his colleagues. He later mentioned that on his way to the meeting a woodcutter had told him how he believed ancient glaciers had formed the landscape, exactly as de Charpentier himself had been similarly convinced. Opponents of the glacier hypothesis, on the other hand, inclined to agree with Lyell's idea that icebergs had previously floated on the sea above Switzerland and, when they melted, erratic rocks had broken free from them and been deposited on the land. In the Alps, mountains are indeed more noticeable than glaciers, so that hypotheses about giant glaciers seemed unlikely, especially since the huge glaciers that actually did exist on Earth at that time still remained unknown to man.

Proclaiming the theory of a large glacier, which had long-since vanished. Jean de Charpentier was not particularly interested in making his views widely known or accepted, but he did want his colleagues to appreciate them. One of these colleagues was a former grammar school student of his, Louis Agassiz (1807–1873), an ambitious biologist who had become chairman of the Swiss Society of Natural Sciences. Agassiz had researched fish fossils in sediment strata and his work had attracted a great deal of attention from geologists because he could differentiate between freshwater and oceanic fish and thus decide whether sedimentary deposits had been formed in lakes or the open sea. In 1836 de Charpentier invited Agassiz to spend some time with him during the summer at his mountain retreat, where he showed him the traces of what he considered evidence that great Alpine glaciers had once covered the land there. Agassiz had listened to de Charpentier's lecture in 1834, but had been sceptical about his glacial hypothesis. Seeing with his own eyes, however, how the landscape had been formed by glaciers pushing ahead of them rubble and striating and smoothing the bedrock, it then became clear to him that the erratic boulders lay in demarcated areas at the mouths of valleys, which glaciers had previously traversed, and were thus not dispersed as haphazardly as might be supposed if they had been transported there by floating icebergs at sea. Moreover, the erratics' sharp edges seemed to indicate they had not been eroded in the tumbling waters of a massive flood. The fact that rocks were found over a much wider area than oceanic fossils also countered the idea that erratics had sunk from icebergs to the ocean bed. Agassiz not only concurred with the ideas of Venetz-Sitten and de Charpentier, but he also became captivated by a new vision (see Carozzi's translation of Agassiz 1967).

2.3   Agassiz, Schimper, and the Birth of the Theory of the Ice Age

Agassiz was a fervent man, imaginative, energetic and indefatigable. His task would not give him any peace, neither by day nor sleepless nights. He now turned all his energy towards glacial research and investigated in detail the movements of the Aar glacier in Switzerland, becoming convinced that glaciers could transport huge boulders. He sent for his old friend and schoolfellow, the botanist Karl Friedrich Schimper (1803–1867), who was also a passionate observer of nature. Schimper encouraged Agassiz, when others tried to dissuade him, and thus began a collaboration which led to Venetz-Sitten and de Charpentier's glacial hypothesis becoming transformed into the theory of the Ice Age. At first the two collaborators pointed out to their colleagues that glaciers had covered the Alps from the Jura Mountains to Bavaria. The glacier then grew so rapidly in their minds, however, that in the end they considered that at an even earlier point in Earth's history a thick glacier had enveloped the northern hemisphere from the North Pole area to as far south as much of the continental mainland of Europe, Asia and North America. This hypothesis of a continental glacier was very daring. No human being lived near the remains of an ice-age glacier in Antarctica and it was still not clear then that one complete ice-shield covered the whole of Greenland. The idea that there had been a period in Earth's history when a glacier had covered land from the North Pole to Germany, had actually been promulgated earlier, in 1832, by the German professor and forester Reinhard Bernhardi (1797–1849), who was familiar with the work of Jens Esmark in Norway, but Bernhardi's publication (1939) had attracted little attention.

Agassiz and Schimper pointed out that the existence of an ancient, enormous continental glacier could only ultimately be explained by the Earth cooling so extensively that an Ice Age had suddenly been unleashed. Since then, the Earth had gradually become warmer once more, but never reaching the temperature prior to the Ice Age, thus in the long term the Earth had cooled overall. This was a revolutionary idea that had not occurred to anyone previously, for it had always been assumed that Earth had been in a continuous cooling state ever since it had come into existence. Lyell had always maintained that, despite a few local and minor fluctuations, Earth's climate had been stable. De Charpentier stated that the large Alpine glaciers had been formed because the mountains had then been higher than in modern times and thus much colder. Agassiz and Schimper put forward yet another more radical idea: that the Ice Age had been a catastrophe for the natural environment and had exterminated all forms of life wherever ice had covered the Earth. Agassiz presented this theory to the Swiss Society of Natural Sciences in 1837, going a step further than Esmark, Venetz-Sitten and de Charpentier, who had proposed that only glaciers in the Swiss Alps and Norway had thickened and extended over nearby areas of land. De Charpentier now considered the young enthusiast to be making reckless assertions about ice-age glaciers covering whole continents, and felt that Agassiz had gone well beyond what his evidence could prove, especially since he had never even been to North America or Asia.

Agassiz became an energetic bearer of these new tidings and published his theory of the Ice Age three years later in 1840 in his book *Études sur les glaciers*, which he dedicated to Venetz-Sitten and de Charpentier. Both they and Schimper, however, felt that their contributions were given scant acknowledgement in Agassiz's book (1840, 1967), and they never forgave him for this. De Charpentier had himself been working for a long time on a book and considered he had the right to publish his own findings first, and thus Agassiz's dedication of his book to him

Fig. 2.7 Karl Friedrich Schimper's diagram showing how temperatures might have fluctuated on Earth and thus caused the sudden onsets of ice ages. From Agassiz (1840)

(which had been published a few months ahead of de Charpentier's work) did not make up for this. Furthermore, Agassiz had included Schimper's diagram in his publication, showing how fluctuating temperatures on Earth could have brought about ice ages, without even citing his original source (Fig. 2.7).

Schimper had become the first to put forward the concept of the Ice Age, *Eiszeit*, in 1837, and he was a much more likely scholar than Agassiz to leap, without evidence, from the glacial hypothesis to the Ice Age theory. Schimper was a romantic scientist who was searching for a comprehensive, universal and perfect coherence in the natural world. He was looking for a complete picture, without closely examining its individual pieces, as he created the whole by fitting them together, as in a jigsaw puzzle; the theory of the Ice Age was a vision similar to a model he was looking for to illustrate the pattern of plant life. Agassiz, on the other hand, was more down-to-earth, and his conclusions always strictly adhered to factual evidence; he did not regard his results as a theory, but as established facts. Later in his life, and using a similar tone to that of his mentor, the French pioneer in palaeontology Georges Cuvier (1769–1832), who had always scoffed at romantic scientists, Agassiz wrote scornfully of Schimper's contribution, claiming that Schimper had merely put forward some speculations. It is now an impossible task to distinguish exactly between Agassiz's original contribution to the theory of the Ice Age, on the one hand, and Schimper's on the other, but it is clear that both of them made such vital contributions that neither of them could have put forward the theory without the work of the other. Indeed, it is a needless simplification to attribute the theory of the Ice Age to a single man, as is so often done. Nonetheless, it should be pointed out that if the pioneers of the Ice Age theory had restricted themselves to only the obvious facts, they would not have advanced knowledge beyond the stage it had reached in 1830. Neither Agassiz nor Schimper attempted to explain the cause of the Ice Age, and indeed they did not completely understand all the implications of the Ice Age hypothesis, as will be noted later.

Difficulties of promulgating the theory of the Ice Age. Agassiz preached the Ice Age theory with religious conviction and eloquence. Sheets of ice had destroyed all life forms, buried mammoths, covered inland lakes, oceans, and large tracts of land, so that the silence of death reigned there: rivers ceased to flow, the Sun rose over a frozen Earth, the northern winds howled, and deathly-deep crevasses in the ocean of ice groaned and rumbled. Agassiz considered the frozen remains of whole mammoths, which had been discovered in Siberia, as illustrating two points: how rapidly the Ice Age had descended and what kind of catastrophe had caused the mammoths' extinction. There had been previous hypotheses that a gigantic flood

Fig. 2.8 Complete frozen mammoths can be found in Siberia. Had a catastrophic flood swept them onto this tundra zone, or had an ice age been suddenly unleashed upon them? Model of a mammoth in the Parc de la Ciutadella in Bercelona, Spain

had swept the mammoths along great rivers from their southern climes all the way north to Siberia, where they froze to death, for indeed the land there was uninhabitable because of the cold and there was nowhere near enough forage for such a large land animal (Fig. 2.8).

Georges Cuvier had in 1796 previously come to the conclusion from his study of the mammoths' skeletons that they were an extinct species unrelated to African elephants. Agassiz had now presented the hypothesis that they had become extinct because of an ice age and not due to a catastrophic flood. The son of a pastor, Agassiz felt that this could well accord with his belief in God and even be cited as God's direct intervention in history and His creation of a new world following catastrophes. Ice instead of water could hardly be seen as seriously contradicting the words of the Bible. The Ice Age completely destroyed all life, and when it came to an end new species of life forms immediately arose without being connected to any predecessors on Earth, and they would remain unchanged until they too became extinct. Agassiz thus supported the stance of his mentor Cuvier, as opposed to the theory of evolution of Jean-Baptiste Lamarck (1744–1829) that species developed into other species. Agassiz considered his and Cuvier's stance not as a theory but a fact! The existence of fossils in sediment strata was explained in the 18th and 19th centuries by natural catastrophes that had swept over the Earth and destroyed life again and again, making species extinct, and that life had then been revived once more by an intelligent instigator who had a continuing interest in maintaining his creation. This was how life developed through natural catastrophes. Agassiz believed that all connections between life forms before and after the Ice Age had been severed and he never accepted the theory of the origins of species by Charles Darwin (1809–1882).

The response to Agassiz's message was a mixture of apathy and denial. More was clearly needed to overturn the ideas then current and to introduce new ones. Agassiz was a tireless herald of the theory of the Ice Age and gained a tremendous victory when he managed to persuade an important adversary to support his views: William Buckland (1784–1856), Professor of Geology at the University of Oxford in England. Buckland (1836) had been a convinced spokesman for the theory of the catastrophic flood, which had once inundated the entire world, and believed it explained the irregular distribution of rocky debris in the British Isles. Being a geologist, his contribution to supporting the text of the Bible had increased respect for geology within the conservative British society; one of his publications, for instance, was titled: *Geology and Mineralogy, Considered with Reference to Natural Theology*. It was evident to the geologist, nonetheless, that the supposed dispersal of rocks from a catastrophic flood could not be specifically attached to one area, or be an indication of one forty-day Biblical flood.

Daring scientific theories often greeted with scepticism. In 1838 Buckland travelled around Switzerland with Agassiz and saw the polished and striated rocks and moraines that Agassiz considered traces of ancient glaciers. Buckland was not convinced, however, until in 1840, when he travelled with Agassiz in Britain and saw for himself similar landscapes in Scotland, Wales and northern England. From then on, Buckland became a sincere spokesman for the theory of the Ice Age, as this explained the existence of sediment layers in large areas of Britain far better than the Great Flood. Buckland then showed these landscapes to Lyell which resulted in the latter also agreeing with the Ice Age theory, though he did not become its spokesman. Indeed, Lyell never believed that ice ages had destroyed all life on Earth and, like his predecessor James Hutton, always criticised theories founded on the Biblical version of the creation of the world, which presented catastrophes as an explanation for the existence of fossils of extinct life forms. Both Lyell and Hutton maintained that fossils had been buried in sediment strata which would have been layered at exactly the same pace as in contemporary time. Thus the present would be the key to the past. Past events in geological history could be explained through processes which were still ongoing and which always obeyed and revealed the same laws of nature; processes and natural laws remaining constant are the core principles of uniformitarianism. Hutton's theories, however, had been more or less dismissed because they implied that Earth was much more than 6000 years old, contradicting a belief the ruling social powers strictly adhered to in the 18th century.

British geologists still maintained the old opinions, therefore, and could not envisage large glaciers in Wales or anywhere else in the British Isles. The mountains were simply not high enough in order to help create considerably large glaciers, and inland lakes were too big to have been buried under the ice of small glaciers. British scientists could agree to earlier glaciers in the Alps once being larger than they were currently, for indeed it was clearly visible towards the end of the Little Ice Age that the Alpine glaciers were variable and had recently advanced significantly. On the other hand, it was extremely difficult for people to imagine gigantic continental glaciers many kilometres thick and thousands of kilometres in length. Moreover, physicists were unfamiliar with glaciers and could not understand how they could move, let alone creep up mountainsides. Scientists did not then know that a glacier came into existence through the accumulation of snow, which then travelled downwards from its accumulation zone towards its snout. Agassiz was more than convinced, however, that glaciers had pushed up moraines, transported large boulders, and polished and grooved the bedrock over which they traversed; uncertainty as to how exactly this was done did not alter the fact! Pioneers of the glacier hypothesis thought it enough to know this, they did not need to understand it. Indeed, none of the original spokesmen of the Ice Age theory then understood the connection between glaciers and climate. Schimper and Agassiz believed that colder temperatures had caused an ice age, and that was sufficient. Coldness on its own is not enough for the formation of glaciers, it also needs snow: i.e. it is a combination of cold temperatures plus precipitation. Nor was it enough to suggest that mountains had been higher when glaciers were formed, as de Charpentier had maintained, because if the climate had remained unchanged the ice would have melted as it reached the lowlands. What was still needed was an understanding that glaciers had been formed by cooling weather conditions and an increasing accumulation of snow.

A sceptical reception of radically new scientific ideas is nothing new, for naturally such ideas had to be tested and proved. But the reluctance to accept them was also connected to the fact that many highly-respected scientists in the 19th century had gained their rank and reputation in a conservative society because their explanations of nature had strengthened the theories on how God had created the world and all its living creatures. Opposition to the Ice Age theory, however, was not primarily because it was not in accord with the Bible's narrative and the academic works of scholars on the creation of the world, for there was already in existence an opposition to the policy of religiously explaining nature exclusively from the Holy Scriptures. All Agassiz and Schimper had done was to suggest an Ice Age had been unleashed instead of a Great Flood, but scholars were so fixed in their opinion that only flowing water could transport rocks and debris that it took the foremost natural scientists of that time three decades to bring others round to accepting the indications of ice ages as indisputable facts, even though no explanation had been found as to their cause.

In the middle of the 19th century it became clear that a huge glacier covered all of Greenland, and only then did many scholars finally come to agree that continental Europe might also have been buried under glaciers. Convincing indications came to light decades later that erratic boulders and bedrock striation in Scotland (Geikie 1874) had been caused by glaciers that had extended all the way from Scandinavia. The Swede Otto Torell (1828–1900) later traced the expansion of glaciers from Scandinavia to northern Germany. He had previously travelled around Iceland and been convinced that the island had been covered by an ice-age glacier. Moreover, awareness of the enormous expanse of the ice sheet on Antarctica only finally became common knowledge later in the 19th century.

As for Agassiz, he immigrated to the United States in 1846, became a professor of zoology at Harvard University in New England in 1848, and made a tremendous impact on the development of teaching and research in the natural sciences in America. He returned to researching fish fossils, but he also investigated the expansion of an ice-age glacier in the New World, eventually even going so far as to claim that glaciers had covered all the continents of the Earth and had decimated all forms of life. God had then created the world once again.

2.4 Locating and Dating the Ice-Age Glacier

Research was now concentrated on investigating the extent of the ice-age glacier by marking the limits of the furthest glacial moraines. This was done by trying to trace where the ice had been originally transported from and by examining the directions of glacial striations and the location and origins of erratic boulders. This was mostly completed around 1875 when it seemed clear that one continuous glacier had covered all of Scandinavia and expanded towards more central latitudes, even reaching as far south as southern England (London) and the English Channel, northern Germany and Poland, the western parts of Siberia, and to Canada and far south on the North American continent. In the northern part of Alaska, the snowfall had not been sufficient to form a glacier even though its mean temperature was low. One continuous glacier had not reached all the way from the North Pole, as the pioneers of the Ice Age theory had believed. In the tropics, glaciers were larger and more numerous than they are today. At least a half of Earth's dry land was covered by ice and snow in one way or another: ice caps carpeted about 30 % $(40,000,000\ km^2)$, permafrost about 20 % $(27,000,000\ km^2)$, and sea ice was present on half of the Earth's oceans (Fig. 2.9; Flint 1971; Denton and Hughes 1981; CLIMAP 1981; Clark et al 1996; Bradley 1998). Ice sheets were 3 km thick over a wide area and at their centres there was so much water stored in glacial ice that, at the high point of the last glacial period, sea levels were 120 m lower than they are now. The average thickness of the ice can be calculated from the extent of the glaciers and the level of the sea. There are still remnants of this ice-age ice in Greenland and Antarctica.

Land masses during the last glacial period were different to what they are today. There were terrestrial bridges between England and continental Europe, and settlers crossed to North America over another land bridge between eastern Siberia and Alaska. Land animals inhabited wetlands where the North Sea and English Channel are now situated. The ice-age glacier advanced over land that had previously been under the sea, scraping up and bulldozing onto land shells from shallow waters, just like erratics, a long way inland in Germany, Scotland, England, and even New

| | Continental ice sheets | | Sea ice | | Ocean | | Dry land |

Fig. 2.9 Ice covering the northern hemisphere at the high point of the last glacial period about 21,000 years ago; sea ice is depicted at its greatest extent at the end of winter (Flint 1971; Denton and Hughes 1981; CLIMAP 1981; Clark et al. 1996; Bradley 1998)

England in North America. The Scot James Croll (1821–1890) drew attention to this in 1865, for it had previously been thought that the presence of shell and fish fossils in sediment layers on dry land proved that these areas had previously been suboceanic. Croll's contribution (1864, 1875, 2012) to theories on climate change will be referred to again later.

So gigantic was the overburden of ice-age glaciers that they compressed the Earth's crust, pushing it even further below the already lowered sea level. Eroding seas thus scoured far inland from the coast, and relics of oceanic life can be found in the hinterland at a great distance from the shoreline. When the ice age ended, the water that had been frozen in glaciers was once more discharged into the oceans and seas, which immediately flooded over dry land. The habitat of mammoths between England and the Netherlands became the North Sea. Coral that had been formed on shorelines in southerly climes are now 120 m deep underwater. Sea also swept over land that had been previously compressed by glaciers, but the Earth's crust quickly began to uplift again, once the weight of the ice had been removed, and coastal regions began to arise slowly out of the ocean. Where there had once been a thick glacier, ancient tidelines are now visible far inland and high above the ocean, indicating an elevation of land far greater than the rise of the sea level due to glacial melting. Fossils of oceanic organisms can thus be found in sediment layers at the same height on dry land in many countries.

Palaeontological research also made it clear that even though organisms disappeared wherever a glacier covered land, ice ages had not totally decimated life, for plants and animals had retreated before the ice, moving to more viable habitats and adjusting to climate changes. Glacial periods became a catalyst for changes in flora and fauna, and indeed modern man developed considerably during the last ice age. Organisms were under pressure from the various effects of climatic changes, but such a slow development was completely unlike the theory of natural catastrophes held by Agassiz. It is now believed that the mammoths he believed had been frozen to death, when an intense and mortal cold had been unleashed, had actually fallen through the thin ice of the tundra that had begun to thaw during the end of the last glacial period. They had drowned in cold water and then floated downriver. Climate warming had tolled their death knell, not the ice age. The extinction of mammoths

Fig. 2.10 Gerard Jakob de Geer (1858–1943), Swedish geologist who discovered the chronology of varves

and other large mammals that had multiplied very slowly during the ice age is still an unsolved riddle, nonetheless. Many causes for this extinction have been considered, including climate change, transformed ecosystems, epidemics, and the excessive hunting of tribes that had crossed over the land bridge of the mountainous Aleutian Isles from Asia to the new world, America, when the sea level was lower during the last glacial period.

Once the enormous effects of the ice age had become evident, geologists then tried to evaluate how long ago it had come to an end. The American Grove Karl Gilbert (1843–1918), using the same hypothesis as Hutton, that geological processes of the present were the key to the past, investigated how long it would take the Niagara Falls to erode its edges and then, by applying the same erosion speeds since the ice-age glacier had disappeared from the area, he calculated that the gorge beneath the falls had been gouged out during a period of about 7000 years (Gilbert 1890). The Swede Gerard de Geer (1858–1943; Fig. 2.10) then discovered a more exact way to determine the age of deposits when he found regular annual layers of sediment (varves, Fig. 2.11) that glacial rivers had dispersed around the ancient Baltic Sea, which had previously covered a large part of Sweden (de Geer 1912). In summer a thicker, rougher sediment was deposited and in winter this was super-

Fig. 2.11 Holocene sediment strata in the banks of the Jökulsá á Dal preserve a history of climate change. This is now all submerged under the Hálslón reservoir (Chap. 8.8.5). HB, August, 2002

imposed with a thinner more refined deposit, so that distinct varves were created. It has proved possible to count these layers 12,000–13,000 years back in time and so it became clear how long ago the meltwaters had been discharged from the glaciers. The oldest part of the ice-age glacier in Scandinavia had retreated rapidly 12,000 years ago, although all the ice had not disappeared completely until 6000 years ago.

2.5 Traces of Many Glacial and Interglacial Periods in Sediment Layers

In the 1850s and 1860s, sediment strata visibly revealed how ice-age glaciers had advanced and retreated many times. Beneath lifeless remnants of an ice age, sediment layers could be found with fossilised plants and animals that had lived in a warmer climate and then been buried under ground moraine. This indicated that there had not been one continuous ice age, but rather a series of cold periods during which glaciers had lain over the land. These cold periods were interrupted by short warmer interludes during which the glaciers melted enough to enable flora and fauna to inhabit land previously covered by ice.

By the beginning of the 20th century, geologists had already distinguished four separate advances of a glacier far south on the continents. In Europe these periods of glacial advances were named after rivers in which traces of their progress were visible: Günz (the oldest), Mindel, Riss and Würm (Penck and Brückner 1909); in North America they were named after states: Nebraska, Kansas, Illinois and Wisconsin (in order of age). Geologists traced their deposit layers far back in time according to such characteristics as chemical composition and size of grains. Fossils of flora and fauna bore witness to the climate when the layers of sediment came into existence, and microscopic windborne pollen grains revealed the composition of the contemporary vegetation. All discussion on this topic was marked by uncertainties because geologists found it difficult to date sediment deposits and remnants of glaciers, for varves and annual rings in tree trunks were only of use after the last glacial period ended. Glaciers had destroyed older deposits in previous expansions. At the end of the 1930s, however, there was a watershed in the study of ice-age remnants when Willard Libby (1908–1980) at the University of Chicago discovered that the age of seashells and remains of vegetation in deposit layers could be determined by measuring their levels of radioactive carbon (^{14}C; Libby 1952); at first it was only possible to date them back to 40,000 years ago, but later, due to improved measuring techniques, these objects could be dated back to 75,000 years ago. This resulted in a huge increase in knowledge, not only of the last cold period of the ice age, which reached its peak about 18,000 years ago, but also of the late glacial period that succeeded it, and finally of the present contemporary time period, the Holocene. This history will be dealt with in more detail later, but first let us return to the Ice Age theories.

A Radiocarbon Clock Measures the Age of Terrestrial Layers

Dating sediment layers through the measuring of radioactive materials opened up new methods for researching ancient climates. When organisms die, a radioactive clock begins to tick that measures the time from their decease. The radioactive isotope carbon-14 (^{14}C) has a half-life of 5730 years and it decays exponentially, thus by measuring its current levels in animal and plant remains, their age can be deduced if the original level of the organisms' carbon content at death is known. The radioactive isotope carbon-14 spreads evenly throughout the upper atmosphere because of the constant impact of cosmic rays on nitrogen, and it binds with oxygen to form radioactive carbon dioxide ($^{14}CO_2$), which then mixes with ordinary carbon dioxide in the atmosphere (CO_2 with the isotopes ^{12}C and ^{13}C). Plants then absorb this carbon dioxide (CO_2), combined with hydrogen (H), through photosynthesis to form sugar, which is then ingested by animals feeding on the vegetation, and this is how carbon-14 enters the food chain of all living organisms. When they die, this photosynthesis ceases and the radioactive clock begins to tick.

The proportions of $^{14}C/^{12}C$ are measured in the remains of once living organisms. After 40,000 years, 7 half-life periods of radioactive carbon, there is only 0.8 % of its former level left, 0.006 % after 75,000 years (13 half-life periods); it has not proved possible to measure lesser amounts. Many things can impair the clock's accuracy once it has been activated, however. Material containing new or old radioactive carbon has perhaps been added to the remains of the organism in question, e.g. by entering its plant roots or by its being soaked in carbon-rich waters. The levels of carbon-14 have also varied in the Earth's atmosphere over thousands of years. When compared to its current level, if the carbon-14 had been higher at the time of the organism's death, its carbon dating will now be too low. A comparison of the ages of tree rings reveals the necessity of adding 700 years to an age calculated to be about 6000–9000 years old, because the carbon-14 level was greater then, than now. In a 200-year period about 12,000 years ago, and shortly after 10,000 years ago, it is impossible to date things accurately. Willard Libby received the Nobel Prize for chemistry in 1960 for his research into carbon dating (Fig. 2.12).

It later became possible to calculate the age of much older objects in the geological history of the Earth, inorganic volcanic rocks, for example, through radioactive materials that have longer half-lives than radioactive carbon. Radioactive kali (^{40}K) has a half-life between 1.25×10^9 years and then becomes argon (^{40}Ar); it is assumed that during a volcanic eruption minerals are without argon. Using the uranium series (Th^{230}, U^{234} and U^{235}) of radiometric dating, ages between thousands and hundreds of millions of years can be calculated.

Once the dating of objects through radioactive material had been established, it was possible to correlate such data with the results of other methods involving the analysis of tephra layers, pollen dispersal, and amino-acids (decay of protein after death of an organism, which is actually related to temperature). Furthermore, geologists learned to utilise magnetic directions when analysing layers of rock (Fig. 2.13). The Earth's magnetic field has constantly been changing because the location of the magnetic poles has slowly oscillated, though considerably more during Earth's geological history. During lava flows, minerals adjust themselves to the direction of the magnetic field; so too do iron-rich particles in deposits such as drifting soil and volcanic ash. Magnetic directions are thus one of the features of Earth's strata, and by calculating the magnetic direction of rocks and dating them with the K-Ar method, changes in the directions of the Earth's magnetic fields through the ages have been recorded. Furthermore, the specific turning points of Earth's sudden geomagnetic reversals (the last being the Brunhes/Matuyama reversal about 780,000 years ago) can be utilised as a clear reference point in determining the historical layers of geological strata.

Fig. 2.12 Willard Frank Libby (1908–1980), the American physicist who discovered how to date objects by measuring their levels of radioactive carbon, a method since used in archaeological and geological sciences

Fig. 2.13 The basalt mountains of the western fjords of Iceland preserve 15 million years of history. One volcanic eruption after another piled up thick layers of lava. Strata of soil mixed with river and lake sediments were formed on top of these layers, which are now visible as thin layers of clay and sandstone between the strata of lava. Over a wide area they preserve the fossilised remains of plants along with seeds, pollen, leaves and tree trunks, as well as lignite from carbonated peat, all from when it was warm and moist during the Tertiary period and the land was covered by deciduous forests. Over the last three million years, ice-age glaciers have been eroding these geological strata. Looking southwestward, nearest in the photograph is Ísafjarðardjúp fjord, then Skötufjörður and Hestfjörður, on the other side of which Mt. Hesturinn stands guard. Near the top right corner is Mt. Kaldbakur, to the south of Dýrafjörður. In the far distance on the right are the jagged peaks to the east of Kaldbakur that had broken through the ice cover of the last glaciation (Chap. 7). HB, March 1999

2.6 Linking Climate Change to Earth's Orbital Trajectory: Adhémar, Croll and Milankovitch

In 1842, just two years after Agassiz published his book on the Ice Age theory, the French mathematician, Joseph-Alphonse Adhémar (1797–1862), put forward the idea that it was periodical oscillations in solar radiation, in correlation with Earth's orbit in space, which caused ice ages (Fig. 2.14). Adhémar reiterated that the Earth had seasons because its axis of rotation was at an angle of 23.4° in its elliptical orbit around the Sun (Adhémar 1842). Winter was thus a week shorter than summer in the northern hemisphere because the Earth was nearest to the Sun during the winter and moved faster in its orbit due to gravitational forces; daylight in spring and summer was 168 h longer than from autumn until the end of winter. This, he

Fig. 2.14 The adversaries, sunlight and snow, compete with each another, the gleaming brow of Þórisjökull reflecting the sun's rays. Lake Þingvallavatn and Mt. Hengill in the far distance to the left (Chap. 5). HB, September 1999

believed, explained why there was currently no ice age in the northern hemisphere. It was the direct reverse in the southern hemisphere, where there was a week longer of darkness than daylight and thus a huge sheet of ice—though no one then had actually seen Antarctica.

Adhémar then pointed out that because of long-term changes in the direction of Earth's axial tilt in space, the seasons changed according to its orbit around the Sun. He put forward the hypothesis that due to Earth's precessional cycles its polar regions had large ice sheets in turn. About 13,000 years ago, Earth had been nearest to the Sun in summer in the northern hemisphere, interglacial periods had been short, and ice sheets had covered the northern hemisphere all winter long, while the southern hemisphere had been free of ice. There was little support for Adhémar's hypothesis. The great natural scientist Alexander von Humboldt (1769–1859) did not believe that the differing distribution of solar radiation during the seasons could explain ice ages. He pointed out that Earth's hemispheres received the same sum total of solar radiation over the year as a whole; what was lacking in solar radiation in one half-year was compensated for in the other. Nor does Adhémar's hypothesis actually explain the existence of an ice sheet at Antarctica, either. It came into being 30–35 million years ago with the cooling of the early Oligocene period, reaching its full extent 10 million years ago, and has remained little changed ever since. Isolated from warm oceanic currents, it rises high above the sea and reflects almost all the solar radiation it receives and so maintains an intense coldness.

Indications appearing in the 1850s, that there had been glacial and interglacial periods in turn, helped maintain a continuing search for the causes of fluctuations in climate. Approximately two decades after the publication of Adhémar's work, James Croll pointed out that in addition to axial precession, eccentricities in Earth's orbit could also cause regular oscillations in the levels of solar radiation that reached Earth. Croll believed that the combined effect of Earth's orbital eccentricities and precession could explain the appearance of ice ages. The level of solar radiation is fairly equal all year when Earth's orbit around the Sun is almost circular, but varies according to seasons when it is mostly elliptical. When the inclination of Earth's axis is tilted away from the Sun and its orbit is at its furthest distance from it, the period of darkness increases by five weeks (36 days) from when it is in a circular orbit. Croll believed that, in such circumstances, cold winters over thousands of years could produce ice ages. Even though the hemispheres received the same amount of solar radiation over the whole year, small irregularities in Earth's orbit could still affect the seasons just enough to produce glacial periods. When the orbit is mostly elliptical, Earth's hemispheres take it in turn to bear ice-age glaciers for a period of around 11,000 years, and regular interglacial periods would exist when the orbit was almost circular.

Scientists generally felt that fluctuations in solar radiation caused by these influences would be too small to have an impact on climate. Croll's main idea, however, was that the influence of small variations in solar radiation could be amplified by their multifarious effects on air flow patterns and ocean currents influencing changes in the transportation of humidity in the air and sea from the equatorial to the polar regions. An increased difference in temperature between the equatorial and polar regions would strengthen moisture-laden winds and snowfalls that reached northern areas. Once a glacial period has begun, it starts to grow in magnitude, because as glaciers expand there is a greater reflection of solar radiation from Earth, reducing its energy gain, and thus temperatures continue to fall. Croll was thus the originator of the idea that scientists are now concentrating on, i.e. that astronomical causes could instigate, rather than fully explain, the fluctuations between glacial and interglacial periods.

Croll did not know how long the cycles between axis inclinations could be, and nor could he evaluate their effects on solar radiation; other things being equal, however, glacial periods usually begin when Earth's axial obliquity is small, the polar regions receive the least solar radiation, and the difference in temperature is at its greatest between the polar and equatorial regions. We now know that this cycle lasts for 41,000 years (Fig. 2.15).

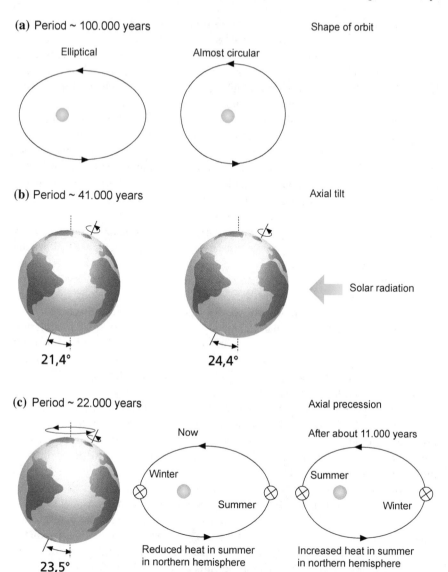

(a) Period ~ 100.000 years   Shape of orbit

Elliptical   Almost circular

(b) Period ~ 41.000 years   Axial tilt

Solar radiation

21,4°   24,4°

(c) Period ~ 22.000 years   Axial precession

Now   After about 11.000 years

Winter   Summer

Summer   Winter

23,5°   Reduced heat in summer   Increased heat in summer
in northern hemisphere   in northern hemisphere

Fig. 2.15 Astronomical causes of climate change

Hypotheses on Astronomical Causes of Climate Change Clarified

From the mid-19th century onwards, hypotheses emerged about how the regular changes in Earth's orbit around the sun and the direction of its axial rotation could affect the levels of solar radiation that reached Earth's atmosphere and could even cause ice ages. The Serb Milutin Milankovitch (1879–1958) is best known for his theories concerning this matter (Fig. 2.16). About 175 years after Johannes Kepler (1571–1630) had explained how Earth's orbit around the Sun is elliptical, the mathematician and astronomer Joseph-Louis Lagrange (1736–1813), who was born in Italy, but spent most of his career in France and Germany, discovered that its orbit oscillates between being elliptical and almost circular. This was caused by the varying and accumulative effects of other planets' gravity in our solar system; they all lie in the same orbital plane and circle the Sun at varying lengths of time, so that the gravitational effects between them change accordingly. When Earth's orbit round the Sun is almost circular, the levels of solar radiation on the planet's surface are even all year round, but when its orbit is mostly elliptical, the seasons vary in length. The limiting effect on solar radiation that reaches Earth is minimal (less than 0.7 % over the year), but there are great fluctuations, nonetheless, because this radiation energy is unevenly dispersed over the seasons and longitudes due to the direction and tilt of Earth's axis changing in the course of time.

During its orbit around the Sun, Earth's axis tilts at an angle of 23.4° from an imaginary vertical line through the centre of the ellipsis. In winter, Earth's northern part tilts away from the sun, but in summer it tilts towards it, so that the midnight sun reaches the Arctic Circle at 66.6°N. The tilt of Earth's axis oscillates, on the other hand, over a period of 41,000 years between 22.0° and 24.5°. The Arctic circle crossed Hrísey island in Eyjafjörður 5000 years ago, but it now crosses Grímsey, off the northern shore of Iceland, and is still moving northward at about 14 m a year, though its movements fluctuate. With an increased axial tilt, the Sun reaches higher in the sky and summers are warmer, but conversely winters are darker and colder; the lengths of seasons increasingly vary.

The direction of Earth's axis in space also changes and this also affects the lengths of seasons; the solstices (when the Earth is nearest to and furthest from the sun) and the equinoxes are moved on the orbital ring. Earth thus rotates like a spinning top so that the axis completes a whole circle in 26,000 years, and its spinning pattern can be visualised as outlining the shape of a cone. Due to changes in the Earth's orbit, mentioned above, the axial tilt is reckoned from the Earth's perihelion point (nearest Sun) every 22,000 years. Thus the seasons have moved during the Earth's orbit round the Sun by almost three months since the time of the Ancient Egyptians about 5000 years ago. About 11,500 years ago, the perihelion was in June, when the southern hemisphere faced the Sun, and the aphelion in January, and winters were longer in the northern hemisphere than in the southern. About

4000 years ago, Thuban in the Draco constellation was the Pole Star, and in 12,000 years' time it will be Vega in the Lyra constellation.

Astronomers in ancient times discovered Earth's precession (changes in axial direction). They had studied the fixed stars of the night skies and their relative positions to the Sun during various seasons for centuries. In the 2nd century before the birth of Christ, the Ancient Greek astronomer Hipparchus (ca. 190–125 BC) realised that Earth's axis no longer pointed in the same direction into space as it had done during the time of Timocharis's observations 150 years previously. Though the Earth's axis is fairly stable and points almost directly towards the star we call the Pole Star, its direction changes in space over a long period of time (the axial tilt itself still remains the same during the orbit round the Sun, 22.0°–24.5°). It was not until Isaac Newton (1642–1727) put forward his laws of gravity that this precession was explained. The Earth rotates on its axis like a spinning top as it orbits the Sun. Its mass is irregularly dispersed around the globe, which is not a regular circle. The gravitational pull of both the Sun and Moon is around its bulging equatorial centre as they attempt to correct its axis. Because Earth is continuously revolving around itself, the centrifugal force causes it to sway from side to side and create a cone-shaped spin. We have all seen how a spinning top lies flat on the floor once it has ceased to spin. We know the effects of centrifugal force from observing gyrocompasses, bicycles, and aeroplane propellers. If Earth's axis was at right angles in its orbit around the sun, then every day of the year would have twelve hours and there would be no seasons and an everlasting, freezing cold in the polar regions. It is believed that early in the globe's history a planet the size of Mars might have collided with Earth and nudged its axis from its vertical position. Let's be thankful for that chance encounter.

Milankovitch's numerical models of solar radiation reaching Earth. Croll's calculations indicated that 250,000–80,000 years ago Earth's orbital pattern and axial precession had coincided to have a maximum effect on the levels of solar radiation, and he believed that this was the time span when warm and cold periods alternated regularly between the northern and southern hemispheres, each period lasting about 10,000 years. There would not have been a glacial period since then. But although geologists believed that they saw signs of a glacial period about 10,000 years ago, they could find no evidence that cold and warm periods had alternated between the southern and northern hemispheres, and thus support for Croll's ideas had dwindled by the end of the 19th century. Two decades after his death, however, the Serb Milutin Milankovitch (1879–1958) took up the thread and wove it into the detailed calculations that the German Ludwig Pilgrim made (1904) concerning the collective effect of orbital eccentricities and axial obliquity on the movement of equinoxes. With extensive calculations, Milankovitch illustrated the influence of all three factors (axial precession, orbital variation, and axial tilt) on the

Fig. 2.16 (From *left* to *right*, *top* to *bottom*) James Croll (1821–1890), Scottish geologist. Joseph-Louis Lagrange (1736–1813), Italian-French mathematician and astronomer. Milutin Milankovitch (1879–1958), Serbian engineer. Wladimir Köppen (1846–1940), Russian-German meteorologist. Alfred L. Wegener (1880–1930), German geophysicist. Oswald Heer (1809–1883), Swiss palaeontologist

levels of solar radiation over the last million years. He calculated the temperature on specific places on Earth for the entire year for the last 130,000 years; he evaluated the levels of radiation at the furthest reaches of the atmosphere and then how it wanes on its way to Earth's surface. Previously, only the overall influence of this radiation had been calculated for the southern and northern hemispheres. A convincing correlation emerged between the temperatures in many parts of the world, although heat levels calculated near the equator were a little too high, and those in northern latitudes a little too low. This was to be expected as the calculations did not take into account heat fluxes in the oceans and winds (Milankovitch 1930). The greatest collective influence of all three factors would be the equivalent of summer solar radiation being reduced at 65°N to what it is now at 77°N; i.e. solar radiation in the middle of Iceland would be equal to that on the southern part of

Svalbard. Milankovitch believed that although axial tilt oscillated only about one degree within a 41,000-year cycle, it would nonetheless have a tremendous impact on the differing temperatures between winter and summer. With an increase in axial tilt, solar radiation in the northern hemisphere would increase in the summers, but conversely decrease in the southern hemisphere; there would then be the greatest difference in temperature between summer and winter. The effect of the cycle on Earth's axial tilt would be greatest in the northern hemisphere, because that is where the largest land areas of the world are upon which snow could accumulate.

Supportive meteorologists, doubting geologists. Just like Croll, Milankovitch believed that cold winters were the cause of glaciation, but the Russo-German meteorologist Wladimir Köppen (1846–1940) and his son-in-law Alfred L. Wegener (1880–1930) pointed out that it was summer temperatures rather than winter ones that mostly dictated the accumulation of snow and formation of glaciers. With reduced solar radiation during the summer and colder weather, melting would be less, so that a surplus of the winter snowfall would survive the summer. Solar radiation would then be reflected into space at an increasing rate, temperatures would continue to fall, glaciers would expand, and then a glacial period would begin. Milankovitch (1920, 1930, 1941) presented a model of how a growing reduction of the levels of solar radiation reaching the Earth's surface would lower the snowline and cause perennial snow to be more widely dispersed. At Köppen's request, Milankovitch calculated the temperature over the last 600,000 years and they believed that their model predicted four glacial periods when solar radiation was at its minimum in the northern hemisphere: 400,000, 300,000, 250,000 and 125,000 years ago. Köppen and Wegener published these predictions in *Die Klimate der geologischen Vorzeit* (1924), and believed they correlated well with the conclusions of Austrian geologists on four glacial periods in the Quaternary period in the Alps and in North America. But it was then discovered that the geologists' dating methods proved to be unreliable. Milankovitch's calculations had also predicted many more cold periods than traces of which could be found in sediment layers, and the date that had been attained with radiocarbon (^{14}C) and U-Th measurements did not match his predictions on cold periods. Geologists also found remains of interglacial vegetation in 25,000-year-old peat in Europe and North America, whereas Milankovitch's model predicted that solar radiation in the summer had then been at a low level and the summers thus supposedly cold. About 7000 years later, the ice sheets of the last glacial period had actually been at their largest extent ever. Milankovitch then put forward the hypothesis that the reaction time of glaciers could be 5000 years and that a 2000-year margin of error could also be assumed. Fewer and fewer scientists supported Milankovitch's theories, though he himself believed in them until the day he died in 1958. Within two decades of his death, however, data emerged concerning global climate variations that was deemed to confirm the main gist of his theories.

2.7   Reading Climate History in Suboceanic Strata and Glacial Ice

Until beyond the mid-20th century, climate history was deducted from glacial remains from many countries and the evidence assembled to create one intermittent narrative thread. With each new advance, ice-age glaciers and glacial rivers had wiped out traces of earlier glaciers both on land and on shallow sea beds. There were some regions, however, beyond the furthest extents of glaciers that revealed a long, continuous history of climate. The glacial deposits in the Rhine delta in the Netherlands and the loess in central China, for example, showed that cold, ice-age winds had blown from the high-pressure area of the Tibetan plateau. It had become evident that climate changes had been simultaneous over the whole globe.

In the 1960s, however, a radical new knowledge of ancient climates, far distant from glaciers, emerged from deep-sea sediments where glaciers had never disturbed the ocean floor. Physicists had by then discovered methods to calculate temperatures of oxygen isotopes in tiny fossilised calcareous foraminifera shells, which had sunk to the ocean floor for thousands of years and been buried there in mud. The

Fig. 2.17 Harold C. Urey (1893–1981), American physicist

Photographed by Fritz Eschen, ca 1962

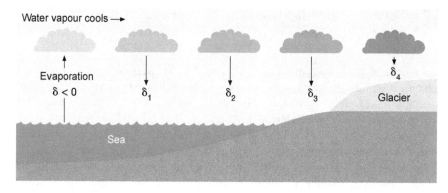

Fig. 2.18 Oxygen isotopes in glacial ice record atmospheric temperatures. In water (H_2O) there are both heavy (^{18}O) and light (^{16}O) molecules of oxygen, and the ratio between their numbers in a sample of water depends on the temperature when the vapour condenses and produces liquid water. Precipitation thus bears a temperature marker; the ratio of the number of heavy to light oxygen isotopes records annual fluctuations in atmospheric temperature as well as long-term climate changes. The explanation lies in the fact that when water vapour condenses there will be an enrichment of heavier isotopes (^{18}O) to lighter ones in condensed water and the precipitation that falls to Earth. Comparatively more lighter isotopes (^{16}O) will be remaining in the air masses and with continuing cooling of the air and condensation the precipitation will gradually contain still relatively more lighter isotopes; an increase in coldness and a higher percentage of lighter isotopes go together. Changes in the ratio of isotopes ($R = {}^{18}O/{}^{16}O$) are described as a deviation from the mean value in the sea (a standard value R_0), so that $\delta = 10^3 (R - R_0)/R_0$. The number of heavy isotopes decreases during evaporation from the sea (are depleted) and thus δ is always a negative size. In winter, more light oxygen falls in precipitation than in summer, and this applies also to both cold and warm climate periods. When glaciers expand on land, the water accumulated has a relatively large amount of light oxygen (^{16}O), while in the sea there is an increasing amount of heavy oxygen isotopes (^{18}O). With evaporation from the sea, proportionately more water with light oxygen isotopes ($H_2^{16}O$) than with heavy ones ($H_2^{18}O$) enters the atmosphere and is borne onto the glaciers when precipitation accumulates there as snow. In cold periods, glaciers preserve even larger amounts of light oxygen isotopes (^{16}O), and there is thus proportionately more numerous lighter isotopes (^{16}O) remaining in the atmospheric water vapour sum of heavy oxygen isotopes in the ocean (^{18}O)

shells covered fauna and flora that had once floated on the ocean surface or above the seabed. Besides being preserved in seashells, temperatures are also recorded in the oxygen and hydrogen isotopes in ice cores retrieved from the glaciers of Greenland and Antarctica.

Oxygen and Hydrogen Isotopes in Water Reveal Air Temperature

Urey (1947; winner of the Nobel Prize for chemistry in 1934; Fig. 2.17) discovered in 1946 how oxygen isotopes (^{16}O and ^{18}O) could be utilised as thermometers because the ratio of these two isotopes in seashells and glacial ice is related to their temperature when they are formed and begin to absorb oxygen. Isotopes with a low atomic mass (^{16}O) evaporate more easily from oceans than those with a high mass, so that a higher percentage of heavy isotopes (^{18}O) remains in the sea. Vapour with light oxygen isotopes is borne through the atmosphere and falls as precipitation on ice sheets (Fig. 2.18). Thus, during glacial periods, a large amount of light oxygen isotopes accumulates as snow on land, while conversely the sea has a relatively higher number of heavy isotopes. The ratio of the heavy and light oxygen isotopes in ice cores thus reveals the temperature of the atmosphere, whereas seashells reveal how much of the Earth's water is stored in ice sheets on land.

The ratio of heavy and light isotopes (called δO) in ocean sediment cores is a much clearer measure of atmospheric temperatures than one might suppose, because in deep oceans temperatures change very little and very slowly. When glaciers melt, the lighter isotopes return to the sea where its ratio will increase while the ratio of heavier ones will decrease. Similarly, hydrogen isotopes in water (protium ^{1}H and deuterium ^{2}H) also provide a means of measuring temperature.

2.8   The Emergence of a Complete Picture of Past Global Climate Variations

Within two decades of Milankovitch's death in 1958, as mentioned above, climate records emerged from deep oceanic deposits, and later from glaciers, that once again made his theories increasingly credible. It was immediately apparent that the four glacial periods previously described, and that were known about at the beginning of the 20th century, are the last in a series of many glacial periods which reach back millions of years in time. In 1971 a continuous core of deposits from the Pacific Ocean illustrated 19 oscillations from glacial to interglacial periods over the last 700,000 years, and this indeed seemed to support Croll's main hypothesis that the main causes of regular climate changes were 100,000-year fluctuations in the Earth's orbit around the Sun. Two year later, in 1973, three of these fluctuations (100,000, 41,000, and 22,000 years ago) were visible in a core from the Indian Ocean that reached back as far as 450,000 years ago, and in 1976, a complete picture of the series of glacial periods appeared in a deep core from the Pacific that spanned the entire 2.5 million years of the ice age (Hays et al. 1976; Imbrie and Imbrie 1979). The oxygen isotopes in this core revealed about 50 cold periods and between them short interglacial periods that lasted for 10–20,000 years each. In the first part of the ice age, the cold periods lasted for 40–50,000 years, but for the last million years they have been maintained for about 90,000 years, have been much colder than previously, and their glaciers larger than in earlier cold periods (Figs. 2.19 and 2.20). An ice core from Antarctica (Vostok, 3623 m long) goes back 420,000 years in time and shows the last four glacial periods that reached their high points 335,000, 245,000, 135,000 and 18,000 years ago. Ice cores from the Greenland ice cap (GISP2, 3053 m long, and GRIP, 3029 m long; Figs. 2.21 and

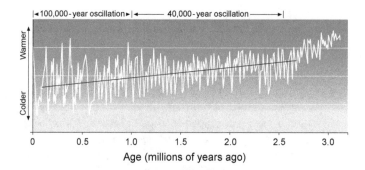

Fig. 2.19 Ice-age climate oscillations. Oxygen isotopes in fossil shells of foraminifera in ocean sediment cores reveal great and frequent changes in climate over the last 1.3 million years. They expose fluctuations in ocean isotopes during the ice age. The climate has nonetheless grown colder over a long period during this time (see direct line) and it has been at its coldest over the last 750,000 years. More than 2.75 million years ago there were few glaciers on Earth, but they then began to form and disappear in a regular ca. 41,000-year cycle. This changed about 900,000 years ago, however, and glaciers have since appeared and departed in 100,000-year intervals, while at the same time becoming much larger than ever before. ("Ice Age" website; Emiliani 1955, 1966; Shackleton 1967, 1984; Raymo 1994; Marchant and Denton 1996; Maslin et al. 1998; Shackleton et al. 1990; Berger 1977a, b, 1978; Berger and Loutre 1991; Geirsdóttir 1990; Geirsdóttir and Eiríksson 1996; Geirsdóttir et al. 2007; Haflidason et al. 2000)

2.22) clearly reveal an interglacial period that ended about 125,000 years ago as well as all of the last glacial period. There was, in fact, no continuous cold period, for 23 interglacial periods interrupted it, though none of them have been as stable as over the last 10,000 years, during which average temperatures have only fluctuated about 2–4 °C.

The frequency analysis of the climate records from the deep-sea sediment and glacial ice cores reveal that climate fluctuations are in accordance with the predictions of Milankovitch's model of the collective influence of Earth's orbit, axial precession and tilt on solar radiation at various points of latitude and in different seasons. The analyses indicate that fluctuations in levels of solar radiation instigate long-term climate changes. Radiation dictates whether snow survives the summer and continues to accumulate in order for the glacial period to set in, and the latter ends when radiation consistently increases once more in northern regions. Milankovitch's theory does not explain, however, why there is a glacial period simultaneously in both hemispheres, or why they come to an end so abruptly, or why the atmospheric temperature can increase without warning about 5–10 °C within a single decade. There must be other factors beyond those in his theory to explain how the amplifying effects of solar radiation can influence and maintain climate changes. Ocean currents, ice caps on land, sea ice, and the atmosphere's circulation patterns, all play a part in distributing heat around the globe, from the equator to the polar regions. The complex effects of ice are important and include glacial surges and the discharging of glacial meltwater into the oceans, which lower the salinity of seawater, and the expansion of sea ice and snow cover that increases

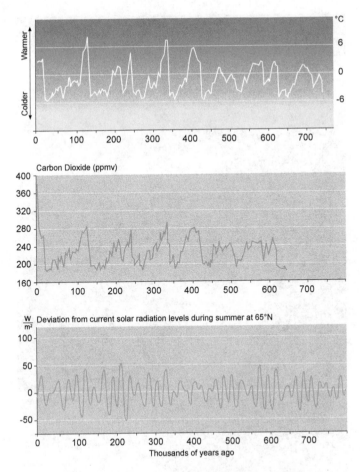

Fig. 2.20 Connection between solar radiation, carbon dioxide and atmospheric temperature. Oxygen isotopes in 3-km-long ice cores from Antarctica (from Dome-Concordia in 2004) have recorded climate changes over the last 740,000 years, from a total of eight glacial periods (*top graph*). It became increasingly colder during each cold period, though temperatures fluctuated somewhat. It then grew warmer very rapidly for a few thousand years, but this warm period did not last long. The graph shows deviations from the mean value (shown as 0 °C) at the end of the last glacial period. Temperatures fluctuated in accordance with carbon dioxide levels in glacial air bubbles (*middle graph*). The level of carbon dioxide was about 180–190 units (ppmv, parts per million by volume) during glacial periods, and 280–300 units during interglacial periods. At the beginning of the Industrial Revolution at the end of the 18th century, the level was 280, but it is now (2015) around 400 and has never been as high as this over the last 750,000 years. It is thus predicted that, by the end of the 21st century, the level of carbon dioxide could be between 530 and 980 units, depending on how much pollution mankind generates. Long-term climate changes are believed to be connected to fluctuations in levels of solar radiation caused by changes in the Earth's orbit around the Sun, and the tilt and direction of the Earth's axis, see e.g. the increased levels of solar radiation at the beginning of the interglacial periods 12,000, 130,000, 220,000, 285,000 and 350,000 years ago (*bottom graph*). (EPICA 2004; Petit et al. 1999; Berger 1977a b, 1978; Hays et al. 1976; Raynaud et al. 1993; Barnola et al. 1987; Heinrich 1988)

Fig. 2.21 Sigfús J. Johnsen (1940–2013), glaciologist, drilling for ice cores on the Greenland ice sheet

the reflection of solar radiation. It is believed that climate changes begin in the northern hemisphere and, if they endure long enough, the cycle of oceanic currents begins to have an effect in the southern hemisphere. Fluctuations longer than 2000 years can be seen in both hemispheres. One of the features of long-term fluctuations is that cooling is a slow process while warming is sudden and rapid (a theory associated with the Dane Willi Dansgaard [1922–2011, Fig. 2.23] and the Swiss Hans Oeschger [1927–1998]). This serrated pattern of climate fluctuation is believed to be caused by the interaction of ocean currents and glacial ice caps, which form much more slowly than they disintegrate. Some sudden changes take place simultaneously over the whole globe, while other fluctuations, especially the smaller and more frequent ones, are not always synchronous everywhere on Earth. These sudden variations are believed to be connected to rapid changes in ocean

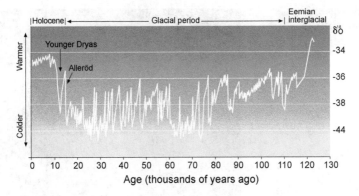

Fig. 2.22 Temperature fluctuations from the last glacial period until the present. Oxygen isotopes in 3-km-long ice cores from the Greenland ice sheet have recorded climate changes over the last 125,000 years. The last glacial period began about 120,000 years ago, but it did not, indeed, become really cold until about 40,000 years later, and it was coldest during the period's final 20–30,000 years. Twenty-three sudden fluctuations of climate have been distinguished during the period and are named after two glaciologists as Dansgaard-Oeschger events. The events are believed to have been a result of changes in the currents of the North Atlantic. Each event began with an increasing coldness, but finally ended with a sudden warming, until it cooled again and the same pattern was repeated. Each group of Dansgaard-Oeschger events is called a Bond-event and it ends with glaciers surging into the sea, dispersing icebergs and sediment layers. The sediment settles in layers named after the German paleo-oceanographer Hartmut Heinrich. The glacial period finally ended around 14,700 years ago, when the warm Bølling period began. Around 14,000 years ago there came a short 300-year cold period (Older Dryas), but at the end of this the climate became rapidly warmer during the Allerød period, until a cold period returned about 12,900–11,500 years ago (Younger Dryas). Finally, the present, stable climate of modern times emerged, with a temperature 9 °C warmer than during the coldest point of the glacial period. (Dansgaard et al. 1984; Dansgaard et al. 1985; Bond et al. 1993; Grootes et al. 1993; NorthGRIP members 2004; Johnsen et al. 1998; Andrews et al. 2000; Eiríksson 1980, 2008; Ingólfsson et al. 1997; Larsen and Eiríksson 2008; Símonarson 1979; Símonarson and Leifsdóttir 2007; Maclennan et al. 2002; Ingólfsson 1984, 1985, 1987a, b, 1988, 1991; Pétursson 1986; Hubbard et al. 2006)

currents, changes in the North Atlantic being especially important for the climate all over the world. The boundaries between cold polar and warm southern ocean currents can move and may be likened to a door that opens during interglacial periods to allow warm seawater and air to flow far into the northern oceans. This door is closed during cold periods, but in-between times it is half ajar. Thus sudden climate changes are believed to be connected to glacial waters being borne into the North Atlantic when either the gigantic lakes at the margins of the ice sheets of the last glaciation were emptied, or when an armada of icebergs suddenly floated out to sea during glacial surges. Freshwater then floated around the ocean, cooling the sea and diluting it so that it froze and disrupted the heat flux around the globe. This will be discussed in more detail below.

On the ice sheets of Greenland and Antarctica, not only can the atmospheric temperature be detected, but also the composition of the atmosphere and thereby

Fig. 2.23 Willi Dansgaard
(1922–2011), Danish
glaciologist

changes in the levels of so-called greenhouse gases and air pollution. Results
indicate that changes in solar radiation first instigate warming, and this is then later
amplified by increasing levels of greenhouse gases. A change from a glacial to an
interglacial period has always been followed by an increase in carbon dioxide. This
link between changes in temperature and the level of greenhouse gases is believed
to be connected to the temperature of the oceans and the exchange of gases between
sea and air.

Surges of Ice-Age Glaciers Dispersed Sediment Layers on the Seabed
The German paleo-oceanographer Hartmut Heinrich (1988) found six thin,
white stone deposits in sediment strata from the North Atlantic. The layers
were thickest nearest the Hudson Bay in northern Canada (0.5 m) and they
contained grains of limestone originating from the eastern part of Canada.
They had sunk to the seabed from the enormous amount of ice-rafted detritus
during surges of the Laurentide ice sheet at intervals of 7–10,000 years about
10,500–70,000 years ago; the layers are called H1 (16,500 years old, shortly
before the ice-age glacier began to recede), H2 (23,000 years old, at the high
point of the glacial period), H3 (29,000 years old, when it was quite warm)
and then H4 (37,000 years old), H5 (51,000 years old) and H6
(~70,000 years old). It then emerged that sediment deposits from Iceland
and northwest Europe had also been deposited there, which indicated that ice
sheets on the other side of the Atlantic had surged at the same time. The
causes of these surges are believed to be connected to either climate change

(Denton's model, which is believed to explain why ice sheets in different parts of the world surged simultaneously), or irregularities in the flows of the ice sheets (MacAyeal's model, whereby a glacier becomes so thick that its base had warmed to freezing point and it surged forward, thinning out, freezing again at the base and thickening once more before surging again).

2.9   Climate Fluctuations During the Last Glacial Period

There was an ice-cold glacial period about 130,000 years ago, which was then followed by an interglacial period that reached its warmest climax about 125,000–120,000 years ago. Remains of vegetation and oxygen isotopes indicate that it had been 2 °C warmer on Earth than it is today; glaciers had correspondingly been smaller and the sea level 5–8 m higher, probably because the western part of Antarctica had been free of ice. There is a layer of oceanic sediment from this interglacial period at the bottom of strata in Elliðavogur Bay near Reykjavík (Einarsson 1968). Then, 118,000 years ago, there was a sudden cooling and 3000 years later the Würm/Wisconsin glacial period began. Mankind's predecessors then lived far from glaciers in the warm regions of Africa and in caves in Asia. The climate history of this glacial period is well known. The last remnants of ice caps, which had begun to be formed at the beginning of this period, disappeared from Europe and North America about 8000–8500 years ago, and traces of this glacial period are still visible on Earth. Most of the traces of older glacial periods, however, have been destroyed by these later glaciers.

The cold period that began about 118,000 years ago was not continuous but interrupted six times, so that there was some temporary warming and glaciers retreated for a while, but at the end it grew increasingly colder until its lowest temperature was reached about 18,000 years ago. The cold period was rather mild at first, but it suddenly cooled about 75,000 years ago and then the real glacial period first began. Conglomerate rock and then lignite were deposited over the sediment strata in Elliðavogur Bay. Finally, Reykjavík dolerite lava flowed over these layers, and this prevented glacial erosion from reaching and destroying the strata. There was extreme cold for 15,000 years. Ice covered northern Europe and the sea dropped to 100 m below its current level. But then a rather warm 45,000-year period followed and vegetation grew, and animals and mankind proliferated over a wide area. Glaciers still maintained themselves, though, and the sea level may have been about 50 m lower than it is today.

Neanderthal man had disappeared about 30,000 years ago and Cro-Magnons were ascendant all over Europe. In small groups, and continually on the move, they adjusted to their environment and developed technical skills and artistic creativity, as can be seen in the cave drawings in southern France and Spain (Figs. 2.24 and 2.25). But then, around 25,000 years ago, the temperature cooled significantly and

Fig. 2.24 An ice-age man 15,000 years ago, clad in animal skins, bringing some wood to his home, a shelter made from mammoth bones. American Museum of Natural History

Fig. 2.25 An approximately 14,000-year-old chalk drawing of a bison in a cave in Lascaux in southern France

glaciers began to expand rapidly in continental Europe and North America, and there are many indications that they reached their greatest size in the regions further north, on Svalbard, and in Canada and Greenland. A glacier up to 3 km thick covered all mainland Europe and the sea level was 120 m lower than it is currently. A third of the world's land mass was then covered in ice, and from the ice sheets blew constant, dry, cold katabatic winds across the frozen tundra to as far south as the Mediterranean Sea. The earth in the forefields of ice sheets was then bound in permafrost, just as it is now in Siberia and the highlands in northernmost Scandinavia.

Homo sapiens was now the only humanoid species on Earth. Bones of large mammals have been found in caves from this time along with artistic petroglyphs. Mean winter temperatures were about 20–25 °C lower than they are today, especially in the northern parts of the globe, and the weather was drier. At the equator, the average temperature may have been 3 °C colder, but by then the high point of the glacial period had been attained and ahead lay 8000 years of a late glacial period.

Earth began to be warmer around 17,000 years ago (when solar radiation increased according to Milankovitch's calculations), and glaciers on land shrank and seasons of vegetation lengthened, though vegetation on the steppes and tundra were slow to recover. A land bridge over the Bering Strait to North America became ice-free and peoples from Asia crossed over it, settling and moving south along the western side of the continent, west of the Laurentide ice sheet. Temperatures then rose sharply around 15–14,000 years ago. In the British Isles the summer heat rose by an average of 8–10 °C. It is believed that half of the warming from the complete glacial period of about 14,600 years ago had occurred in one decade. During this blooming Bølling interstadial period forests spread rapidly over northern regions, and pollen grains, insect and marine fossils, and glacial ice cores all bear witness to this warming. Glaciers shrunk at a much faster rate than they had previously formed and expanded, and the sea level rose rapidly, 5 m in a century.

The glacial period did not surrender without a fight. There was then a sudden end to the retreat of ice-age glaciers about 14,000 years ago, and a five-century cooling episode succeeded. Glaciers began to advance once more, pushing up moraines. This glacial period is called the Older Dryas after the alpine and tundra plant, the fossilised traces of which can be found in sediment strata from this time (Fig. 2.26).

This cold period ended suddenly when, within a few decades, temperatures rose by about 10 °C on the shores of the North Atlantic, and there are even examples of 16 °C warming and a doubling of precipitation. The warm Allerød interstadial period succeeded. Forests spread out over a great part of Europe, reindeer, musk-oxen and mammoths moved northward, where there were still tundra, and the hunters followed them. There was still dry land between England and France and large parts of the North Sea were not as yet underwater. Meltwater first flowed from the Laurentide ice sheet south to Mississippi and the Bay of Mexico, also down the Hudson River, and finally eastward through the Gulf of St. Lawrence into the North Atlantic Ocean. Huge, ice-dammed lakes were formed at the margin of the

Fig. 2.26 Two sudden cold episodes at the end of the last glacial period are named after the white dryas plant (*Dryas octopetala*) because its fossilised leaves are found in sediment layers from this time. The first cold episode (the 500-year long Older Dryas) began about 12,000 years ago and the later episode about 11,000 years ago (the ca. 1200-year-long Younger Dryas). Þóra Ellen Þórhallsdóttir

retreating glacier in North America, the largest being Lake Agassiz, which, at its greatest extent around 13,000 years ago, had covered much of Manitoba, western Ontario, northern Minnesota, eastern North Dakota, and Saskatchewan, or as much as 440,000 km^2, larger than any currently existing lake in the world (including the Caspian Sea). Finally, about 11,000 years ago, the lake burst out from beneath the glacier's margin and for a few months freshwater flowed through the Gulf of St. Lawrence into the Labrador Sea, where it floated on the surface of the more saline Gulf Stream.

But suddenly, 11,500 years ago, this warm period ended. An ice-age cold descended and within a hundred years a 1200-year-long cold episode had begun called the Younger Dryas period. The temperature in northwestern Europe dropped by about 8–10 °C, and long winters maintained at 20 °C frost. There were snow-falls from September until May. Cold, arctic airstreams swept over the country, glaciers advanced about 30–40 km in western Norway and Iceland, and formed once more in the highlands of Scotland. Treelines descended and retreated south-ward. The boundary between cold and warm ocean currents moved as far south as it had been 18,000 years ago, and sea ice encroached over a large area of the North Atlantic.

Did glacial meltwater turn off the heating system of northern regions? It has long been a puzzle as to what caused the reversal into an ice-age cold during the

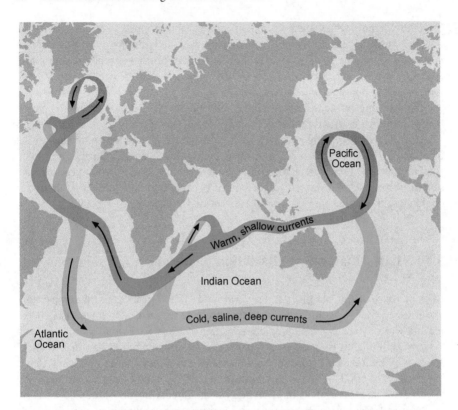

Fig. 2.27 Heating system of the world's oceans. The heating system of the world's oceans carries warmth from the tropics to the North Atlantic and Western Europe. Air currents drive the warm Gulf Stream and North Atlantic current to northern regions where its warmth is met and absorbed by the cold atmosphere. Cold and saline seawater sinks to the northeast of Iceland and then flows southward as a deep oceanic current picking up once again the warmth of the heating system, also called the oceans' conveyor belt of warmth. (Broecker 1965, 1991, 2003; Broecker and Denton 1989, 1990; Bianchi and McCave 1999; Broecker et al. 1968, 1999; Dickson et al. 1988; Häkkinen and Rhines 2004; Hanna et al. 2006; Hátún et al. 2005)

Younger Dryas period; the answer was not to be found in Milankovitch's theory, for solar radiation levels were high at this point in time. It is now believed that the huge flow of freshwater from melting ice sheets had so diluted the saline content of the sea that ocean currents changed course, thereby disrupting the oceanic heat flux into the northern regions (Fig. 2.27). Previously, warm and salt-laden surface seas flowed from the south into the northern oceans and, once this heat had been absorbed into the atmosphere, the cold, saline and denser seawater sank and became deep ocean currents that flowed to the South Atlantic and Antarctica and then into the Pacific and Indian Oceans. The seawater in the northern oceans around Iceland and in the Labrador and Greenland Seas had now mixed with freshwater, so that it no longer sank to form a deep oceanic current, while the freshwater floated above the saline sea and then froze, covering the ocean area with ice. Sea ice confronted the Gulf Stream, which then veered off course and stopped bearing heat to the northern regions. The conveyor belt of heat that had nurtured the warming of northern regions for 3000 years came to a halt, heat no longer reaching the North Atlantic or warming polar air masses. With new courses for ocean currents, the tracks of low-pressure weather systems also moved in the North Atlantic. The sudden warming of the Allerød period had in this manner turned off the conveyor belt of heat into northern regions. It is this kind of sequence of events that modern scientists are now bearing in mind when considering global warming due to the increased greenhouse effect and the rapid melting of glaciers. Glaciers are much smaller now than they were at the end of the last glacial period 11,000 years ago, so scientists are not afraid the conveyor belt of oceanic currents will cease, though it could slow down, causing the sea to cool around the shores of Iceland.

2.10 Late Quaternary Glaciation of Iceland

At the climax of the last glacial period, around 20,000 years ago, the temperature was lowered by 5–6 °C in Iceland and the snowline reached right down to sea level. Iceland was covered by one complete ice sheet that extended in all directions from the central highlands right down into the sea and over the offshore continental shelf. Judging by the tuyas that rose through and above the ice, it can be estimated that it had been all of 1000–1500 m thick in the centre of the country and had compressed the land by about 300–500 m. The highest mountains on the shores of western, northern and eastern Iceland might also have risen above the ice as nunataks. The ice sheet was thinnest along the shorelines, and land compression correspondingly less there too. Both on land and on Iceland's offshore oceanic bed, there is visible evidence of this ice sheet's erosive powers and the sediments which were borne with it: striation in rocks, troughs gouged into malleable deposits, channels through which water had flowed beneath the ice, sediment strata, erratic boulders and moraines. The furthest moraines can be found between 50 and 120 km from Iceland's current shoreline. In all, the land-based ice covered an area twice the size of what Iceland is today. It is still not totally clear, however, where the exact boundaries of this ice mass was at the climax of the last glacial period. The sub-oceanic troughs end at the edge of the continental shelf (see Chap. 3, Fig. 3.3). Beneath the ice shelves, sediments accumulated on the ocean floor at the mouth of rivers gradually forming deltas. Repeated glacial surges had increased sediment loads and jökulhlaups had also borne sediment from subglacial volcanic eruptions.

It is estimated that the ice sheet had been divided into 30–40 ice streams, each some 2 km wide, and they had borne most of the ice into the sea, the ice being transported much more rapidly than surrounding matter. Water had lubricated the soft sediments beneath the ice streams and basal movement had been much more rapid beneath the ice than at its margins. Meltwater resulting from geothermal heating at the glacial bed could also have facilitated sliding at certain places. These major ice flows of the central mass gouged out the main valleys, fjords and bays of

Iceland and harrowed out and dissected the continental shelf with deep troughs, right to its very edge.

Ice continuously crept forwards from land out into the ocean, where flat ice shelves began to float on the surface of the sea. At the furthest edges of the continental shelf, however, ice bergs broke off from the ice mass and floated away until they melted. Further away they would also meet very thin sea ice. For a long time there was a balance between the snow accumulation on the main ice sheet, its high plateaus, and the ice streams down to the floating shelves, where the ice calved into the sea. In some places the ice shelves would collide with islands, which would then form a buttress against further advancement out to sea.

The shrinking of Iceland's ice sheet. When it began to grow warmer in the northern hemisphere, around 17–15,000 years ago, the mass balance of the ice sheet on land was gradually reduced and the ice streams into the ocean abated. The ice shelves became thinner and ice resting on the sea bed gradually uplifted and the grounding line of ice finally moved inland. This resulted in a reduction in offshore buttressing against advancing ice streams, so that ice could flow more rapidly to the coast, where it calved into the sea. The climate then suddenly grew even warmer 15–14,000 years ago at the beginning of the Bølling Period. This mild interlude is named after Kópasker in northeastern Iceland, as oceanic sediments with fossilised shells from this time can be found there beneath ground moraine.

Meltwater flowed in increasing volume along glacial beds and this increased the basal sliding of land-based ice into the sea. Warm, southern oceanic currents also flowed into the deep offshore troughs and melted the ice from below. The sea levels also rose about 20 m in 500 years (4 cm per annum) because of the rapid melting of ice sheets in North America (Laurentide), Scandinavia and the Barents Sea. This began a chain of events which would not end until all the ice shelves had collapsed. It appears that all the ice mass had disappeared from Iceland's continental shelf just over 13,000 years ago and had withdrawn up onto land by about 12,500 years ago. In spite of a short cold period (the Older Dryas, about 14–13,000 years ago) the collapse of the ice shelves continued because the ice streams supplying them had become unstable. In Iceland the Older Dryas period is named after the Álftanes peninsula, where glacial moraines from that time are to be found, as well as in the Melasveit district. During the warm Allerød period, called the Saurbær period, after a farm in Dalir County, where shells can be found in oceanic sediment strata, the glaciers retreated far inland, the seas following them, and so came into existence the well-known Fossvogur sediment layers on top of the glacially striated Reykjavík dolerite. Finally the glaciers advanced during the Younger Dryas period, called the Búði stage in Iceland, after a waterfall in the river Þjórsá. The glacial moraines on Rangárvellir plain, which stretch southeastward to Keldur, are from this time.

When the ice melted at the end of the last glacial period, the sea levels of the world's oceans rose rapidly and followed the shrinking glacial margins inland, inundating the continental shelves and flowing over lowlands previously covered by glaciers. The waves of the Atlantic penetrated shorelines below coastal cliffs. When the large ice masses of the northern hemisphere had completely melted, there was no further rise in sea levels, but the Earth's crust began to uplift after the overburden of the ice had been removed, and land was thus uplifted from the sea along the coasts of Iceland. It is believed it could have risen as fast as 10 cm a year, and it is estimated that it had reached its maximum height around 9000 years ago. Glaciers only then existed in the central highlands and in southeastern Iceland. For further details see Kjartansson (1939), Einarsson (1968), Hjartarson and Ingólfsson (1988), Hjartarson (1989), Sveinbjörnsdóttir and Johnsen (1990), Ingólfsson (1991), Ingólfsson and Norðdahl (1994), Ingólfsson et al. (1997), Eiríksson et al. (2000), Harðardóttir et al. (2001), Geirsdóttir et al. (2007).

2.11 Holocene Climate and Glaciers

Around 11,500 years ago, the extreme cold of the last glacial period finally gave way and the so-called Holocene period began and is still continuing (Fig. 2.28). In around 50 years, the climate was sharply transformed from the Younger Dryas period into a warm and humid Pre-Boreal age. Earth has remained relatively warm ever since, although climate and vegetation have fluctuated, so that the period might be divided into a few vegetation periods correlating with air temperatures and moisture (Table 2.1; Hallsdóttir 1990; Hallsdóttir and Caseldine 2005). The floating surface layer of freshwater on the sea had gradually become thinner, summer

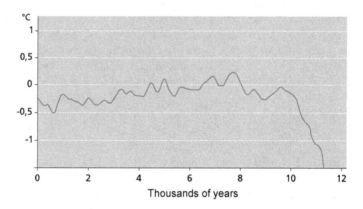

Fig. 2.28 Temperature levels from the end of the last glacial period. Earth's average temperature at the end of the last glacial period 11,000 years ago, calculated as a deviation from the average of the 20th century; data compiled from suboceanic sediment layers along with ice core readings from the Greenland and Antarctica ice sheets. The climate rapidly grew warmer 11,000 years ago and then maintained a stable average for the next 6000 years, during which most of the glaciers disappeared from Iceland. It then cooled for the next 3000 years, and glaciers expanded once again. ("Holocene Temperature Variations 2008; Website; Dansgaard et al. 1969, 1984; Denton and Karlen 1973; Oeschger and Langway 1989; Masson-Delmotte et al. 2005; Vinther et al. 2006; Bond et al. 1997, 2001; CAPE Project Members 2001; COHMAP Members 1998; Chapman and Shackleton 2000; Andrews et al. 2001; Eiríksson et al. 2000; Kaufman et al. 2004; Zielinski et al. 1994; Guðmundsson 1997; Ingólfsson and Norðdahl 1994; Ingólfsson et al. 1995)

Table 2.1 Post-glacial time divided into climate periods according to vegetation patterns in Iceland (Hallsdóttir and Caseldine 2005; Hallsdóttir 1990, 2008, personal communication 2009)

Period	Age in carbon-dated years	Climate and vegetation
Pre-Boreal	10,000–9000	Cool and dry. Plants began to grow, first resilient grasses and herbs and later bushes, heather and species of wood. Moors began to bear vegetation
Borea birch	9000–8500	Juniper spread out, then dwarf-birch, and finally trees took root
Atlantic	8500–5000	Warm and moist. Birch trees proliferated and reached their greatest extent in modern times, about 6000 years ago
Sub-Boreal	5000–2500	Cooling. Birch and brushwood receded in the lowlands and on moors while marshlands increased. Birch trees receded and advanced in turn, their last expansion probably being about 6000 carbon-dated years ago
Sub-Atlantic	2500-present	Increasingly cooler and damper, increase in marshland areas
	1500-present	After the settlement of Iceland there were major changes in the island's vegetation. Forests and brushwood disappeared, grasslands spread out along with arable grass and weeds. Soil erosion soon began

warmth had melted sea ice, and saline seawater began to sink at the margins of ice and form deep ocean currents thus reviving the heat flux from the equator to the polar regions. The heating pump of the North Atlantic, which had been turned off for over 1000 years, had once again been reactivated. The cold episode of the Younger Dryas period ceased as suddenly as it began, and now permanently. The expansion of ice sheets in North America during this millennium-long cold episode could have directed meltwater southward to the Mexico Bay once more instead of letting it flood into the North Atlantic.

Tremendous changes occurred in climate, vegetation and animal life. Pollen grain analyses of sediment deposits reveal that tundra disappeared and birch trees took root in land vacated by retreating glaciers. The glacial shield over Iceland mostly disappeared within 1000 years. It retreated rapidly from the lowlands and around 9500 years ago there was only an ice cap over the central highlands and the southeastern part of the country; 1000 years later even that had gone. Continuous sea ice no longer encroached on its shorelines. About 8000 years ago, Iceland was free of glaciers on the Tungnaáöræfi wilderness, from where the Þjórsárhraun lava field emanated and flowed south to the sea. The glacial shield over northern Europe and North America disappeared within 3000 years. Great pine and deciduous forests spread out over the northern parts of continental Europe and America and large mammals such as mammoths roamed the land, although they were soon wiped out by hunters. Until this moment, mammals had survived many fluctuations from one cold to warm period after another. About 8–10,000 years ago, mankind learned how to grow and utilise wheat and corn, and a static agricultural mode of life

Fig. 2.29 Cross section of the trunk of a coast redwood tree (Sequoia sempervirens) from California which the United States of America presented to Iceland in 1974 to mark the 1100th anniversary of the island's settlement. Marked on the tree rings are the years of important events in Iceland's history: the settlement (874), the establishment of the Althing (930), the acceptance of Christianity (1000), the end of the Commonwealth (1262), the Reformation (1550), the Skaftá fires volcanic eruption (1783), the resurrection of the Althing (1843), and the establishment of the Republic of Iceland (1944). The cross section is about 2 m in diameter

replaced that of the peripatetic hunter-gatherers. Birch woods spread out over Iceland 8–9000 years ago, and this time span is thus called the Birch period in Iceland and the previous millennium was called the Birch-free period. Glaciers retreated rapidly, although the most active still managed to push forward some marginal moraines during this time.

Vegetation Tells the Story of Climate History
It is possible both to calculate the age of trees and to analyse the climate changes during their growth by counting the annual rings and measuring their width (Fig. 2.29). Annual rings are formed in tree trunks while they are

growing in the warm and damp times of the year. Trees are in hibernation during cold and dry periods. Each year adds another ring, and its width depends on the warmth and precipitation of that year. The paler part of the ring is made from larger cells that form at the beginning of each growing season, and the darker parts of the ring are from smaller cells from the late summer. A tree does not have to be cut down in order to count its rings, for a 0.5-cm-thick core sample from a borehole is sufficient for this purpose, and its extraction does not harm the tree. The climate history of Europe and North America has been traced back thousands of years over a wide area by examining tree rings and treelines; in some places this history has been traced as far back as Roman times, and in a few special locations, even up to 6000 years ago.

The history of vegetation has been determined in historical times by analysing the species of pollen grains dispersed by plants that have been preserved in marshes and lakes. Each annual pollen rain records the history of vegetation, from which may be drawn conclusions about climate history. The age of vegetation can be determined through radiocarbon dating and tephra layers. Thus it can be seen when birch trees grew on land and when they were subsequently driven into marshes, when moss began to proliferate, and when treelines moved higher or lower and different species of grass multiplied. Vegetation can also be affected by volcanic activity, the encroachment of humans and animals, and wind erosion. Pollen analysis can thus be utilised in many ways in studying and researching natural history.

With the rapid shrinking of glaciers, huge amounts of meltwater were discharged into the sea. Powerful rivers bore jökulhlaups that gouged out great canyons. Around 8800 years ago, a large inland lake covering the centre of Jamtland in Sweden emptied and deposited a thick layer of sediment that the Swede de Geer used as a reference point in his dating of varves. All over the globe the sea level rose and seawater flooded over lowlands, hewing new shorelines along the coast and submerging land bridges. An ocean swept through the English Channel 8500 years ago, finally dividing Britain from continental Europe, and the southern part of Scandinavia also disappeared under water. About 8200 years ago, two large lakes emptied at the margins of the ice sheet in North America on the Hudson Bay causing the sea level to rise considerably. As a result, the North Atlantic cycle slowed down and a short-lived cold episode was unleashed, temperatures falling by as much as 6 °C. Around 7800 years ago, the atmospheric circulation began once more with low air pressure over Iceland and high pressure over the Azores. Strong westerly winds bore heat and moisture over the Atlantic to Europe. This period, from 8000 to 5000 years ago, has thus since been called the Atlantic period, though it is also named the Alder-elm-lime period in Scandinavia, while in Iceland birch trees proliferated. About 5600 years ago, the Mediterranean (Marmara Sea) rose to fill up the Bosphorus and within a few years it had torn a pass through the earthen dam and rushed through along a 30-km-broad channel into the Black Sea, raising its water level by 150 m; until then the Black Sea had been a freshwater ocean, but it has been a saltwater one ever since. Its water levels rose 15 cm per day. These catastrophic events could have been the origins of the story of Noah's Flood.

Features of post-glacial times: taking the rough with the smooth. Land that glaciers had compressed rose slowly out of the sea once the overburden of ice was reduced, and tidelines can now be seen dozens of metres above the current sea surface. They rise toward areas which the central ice sheet covered and are highest where the ice-age glacier was thickest, in the basin of the Baltic Sea and by Hudson Bay on the northern coast of Canada, 300 m above today's sea level. The Earth's crust rose so quickly in Iceland, once the weight of the glaciers had been reduced, that 2000 years after the ice-age glacier had dissolved, 9000 years ago, the land had risen to its fullest extent. There was still a huge amount of ice bound up in glaciers all over the world and so the sea level was then 20 m lower than it is today.

Research in Europe indicates that the warmest period had been about 6000 years ago, the average temperature being around 2–3 °C higher in the northern hemisphere than it is today. Although there had been ample moisture in the air, there were probably no glaciers in Iceland except on its highest peaks; the snowline is believed to have been at a height of about 1400 m in southern Iceland. The ice sheet in North America finally disappeared, the ocean was about 3 m higher than it is now, and there was little sea ice in northern oceans. Highly civilised cultures were then developing along the banks of the great Indus, Nile and Hwang Ho rivers. A cultured society was flourishing 5000 years ago in Sumer in Mesopotamia, between the Euphrates and Tigris rivers. The Pyramid of Cheops was constructed around 4600 years ago and a bit younger is the 4500-year old bristlecone pine (*Pinus aristata*) in California, the oldest organism still with us today, and which shared its childhood on Earth with the last of the mammoths (although a few of the latter are believed to have survived on the arctic islands north of Siberia until about 2000 years ago).

Around 5500 years ago it gradually began to cool, as changes in vegetation bear witness. The Sub-Boreal period began in Scandinavia, although in Iceland birch trees were giving way to increasingly encroaching marshlands. The frozen Copper-Age man, widely known as Ötzi the iceman, who was found in 1991 at an altitude of 3200 m in the Tyrol on the borders of Austria and Italy, died of exposure there 5300 years ago and was preserved in ice until our times (Fig. 2.30). High, mountain glaciers that had survived the warm and humid period from 8000 to 5000 years ago, then began to expand 4500 years ago. There are ancient Egyptian sources on the distress and hardship, caused by the climate fluctuation of 4000 years ago, especially drought. There was tremendous air pollution from the eruptions of Hekla 4200 and 3100 years ago, possibly followed by a few years of hunger in many parts of the world. Mycenaean civilisation flourished just over 1000 B.C. on the Peloponnesian peninsula, to be followed by the Hellenic civilisation for the next nine centuries. The climate then began to cool even more, however, 3000–2500 years ago, and precipitation increased. A cool and damp time began in Europe at the beginning of the Iron Age in Scandinavia, where oak receded and spruce

Fig. 2.30   The body of a Copper-Age man was found in the autumn of 1991 at a height of 3200 m near Ötztal on the border of Austria and Italy. His mummified corpse had been frozen there for 5300 years, or from the end of the Atlantic period. Ötzi the iceman, as he has been called, is now preserved in the Tyrol Museum in Bolzano, Italy. *Photo* Austrian police

proliferated, while in Iceland wetlands became even more widespread. The climate then entered the phase that it has for the most part followed ever since, although there have been some fluctuations. It may well have been from this time that the eldest narratives could have emerged of freezing cold winters, when a wolf swallowed the Sun, Moon and stars. Glaciers, which had not been particularly large at the time, began to expand and have remained in existence ever since. In Iceland it is believed that the glaciers on the highest mountains converged to form the great ice caps of Vatnajökull, Langjökull and Hofsjökull. Glacial ice covered vegetated moorlands and wooded mountainsides and finally hid from sight the large marshes in the valley basins. Birchwood branches and slabs of peat from this era are still emerging on outwash plains from beneath Vatnajökull's outlet glaciers, especially Skeiðarárjökull and Breiðamerkurjökull.

There was a warm climate for centuries during the age of Hellenic civilisation in Greece, the Celts in the British Isles, the Gauls in central Europe, and then later the Romans and their empire, right up to the end of the Middle Ages. In the rain forest belt, the civilisation of the Mayan Indians was at its peak from 300 to 900 A.D. The Middle Ages were warmest from 800 to 1200 A.D. with an average temperature 0.5–2.5 °C above that of the beginning of the 20th century. Human life and activity flourished around the North Atlantic. Monks sailed to Iceland in search of a sanctuary, longships appeared, and a Nordic people claimed and settled Iceland and

Greenland and reached as far as Vinland, or North America. Glaciers were lean and sailing routes were much less disrupted by sea ice. Crusades and Gothic cathedrals bear witness to a time of prosperity and expansion in medieval Europe. From tree rings, the position of treelines in the Alps, and written sources, it can be deduced that forests, cornfields, and vineyards reached further north and up to 200 m higher on mountainsides than they do today.

2.12 Climate and the Glaciers of Iceland Over the Last Millennium

During the first centuries of the settlement of Iceland and the Icelandic Commonwealth (ca. 870–1262 A.D.), the climate was mild, though there had been occasional severe winters. Sea ice seldom appeared and glaciers were smaller than they are now. Natural scientists believe that, during the first centuries when Icelanders were settling the island, three quarters of the land had vegetation and just under half of the land was covered by trees. In those days, Kjölur, at an elevation of almost 700 m between the Langjökull and Hofsjökull ice caps, and the Möðrudalsöræfi wilderness had both been covered with vegetation, but there has been little work done on trying to evaluate how widespread vegetation was in the central highlands between these two glaciers. There is much to indicate that, during the first centuries of the settlement, the highland routes of Iceland were frequently travelled and the interior utilised, e.g. the Þjórsárver wetlands. Today the land is drastically changed; just less than a fourth of Iceland has vegetation, only 1 % is covered by trees, and the highest continuous vegetation limit in the central highlands is now 100–200 m lower than at the time of settlement. Only in a very few places is there any significant continuous vegetation above a height of 630 m, but there is a considerable amount in the central highlands at around 500–600 m.

At the end of the 12th century, the warm period of the North Atlantic during the Middle Ages drew to a close as winters often became cold and damp, westerly winds piled up snow, and summers were cool. Glaciers grew and extended further than had previously been known. There was not one continuous cold period, however, but rather unpredictable weather conditions with irregular fluctuations in temperature and precipitation. These periods repeatedly oscillated, few lasting longer than for twenty-five years, with very cold winters being followed by a few years of frequent currents of mild air crossing the North Atlantic. Climate fluctuations did not come in slow stages but rather suddenly, and differently, from year to year and from one area of the country to another. Tree rings and ice cores reveal that cooling began in Greenland and other northern regions around 1200. It is possible the Gulf Stream had changed direction, as had first been visible in agriculture and in the sea journeys of Greenlanders and Icelanders. Harbours froze and contact between the Nordic peoples in Greenland and the rest of Scandinavia and Europe ceased, the settlers starving, dying out, or blending with the Inuit natives.

Pope Alexander stated in a letter dated 1492 that very few ships sailed to Greenland because of the ice. The appearance of sea ice off the coasts of Iceland became frequent and with them came cold summers. Sea ice drifted southward along the eastern fjords and surrounded Iceland annually, months on end, for many years. Iceland became isolated from the rest of the world. Worsening weather conditions had already played a large part in the cessation of grain farming (barley) in Iceland, first in the north and then the south, though wheat was grown in southern Iceland until well into the 16th century. At the end of the 13th and beginning of the 14th century, there were many years of crop failure and starvation in Europe. Around 1400 the weather became stormier than previously and remained rather cool, but then grew even colder as the 16th century progressed. In the latter part of the 16th century, drought harried the inhabitants of North America from Mexico to Canada. We do not know if this coldness was only in the North Atlantic, but late in the 17th century it is known for certain that glaciers began to advance over the entire world, in the Alps, in the Andes, and in China and in New Zealand. It was coldest in Asia during the 17th century, and in North America in the 19th century.

In Iceland it is believed the 17th, 18th, and 19th centuries were the coldest since the settlement (Fig. 2.31). A lowering of the average temperature during the growing seasons influenced both the speed and magnitude of production. The livestock of farms was reduced, because in times of starvation there was a gradual long-term shortage of winter fodder. The cold period from 1550 to 1900 was a

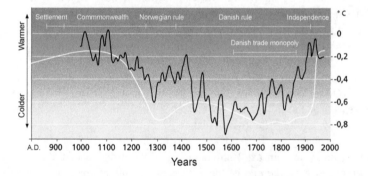

Fig. 2.31 Iceland had a warm climate during its Settlement and Commonwealth periods (870–1262). Oxygen isotopes in ice cores from the Greenland ice sheet indicate that temperatures there (*black line*) had fluctuated in a similar way as they are believed to have done in Iceland (*white line*, graph without numerical value; Þórarinsson 1974). It was coldest during the 14th century and from the end of the 17th until the beginning of the 20th century. Not only Iceland's environment was shaped by climatic fluctuations during these cold centuries, but also the history and societies of Europe, from the medieval Renaissance to the Enlightenment, from the French and Industrial revolutions to modern times. (Dansgaard 2000, 2005; Johnsen et al. 1995; Mann et al. 1999; Þórarinsson 1974; Sveinbjörnsdóttir and Johnsen 1990; Sveinbjörnsdóttir 1993; Geirsdóttir et al. 2009; Harðardóttir et al. 2001; Crowley 2000; Masse et al. 2008; Shindell et al. 2001; Sicre et al. 2008; Thompson et al. 1993; Ogilvie 1986, 1992; Ogilvie and Jónsson 2001; Finnsson 1970; Thoroddsen 1916–1917; Bergþórsson 1969; Guðmundsson 1912–22, 1948; Storm 1888; "1000 Year Temperature Comparison" 2005)

difficult time for vegetation and it was very susceptible to the effects of volcanic activity and human habitation. The destruction of soil and vegetation increased in the central highlands and caused additional livestock grazing in the lowlands, which, together with erosion, brought about the abandonment of farms. Avalanches became more common and they also destroyed vegetation and transported mud and debris over cultivated land. The migrations of fish were also diverted due to changes in the biological environment of the ocean and this also increased Icelanders' difficulties. A cold sea drove fish away from the shore, thus catching them meant going further out to sea for longer periods. Cod often disappeared from Icelandic fishing grounds altogether as it thrives badly in very cold waters.

The expansion of glaciers in Iceland. The climate changes from the 13th to the end of the 19th centuries had the most effect on countries with a maritime climate, because when temperatures drop in areas with wet weather systems, an increasing part of precipitation falls as snow. Snow-free areas also grew fewer because of cold summers. From the middle of the 16th until the end of the 19th centuries, it is believed the snowline fell to a height of 700–800 m to the south of Vatnajökull, 300 m below what it usually was during the 20th century. On peninsulas and promontories in northern Iceland, the snowline could have been as low as 800 m, and in the western fjords even below 600 m above sea level. Further inland, however, the snowline in northern Iceland has often been above 900 m and in the central highlands at about 1100 m while to the north of Vatnajökull it had been even higher because there was little precipitation there during all of this period. In depressions and hollows all over the country, snow has lain for years below the snowlines described above. Vegetation often appeared late from under snow cover and, during most years, snow settled high up on mountain tracks between glaciers before winter returned, although vegetation in the highlands could survive up to an elevation of 600 m. This has been proved by sources from the 18th century, such as the ruins of the outlaw Fjalla-Eyvindur's hut in Eyvindaver and bridle paths at a height of 600 m in Arnarfellsmúlar and Þúfuver (south of Hofsjökull). Permafrost had increased in the highlands, on the other hand, and is even believed to have reached as low as 400 m above sea level.

Glaciers probably began to advance just before the end of the 12th century, but with increasing power in the 16th century, and the progress of the largest glaciers was unimpeded until the end of the 19th century. While creeping forwards, these glaciers destroyed moraines from the Sub-Atlantic period. In Norway and the Alps, glaciers expanded from 1200 to 1550, but the first sources relating to advancing glaciers in Iceland are to be found in *Íslandslýsing* ('Account of Iceland') by Oddur Einarsson (1559–1630) from about 1590 (Einarsson 1971). Glaciers then progressed considerably from 1600–1620 and 1640–1650. Glacial lakes grew larger too, encroaching on, flooding and destroying farmland. Glaciers advanced the most in the dampest and coldest areas of the country, along the south coast and in the western fjords. In the first part of the 18th century, the outlet glaciers from Drangajökull on the western-fjords peninsula had spread out extensively, forcing five farms to be abandoned. Around 1700 the settlement farms of Breiðá and Fjall disappeared under Breiðamerkurjökull in the south, which had by then advanced

15 km over Breiðamerkursandur plain. The Breiðá farm was abandoned in 1698 and the farmstead buildings were in ruins and virtually covered by the glacier by 1712. The Fjall farm was probably abandoned before 1694 and had disappeared under the glacier (Fjallsjökull) by early in the 18th century. Árni Magnússon reported changes in Sólheimajökull glacier in the years 1702–1712 and described the advance of outlet glaciers from Drangajökull (Magnússon 1955). In Eggert Ólafsson and Bjarni Pálsson's (1981) *Ferðabók* ('Travelogue') there is mention of glaciers becoming larger, and around the middle of the 18th century they gave an account of how a glacier expanded on Skarðsheiði and how a new glacier had come into existence. This glacier was a quarter of a square mile when Sveinn Pálsson (1945, 2004) saw it on 27 August 1791. Due to the cold, glaciers began expanding again in Iceland in the years 1810–1860, as did glaciers in the Alps and elsewhere in the world, e.g. in Alaska and Karakoram in the Himalayas. Steep, fast-moving valley glaciers in Öræfi attained their greatest expansion from 1750 to 1850. The country's largest glaciers, on the other hand, continued to advance until the end of the 19th century, the glaciers in Iceland then being larger than they had ever been since the last glacial period had ended; indeed Vatnajökull had become larger than it had been for 8–9000 years. It is believed that the main outlet glaciers from Vatnajökull had extended 10–15 km further than they had been during the time of the settlement and that by 1900 glaciers had covered, in all, about 15 % of the country. Breiðamerkurjökull was just under 300 m short of reaching the sea. The land was compressed by the tremendous overburden of ice and tidelines rose higher due to this subsidence. Evidence of this can be seen in Hornafjörður in southeastern Iceland and will be discussed in more detail in the chapter on Vatnajökull.

Was the Little Ice Age a Result of a Minimum Number of Sunspots or Increased Volcanic Activity?
The cooling of the Little Ice Age has been connected to the fact that there were hardly any sunspots over a 70-year period from 1645 until 1715 (a theory named after the Englishman Walter Maunder [1851–1928]), and prior to that there had been a minimum amount of sunspots for a period twice as long (1420–1570; associated with the German astronomer Gustav Spörer [1822–1895]). No convincing physics-based explanation has been advanced to explain the link between sunspots and weather, however. Normally, there are 11-year oscillations in the Sun's activity, visible on Earth as sunspots. A higher number of high-speed particles then reach the outer layer of the Earth's atmosphere, or stratosphere, and northern lights are seen more frequently. The total solar energy varies about 0.1 % but the ultraviolet part of the sun's light about 20 %. The stratosphere completely absorbs this ultraviolet part of solar radiation and this is where any effect on the inner layer of Earth's atmosphere, the troposphere, would have to come from, although no one has yet worked out how this could cause fluctuations in climate. There are those who believe there are statistical calculations, however, which do

indicate that the number of sunspots can indeed be correlated with changes in precipitation and temperatures on Earth.

As well as the low number of sunspots, the Little Ice Age has also been linked to high levels of volcanic activity in the period from the 14th to the 19th centuries. In major volcanic eruptions, hundreds of cubic kilometres of ash and volatile materials, such as sulphur dioxide, have been borne through the troposphere and up into the stratosphere where they have mixed with water vapour, so that microscopic drops float for years on the outer edges of the troposphere, reflecting solar radiation and thus reducing the amount of solar energy reaching the Earth. This haze, when cooling, descends into the troposphere and changes the atmospheric circulation around the globe. This cooling of the Earth is rather short-lived, however, because the winds high in the atmosphere blow rapidly around the world and engulf the floating drops so that they eventually fall to Earth as precipitation. The best-known volcanic eruptions that are believed to have influenced weather conditions all over the world were Vesuvius in Italy (79 A.D.), the Laki fires in Iceland (1783), Tambora (1815) and Krakatoa (1883) in Indonesia, El Chichón in Mexico (1982), and Pinatubo in the Philippines (1991).

2.13   Climate in the 20th Century and the Future of Iceland's Glaciers

There was a worldwide increase in warming after the middle of the 19th century, and at the beginning of the 20th century the cold period that had lasted from about 1300 in northern Europe had finally ended (Fig. 2.32). A period succeeded with a temperature on average 1–2 °C warmer than it had been for seven centuries.

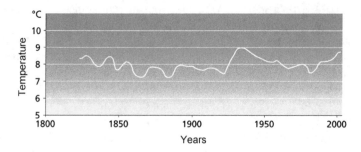

Fig. 2.32 Outline of average temperatures in Iceland since 1820. Data from Stykkishólmur indicates a running 11-year average (with triangular weight). (IMO data 2009; Eyþórsson 1950; Eyþórsson and Sigtryggsson 1971; Sigtryggsson 1972; Sigfúsdóttir 1969; Einarsson 1984, 1991, 1993; Jónsson 1991; Hanna et al. 2004)

Vegetation expanded and new species of birds began to settle and breed in Iceland. The greatest warming in the North Atlantic was in the 1920s and was connected to low-depression systems reaching further north than previously. The warmest period in Icelandic history was from 1930 until 1960 and almost all glaciers retreated except for a few surging ones. From around 1940 the temperature started to drop once again, and during the 1950s and 1960s it was on average 1 °C colder over the summer months than in the previous years of the Second World War. With this cooling, there was a reduction in summer meltwater from the glaciers and a greater percentage of precipitation fell as snow. The retreat of glaciers in Iceland gradually slowed down and over the next thirty years or so (1970–1995) most glaciers either remained stationary or advanced little because of the increasingly colder weather. It has been suggested that the causes of this cooling after 1940 may have been air pollution, especially due to smog from coal-burning, which had reduced the levels of solar radiation reaching Earth. By the end of the 20th century, Earth began to get warmer very quickly, especially after 1995, and glaciers have retreated all over the world; the effect of air pollution on atmospheric radiation now weighs more in importance than the dispersal of solar radiation on its way to Earth. It is also believed, however, that some oscillations might be caused by irregular fluctuations in the transporting of heat by air and ocean currents over the last decades.

Although changes in temperature levels over the last 1000 years may be connected to fluctuations in solar radiation and volcanic activity, the influence of mankind after 1850 is still considered the main cause of global warming: i.e. an enlarged absorption of terrestrial radiation in the atmosphere due to an increased concentration of greenhouse gases (IPCC 1995, 2007, 1990). The Earth is now warmer than it has been for 90 % of its 4600-million-year history due to two centuries of atmospheric pollution. Further warming is rated likely to continue for several hundred years, even if there were to be a reduction of so-called greenhouse gas emissions. Glaciers here in Iceland are now smaller than they have been since the 17th century, and if the climate change forecasts prove correct (i.e. +2 °C warmer per century), most of the country's main glaciers will disappear within the next 100–200 years (Jóhannesson 1997; Jóhannesson et al. 2004, 2006b, 2007; Jóhannesson et al. 2006a; Bergström et al. 2007; Björnsson and Pálsson 2008). Runoff waters from areas now covered by glaciers will increase around 25–50 % over the next 30–100 years before being reduced, and once glacial water reserves have drained away the only source of water will be from precipitation. Increasing meltwater flows and changes in waterways will have a great impact on the design and construction of bridges and roads and the operating of hydro-electric power stations that utilise water from the central highlands of Iceland.

Fluctuating ice ages in the future. We are living in a warm period of an ice age with exceptional circumstances in the history of the Earth because warm periods have only taken place during 10% of the last 2,000,000 years. The last time the Earth was as warm as it is today was 120,000 years ago. The ice age continues, nonetheless, and cannot be considered to have ended until ice has mostly disappeared from the world's mountain ranges and polar regions. Glaciers currently cover almost 11 % of the Earth's land surface (15,000,000 km^2) and in addition there is an equally large area of frozen tundra on Earth in North America and Eurasia (14,000,000 km^2). During winter, sea ice covers around 12 million km^2 of ocean in the northern hemisphere, and 20,000,000 km^2 around Antarctica during the summer (when it is, of course, winter in the southern hemisphere). In comparison, the whole of the Greenland ice cap is 1.7 million km^2. About 90 % of all the freshwater of this Earth, the blue planet, is now frozen in glaciers, most of it in Antarctica and Greenland.

According to Milankovitch's theories, there lies ahead of us a series of 50 glacial periods that will endure for about 100,000 years before there is any let-up. The Earth will gradual cool due to its orbit and axial tilt, even though precession would still promote warming. A combination of these factors will cause a gradual cooling and a move towards a cold period after 10–15,000 years, and after 20–25,000 years a glacial period will begin which will reach its peak after 60,000 years (Berger 1977b, 1978). Mankind cannot avoid having to deal with a glacial period. Cities in the northern hemisphere will then be covered in ice, crushed, pulverised and bulldozed by glaciers. A massive flow of refugees southward will follow with the consequent fight for survival and danger of wars. Human life will only be a shadow of what it is today in a modern, technological civilization. The question can thus be raised as to whether polluting the atmosphere could in fact delay this cooling of Earth and that global warming from the greenhouse effect might therefore be welcomed? Experience of how a sudden warming during the Younger Dryas period led to the shutting down of the conveyor belt of heat to northern regions, however, raises fears that an increasing greenhouse effect could actually speed up the coming of a cold period. A tremendous melting of glaciers would change the saline content of seas and the routes of ocean currents in the North Atlantic and bring about a cooling of the whole northern hemisphere; there would be severe weather around the North Atlantic, in Greenland, Iceland, the British Isles and Scandinavia. We can no longer rely on an increase in solar radiation, as in the beginning of this post-glacial, or Holocene period. But all of this is difficult to predict, for scientists still do not fully understand the causes of sudden climate changes (Alley 2000, 2004; Masse et al. 2008). On the other hand, we do know from the climate history of the Earth, that climate has often suddenly been jolted out of an equilibrium when a slow-working change has pushed Earth's climate system over a certain threshold. Effects are then amplified and continue for centuries; and it appears that ocean currents in the North Atlantic have the final say in this. It is thus ominous news that ocean currents southward between Scotland and the Faroe Isles have slowed by about 20% since the middle of the 20th century (Häkkinen and Rhines 2004; Hátún et al. 2005). Being at the centre of North Atlantic oceanic currents, Iceland is very sensitive to climate change and it has occurred there in the past more rapidly than in most other countries.

Note: The information in this chapter has been gathered from many sources, especially works by Alley (2002, 2004), Agassiz (1967), Fagan (2000), Grove (1988), Flint (1971), Imbrie and Imbrie (1979), Lamb (1966, 1995), Broecker (1966, 1975, 2005), Ruddiman (2001), Dansgaard (2000, 2005) and IPCC (1990, 1995, 2007).

References

Adhémar, J. (1842). *Révolutions de la mer*. Paris: Carilian-Goeury & Dalmont.

Agassiz, L. (1967) [1840]. *Études sur les glaciers*. Neuchâtel: Joseph Bettannier. English edition: Agassiz, L. (1967) *Studies on glaciers. Preceded by the discourse of Neuchâtel (1837)*. (A. Carozzi, Trans.) New York: Hafner.

Alley, R. (2000). *The two-mile time machine: Ice cores, abrupt change, and our future*. Princeton: Princeton University Press.

Alley, R. (2004). Abrupt climate change. *Scientific American*. November. pp. 62–69.

Andrews, J., Harðardóttir, J., Helgadóttir, G., et al. (2000). The N and W Iceland shelf: Insights into LGM Ice extent and deglaciation based on acoustic stratigraphy and basal radiocarbon AMS dates. *Quaternary Science Reviews, 19*, 619–631.

Andrews, J., Helgadóttir, G., Geirsdóttir, Á., & Jennings, A. (2001). Multicentury-scale records of carbonate (hydrographic?) variability on the Northern Iceland margin over the last 5000 years. *Quaternary Research, 56*, 199–206.

Barnola, J., Raynaud, D., Korotkevich, Y., et al. (1987). Vostok ice core provides a 160,000 year record of atmospheric CO_2. *Nature, 329*, 408–414.

Berger, A. (1977a). Support for the astronomical theory of climatic change. *Nature, 269*, 44–45.

Berger, A. (1977b). Long-term variation of the earth's orbital elements. *Celestial Mechanics, 15*, 53–74.

Berger, A. (1978). Long-term variations of caloric insolation resulting from the Earth's orbital elements. *Quaternary Research, 9*, 139–167.

Berger, A., & Loutre, M. (1991). Insolation values for the climate of the last 10 million years. *Quaternary Science Reviews, 10*, 297–317.

Bergström, S., et al. (2007). *Impacts of climate change on river runoff, glaciers and hydropower in the Nordic area*. Joint final report from the CE hydrological models and snow and ice groups. CE Rep. 6, The CE Project. Reykjavík.

Bergþórsson, P. (1969). An estimate of drift ice and temperature in Iceland in 1000 years. *Jökull, 19*, 94–101.

Bernhardi, R. (1939). An hypothesis of extensive glaciation in prehistoric time. In K. Mather & S. Mason (Eds.), *Source book in geology* (pp. 327–328). New York: McGraw-Hill.

Bianchi, G., & McCave, I. (1999). Holocene periodicity in North Atlantic climate and deep-ocean flow south of Iceland. *Nature, 397*, 515–517.

Björnsson, H., & Pálsson, F. (2008). Icelandic glaciers. *Jökull, 58*, 365–386.

Bond, G., Broecker, W., Johnsen, S., et al. (1993). Correlations between climatic records from North Atlantic sediments and Greenland ice. *Nature, 365*, 143–147.

Bond, G., Kromer, B., Beer, J., et al. (2001). Persistent solar influence on North Atlantic climate during the Holocene. *Science, 294*, 2130–2136.

Bond, G., Showers, W., Cheseby, M., et al. (1997). A pervasive millennial-scale cycle in North Atlantic Holocene and glacial climates. *Science, 278*, 1257–1266.

Bradley, R. (1998). Paleoclimatology: Reconstructing climates of the Quaternary. *International geophysics* (Vol. 64). San Diego: Harcourt Academic Press.

Broecker, W. (1965). Isotope geochemistry and the pleistocene climatic record. In H. Wright Jr. & D. Frey (Eds.), *The quaternary of the United States* (pp. 737–753). Princeton: Princeton University Press.

Broecker, W. (1966). Absolute dating and the astronomical theory of glaciation. *Science, 151*, 299–304.

Broecker, W. (1975). Climatic change: Are we on the brink of a pronounced global warming? *Science, 189*, 460–463.

Broecker, W. (1991). The great ocean conveyor. *Oceanography., 2*(2), 79–89.

Broecker, W. (2003). Does the trigger for abrupt climate change reside in the ocean or in the atmosphere? *Science, 300*, 1519–1522.

Broecker, W. (2005). Global warming: Take action or wait? *Jökull, 55*, 1–16.

Broecker, W., & Denton, G. (1989). The role of ocean-atmosphere reorganizations in glacial cycles. *Geochimica et Cosmochimica Acta, 53*(10), 2465–2501.

Broecker, W., & Denton, G. (1990) What drives glacial cycles? *Scientific American.* January. pp. 48–56.

Broecker, W., Thurber, S., Goddard, J., et al. (1968). Milankovitch hypothesis supported by precise dating of coral reefs and deep-sea sediments. *Science, 159,* 297–300.

Broecker, W., Sutherland, S., & Peng, T. (1999). A possible 20th-century slowdown of southern ocean deep water formation. *Science, 286,* 1132–1135.

Buckland, W. (1836) *Geology and mineralogy, considered with reference to natural theology* (vols. 2). London: William Pickering.

CAPE Project Members. (2001). Holocene paleoclimate data from the Arctic: testing models of global climate change. *Quaternary Science Reviews, 20,* 1275–1287.

Chapman, M., & Shackleton, N. (2000). Evidence of 550-year and 1000-year cyclicities in North Atlantic circulation patterns during the Holocene. *Holocene, 10*(3), 287–291.

Clark, P., Licciardi, J., et al. (1996). Numerical reconstruction of a soft-bedded Laurentide ice sheet during the last glacial maximum. *Geology, 24,* 679–682.

CLIMAP Members. (1981). Seasonal reconstruction of the Earth's surface at the last glacial maximum. In *Map and chart series* MC-36. Boulder: Geological Society of America.

COHMAP Members. (1998). Climatic changes of the last 18,000 Years: Observations and model simulations. *Science, 241,* 1043–1062.

Croll, J. (1864). On the physical cause of the change of climate during geological epochs. *Philosophical Magazine, 27,* 121–137.

Croll, J. (1875). *Climate and time in their geological relations.* London: Daldy, Isbister.

Croll, J. (2012). *Climate and time in their geological relations.* Cambridge, UK: Cambridge University Press.

Crowley, T. (2000). Causes of climate change over the past 1000 years. *Science, 289,* 270–277.

Dansgaard, W. (2000). *Grønland i istid og nutid.* Copenhagen: Rhodos.

Dansgaard, W. (2005). *Frozen annals: Greenland ice sheet research.* Copenhagen: Rhodos.

Dansgaard, W., Johnsen, S., Clausen, H., et al. (1984). North Atlantic climate oscillations revealed by deep Greenland ice cores. In J. Hansen, & T. Takahashi (Eds.), *Climate processes and climate sensitivity* (pp. 288–298). *Geophysical monographs,* 29.

Dansgaard, W., Johnsen, S., & Møller, J. (1969). One thousand centuries of climatic record from Camp Century on the Greenland ice sheet. *Science, 166,* 377–380.

Dansgaard, W., Oeschger, H., & Langway, C. (Eds.). (1985). *Greenland ice core. Geophysics, geochemistry, and the environment. Geophysical monographs.* Washington: American Geophysical Union.

De Geer, G. (1912). A geochronology of the last 12,000 years. In *Congress geological international Stockholm 1910.* C.R. pp. 241–253.

Denton, G., & Hughes, T. (Eds.). (1981). *The last great ice sheets.* New York: Wiley.

Denton, G., & Karlen, W. (1973). Holocene climatic variations: Their possible causes. *Quaternary Research, 3,* 155–205.

Dickson, R., Meincke, J., Malmberg, S., et al. (1988). The 'great salinity anomaly' in the Northern North Atlantic 1968–1982. *Progress in Oceanography, 20,* 103–151.

Einarsson, M. (1984). Climate of Iceland. In H. Van Loon (Ed.), *Climates of the oceans* (pp. 673–697). Amsterdam: Elsevier.

Einarsson, M. (1991). Temperature conditions in Iceland, 1901–1990. *Jökull, 41,* 1–20.

Einarsson, M. (1993). Temperature conditions in Iceland and the eastern North-Atlantic region, based on observations 1901–1990. *Jökull, 43,* 1–13.

Einarsson, O. (1971) [1585]. *Íslandslýsing - Qualiscunque descriptio Islandiae.* (S. Pálsson, Trans.). Reykjavík: Bókaútgáfa Menningarsjóðs.

Einarsson, Þ. (1968). *Jarðfræði. Saga bergs og lands.* Reykjavík: Mál og Menning.

Eiríksson, J. (1980). Tjörnes, North Iceland: A bibliographical review of the geological research history. *Jökull, 30,* 1–20.

Eiríksson, J. (2008). Glaciation events in the Pliocene-Pleistocene volcanic succession of Iceland. *Jökull, 58*, 315–329.

Eiríksson, J., Knudsen, K., et al. (2000). Late-glacial and Holocene palaeoceanography of the North Icelandic shelf. *Journal Quaternary Science, 15*, 23–42.

Emiliani, C. (1955). Pleistocene temperatures. *Journal of Geology, 63*, 538–578.

Emiliani, C. (1966). Paleotemperature analysis of Caribbean cores P6304-8 and P6304-9 and a generalized temperature curve for the past 425,000 years. *Journal of Geology, 74*, 109–126.

EPICA community members. (2004). Eight glacial cycles from an Antarctic ice core. *Nature, 429*, 623–628.

Esmark, J. (1824). Bidrag til vor Jordklodes Historie. *Magazin for Naturvidensaberne, 2*(1), 29–54.

Eyþórsson, J. (1950). Hitafarsbreytingar á Íslandi. *Náttúrufræðingurinn, 20*, 67–85.

Eyþórsson, J., & Sigtryggsson, H. (1971). The climate and weather of Iceland. In *The zoology of Iceland* (1 (3)). Copenhagen: Munksgaard.

Fagan, B. (2000). *The little ice age: How climate made history 1300–1850*. New York: Basic Books.

Finnsson, H. (1970). In J. Eyþórsson & J. Nordal (Eds.), *Mannfækkun af hallærum*. Reykjavík: Almenna bókafélagið.

Flint, R. (1971). *Glacial and quaternary geology*. New York: Wiley.

Geikie, J. (1874–94). *The great ice age* (1ˢᵗ ed.). London: Isbister, 1874; (2ⁿᵈ ed.). London: Daldy, Isbister, 1877; (3ʳᵈ ed.). London: Stanford.

Geirsdóttir, Á. (1990). Diamictites of late Pliocene age in western Iceland. *Jökull, 40*, 3–25.

Geirsdóttir, Á., & Eiríksson, J. (1996). A review of studies of the earliest glaciations in Iceland. *Terra Nova, 8*, 400–414.

Geirsdóttir, Á., Miller, G., & Andrews, J. (2007). Glaciation, erosion, and landscape evolution of Iceland. *Journal of Geodynamics, 43*, 170–186.

Geirsdóttir, Á., Miller, G., Þórðarson, Þ., & Ólafsdóttir, K. (2009). A 2,000 year record of climate variations reconstructed from Haukadalsvatn, West Iceland. *Journal of Paleolimnology, 41*, 95–115.

Gilbert, G. (1890). Lake Bonneville. *U.S. geological survey, monograph* 1 (pp. 1–438). Washington: USGS.

Grootes, P., Stuiver, M., White, J., et al. (1993). Comparison of Oxygen isotope records from the GISP2 and GRIP Greenland ice cores. *Nature, 366*, 552–554.

Grove, J. (1988). *The little ice age*. London & New York: Methuen.

Guðmundsson, H. (1997). A review of the Holocene environmental history of Iceland. *Quaternary Science Reviews, 16*, 81–92.

Guðmundsson, P. (Ed.) (1912–22, 1948) *Annáll nítjándu aldar*. Akureyri: Hallgrímur Pétursson.

Haflíðason, H., Eiríksson, J., & Van Kreveld, S. (2000). The tephrochronology of Iceland and the North Atlantic region during the Middle and Late Quaternary: A review. *Journal of Quaternary Science, 15*, 3–22.

Häkkinen, S., & Rhines, P. (2004). Decline of subpolar North Atlantic circulation during the 1990s. *Science, 304*, 555–559.

Hallsdóttir, M. (1990). Studies in the vegetational history of north Iceland. A radiocarbon-dated pollen diagram from Flateyjardalur. *Jökull, 40*, 67–81.

Hallsdóttir, M., & Caseldine, C. (2005). The Holocene vegetation history of Iceland, state of the art and future research. In C. Caseldine, A. Russell, J. Harðardóttir, et al. (Eds.), *Iceland—modern processes and past environments 5* (pp. 319–334). Amsterdam: Elsevier.

Hanna, E., Jónsson, T., & Box, J. (2004). An analysis of Icelandic climate since the nineteenth century. *International Journal of Climatology, 24*, 1193–1210.

Hanna, E., Jónsson, T., Ólafsson, J., & Valdimarsson, H. (2006). Icelandic coastal sea-surface temperature records constructed: Putting the pulse on air-sea-climate interactions in the northern North Atlantic. Part I: Comparison with HadISST1 open ocean surface temperatures and preliminary analysis of long-term patterns and anomalies of SSTs around Iceland. *Journal of Climate, 19*, 5652–5666.

Harðardóttir, J., Geirsdóttir, Á., & Sveinbjörnsdóttir, Á. (2001). Seismostratigraphy and sediment studies of Lake Hestvatn, southern Iceland: Implications for the deglacial history of the region. *Journal of Quaternary Science, 16*(2), 167–179.

Hátún, H., Sandø, A., Drange, H., et al. (2005). Influence of the Atlantic subpolar gyre on the thermohaline circulation. *Science, 309*, 1841–1844.

Hays, J., Imbrie, J., & Shackleton, N. (1976). Variations in the Earth's Orbit: Pacemaker of the ice ages. *Science, 194*, 1121–1132.

Heinrich, H. (1988). Origin and consequences of cyclic ice rafting in the northeast Atlantic Ocean during the past 130,000 years. *Quaternary Research, 29*, 142–152.

Hjartarson, Á. (1989). The ages of the Fossvogur layers and the Álftanes end-moraine, SW-Iceland. *Jökull, 39*, 21–31.

Hjartarson, Á., & Ingólfsson, Ó. (1988). Preboreal glaciation of Southern Iceland. *Jökull, 38*, 1–16.

"Holocene Temperature Variations" (2008) http://en.wikipedia.org/wiki/File: Holocene temperature variations. Accessed September, 1.

Hubbard, A., Sugden, D., Dugmore, A., et al. (2006). A modelling insight into the Icelandic last glacial maximum ice sheet. *Quaternary Science Reviews, 25*, 2283–2296.

Hutton, J. (1959). *Theory of the earth* (Vols. 2) (Edinburgh: William Greech 1795) New York: Hafner.

"Ice Age." (2009). http://en.wikipedia.org/wiki/Iceage. Accessed June 1.

Imbrie, J., & Imbrie, K. (1979). *Ice ages: Solving the mystery*. Cambridge, Mass. & London: Harvard University Press.

IMO data (2009). www.vedur.is. Accessed March 20.

Ingólfsson, Ó. (1984). A review of late Weichselian studies in the lower part of the Borgarfjörður region, Western Iceland. *Jökull, 34*, 117–130.

Ingólfsson, Ó. (1985). Late Weichselian glacial geology of the lower Borgarfjörður region, western Iceland: A preliminary report. *Arctic, 38*, 210–213.

Ingólfsson, Ó. (1987a). *Investigation of the Late Weichselian glacial history of the lower Borgarfjörður region, western Iceland*. Lundqua Thesis 19. Doctoral dissertation. University of Lund.

Ingólfsson, Ó. (1987b). The late Weichselian glacial geology of the Melabakkar-Ásbakkar coastal cliffs, Borgarfjörður, W-Iceland. *Jökull, 37*, 57–81.

Ingólfsson, Ó. (1988). Glacial history of the lower Borgarfjörður area, western Iceland. *Geologiska Föreningens i Stockholm Förhandlingar, 110*, 293–309.

Ingólfsson, Ó. (1991). A review of the Late Weichselian and early Holocene glacial and environmental history of Iceland. In J. Maizels & C. Caseldine (Eds.), *Environmental change in Iceland: Past and present* (pp. 13–29). Dordrecht: Kluwer Academic Publishers.

Ingólfsson, Ó., & Norðdahl, H. (1994). A review of the environmental history of Iceland, 13,000–9000 year BP. *Journal of Quaternary Science, 9*, 147–150.

Ingólfsson, Ó., Björck, S., et al. (1997). Glacial and climatic events in Iceland reflecting regional North Atlantic climatic shifts during the Pleistocene-Holocene transition. *Quaternary Science Reviews, 16*, 1135–1144.

Ingólfsson, Ó., Norðdahl, H., & Hafliðason, H. (1995). Rapid isostatic rebound in southwestern Iceland at the end of the last glaciation. *Boreas, 24*, 245–259.

IPCC. (1990). *Climate change: the IPCC scientific assessment*. In J. Houghton., G. Jenkins., & J. Ephraums. (Eds.), Cambridge: Cambridge University Press. www.ippc.ch. Accessed July 1, 2008.

IPCC. (1995). *Climate change: The science of climate change*. Cambridge: Cambridge University Press. www.ippc.ch. Accessed July 1, 2008.

IPCC. (2007). *Climate change: The science of climate change*. Cambridge: Cambridge University Press. www.ippc.ch. Accessed July 1, 2008.

Jóhannesson, T. (1997). The response of two Icelandic glaciers to climatic warming computed with a degree-day glacier mass balance model coupled to a dynamic glacier model. *Journal of Glaciology, 43*(143), 321–327.

Jóhannesson, T., Aðalgeirsdóttir, G., et al. (2004). Impact of climate change on glacier in the Nordic countries. In *Nordic project climate, water and energy* (CWE). Website: http://www.os. is/cwe. Accessed June 3, 2005.

Jóhannesson, T., Aðalgeirsdóttir, G., Björnsson, H., et al. (2006a). The impact of climate change on glaciers and glacial runoff in the Nordic countries. In *Proceedings of the European conference of impacts of climate change on renewable energy sources*. Reykjavík: Orkustofnun.

Jóhannesson, T., Aðalgeirsdóttir, G., Björnsson, H., et al., (2006b). Mass balance modelling of the Vatnajökull, Hofsjökull and Langjökull ice caps. In *Proceedings of the European conference of impacts of climate change on renewable energy sources*. Reykjavík: Orkustofnun.

Jóhannesson, T., Aðalgeirsdóttir, G., Björnsson, H., et al. (2007). *Effect of climate change on hydrology and hydro-resources in Iceland*. Final report of the VO-project. OS-2007/011.

Johnsen, S., Clausen, H., Dansgaard, W., et al. (1998). The δ18O record along the Greenland Ice Core Project deep ice core and the problem of possible Eemian climatic instability. *Journal of Geophysical Research, 102*, 26397–26410.

Johnsen, S., Dahl-Jensen, D., Dansgaard, W., & Gundestrup, N. (1995). Greenland palaeotemperatures derived from GRIP bore hole temperature and ice isotope profiles. *Tellus, 47B*, 624–629.

Jónsson, T. (1991). Ný meðaltöl veðurþátta, 1961–1990. *Jökull, 41*, 81–87.

Kaufman, D., et al. (2004). Holocene thermal maximum in the western Arctic (0–180°W). *Quaternary Science Reviews, 23*, 529–560.

Kjartansson, G. (1939). Stadier i Isens Tilbagerykning fra det sydvestislandske Lavland. Skuringmerker. En isdaemmet sö. Marine dannelser. Postglacial tektonik. *Meddelelser fra Dansk Geologisk Forening, 9(4)*, 426–458.

Köppen, W., & Wegener, A. (1924). *Die Klimate der Geologischen Vorzeit*. Berlin: Gebrüder Borntraeger.

Lamb, H. (1966) Climatic fluctuations. In H. Flohn (Ed.), *World survey of climatology*. 2. *General climatology* (pp. 173–249). New York: Elsevier.

Lamb, H. (1995). *Climate, history and the modern world*. London: Routledge.

Larsen, G., & Eiríksson, J. (2008). Late Quaternary terrestrial tephrochronology of Iceland— frequency of explosive eruptions, type and volume of tephra deposits. *Journal of Quaternary Science, 23*, 109–120.

Libby, W. (1952). *Radiocarbon dating*. Chicago: University of Chicago Press.

Lyell, C. (1830–1833) *Principles of Geology*. 3 vols. London: John Murray.

Lyell, C. (1865). *Elements of geology*. London: John Murray.

Lyell, C. (1990–1991). *Principles of geology*. Chicago: University of Chicago Press.

Maclennan, J., Jull, M., McKenzie, D., et al. (2002). The link between volcanism and deglaciation in Iceland. *Geochemistry Geophysics Geosystems, 3*, 1062.

Mann, M., Bradley, R., & Hughes, M. (1999). Northern hemisphere temperature during the past millennium. *Geophysical Research Letters, 26*, 759–762.

Marchant, D., & Denton, G. (1996). On the structure and origin of major glaciation cycles; 2. The 100,000-year cycle. *Marine Micropaleontology, 27*, 253–271.

Maslin, M., Li, X., et al. (1998). The contribution of orbital forcing to the progressive intensification of northern hemisphere glaciation. *Quaternary Science Reviews, 17*, 411–426.

Masse, G., Rowland, S., Sicre, M., et al. (2008). Abrupt climate changes for Iceland during the last millennium: Evidence from high resolution sea ice reconstructions. *Earth and Planetary Science Letters, 269*, 564–568.

Masson-Delmotte, V., Landais, A., Stievenard, M., et al. (2005). Holocene climatic changes in Greenland: Different deuterium excess signals at GRIP and NorthGRIP. *Journal of Geophysical Research, 110*(D14102), 2005. doi:10.1029/2004JD005575.

Milankovitch, M. (1920). *Théorie mathématique des phénomènes thermiques produits per la radiation solaire*. Paris: Gauthier-Villars.

Milankovitch, M. (1930). Mathematische Klimalehre und astronomische Theorie der Klimaschwankungen. In W. Köppen & R. Geiger (Eds.), *Handbuch der Klimatologie* (pp. 1–176). I (A). Berlin: Gebrüder Borntraeger.

Milankovitch, M. (1941). *Kanon der Erdbestrahlung und seine Andwendung auf das Eiszeitenproblem.* Belgrade: Königliche Serbische Akademie. English translation: *Canon of insolation and the ice-age problem.* Washington: Department of Commerce and National Science Foundation (1969).

North Greenland Ice Core Project (NorthGRIP) members. (2004). High-resolution record of Northern Hemisphere climate extending into the last interglacial period. *Nature, 431,* 147–151.

Oeschger, H., & Langway, C. (1989). (Eds.), The environmental record in glaciers and ice sheets. *Report of the Dahlem workshop on the environmental record in glaciers and ice sheets, Berlin, March 13–18, 1988.* New York: Wiley.

Ogilvie, A. (1986). The climate of Iceland 1701–1784. *Jökull, 36,* 57–73.

Ogilvie, A. (1992). Documentary evidence for changes in the climate of Iceland, AD 1500 to 1800. In R. Bradley, & P. Jones (Eds.), *Climate since AD 1500* (pp. 93–117). London: Routledge.

Ogilvie, A., & Jónsson, T. (2001). "Little ice age" research: A perspective from Iceland. *Climatic Change, 48*(1), 9–52.

Ólafsson E., & Pálsson, B. (1981). *Ferðabók Eggerts Ólafssonar og Bjarna Pálssonar um ferðir þeirra á Íslandi árin 1752–1757* (S. Steindórsson, Trans.) Reykjavík: Örn og Örlygur.

Pálsson, S. (1945). In J. Eyþórsson, P. Hannesson, & S. Steindórsson. (Trans. and Eds.), *Ferðabók Sveins Pálssonar. Dagbækur og ritgerðir 1791–1794* (423–552). Reykjavík: Snælandsútgáfan. (Thesis on glaciers).

Pálsson, S. (2004) [1795]. Williams, R., & Sigurðsson, O., (Trans.) *Draft of a physical, geographical, and historical description of Icelandic Ice Mountains on the basis of a journey to the most prominent of them in 1792–1794.* Reykjavík, Hið íslenzka bókmenntafélag.

Penck, A., & Brückner, E. (1909). *Die Alpen im Eiszeitalter.* Leipzig: Tauchnitz.

Petit, J., Jouzel, J., Raynaud, D., et al. (1999). Climate and atmospheric history of the past 420,000 years from the Vostok Ice Core, Antarctica. *Nature, 399,* 429–437.

Pétursson, H. (1986). *Kvartærgeologiske undersøkelser på Vest-Melrakkaslétta, Nordøst-Island.* Doctoral dissertation. University of Tromsø.

Pilgrim, L. (1904). Versuch einer rechnerischen Behandlung des Eiszeitenproblems. *Jahreshefte für Väterlandische Naturkunde in Württemberg.* 60.

Raymo, M. (1994). The initiation of northern hemisphere glaciation. *Annual Reviews of Earth and Planetary Sciences, 22,* 353–383.

Raynaud, D., Jouzel, J., Barnola, J., et al. (1993). The ice record of greenhouse gases. *Science, 259,* 926–934.

Ruddiman, W. F. (2001). *Earth's climate past and future.* New York: W.H. Freeman.

Shackleton, N. (1967). Oxygen isotope analyses and Pleistocene temperatures re-assessed. *Nature, 215,* 15–17.

Shackleton, N. (1984). Oxygen isotope evidence for Cenozoic climatic change. In P. Brenchley (Ed.), *Fossils and climate* (pp. 27–34). Chichester: Wiley.

Shackleton, N., Berger, A., et al. (1990). An alternative astronomical calibration of the lower Pleistocene time based on ODP Site 677. *Transactions of the Royal Society of Edinburgh: Earth Sciences, 81,* 251–261.

Shindell, D., Schmidt, G., Mann, M., et al. (2001). Solar forcing of regional climate change during the Maunder Minimum. *Science, 294,* 2149–2152.

Sicre, M., Jacob, J., Ezat, U., et al. (2008). Decadal variability of sea surface temperatures off North Iceland over the last 2000 years. *Earth Planet Science Letters, 268,* 137–142.

Sigfúsdóttir, A. (1969). Temperature in Stykkishólmur 1846–1968. *Jökull, 19,* 7–10.

Sigtryggsson, H. (1972). An outline of sea ice conditions in the vicinity of Iceland. *Jökull, 22,* 1–11.

Símonarson, L. (1979). On climatic changes in Iceland. *Jökull, 29,* 44–46.

Símonarson, L., & Leifsdóttir, Ó. (2007). Early Pleistocene molluscan migration to Iceland, Palaeoceanographic implication. *Jökull, 57,* 1–20.

Storm, G. (Ed.). (1888). *Islandske Annaler indtil 1758*. Christiania: Grøndahl & Søns.

Sturluson, S. (1987) [12th century]. *Edda*. A. Faulkes, (Trans. and Ed.) London: J. M. Dent.

Sveinbjörnsdóttir, Á. (1993). Fornveðurfar, lesið úr ískjörnum. *Náttúrufræðingurinn, 62*, 99–108.

Sveinbjörnsdóttir, Á., & Johnsen, S. (1990). The late glacial history of Iceland. Comparison with isotopic data from Greenland and Europe, and deep sea sediments. *Jökull, 40*, 83–96.

Thompson, L., Mosley-Thompson, E., Davis, M., et al. (1993). Recent warming: Ice core evidence from tropical ice cores with emphasis on Central Asia. *Global and Planetary Change, 7*, 145–155.

Thoroddsen, Þ. (1916–1917). *Árferði á Íslandi í þúsund ár*. Copenhagen: Hið íslenska fræðafélag.

Urey, H. (1947). The thermodynamic properties of isotopic substances. *Journal of the Chemical Society, 11*, 562–581.

Venetz-Sitten, I. (1861). Mémoire sur l'Extension des Anciens Glaciers, Renfermant Quelques Explications sur Leur Effects Remarquables. *Nouveaux Mémoires de la Société Helvétique des Science Naturelles*. 18. Zurich.

Vinther, B., Clausen, H., Johnsen, S., et al. (2006). A synchronized dating of three Greenland ice cores throughout the Holocene. *Journal of Geophysical Research, 111*, Issue D13102.

"1000 Year Temperature Comparison." (2005). Wikimedia file. Accessed August 4, http://commons.wikimedia.org/wiki/File:1000_Year_Temperature_Comparison.png

Zielinski, G., Mayewski, P., Meeker, L., et al. (1994). Record of volcanism since 7000 B.C. from the GISP2 Greenland ice core and implications for the volcano-climate system. *Science, 264*, 948–952.

Þórarinsson, S. (1974). Sambúð lands og lýðs í ellefu aldir. In S. Líndal (Ed.), *Saga Íslands* (Vol. 1, pp. 29–97). Hið íslenska bókmenntafélag: Reykjavík.

Chapter 3
Iceland

'... for there are vast and boundless fires, overpowering frost and glaciers, boiling springs and violent ice-cold streams.' The King's Mirror from the mid-13th century (Larson 1917, p. 131).

Abstract Iceland is located in the North Atlantic Ocean, just below the Arctic Circle, and is a meeting point of warm and cold oceanic and atmospheric currents. The island's rugged terrain, heavy precipitation and relatively low air temperatures have thus produced large ice caps that cover about 10 % of the country (10,500 km^2, 3400 km^3). These glaciers helped form the country's mountainous landscape, interspersed with valleys and fjords, and their meltwaters are the sources of many of the country's main rivers. A brief overview of Iceland's main glaciers follows, with their locations, altitudes, snowlines, meteorological conditions, mass balances, and discharge of meltwaters. About 60 % of Iceland's ice masses cover active volcanoes, and thus subglacial volcanic eruptions and jökulhlaups are frequent. The close and centuries-long cohabitation of Icelanders with ice masses has resulted in detailed observations, maps and digital models of glaciers' surfaces, bedrock topographies, subglacial watercourses and geomorphological effects of glacier-volcano interactions. A brief history of the advances and retreats of Iceland's major glaciers, including surges, is then presented from the time of settlement until the present day. Finally, some predictions as to future developments of the country's two largest glaciers, Vatnajökull and Langajökull, are discussed. Iceland is promoted as an ideal location for research into glacier-volcano interaction and temperate glaciers, for increased knowledge in the latter could prove invaluable for predicting responses of ice masses to climate change, including even the consequences of global warming for the polar ice caps.

© Atlantis Press and the author(s) 2017
H. Björnsson, *The Glaciers of Iceland*, Atlantis Advances in Quaternary Science 2, DOI 10.2991/978-94-6239-207-6_3

3.1 'Iceland Is One Great Mountainous and Glacier Mickle Land.' (Stefánsson 1957, p. 1)

The island of Iceland lies in the North Atlantic near the Arctic Circle, though due to the warmth of the Irminger Current its climate is rather mild with little variation in seasonal temperatures (Figs. 3.1 and 3.2). The mean winter temperature is almost 2 °C along the southern coast, although the mean temperature of the area's warmest month is only 11 °C, and thus the annual mean temperature is about 5 °C. Iceland's northern coasts are influenced by the cold East Greenland Current, which some-times brings sea ice to its shores. In the central highlands, permafrost can be found above altitudes of 550–600 m. Heavy snowfalls are common when low-pressure systems cross the North Atlantic, where wind and ocean currents from the tropical and arctic regions collide, and this snow accumulates on mountains. About 10 % of Iceland is now covered by glaciers and they are temperate and dynamically active (Fig. 3.3). They not only respond quickly to climate changes, but they are also long-term reservoirs of ice and the source of meltwater that flows into the country's principal rivers, some of which have been harnessed to produce hydro-electric power. Up until the late 20th century, ice sheets in Iceland had covered previously unexplored terrains, including active volcanoes, geothermal areas and subglacial lakes. There are subglacial volcanic eruptions and frequent jökulhlaups from geothermal areas that have time and again threatened local settlements, destroyed

Fig. 3.1 Iceland viewed from space (MODIS)

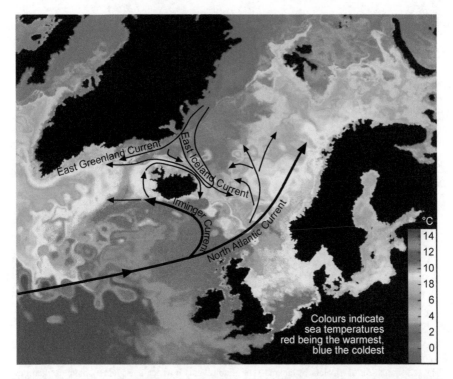

Fig. 3.2 Ocean currents—Location of Iceland in the North Atlantic. Iceland is a meeting point of warm and cold ocean currents. The warm North Atlantic Current (the northern branch of the Gulf Stream), meets resistance at an underwater ridge between Iceland and Scotland and veers westward along the country's southern shores before flowing clockwise around it. Iceland can thank this current for its mild, maritime climate. Along the southern coast, the sea temperature during winter is about 6 °C and the air temperature on average about 2 °C, though it can be 11 °C during the warmest months of the year. From the northern Greenland Sea comes a cold current that diverges to the north of Iceland into the East Greenland Current and the East Iceland Current, and which sometimes brings sea ice to Iceland's shores. Off the western fjords, the northern coastline, and the eastern fjords, these currents constrict the warm arm of the Irminger Current that encircles Iceland (National Oceanography Centre 2009)

vegetation, torn up roads and even temporarily diverted fish migrations away from coastal waters. About 60 % of Iceland's glacial areas cover active volcanoes.

Iceland's rugged terrain, lowland areas and shallow coastal sea beds have all been moulded by glacial erosion and glacial river sediment deposits from both the last glacial and the current Holocene periods. Glaciers have carved out an Alpine landscape typified by cirques, jagged mountain peaks, broad lowland expanses, and long U-shaped valleys and narrow fjords. The effects of glacial river activity can be seen in deep, eroded canyons and in outwash plains and deltas. The country's main agricultural areas in southern and western Iceland are cultivated on these plains and their ground moraine (Figs. 3.4, 3.5, 3.6 and 3.7). Palagonite rock formations were created by volcanic eruptions under glaciers which then later eroded.

Fig. 3.3 Location of glaciers in Iceland. The map reveals how ice-age glaciers have gouged one deep fjord after another out of Iceland's continental shelf. Around 10 % of Iceland is covered by glaciers (10,500 km^2 in 2014) that are active and respond rapidly to changes in climate and have a marked effect on their environments. They contain about 3400 km^3 of water, the equivalent of a 33-m-thick layer of ice spread evenly over the whole country; if it were to melt, the world's sea level would rise about 1 cm. Glaciers are the greatest reservoirs of freshwater in Iceland, holding the equivalent of the total precipitation of the last twenty years. It would take the river with the greatest volume in Iceland, Ölfusá (with an average flow rate of 400 m^3 s^{-1}), just over 300 years to transport all that stored water out to sea. Double the continuous maximum flow of the huge jökulhlaup over Skeiðsarársandur in November 1996 would be needed to discharge it all into the sea in one year. More than 99 % of Iceland's freshwater is stored in glaciers. Picture preparation: Finnur Pálsson

Fig. 3.4 Icelanders have lived next to glaciers for centuries. Barley and wheat are currently cultivated at Þorvaldseyri in the foothills of Eyjafjöll (Chap. 5). HB, 2005

Fig. 3.5 One and the same photograph capturing a glacier, outwash plain, lava field, glacial river, grassy fields and brushwood copse. Hrífunes at the junction of the Tungufljót and Hólmsá rivers, with Mýrdalsjökull in the distance (Chap. 5). HB

Due to the tremendous influence they have on their environment, glaciers in Iceland have long been the centre of research, resulting in an accumulation of much knowledge about glaciovolcanic activity and meltwater floods. More recently, a lot more data has been systematically collected on glaciers, glacial rivers and subglacial volcanic activity, while meteorological observations on glacial surfaces have linked climate to their mass balances. With all this glaciological knowledge, numerical models of mass balance and ice flow have been created and correlated to evaluate the response of glaciers and their runoff waters to climate change, both in the past and in the possible, foreseeable future.

3.2 Influence of Climate and Landscape on Location of Glaciers

The southerly winds bearing precipitation to Iceland are all to blame for the location of its main glaciers. The average annual precipitation is more than 4000–5000 mm (max. 7000 mm) on the highest flanks of Vatnajökull and Mýrdalsjökull (above 1300 m) and on Hofsjökull and Langjökull it can be 3500 mm per annum, for

Fig. 3.6 Norðurjökull, an outlet of Langjökull (Chap. 6), previously calved into the lake Hvítárvatn (420 m above sea level). The glacier has retreated so much, however, that since 2009 it has no longer reached the lake. An abundant variety of vegetation grows in Karlsdráttur, the remains of a vast swathe of vegetation that covered all of central Iceland centuries ago. HB, 2000

Fig. 3.7 The ring road around Iceland now crosses a narrow barrier beach between the sea and Jökulsárlón lagoon on Breiðamerkursandur (Sect. 8.5). HB, 15 May 1995

Iceland's largest glaciers are indeed in the south and the central highlands (Fig. 3.8). At the summits of the largest ice caps, the mean temperature is just below or about freezing point the whole year round and so most precipitation falls as snow (Fig. 3.9). The summer's mass balance is usually negative in the central area of Vatnajökull, but it can be just a little positive (up to 0.5 m) when persistent cold spells with northerly winds repeatedly bring replenishing layers of snow to the glacier, which reflects most of any incoming solar radiation. Summers in the Icelandic highlands are cool, so on the highest reaches of the glaciers there is only 10–20 days of melting each year, although the ablation period lower down their slopes usually lasts from three to four months (from June to mid-September).

The southernmost outlet glaciers from Vatnajökull and Mýrdalsjökull are not far from the coastline and creep down to a height of 100 m (lowest 20 m) and their winter mass balances are even slightly negative, for although annual precipitation is as much as 1500 mm, it mostly falls as rain, so that on the whole the glacier is losing ice during all the seasons of the year. In summer, a typical loss on the whole is about 9 m (water equivalent) at the margins of Vatnajökull (Breiðamerkurjökull and Skeiðarárjökull creep down to a height of 100 m above sea level). On glacial termini in central Iceland, which descend as low as 600–800 m, the summer mass

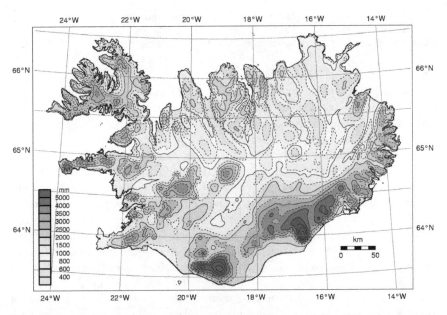

Fig. 3.8 Precipitation in Iceland. Mean annual precipitation from 1971–2000. On average, the annual precipitation in Iceland is about 2000 mm, but on the southern parts of Vatnajökull and Mýrdalsjökull it is more than 4000 mm, and on Öræfajökull and the highest points of Mýrdalsjökull it can reach a maximum of 7000 mm. On the highest domes of the main ice caps, it snows every month of the year. Where there are settlements and communities, and up on the moors, the snow melts, but above glacial equilibrium lines the summers are not warm enough to eliminate all the snow, so it accumulates there every year (Crochet 2007; Crochet et al. 2007; Sigfúsdóttir 1964, 1975; Rögnvaldsson et al. 2004, 2007; Rögnvaldsson and Ólafsson 2005)

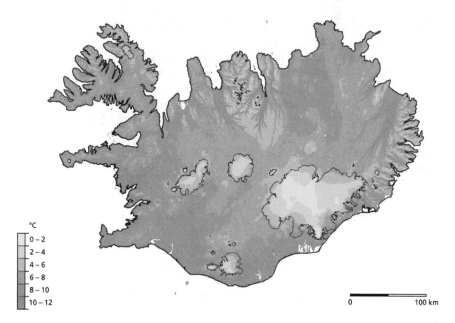

°C
0 – 2
2 – 4
4 – 6
6 – 8
8 – 10
10 – 12

0   100 km

Fig. 3.9 Mean temperatures in Iceland in July. Summers are cool and the winter snow does not fully melt away on glacial domes and the highest mountains. The highest mean temperature of the summer months is 11 °C on the southern coast and 8–9 °C on the northern coast (IMO 2009)

balance is from −4 to −6 m. In the southern regions of the eastern fjords, there is considerable precipitation and there are two small carapace glaciers, Þrándarjökull and Hofsjökull, even though the mountains there are not at a high elevation. Further north, the precipitation becomes less and there are just a few small glaciers, e.g. in the Dyrfjöll mountains and on Fönn.

Snow accumulation and reduced melting on the shadowy lee sides of mountains. In the central highlands of Iceland there are a good number of steep mountain peaks that reach a height of 1400 m above sea level and which have small glaciers on them. In the driest areas far inland, further north, the annual precipitation is only 400–700 mm, so that the snowline rises above 1600 m in the rain shadow north of Vatnajökull. Along the northern coast, the snowline falls to a height of 1100 m. On the broad Tröllaskagi peninsula there are about 100 small cirque glaciers, which mostly face north, crouching above main valleys and around the highest peaks, reaching an elevation of 1300–1500 m. Mountain peaks here receive as much as 3000 mm precipitation per annum, mostly from northerly winds. Spindrift increases snow accumulation in the corries, and many glaciers are hidden in narrow valleys where they receive little sunlight. There are also a few glaciers on Mt. Kaldbakur, between Eyjafjörður and Skjálfandi Bay, and on the mountains above Flateyjardalur.

In the highlands of the western fjords the annual precipitation reaches 3000 mm and the snowline there is the lowest in the country, at 600–700 m (Fig. 3.10). The plateau of the western fjords is from 700 to 900 m high, so there are a number of

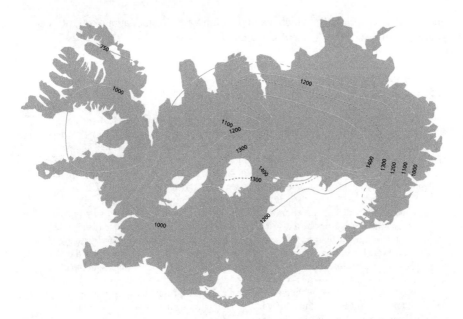

Fig. 3.10 The average height of the snowline at the end of summer in Iceland (glaciation limits in m above sea level). The snowline can be visualised as a dome that covers the country and which is penetrated by mountains where all the winter snow does not melt during the summer but piles up and accumulates there for years, gradually becoming glaciers. The snowline is at an elevation of about 1100 m along the southern parts of Vatnajökull and Mýrdalsjökull, where the precipitation over large areas is up to 4000 mm per annum; the snowline is even lower over the eastern flanks of Öræfajökull, which have the wettest maritime climate of southern Iceland—though it is at least the warmest part of the country too. Map by Ahlmann (1937) Further north from the south and southeast, precipitation lessens and the snowline rises inland. It reaches up to 1150 m on the northwestern part of Mýrdalsjökull and the southern sides of Hofsjökull and Langjökull; on the northern parts of these glaciers it attains a height of 1200–1300 m. In the rain shadow north of Vatnajökull, the driest part of the country, the annual precipitation is only about 300 mm, so that it is the region nearest to having a continental climate in Iceland. The snowline rises to between 1350 and 1400 m in Dyngjufjöll, and then up to 1700 m even further north, which is why Mt. Herðubreið (1682 m) has no glacier. Precipitation increases once more near the northern coastline and temperatures cool so that the snowline descends again, reaching as low as just 900–1100 m in the shadowy mountainsides between Skagafjörður and Eyjafjörður and between Eyjafjörður and Skjálfandi Bay (Chap. 7). On the southern side of Snæfellsjökull, on the west coast, the snowline is 1100 m high, but it is 200 m lower on its northeastern slopes (Sect. 7.5). In cold years, snow cover may have extended as far down as 600 m on the Snæfellsnes peninsula by the end of summer. The snowline is at its lowest in Iceland on Drangajökull in the far northwest, 700 m on its western side, but 500 m on its eastern side, which faces the cold and dampness of the Northeast Greenland Current. In the northern parts of the western fjords there are often years when firn snow patches can be found at a height of 300–400 m at the end of summers

snow-filled corries and about 10 small cirque glaciers at a height of 600–700 m above sea level. The crest of Drangajökull glacier covers the plateau of the northernmost part of the western fjords. It is the northernmost glacier in the country and its outlet glaciers descend to below a height of 200 m. Along with Kvíárjökull and Breiðamerkurjökull, which extend southward from Vatnajökull, it reaches nearest to the sea of all Iceland's glaciers.

Account of Iceland (from ca. 1590)

Thus there are not just the highland mountain ranges that stretch more or less continuously from west to east through the centre of the island, covered by glaciers and snow which never melts, but also, rising here and there all over the land, peaks and ridges or knolls of the highest mountains, all dressed in this horrifying and drab apparel, where the denuding power of the sun's rays hardly ever shine. To this unprecedented snowfall that has collected over the many centuries of long ago, increasing little by little, there is then added some gruffness from the heavens and an amazing unbridledness and turbulence of the air, caused in some part by precipitation and in some part by a dry and excessive coldness. The sky has thus seemed to me to be rigid with cold for weeks on end, as if it was made of iron or copper. Moreover, there then comes an unremitting fog and mist that often hides the sky from view and closes out the sun's delightful journey across the heavens. Sometimes it is as if this fog is so dense that it sinks deep down into the valleys, settling over people's heads, never moving away, but then again it sometimes just scurries away at amazing speed in a flurry of wind and weather, as if it were just taunting folk. (Einarsson 1971, p. 36)

3.3   The Hidden Terrains of Iceland

Maps of the surface and bedrock topographies of glaciers reveal previously unknown landscapes and formations involving central volcanoes, active volcanic chains, fissure swarms and calderas (Figs. 3.11 and 3.12). New maps of glacial surfaces have revealed important data about their shapes, altitudes and inclines. They also portray the outlines of ice cauldrons wherever subglacial lakes are located in geothermal areas beneath glaciers. Individual catchment areas have been delineated, glaciers' mass balances calculated, and the discharge of runoff meltwater into the river systems estimated.

In many parts of Iceland only 10–20 % of glaciers' beds are above the present snowline. It is thus clear that the country's largest ice caps are maintained by their own altitude. Even though the Vatnajökull ice cap rests on high land at 600–800 m above sea level, and 88 % of its bed is above 600 m, only 20 % of this bed is above 1100 m, which is the height of the snowline in southern Iceland. Long ago, glaciers succeeded in forming on six mountain ranges within the boundaries of Vatnajökull rising to elevations of between 1200 and 2000 m: Grímsfjall (1719 m), Bárðarbunga (2009 m), Kverkfjöll (1936 m), Öræfajökull (2110 m), Esjufjöll (1522 m) and Breiðabunga (1200 m). A further example of how ice alone can raise an ice cap above the glaciation line is Mýrdalsjökull, which rests on a huge volcanic centre with

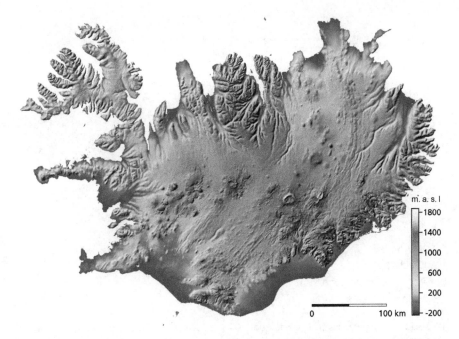

Fig. 3.11 Iceland without glaciers. Were glaciers to be removed from Iceland, we would be presented with undulating sandy wastelands, sharp outlines of valleys, straight lines of palagonite ridges, precipitous promontories, bald shield volcanoes, deep calderas, and long trenches reaching below sea level. Numerous lakes would take the place of some glaciers. Illustration by Finnur Pálsson

rims 1300–1380 m high around a 650–750 m deep caldera. This means that only 10 % of the bed of Mýrdalsjökull rises above the snowline at 1100 m above sea level. Hofsjökull glacier also covers a huge volcanic centre with rims 1300–1650 m high around a caldera that drops to a height of around 980 m. Around 20 % of the bed of Hofsjökull reaches above 1200 m, but only 11 % of it is higher than 1300 m. Langjökull glacier covers a 50-km-long mountain range that only rises to a height of 1000–1250 m, so that only 5 % of its bed reaches above 1200 m.

'Almost all of Vatnajökull then disappeared ...'

There is no doubt that there was a warm weather period that succeeded the last glacial period—as elsewhere in the Nordic countries. It is probable that the annual temperature here in Iceland had been 2–3 degrees higher than now. Almost all of Vatnajökull then disappeared except for the ice caps that remained on Öræfajökull, Bárðarbunga, Kverkfjöll and Esjufjöll.

(Eyþórsson 1951, p. 5)

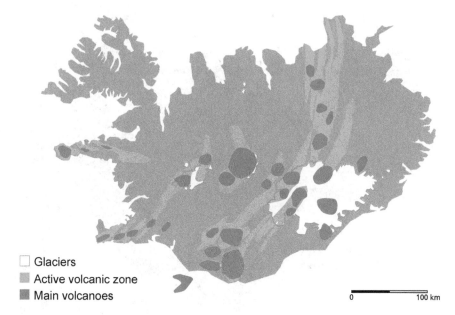

☐ Glaciers
▨ Active volcanic zone
▨ Main volcanoes

0   100 km

Fig. 3.12 Volcanic zones in Iceland. About 60 % of Iceland's glacial areas lie over its active volcanic zones. During the last glacial period, a glacier covered the entire country and palagonite rock formations were created along this volcanic zone. The western part of the zone enters Iceland at the Reykjanes peninsula and continues under Langjökull and Hofsjökull. The eastern branch of the zone comes from the Vestmannaeyjar islands through Mýrdalsjökull and then on to the western part of Vatnajökull and up through the northeast of Iceland. In subglacial eruptions, palagonite ridges and tuyas are formed and jökulhlaups rush out to sea, gouging out huge river canyons and dispersing sediments over outwash plains (Kjartansson 1960, 1962, 1965; Sæmundsson 1977, 1979, 1980; Einarsson and Sæmundsson 1987; Jóhannesson and Sæmundsson 1989; Jóhannesson et al. 1990; Jóhannesson 1994; Sigurðsson 2000; Einarsson 1991, 1994; Larsen 2002; unpublished material from project of IESUI and University of Toulouse). With ice melting due to continuous geothermal heat in some of the main volcanic centres, meltwater accumulates in glacial lakes, which then burst forth in jökulhlaups and can threaten populated areas

3.4   Mass Balances and Runoff Meltwaters of Icelandic Glaciers

Surface maps of glaciers indicate the main direction of ice flows down inclines and are used in delineating the sources of the principal glacial rivers that flow from them. From around 1990 onwards, annual mass balance measurements have been made on all of the main ice caps of Iceland: Hofsjökull (since 1987/88), Vatnajökull (1990/91), Langjökull (1996/97) and Drangajökull (2004/05), and some measurements have also been taken on smaller glaciers. This mass balance data provides a basis for calculating how much meltwater drains from glaciers into the country's river systems. Bedrock maps are also used in conjunction with surface maps to determine the boundaries of the glacial catchment areas from which meltwater flows into the rivers.

Vatnajökull's winter balance was generally highest in the early 1990s, a minimum being recorded in 1996–97 and a maximum in 2003, but it has since slowly declined. Summers were colder during the first part of the 1990s, as can be seen in the lower summer ablation levels. The high level of summer melting in 2000, on the other hand, was a result of warm and windy weather continually blowing over the glacier. The annual net balance on Vatnajökull remained positive annually from 1991 to 1994, was more or less even in the years 1994–95 (income and expenditure equal), but has remained negative ever since. From 1995–96, Vatnajökull has, on average, lost the equivalent of 0.8 m a year, as a loss measured uniformly over the whole glacier. Vatnajökull's total loss of mass from 1994 to 2006 was 9.2 m (water equivalent) or 84 km^3 (which is six times the average winter balance), so that the glacier has lost 2.7 % of its total mass in one decade. In addition to this surface melting, constant geothermal activity and occasional volcanic eruptions at the base of Vatnajökull have also melted about 0.55 km^3 a year in the 1990s, which equals only 4 % of the total surface ablation. The volcanic eruption in Vatnajökull in October 1996, however, alone melted around 4.0 km^3 of ice. This calculation relates to Vatnajökull in its entirety, for the role of volcanic eruptions is greater, of course, when considering the catchment areas around the volcanic centres.

On most glaciers, 55–65 % of the surface is above the equilibrium line in the years when the mass balance is around zero. In the years 1992–2007, the accumulation zone of Vatnajökull varied from 20 to 70 % of the total surface area of the glacier, its equilibrium line shifting in altitude from between 200 and 400 m, and the year's net mass balance was from 1 to −1 m. If the equilibrium line moved about 100 m in elevation, the net mass balance of Vatnajökull changed around 0.7 m per annum.

In years when income and expenditure were equal in terms of mass balance, the meltwater runoff from Vatnajökull was about 60 l s^{-1} km^{-2} (60 litres per second for every square kilometre, averaged over the entire glacier for the whole year) but is reduced to half this volume when the glacier's mass balance is positive (the early years of the 1990s). Rainfall on the glacier during the five summer months can add a further 10–20 l s^{-1} km^{-2} to its meltwater discharge.

The mass balance measurements for the Hofsjökull and Langjökull glaciers reveal similar characteristics. The mass balance of Langjökull has remained negative since measurements began there in 1997, and the accumulation zone has lurched back and forth from about 10 to 40 % of the glacier's total surface area. The glacier's net loss over the period 1997–2006 was 12.8 m (13.1 km^3 of ice), or 7 % of the total mass of Langjökull. In years when the mass balance is even (expenditure equals income), the annual turnover rate of Vatnajökull is about 0.4 % of its total volume, while the comparable figure is 0.8 % for Langjökull and Hofsjökull.

3.5 Glacial Meteorology

Every summer several automatic meteorological stations operate simultaneously on Vatnajökull (since 1994) and Langjökull (2001). Radiation levels are measured in situ at the stations while heat fluxes from warm and moist air are calculated from the wind speeds, atmospheric temperatures, and the humidity levels of air currents traversing the glaciers (Fig. 3.13).

As has previously been noted, all the various forms of radiation combine to be the main cause of melting, although heat fluxes in air turbulence can occasionally bring an equal amount of warmth. During the ablation season, radiation usually provides two thirds of the melting energy, and heat from turbulent air currents one third. On the higher reaches of the ice cap, the heat flux brought by winds becomes proportionately less and less. Occasionally, radiation can bring about melting even though there is frost on the glacial surface. When the winter snow disappears from the ablation zone, solar radiation penetrates the glacier to a greater extent because the ice that emerges there is covered by tephra and sand; in these conditions, up to 90 % of solar radiation can penetrate the surface ice, the glacier's albedo (ability to reflect) being reduced to as low as 10 %. The glacier's ablation zone receives most of its radiation in June, its accumulation zone in August. Heat that is borne by winds increases throughout the summer and reaches a high point in August and

Fig. 3.13 Instrument on Grímsfjall covered in hoarfrost at 1700 m above sea level (Sect. 8.3). Finnur Pálsson, electrical engineer, collecting data. The huts of the IGS in the background. HB, 1 October 1996

September. These winds blow across glaciers and become stronger when reaching the steep ablation zones. The air then cools when coming in contact with a melting glacier as the latter absorbs its warmth. Becoming heavier and colder as a result, the winds then rush down the incline. It is only during the passage of powerful storms over the glacier that the impact of these katabatic winds becomes relatively small.

The tephra that falls from volcanic eruptions onto a glacier has a short-lived impact on the accumulation zone, and usually only affects the ablation zone the summer after the eruption. The great melting on Vatnajökull in the summer of 1997 was to a great extent due to the glacier's low albedo level after tephra from the Gjálp eruption in October 1996 had become exposed. Furthermore, a lot of dust, which the jökulhlaup had deposited on Skeiðarársandur in October 1996, had been blown over a large area of the glacier.

3.6   The Dynamics of Iceland's Glaciers

The average surface velocity of a glacier during the summer has been monitored by GPS instruments in most of the locations where glacial mass balance is measured. In addition, velocity maps have been compiled over large areas based on data from satellites (InSAR and SPOT, see Fig. 3.14). Steeply sloping glaciers, whether on hard bedrock or on soft sediment that might deform, seem to move fast enough to maintain a balance with the snow accumulated over the years. Surge-type glaciers, however, have only slight inclines (typically 1.6–4°) and creep too slowly to maintain a balance with the accumulated snow that has fallen on them. The time periods between surges vary from one glacier to another, from a few years to almost a whole century. Moreover, the frequency of surges are neither regular nor bear any obvious connection to the glacier's size and mass balance. In all, there are 26 surge-type glaciers in Iceland, varying in size from 0.5 to 1500 km^2; about 80 surges have been recorded and the glaciers have progressed from just a few metres to up to 10 km. Although some of these glaciers surge at regular intervals, the majority of them do not.

On all of Iceland's major ice caps, surges account for a large part of total mass balance movement of ice through outlet glaciers and have a great influence on the flow of ice and water within glaciers. The surface areas of outlet glaciers expand, ice is transported down from the high ice cap to lower levels where it is warmer, and thus melting increases, and so consequently does the volume of runoff melt-water from the glaciers into glacial rivers. In the 1990s, glacial surges affected a total of ca. 3000 km^2 area of Vatnajökull (38 % of the total ice cap) and about 40 km^3 of ice were transported in surges from the accumulation zone to the ablation zone. This accounted for about a fourth of all ice carried down to the ablation zone

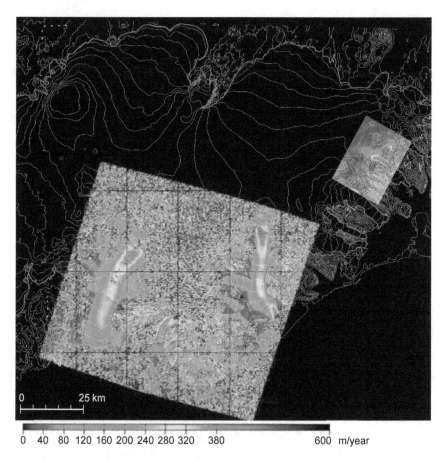

0   40   80 120 160 200 240 280 320   380   600   m/year

Fig. 3.14 Glacier monitoring from space. The movements of Skeiðarárjökull and Breiðamerkurjökull are calculated from satellite pictures (Chap. 8) (Unpublished data form collaborative work with Etienne Berthier, LEGOS, Toulouse, France)

during this period. In some outlet glaciers the role of surges in the mass transportation of ice was even greater. Looking back over the 20th century, surges accounted for at least a tenth of all the ice transported down to glaciers' ablation zones.

Surges increase river sediment loads significantly, especially those containing the most fine-grained fraction. For the first couple of years after a surge from Vatnajökull, the sediment loads in rivers from the glacier have generally reached levels of 7–10 kg/m^3 of water, which is comparable to levels in jökulhlaups. When Brúarjökull, which stretches north from Vatnajökull, surged in the years 1963–1964, the river Jökulsá á Brú transported on average 6.5 kg/m^3 of sediment. In surges that usually last less than a year, a glacier's bed can be denuded by about 1.5 cm, as a calculated equivalent over the whole base area.

Niels Horrebow's Descriptions of Icelandic Glaciers in the Mid-18th Century

These Jokeler are not the highest part of the mountains, there being many near them much higher, yet without snow continually on them. This may probably be owing to the nitrous quality of the earth. There appears a very extraordinary phænomenon [sic] in these places which may rather belong to metaphysical than historical description. (Horrebow 1758, p. 3)

'A large pond or lake, frozen over, very dangerous to pass, and not there the day before...'

It will not be amiss to give a brief account of ... the strange property of these places called Jokells, which increase in bulk, and again diminish and change their appearance almost every day. For instance, paths are seen in the sand, made by travellers that passed the day before. When followed, they lead to a place, like a large pond or lake, frozen over, very dangerous to pass, and not there the day before. This obliges travellers to go two or three English miles round. Then they come again to the very path opposite to that they were obliged to leave. In a few days the interrupted path appears again, all the ice and water having, as it were, vanished. Sometimes travellers are bold enough to venture over the ice rather than go so much about. But it often happens that their horses falling into the great breaks which are sometimes in the ice, it is not in their power to save them. A few days after these very horses are seen lying on the top of the ice, where before was a hole several fathoms deep, but now closed up and frozen. The ice must therefore in this intermediate time melt away, and the water freeze again. Hence it may be concluded, that there is no sure road round and over these mountains, but by thus continually passing and repassing. Sometimes travellers meet with accidents, but not very often. These kind of Jokells are only in Skaftafelds Syssel [sic; Skaftafell County], a fourth part of the country. Hecla and the western Jokells [Snæfelsjökull] are of another kind, and do not change their appearance in this manner. (Horrebow 1758, pp. 3–4)

3.7 Recent Developments and the Importance of Studying Icelandic Glaciers

In the warmest period of post-glacial times, which reached its high point about 7000 years ago, the ice-age glacier disappeared from Iceland and there were probably only small glaciers left on the highest mountains, for example Öræfajökull (Chap. 8). During the so-called Atlantic period, from about 8000 to 5000 years ago, the climate in Iceland was much warmer and drier than it is now, with the mean temperature believed to have been about 2 °C higher than it was in the period 1920–1960. It began to grow considerably colder again about 4500 years ago, but glaciers did not begin to form, expand, or creep down from the highest mountains

until the cooler and damper marsh period (Sub-Atlantic). Some steep valley glaciers, however, could respond quickly, and by around 500 B.C. they had reached
the limits of their greatest expansion since the last glaciation in the Würm period.
The furthest terminal moraines of Kvíárjökull and Svínafellsjökull (outlets from
Öræfajökull) probably came into existence during this period. At the same time,
other glaciers started to expand over the highlands and Iceland's ice caps began to
take on their modern forms. Vatnajökull became one vast ice sheet when glacial
snouts from several mountain ranges coalesced (Öræfajökull, Grímsfjall,
Bárðarbunga, Kverkfjöll, Esjufjöll and Breiðabunga).

From the beginning of the settlement of Iceland (about 874 A.D.) until the 13th
century, the climate was similar to that in the later warm period from 1920 to 1960,
with a mean temperature probably 2 °C lower than when warmest in the
post-glacial period. Thus land that is now for the most part covered by
Breiðamerkurjökull was then cultivated and supported a few farms. The later period
of the great expansion of glaciers, called the Little Ice Age, then began and lasted
from the 14th century to the end of the 19th century. During this time, some outlet
glaciers advanced from 10–15 km, destroying cultivated land and settlements. The
snowline to the south of Vatnajökull crept down from 1200 m (during the early
settlement) to 700 m during the latter part of the Little Ice Age (Fig. 3.15). Steep
outlet glaciers reached their greatest expansion around 1750 (Fig. 3.16), but broad,
flat glacial tongues from the ice caps reached their greatest extensions between
1850 and 1890 (like Breiðamerkurjökull, Sect. 8.5). The glaciers crept over sediment layers, burrowing down into them, the most active glacial snouts from the

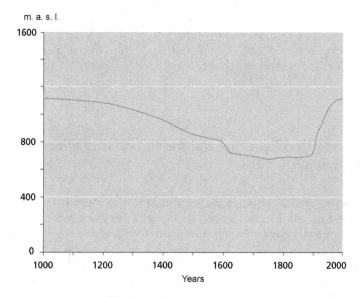

Fig. 3.15 Altitude of firn line on southern edges of Vatnajökull. The altitude of the firn line is
estimated to have lowered about 400 m lower during the Little Ice Age (Þórarinsson 1974, p. 94)

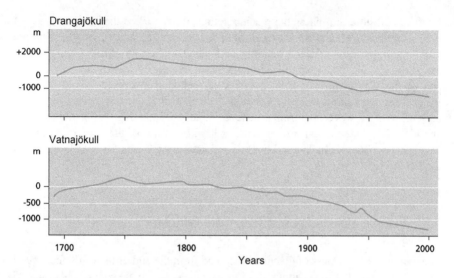

Fig. 3.16 Steep outlet glaciers of the Drangajökull ice cap and southern Vatnajökull reached their furthest extent in the 18th century (Eyþórsson 1935, pp. 121–137; Þórarinsson 1943, p. 47)

southern verges of Vatnajökull gouging down to hard bedrock 200–300 m below sea level. Breiðamerkurjökull, typical of many outlet glaciers advancing south from Vatnajökull, crept forwards in this manner during the Little Ice Age, ploughing away the sediment layers that had accumulated on the coastal lowlands from the last glacial period, which had ended about 10,000 years ago, and creating a 20-km-long trench that is 2–5 km broad and almost 300 m below sea level (Chap. 8).

At the end of the 19th century, however, the glaciers began to retreat, slowly at first, but then more rapidly during the warm period after 1930. The summers, on the other hand, began to grow colder again, so that the retreat of the glaciers slowed down in the 1950s and 1960s and many steep glaciers even began to advance around 1970 (Eyþórsson 1963; Rist (1967–1987; Sigurðsson 2005; Sigurðsson et al. 2007)). After 1985 a warm period led to glaciers rapidly receding once more and all the glaciers of Iceland have shrunk since 1995, though the general withdrawal has been interrupted by a few surges. The retreat of glaciers has gradually become more rapid due to increased summer melting, for no long-term changes have been seen in precipitation. From 1890 onwards, most of Vatnajökull's outlet glaciers have withdrawn about 2–5 km, and the surface area of the glacier as a whole has been reduced by 300 km^3 (about 10 %), and this has contributed to the sea level rising by about 1 mm. Since the warm period began in 1995, the altitude of ablation zones has been lowered by dozens of metres and glacial margins have been reduced by up to $\sim$ 100 m a year. Outlet glaciers on the southern side of Vatnajökull have been especially vulnerable to this warming because many of them have gouged themselves down into the sediment beds beneath them in their advance during the Little Ice Age and their beds are now hundreds of metres below the land in front of their snouts. In addition, glacial lakes have formed in the forefields of these termini as the glaciers retreated from areas they had gouged out below ground level, and this has helped speed up melting and calving at what are now floating margins. As the majority of Iceland's rivers originate from glaciers, this shrinkage has had a significant effect on the country's waterways. In the latter part of the 20th century, glacial rivers were reckoned to provide about 30 % of Iceland's total running water (1500 m^3 s^{-1} of 5000 m^3 s^{-1}). This calculation was only relevant, however, before the increase in summer melting that began in 1995, and glacial drainage might now provide at least one third of the country's running water. River channels have also changed and caused problems for both farmers and the IRCA.

3.8 Future Prospects for Icelandic Glaciers

The total loss of the mass of glaciers in Iceland from the end of the Little Ice Age in 1890 right up to 2013 was about 500 km^3 of water, i.e. an average of about 4 km^3 per year over the 120 years. It must be noted with some concern that recent measurements indicate an escalating rate of shrinking of the country's glaciers. The recent rate of mass loss over the years after 1996 has more than doubled, and is now up to 10 km^3 per year on average (Björnsson et al. 2013; Björnsson and Pálsson 2008; Björnsson 1979; see also Appendix C). Similarly, the average retreat of glaciated areas has increased from about 10 to 30 km^2 per year from the mid-20th century until the last two decades. At this high rate the entire glacier cover of Iceland would be eliminated within 250 to 300 years. In reality, however, the ice masses in Iceland are expected to diminish even faster in a continually warming world, and they will all have disappeared in around 200 years' time. This evaluation is based on models of glacier response to future scenarios of climate change, as detailed below.

The ice flow of glaciers in Iceland has been revealed by numerical models that also show how the mass balances of these glaciers are related to temperatures and precipitation. Moreover, models have proposed precise details as to how this precipitation is currently dispersed over the whole country. Furthermore, meteorologists have predicted how temperature and precipitation levels could probably change in the coming years in Iceland. Behind these calculations are models that have illustrated how weather conditions in various and different parts of the world are determined by processes involving oceans, glaciers and the atmosphere, as well as predictions of probable climate changes on Earth. It has been calculated (as A1B scenario of IPPC) that in the period 2071–2100 the climate may be on average 2.8 ° C warmer and with 6 % more precipitation in Iceland than in the years 1961–1990 (Figs. 3.17 and 3.18). This warming and increased precipitation will vary between seasons, with more of both in the winter than in the summer. It is also predicted that Langjökull will shrink much faster than the two just mentioned, losing about a third of its mass over the next 50 years and disappearing completely within 150 years. The runoff meltwater from glacial melting will increase immediately and reach a maximum after 40–50 years before returning to similar volumes as today for the

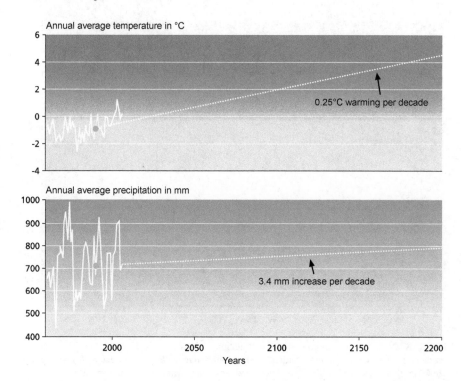

Fig. 3.17 Predicted temperature and precipitation changes in Iceland in the 21st century, IPCC's (2007) A1B scenario. The scenario suggests that the period from 2071 until 2100 will be on average 2.8 °C warmer than the period 1961–1990 and that precipitation will have increased by about 6 % (Rummukainen 2006). Straight lines depict measurements from 1961–2006. Initial reference value is the average from 1981–2000

next 100 years. The flow of runoff waters will increase the most from the lowest parts of Langjökull and Vatnajökull. It is also believed that the distribution of this meltwater will also change according to seasons, some rivers drying up completely, while the flow of others will depend only on precipitation once the glaciers have melted and no longer exist (Fig. 3.19).

Concluding remarks. Iceland is well-suited for research in many fields of glaciology. The hydrology of temperate glaciers, the nature and origins of jökulhlaups and glacial surges, the interaction between glaciers and volcanic activity, and the evaluation of how glaciers on an island in the North Atlantic, just south of the Arctic Circle, can be sensitive to and define climate change, and, of course, the future prospects for Icelandic glaciers themselves, are all important research fields that can be pursued in Iceland.

Research projects in all these glaciological fields are supported by precise data, including compiled maps of the topography of the surfaces and beds of glaciers (using data from radio echo soundings, GPS-measurements and satellites), together with extensive observations on the mass balance of glaciers and recent

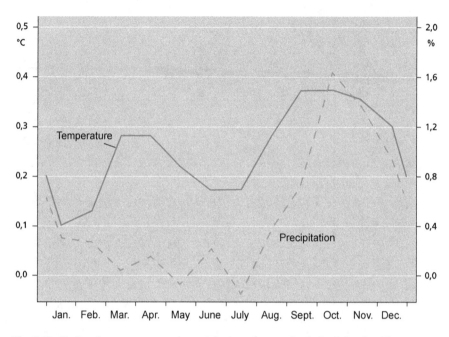

Fig. 3.18 Predicted temperature and precipitation changes in Iceland in the 21st century. Predictions of how average temperatures and precipitation may change per decade, from the period 1961 to 1990 onwards. The predicted changes may vary from one month to another (Rummukainen 2006)

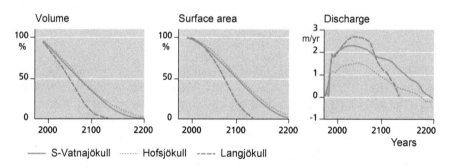

Fig. 3.19 The disappearing glaciers. Predicted volume, surface area, and meltwater discharge until 2200. The runoff is calculated in metres of water spread equally over the glaciers' current surface areas (Jóhannesson et al. 2004, 2006a, b, 2007; Aðalgeirsdóttir et al. 2006; Guðmundsson et al. 2009, 2011; Björnsson and Pálsson 2008)

developments in glaciology and hydrology. All of these scientific fields have been used to gain a growing and comprehensive understanding of the glaciers of Iceland, especially comparatively, in calculating the expansion and reduction of present-day and former glaciers and the influence of climate change on glacial meltwater runoff.

The ice caps in Iceland today are important as models of ice sheets from the glacial period after they became temperate. Knowledge and data accrued from glaciers in Iceland should thus also be of interest internationally in predicting and evaluating what may happen to polar glaciers that are still frozen to their bedrocks.

References

Aðalgeirsdóttir, G., Jóhannesson, T., Björnsson, H., Pálsson, F., & Sigurðsson, O. (2006). The response of Hofsjökull and southern Vatnajökull, Iceland, to climate change. *Journal of Geophysical Research, 111*, F03001. doi:10.1029/2005JF000388.

Ahlmann, H. (1937). Vatnajökull in relation to other present-day Icelandic glaciers. *Geografiska Annaler, 19*, 212–229.

Berthier, E. Unpublished data in collaboration with LEGOS. Toulouse, France.

Björnsson, H. (1979). Glaciers in Iceland. *Jökull, 29*, 74–80.

Björnsson, H., & Pálsson, F. (2008). Icelandic glaciers. *Jökull, 58*, 365–386.

Björnsson, H., Pálsson, F., Guðmundsson, S., Magnússon, E., Aðalgeirsdóttir, G., Jóhannesson, T., et al. (2013). Contribution of Icelandic ice caps to sea level rise: Trends and variability since the Little Ice Age. *Geophysical Research Letters, 40*(1–5), 1546–1550. doi:10.1002/grl.50278.

Crochet, P. (2007). A study of regional precipitation trends in Iceland using a high quality gauge network and ERA-40. *Journal of Climate, 20*(18), 4659–4677.

Crochet, P., Jóhannesson, T., Jónsson, T., Sigurðsson, O., Björnsson, H., Pálsson, F., et al. (2007). Estimating the spatial distribution of precipitation in Iceland using a linear model of orographic precipitation. *Journal of Hydrometeorology, 8*(6), 1285–1306.

Einarsson, O. (1971) [1585]. *Íslandslýsing [Qualiscunque descriptio Islandiae]* (S. Pálsson, Trans.) Reykjavík: Bókaútgáfa Menningarsjóðs.

Einarsson, P. (1991). Earthquakes and present-day tectonism in Iceland. *Tectonophysics, 189*, 261–279.

Einarsson, Þ. (1994). *Geology of Iceland: Rocks and landscape*. Reykjavik: Mál og Menning.

Einarsson, P., & Sæmundsson, K. (1987). Map. Upptök jarðskjálfta 1982-1985 og eldstöðvakerfi á Íslandi. In Þ. Sigfússon (Ed.), *Í hlutarins eðli* (p. 270). Reykjavík: Menningarsjóður.

Eyþórsson, J. (1935). On the variations of glaciers in Iceland I (Drangajökull). Some studies made in 1931. *Geografiska Annaler, 17*, 121–137.

Eyþórsson, J. (1951). Þykkt Vatnajökuls. *Jökull, 1*, 1–6.

Eyþórsson. J. (1963). Variation of Iceland glaciers 1931–1960. *Jökull, 13*, 31–33.

Guðmundsson, S., Björnsson, H., Aðalgeirsdóttir, G., et al. (2009). Similarities and differences in the response of two ice caps in Iceland to climate warming. *Hydrology Research, 40*(5), 495–502.

Guðmundsson, S., Björnsson, H., Magnússon, E., Berthier, E., Pálsson, F., & Guðmundsson, M., et al. (2011). Response of Eyjafjallajökull, Torfajökull and Tindfjallajökull ice caps in Iceland to regional warming, deduced by remote sensing. In *Polar Research*. 30. doi: 10.3402/polar. v30i0.7282. Available at: http://www.Polarresearch.net/index.php/polar/article/view/7282. Accessed March 16, 2016.

Horrebow, N. (1758). *The natural history of Iceland*. (Anderson, J. Trans.) London: Linde.

IMO. (2009). Iceland Meteorological Office website. http://www.vedur.is. Accessed 2009.

IPCC. (2007). *Climate change: The science of climate change*. Cambridge: Cambridge University Press. Website: www.ippc.ch. Accessed 2008.

Jóhannesson, H. (1994). *Jarðfræðikort af Íslandi*. Map. 2. Miðvesturland. Scale 1:250,000. Reykjavík: Náttúrufræðistofnun and Landmælingar Íslands.

Jóhannesson, H., & Sæmundsson, K. (1989). *Jarðfræðikort af Íslandi*. Bedrock map. Scale 1:500,000. Reykjavík: Náttúrufræðistofnun and Landmælingar Íslands.

Jóhannesson, H., Sæmundsson, K., & Jakobsson, S. (1990). *Jarðfræðikort af Íslandi*. Map 6. Suðurland. Scale 1:250,000. Reykjavík: Náttúrufræðistofnun Íslands and Landmælingar Íslands.

Jóhannesson T., Aðalgeirsdóttir, G., Björnson, H., et al. (2004). Impact of climate change on glaciers in the Nordic countries. Report for Scandinavian co-operative project on Climate, Water and Energy (CWE). Website: http://www.os.is/cwe.

Jóhannesson, T., Aðalgeirsdóttir, G., Ahlstrøm, A., Andreassen, L., Björnsson, H., de Woul, M., et al. (2006a). The impact of climate change on glaciers and glacial runoff in the Nordic countries. In S. Árnadóttir (Ed.), *The European Conference of Impacts of Climate Change on Renewable Energy Sources, Reykjavík, Iceland, June 5–6* (pp. 31–38). Reykjavík: NEA.

Jóhannesson, T., Aðalgeirsdóttir, G., Björnsson, H., Bøggild, C., Elvehøy, H., Guðmundsson, Sv, et al. (2006b). Mass balance modeling of the Vatnajökull, Hofsjökull and Langjökull ice caps. In S. Árnadóttir (Ed.), *The European Conference of Impacts of Climate Change on Renewable Energy Sources, Reykjavík, Iceland, June 5–6* (pp. 39–42). Reykjavík: NEA.

Jóhannesson, T., Aðalgeirsdóttir, G., et al. (2007). *Effect of climate change on hydrology and hydro-resources in Iceland*. Final report of the VO-project, OS-2007/011.

Kjartansson, G. (1960). *Jarðfræðikort af Íslandi*. Map 3. Suðvesturland. Scale: 1:250,000. Reykjavík: Náttúrugripasafn Íslands.

Kjartansson, G. (1962). *Jarðfræðikort af Íslandi*. Map 6. Miðsuðurland. Scale: 1:250,000. Reykjavík: Náttúrugripasafn Íslands.

Kjartansson, G. (1965). *Jarðfræðikort af Íslandi*. Map 5. Mið-Ísland. Scale: 1:250,000. Reykjavík: Náttúrugripasafn Íslands.

Larsen, G. (2002). A brief overview of eruptions from ice-covered and ice-capped volcanic systems in Iceland during the past 11 centuries: frequency, periodicity and implications. In J. Smellie & M. Chapman (Eds.), *Ice-Volcano Interaction on Earth and Mars*. London: Geological Society of London Special Publication 202. pp. 81–90.

Larson, M. (Trans.) (1917). *The King's Mirror [Speculum regale. Konungs skuggsjá*. Mid-13th century]. Scandinavian Monographs 3. New York: American-Scandinavian Foundation.

National Oceanography Centre, Southampton University. http://www.noc.soton.ac.uk/omf/. Accessed 2009.

Rist, S. (1967–1987). Jöklabreytingar (Glacier variations in metres). *Jökull*, 17–37.

Rögnvaldsson, Ó., & Ólafsson, H. (2005). The response of precipitation to orography in simulations of future climate. *Croatian Meteorological Journal 40*, 526–529. Proceedings of International Conference Alpine Meteorology, (ICAM), Zadar, 23–27 May 2005.

Rögnvaldsson, Ó., Crochet, P., & Ólafsson, H. (2004). Mapping of precipitation in Iceland using numerical simulations and statistical modeling. *Meteorologische Zeitschrift, 13*(3), 209–219.

Rögnvaldsson, Ó., Jónsdóttir, J., & Ólafsson, H. (2007). Numerical simulations of precipitation in the complex terrain of Iceland—Comparison with glaciological and hydrological data. *Meteorologische Zeitschrift, 16*(1), 71–85.

Rummukainen, M. (2006). The CE regional climate scenarios. Report. *The European Conference of Impacts of Climate Change on Renewable Energy Sources, Reykjavík, Iceland, June 5–6*. Reykjavík: Orkustofnun.

Sæmundsson, K. (1977). *Jarðfræðikort af Íslandi*. Map 7. Norðausturland. Scale 1:250,000. Reykjavík: Náttúrufræðistofnun Íslands og Landmælingar Íslands.

Sæmundsson, K. (1979). Outline of the geology of Iceland. *Jökull, 29*, 7–28.

Sæmundsson, K. (1980). *Jarðfræðikort af Íslandi*. Map 3. Suðvesturland. Scale 1:250,000. Reykjavík: Náttúrufræðistofnun Íslands og Landmælingar Íslands.

Sigfúsdóttir, A. (1964). Nedbör og temperatur i Island: kort orientering med hensyn til de hydrologiske forhold. *Den 4. nordiske hydrologkonferanse: Reykjavík 10.–15. August 1964* (1I2, pp. 1–11). Reykjavík: Raforkumálastjóri, vatnamælingar.

Sigfúsdóttir, A. (1975). Úrkoma á Vatnajökli. *Veðrið, 19*(2), 46–47.

Sigurðsson, H. (2000) The history of volcanology. *Encyclopedia of Volcanoes*. In H. Sigurðsson & B. Houghton (Eds.), et al. (pp. 15–37). San Diego: Academic Press.

Sigurðsson, O (2005). Variations of termini of glaciers in Iceland in recent centuries and their connection with climate. In C. Caseldine, A. Russell, J. Harðardóttir & Ó. Knudsen (Eds.), *Iceland. Modern processes and past environment* (pp. 241–255). Amsterdam: Elsevier.

Sigurðsson, O., Jónsson, T., & Jóhannesson, T. (2007). Relation between glacier-termini variations and summer temperature in Iceland since 1930. *Annals of Glaciology, 42*, 395–401.

Stefánsson, S. (1957) [1746]. Austur-Skaftafellssýsla. In Guðnason, B. (Ed.). *Sýslulýsingar 1744–1749. Sögurit* (28. pp. 1–23).

Þórarinsson, S. (1943). Oscillations of the Iceland glaciers in the last 250 years. *Geografiska Annaler, 25*(1–2), 1–54.

Þórarinsson, S. (1974). Sambúð lands og lýðs í ellefu aldir. In S. Líndal (Ed.), *Saga Íslands* (Vol. 1, pp. 29–97). Hið íslenska bókmenntafélag: Reykjavík.

Chapter 4
History of Glaciology in Iceland

'Might it not be possible, I thought, to imagine that there was a similar element of viscosity in ice ...' (Pálsson, 1945 [1795], p.478)

Abstract Icelanders are shown to have had a greater knowledge and experience of glaciers than most nations, from the time of settlement through to the 18th century, and they were sometimes pioneers in glaciological studies. The first written accounts of glaciers can be traced back to the 12th century, and the connecting of ice masses with climate is dated to the mid-13th century. The first map showing glaciers was drafted in Iceland in 1570, and in 1580 glaciers were reported to be expanding due to a deteriorating climate. In 1695 the Icelandic scholar Þórður Vídalín wrote a thesis in which he hypothesised about how and why glaciers move. By the 1790s, Icelandic natural scientists had categorised the various kinds of glaciers and Sveinn Pálsson had described glacial surges, correctly surmising that ice was a viscous material. A history of the expeditions of many scientists and travellers from Iceland, Scandinavia and Europe, from the mid-19th to mid-20th century, then follows, as they explored and studied Icelandic glaciers and their environments and the effects of subglacial volcanic activity. Since the mid-20th century, the use of motorised vehicles and aircraft, together with the intensified efforts of the Iceland Glaciological Society, have enabled greater access to, and a more accurate mapping of glaciers, as well as the continuous, systematic research of Icelandic ice masses. The meteorological and glaciological data gleaned from this research in Iceland, it is asserted, can now help predict future effects of climate change on both temperate and polar ice masses.

4.1 Accounts of Glaciers at the Beginning of the Settlement of Iceland

During the first centuries of the country's settlement (870–930) and the subsequent Commonwealth (930–1262), Icelanders' knowledge of glaciers was much more extensive than that of many other nations and continued to remain so, despite the dark

© Atlantis Press and the author(s) 2017

H. Björnsson, *The Glaciers of Iceland*, Atlantis Advances in Quaternary Science 2, DOI 10.2991/978-94-6239-207-6_4

Fig. 4.1 Geitlandsjökull at the southernmost end of Langjökull (Chap. 6). Sabine Baring Gould (1863)

Middle Ages, until the end of the 18th century (Fig. 4.1). Many Icelanders originally came from the glacial areas of western Norway such as the Nord, Sogn and Hardanger fjords, and some of them established settlements near the snouts of Vatnajökull, Mýrdalsjökull and Drangajökull. Glaciers had very little influence on the daily lives of Icelanders during the first few centuries of the settlement, for they then extended only a short distance towards the inhabited lowlands and the unbridged rivers flowing from them were seldom a hindrance to movement and communication to a people used to travelling by sea. Ingólfur Arnarson's slaves were unimpeded in their journey across the southern regions of Iceland on their way to finding Reykjavík. Journeys between northern and southern Iceland were undertaken between the main glaciers of the interior through Kaldidalur and along the Kjölur and Sprengisandur highland routes. Icelanders simply had no reason to climb or explore glaciers, though there were mountain trails across Drangajökull and Gláma in the western fjords, and over small glaciers between Skagafjörður and Eyjafjörður in the north, and on Þrándarjökull and Hofsjökull from the head of Hofsdalur in Álftafjörður in the east. The longest trans-glacial journey, however, was taken by itinerant fishermen in northern Iceland, who used to cross Vatnajökull in the 15th and 16th centuries to man the boats of the fishing stations based on the southern coast, and there are even reports of links between the Skaftafell region in the south and Möðrudalur in the northeast.

Not long after the settlement, however, it soon became clear that the glaciers were not all as harmless as they seemed. The early settlers Þrasi at Skógar and Loðmundur at Sólheimar reputedly diverted glacial rivers onto each other's lands through the use of magic until they finally agreed to let the Jökulsá flow south through the centre of Sólheimasandur. Much later it became even more noticeable that cohabitation with glaciers could prove very difficult for the Icelandic nation, not only because of the volcanic activity beneath the glaciers, but also because of their expansion in a cooling climate. Reports and descriptions of disruptive eruptions, of meltwater floods, called jökulhlaups, and other encroachments of glaciers and their waters became more frequent. Volcanic eruptions beneath glaciers caused outburst floods thronged with icebergs to surge over the lowlands on their way to the sea. The glaciers themselves advanced, breaking up land, blocking paths and tracks, and covering home fields and farmsteads. Glacial rivers disrupted and delayed journeys between districts, changed channels from year to year, and burst out from ice-dammed lakes to inundate meadows and local areas. In such close cohabitation with glaciers and their rivers, Icelanders gained a great deal of experience and understanding as to their nature and behaviour.

The very first descriptions of the movement of glaciers are preserved in Saxo Grammaticus' history of the Danes, *Gesta Danorum* (Fig. 4.2), from around 1200, and they are believed to have originated from farmers in the Skaftafell region of Iceland. Ice from the base of the glacier is said to move up onto its surface at its terminus, and this is how it regurgitates everything that has fallen into its crevasses; deep chasms gradually merge and disappear, always returning what they have swallowed up. The corpses of people who have fallen into crevasses are later found on the glacier's surface, where there was no crevasse to be seen and no ice towering above it. As the old Skaftafell aphorism has it: 'The glacier returns what the glacier receives.'

Written Icelandic sources make little mention of ice and glaciers during the first centuries of the settlement, and so a survey of what references there are will be brief. The *Book of Settlements* records various place names referring to glaciers in the 9th century: Jökulsá, Jökuldalur, Jökulfell, Jökulsfirðir, Snæfellsjökull, and Suðurjöklar. In *Egil's Saga* the eponymous hero Egill Skallagrímsson and his companions are amazed at the white colour of a glacial river and so call it Hvítá (Scudder 2002, Chap. 28). Flóki Vilgerðarson gave Iceland its name from the sheets of sea ice along the northern shore. Gnúpa-Bárður from Bárðardalur travelled south over a glacier to the Fljótshverfi district to where warmer winds blew. The famous outlaw Grettir Ásmundarson ascended Geitlandsjökull and walked south from it until he found a grassy valley hidden by glaciers on all sides. It had hot springs and many sheep and was ruled by the half-man, half-giant Þórir and his two daughters. Grettir named the valley Þórisdalur after him and stayed in exile there all winter long (Scudder 2005, Chap. 61). This valley later became the most infamous troll- and outlaw-haunted settlement in Icelandic folklore and right up until late in the 19th century there were stories circulating about grassy and fertile valleys hidden by glaciers over a wide area of the central highlands. The emergence of birch tree remnants and swathes of peat in ground moraine from beneath the outlet glaciers of Vatnajökull, e.g. Skeiðarárjökull and Breiðamerkurjökull, helped maintain this stubborn belief in these secret, fertile valleys. This was the natural science of that time. We are now fairly certain that these woods were smothered by glaciers

Fig. 4.2 Saxo Grammaticus composed the *Gesta Danorum* ('History of the Danes') at the end of the 12th century. The title-page here is from the 1514 edition in Paris, which was also the first full version of this work to be published. NULI

centuries before the birth of Christ and that the wildernesses of Iceland have always been uninhabitable terrain for human beings.

The *Saga of Bárður Snæfellsás* relates a mysterious story of one of Iceland's guardian spirits walking into Snæfellsjökull. He was the son of Mjöll ('new-fallen snow'), the daughter of King Snær ('snow') the Old of Kvænland, and son of King Dumbur, who ruled over Dumbshaf ocean, from whence came northerly winds, sea ice and dark fog. Snæfellsjökull is said to have been formed from a merging of freshly-fallen snow and frozen snow from the northern foggy seas, and this might well be an indication of some understanding of the origins of glaciers by the end of the 13th century (Anderson 1997, pp. 237–38).

The oldest notable explanation of the connection between glaciers and climate, however, is to be found in *The King's Mirror*, from around the middle of the 13th century in Norway, though it is believed that this information originally came from Icelanders (Fig. 4.3). 'As to the ice that is found in Iceland, I am inclined to believe that it is a penalty which the land suffers for lying close to Greenland; for it is to be expected that severe cold would come thence, since Greenland is ice-clad beyond all other lands. Now since Iceland gets so much cold from that side and receives but little heat from the sun, it necessarily has an overabundance of ice on the mountain ridges.' Furthermore, 'in that country the power of frost and ice is as boundless as that of fire,' and there are even descriptions of glacial rivers bursting forth from beneath glaciers' snouts: 'There are also ice-cold streams which flow out of the glaciers with such violence that the earth and the neighbouring mountains tremble; for when water flows with such a swift and furious current, mountains will shake because of its vast mass and overpowering strength. And no men can go out upon those river banks to view them unless they bring long ropes to be tied around those who wish to explore, while farther away others sit holding fast the rope, so that they may be ready to pull them back if the turbulence of the current should make them dizzy (Larson 1917, p. 126, 130). While medieval scholasticism was restricting advances in natural science on the continental mainland of Europe, *The King's Mirror* refers to the curiosity of Icelanders and their exploratory instincts: 'it is also in man's nature to wish to see and experience the things he has heard about, and thus to learn whether the facts are as told or not.' (Larson 1917, p. 140).

4.2   The First Contemporary Documentary Evidence of Glaciers in Iceland

Reliable knowledge of glaciers in Iceland first appeared at the end of the 16th century in *Íslandslýsing* ('Account of Iceland'), written by Bishop Oddur Einarsson (1559–1630) and published in 1590, in which glaciers are said to grow larger during a worsening climate. The book also contains the first ever description of an

Fig. 4.3 Title-page of *Konungs Skuggsjá's* (*The King's Mirror*) 1768 edition. NULI

eruption of Katla and some notable reports of sea ice frequently encroaching onto Iceland's northern shores causing immense cold and the failure of hay harvests (Einarsson 1971, pp. 33–36, 43). Glaciers are clearly marked for the first time on the map of Iceland compiled by the Bishop of Hólar, Guðbrandur Þorláksson (1542–1627), from 1570 (published 1590; Sigurðsson 1978, pp. 9–14), and indeed this is the first map known to have glaciers specifically delineated, being made 15 years prior to the map of the Tyrol and its glaciers by Warmund Ygl (Figs. 4.4,

Fig. 4.4 Map of the northernmost part of Europe, ca. 1570, from the map collection of Abraham Ortelius (1527–1598), *Theatrum Orbis Terrarum*, modelled on Gerhard Mercator's map of 1569. NULI

4.5 and 4.6) Þorláksson's map shows the glaciers of Eyjafjallajökull, Mýrdalsjökull, Sólheimajökull, Baldjökull, Geitlandsjökull, Arnarfellsjökull, Sandjökull and Snæfellsjökull; it also names Gláma, though it is not clear from the map as to whether or not it is considered a glacier (Chap. 7). Vatnajökull is not represented, though glacial rivers are shown flowing from the southeastern highlands. Not all of them are named, but the Jökulsá on Breiðamerkursandur, along with the Almannafljót (now Hverfisfljót), Skaftá, Skjálfandafljót, Jökulsá á Fjöllum and Jökulsá í Fljótsdal rivers are all depicted. The Þjórsá and Hvítá in Árnes County and Borgarfjörður are also named. Þorláksson had little or no knowledge of the central highlands and the southeastern part of the country, so that glaciers, mountains and rivers are shown and named rather randomly, e.g. the river Skaftá is shown as flowing from the Fiskivötn lakes, Jökulsá á Fljótsdal as running into Vopnafjörður, and Áradalur as being near the source of the Jökulsá in Axarfjörður. But Þorláksson did calculate the right latitude for Hólar in Hjaltadalur, however, and was thereby the first man to discover the correct latitude of Iceland. The map also outlined the shape of Iceland much better than previous attempts, though fjords and peninsulas were too serrated; even the southern coastline is indented with too many bays, and the glacial outwash plains are far too small. Iceland was also elongated from west to east in order to be able to show all known landmarks and places in the south and

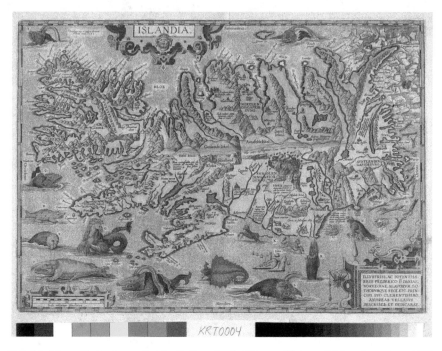

Fig. 4.5 Map of Iceland compiled by Guðbrandur Þorláksson, Bishop of Hólar, from about 1570, published in a 1590 edition of Abraham Ortelius' book of maps. This is the oldest map in the world depicting glaciers. NULI

north, but leaving the north-south axis, including the little known central highlands, far too narrow. Maps of Iceland continued to have this outline throughout the 17th century, and depictions of the coastline were not improved until Danish naval charts appeared in the latter half of the 18th century and the first decades of the 19th century. All the maps from early times showed the glaciers of Snæfellsjökull (Vesturjökull) and Eyjafjallajökull (Austurjökull) as well as Hekla, as they were all well-known landmarks for sailors, their glistening ice caps visible far out to sea, guiding seafarers to Iceland. Vatnajökull, however, is mentioned nowhere on these early maps, for indeed most seafarers did not come within sight of Iceland until approaching the Vestmannaeyjar islands. Even Bishop Þórður Þorláksson's map of 1688, which was crammed with local place names, did not depict Vatnajökull. It is thus quite interesting that the *Atlas Danicus* map compiled by Peder Hansen Resen (1625–1688) over the years 1684–1687 actually has the name Gríms Vatna Jökull on it, but no outlines of the glacier itself.

There was almost no increase in knowledge at all of the topography of the glaciers and central highlands of Iceland from the 12th century until the middle of the 18th century. Most Icelanders were too afraid of meeting ghosts and other monsters to go travelling in the uninhabitable wilderness, never mind ascending glaciers, upon which they risked snow blindness and being swallowed up by

Fig. 4.6 Guðbrandur
Þorláksson (ca. 1541–1627),
Bishop of Hólar, pioneer in
compiling maps of Iceland.
NMI

bottomless chasms, and where ice can catch fire and turn black. Nonetheless there were exceptions. Árni Oddsson (1592–1665) travelled from Vopnafjörður in the far northeast to the meeting of the Alþing at Þingvellir in the southwest in 1618 alone on horseback all along the northern margins of Vatnajökull, the so-called Vatnajökulsvegur highland trail (Árnason 1961, Vol. 2, pp. 124–126; Espólín 1825, Vol. 4, p. 4). Other famous journeys across the highlands include those of the two brave priests, Björn Stefánsson (1636–1717) of Snæúlfsstaðir in Grímsnes and Helgi Grímsson (ca. 1622–1691) of Húsafell, who rode north with supplies and a tent determined to find Þórisdalur (also called Áradalur), if it really existed, and to try and baptise whatever folk they would find there and return them to the fold of Christianity. Folktales about this one-time home of the outcast Grettir the Strong Ásmundarson stubbornly survived and Jón Guðmundsson the Learned (1574–1658) collected and recreated them in his 'Ode to Áradalur' (1936) in which he describes a populous habitation full of outlaws with magical powers, who conjured up thick mists if any overzealously curious travellers approached. The priests found an uninhabited valley enclosed by glaciers that was large, deep and long, but now without any grassy slopes, its mountainsides bare from wind erosion, and with no sign of woods, copses or moss (Gunnlaugsson 1835, 1836, 1953). The farmer Jón Ketilsson (perhaps from Fell in Hornafjörður, b. 1610) is also worth mentioning, for in the middle of the 17th century he was the first to explore the breadth of 'ice mountains'. He went on a two-day expedition onto the eastern part of Vatnajökull, from where he saw north to broad plains in the middle of which was a single mountain with grassy sides, possibly Snæfell (Vídalín 1965a, p. 24, b; Þórarinsson 1965).

4.3   Þórður Vídalín and His Treatise on Ice Mountains

Before the 17th century was over, however, Þórður Þorkelsson Vídalín (1662–1742) emerged as the first Icelander to make serious scientific observations about glaciers. Vídalín was free from all superstitions and was a man ahead of his time. He was a pioneer in the study of natural phenomena, collecting evidence for himself rather than simply relying on the reports of others, though he did refer to sources of what he thought to be the best current thinking on glaciology. His scientific arguments were independent, realistic and focused, even though many of his suppositions have proved incorrect. Vídalín's small treatise on the ice mountains of Iceland from 1695, *Dissertationcula de montibus Islandiae chrystallinis*, is the most instructive work ever to have been produced about glaciers by the end of the 17th century (Fig. 4.7).

Vídalín was the headmaster of the school at Skálholt (1687–1690) in southern Iceland and then later a physician in Austur-Skaftafell County, living in Þórisdalur in the district of Lón in the far southeast. He knew the glaciers of the Skaftafell counties from his visitations throughout his medical practice and from journeys to Sólheimarjökull, where he had been studying for two years under the tutorship of Pastor Oddur Eyjólfsson (ca. 1632–1702) at Holt below the Eyjafjöll mountains. Vídalín saw that glaciers changed shape because they moved, grew larger due to additional snow during winter, and shrank due to melting snow in the warmth of summer; they continued to grow larger in size nonetheless because the cold of winter lasted longer in the mountains than the warmth of summer. He was the first to describe how glaciers might either advance or retreat regardless of what season it might be. '… [ice mountains] rather advance sometimes during the summer, but retreat in the winter. Sometimes they advance in the winter, and move back in the summer.' But glaciers mostly advance when they have 'spewed out mostly fire and water.' Sheer cliffs of ice then tower over everything, between them innumerable deep chasms, although sometimes the blocks of glacial ice are 'low and level, and come in handy as a bridge which travellers can well use when they cannot otherwise cross the waters on both sides.' (Vídalin 1965, pp. 35–36).

Vídalín tried to explain the causes of glacial advances and was the first to maintain the hypothesis that a glacier's movement was due to meltwater collecting in the crevasses during summer and then freezing in winter, causing the glacier to expand and to move slowly down a slope. This in turn created new crevasses in which water then froze, and so on, the glacier advancing, crevasse by crevasse. He said he often heard the frost cracking when the glacial ice burst and that the power of the frost and snow was clearly visible in the way in which rocks appearing from beneath the glacier were soft enough to crumble in his hands. He was aware of the

9

****

II.

Theodor Thorkelsohn Widalins,
gewesenen Rectoris in Skalholt,

Abhandlung
von den

isländischen Eisbergen.

Vorbericht.

Nach demjenigen Begriffe, welchen man dem seligen Herrn Bürgermeister Andersohn, und den dieser vornehme Gelehrte wiederum seinen deutschen Landesleuten beygebracht hat, sollte man wohl nicht glauben, daß es in dem kalten Island Leute gäbe, die etwas besser als ihr Vieh wären, vielweniger aber solche, die ihren Geist über den Pöbel zu erheben, und ihre Vernunft durch schöne, gründliche und nützliche Wissenschaften aufzuklären sucheten. Es würde überflüßig seyn, hier etwas wider diese Schrift zu erinnern, da selbige schon hinlänglich von einem gelehrten Dänen, dem Herrn Horrebow, welcher sich selbst auf königlichen allergnädigsten Befehl ein Paar Jahre im Lande aufgehalten hat, zur Gnüge widerleget worden. Vielleicht wird aber eine kleine Schrift, die wir hier unsern Lesern mittheilen wollen, auch etwas dazu beytragen, daß man sehen könne, wie sehr

A 5   man

Fig. 4.7 Þórður Þorkelsson Vídalín (1662–1742) composed his thesis on glaciers in Latin in 1695. The title-page above is of the German edition from 1754, translated by Páll Bjarnason Vídalín (1728–1759). NULI

origins of glaciers and he described how glacial snouts would transport grooved rocks with sharp edges and how large boulders would fall off the melting ice at the terminus.

Vídalín's ideas concerning the origins of glaciers bore the mark of his textbooks in Copenhagen University, which stated that glacial ice was formed from a mixture of saltpetre and water which flowed up to the surface from underground vaults. As Vídalín wrote: 'The most common view of the beginnings of these ice mountains, and the one most people agree with, was along the lines that they have come into existence through snow that has fallen in the winter and not melted in the summer, because the mountains are always colder than the lowlands, and snow has already become settled there in the autumn and melts later in the spring, and that is how they threaten to spread over the lowlands without end' (Vídalin 1965, pp. 25–26). On the other hand, Vídalín disagreed with the opinion that snow changed into ice, using the argument that on glaciers' summits there was firn snow while in the snouts there was blue and translucent ice, meaning that it must have been formed from water that had frozen. Snow could not become ice without first melting and then becoming water that freezes. He then wrote: 'The water from which this ice is formed probably comes, like other springs, from underground open spaces where seawater reaches along hidden tunnels.... But as soon as it appears on the surface, or just below the ice, it is itself transformed into ice' (Vídalin 1965, pp. 34–35). With saltpetre in the topsoil as a catalyst, water freezes and becomes ice. Vídalín also mentions that there are those who like to believe that glaciers on the lowlands were formed from glacial rivers that had frozen during a flood. 'But if we examine this carefully, we see that these dirty and mud-filled rivers have very strong currents and thus could never, or seldom, come into contact with ice, unless they were smothered in much more snow. Now the ice that contains un-melted snow is never translucent, while the ice that is formed from pure water, on the other hand, is always clear' (Vídalin 1965, p. 29). Vídalín's argumentation has a natural logic to it, but his presuppositions and conclusions are wrong. The local folk of the Skaftafell counties had a much better intuition as to the origins of glaciers than he did, but then they did not have the benefit of a scholastic education, so prevalent in foreign universities, to confuse them.

Vídalín's thesis, written in Latin in 1695, lay unpublished until issued in 1754 in a German translation by Páll Bjarnason Vídalín (1728–1759). In Páll Vídalín's afterword he says that Þórður Vídalín had not explained 'in what manner the ice had risen up and become mountains' (Vídalin 1965, p. 39). He doubts that glaciers are only formed from underground springs, as they do not contain enough water, and indeed water from elsewhere might serve the same purpose. Páll is here referring to meltwater, which flows down to the glacial snout and freezes there every winter, for glaciers are indeed made of annually accumulating layers of ice. Þórður Vídalín's thesis on glaciers was first published in Icelandic in 1965, translated by Gísli Ásmundsson (1906–1990) and with an introduction by the geologist Sigurður Þórarinsson (1912–1983). In 1705, a decade after Vídalín had written his thesis, the Swiss mathematician Johann Jacob Scheuchzer (1672–1733) put forward the same theory as Vídalín's concerning the expansion of frost and the movement of glaciers in his book *Itinera per Helvetiae alpinas regiones facta* (1723) and the theory has since been associated with Scheuchzer even though Vídalín's exposition is much clearer. The frost expansion theory was taken as valid until the 19th century.

4.4 Encroachment of Expanding Glaciers from ca. 1700 Onwards

During the sixty years that Vídalín's thesis lay unpublished, there was little increase in the knowledge of glaciers in Iceland. Inevitably, however, the expansion and encroachment of glaciers into settled, inhabited areas became recorded in written documents. Árni Magnússon (1663–1730; Fig. 4.8) specifically named 21 glaciers in his descriptions on his travels (*Chorographica Islandica*) while compiling his land register during the years 1702–1712, indeed the first reference to Gláma as a glacier is to be found in his work (Magnússon 1955, p. 77). He described the changes in Sólheimajökull (pp. 31–32) and claimed that Öræfajökull was advancing: 'Between the Svínafell and Skaptafell glaciers is Hafrafell, a large and grassy mountain. In days gone by there were bridle paths and sheep round-ups there during the summer. But this Hafrafell is now so enclosed by the stretches of glacier surrounding it, that it is impossible to reach it except on foot and only with great

Fig. 4.8 Árni Magnússon (1663–1730), antiquarian and manuscript collector, wrote a great deal about glaciers in his *Chorographica Islandica.* NMI

difficulty' (p. 17). There used to be a farmstead on Jökulfell mountain, its ruins visible, which provided timber for building other houses, though the largest wood is now gone. Magnússon also cites oral traditions concerning journeys north across Vatnajökull which did not come to an end until after 1500. The story has it that the farmer in Skaftafell also had an estate in Bárðardalur valley and could reach it within a day, even when riding with his wife on one horse. But Magnússon's most important contribution to glaciology is his description of jökulhlaups, or outburst floods, from glacially dammed marginal lakes (Fig. 4.9). He relates how, at Sólheimajökull, ice can block the subglacial tunnels, preventing water from flowing beneath the glacier and compelling it to accumulate in gradually deepening lakes; once the ice dam can no longer resist the growing pressure of the water, the tunnels are forced open and the water bursts out in jökulhlaups. Outburst floods are usually annual events, but they are less powerful when more frequent (Magnússon 1955, pp. 31–32). Such an accurate description of the causes of jökulhlaups from glacially dammed marginal lakes had never previously been recorded anywhere in the world.

The reports of the sheriff Ísleifur Einarsson (1655–1720) at Fell (from 1712) and Sigurður Stefánsson (ca. 1698–1765) (from 1746), describe life in the proximity of glaciers in Austur-Skaftafell County, the abandoned farms in the Öræfi district, and changes caused by volcanic eruptions, jökulhlaups, and encroaching glacial rivers.

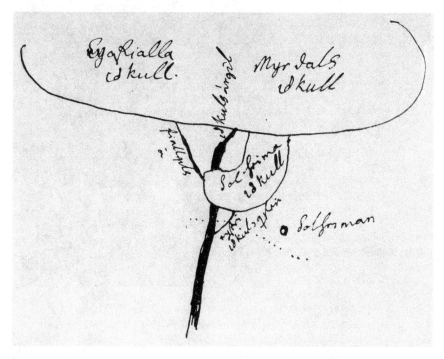

Fig. 4.9 Árni Magnússon's diagram (1955, p. 32) of the souces of the Jökulsá on Sólheimasandur (Chap. 5). NULI

Volcanic eruptions in glaciers are recorded in annals, and there are reports and contemporary descriptions of all the Katla eruptions after 1625. Magnússon listed all the farms that had been destroyed on Mýrdalssandur by volcanic eruptions or jökulhlaups (pp. 25–27). The eruptions of Öræfajökull in 1362 and 1727 have also been described, and annals refer to almost three dozen eruptions in Vatnajökull from 1332 to 1797.

The knowledge and information mediated by Vídalín and Magnússon soon began to appear on maps. The Danish admiral Peter Raben (1661–1727), as Governor of Iceland, instigated a new cartographic survey of the country, and on his Iceland map of 1721, Grímsvötn is first named, though the 1724 map of Hans Hoffgaard (b. 1678), based on Raben's map, only depicts the Öræfajökull glacier in southeastern Iceland. Ten years later, however, in 1734, all the main glaciers of Iceland are depicted on the map of the Norwegian surveyor Thomas Hans Henrik Knoff (1699–1765), which completed the mapping project begun by Raben (Fig. 4.10). One continuous though nameless glacier, now called Vatnajökull, was first shown on Knoff's map, though its outlines were delineated so unclearly that it is impossible to calculate how large it was. Knoff also included Glámujökull on his map, probably influenced by Árni Magnússon. Knoff's map was the greatest

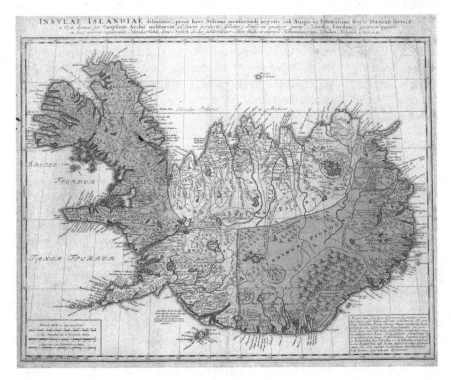

Fig. 4.10 The map of Iceland drafted in 1734 by the Norwegian surveyor Thomas Hans Henrik Knoff (1699–1765). NULI

improvement on the representation of Iceland since the days of Bishop Guðbrandur
Þorláksson, but unfortunately it languished in the Danish archives for more than
two decades (Sigurðsson 1978, p. 143). This is because the maps had been compiled at the instigation of merchants, who held monopolies in the Iceland trade, and
the Danish General Staff, and therefore the Danish government viewed the dia-
grams of the Icelandic coastline and harbours as state secrets. Knoff's map was
finally published and appeared before the public later in the 18th century, along
with three important accounts of Iceland (Sigurðsson 1978).

4.5   The Enlightenment and Knowledge of Glaciers in the 18th Century

Scientific knowledge increased during the 18th-century Enlightenment. The interest
of the authorities in Copenhagen in regenerating Iceland grew, and in the latter part
of the century there was an improvement in exploring and evaluating the country's
natural history and resources. Knoff's map of Iceland was made available to the
public in 1752 as part of the *Natural History of Iceland* by Niels Horrebow (1712–
1760) a Danish lawyer who composed the book at the end of his sojourn in Iceland
1749–1751 under the auspices of the Royal Danish Academy of Sciences and
Letters (Fig. 4.11). Horrebow wrote knowledgeable comments on glaciers and
described the climate in Iceland better than had been done previously, portraying a
maritime climate in which the summer temperatures are low, but the winters much
warmer than might be expected, if the country's global location alone were to
determine its temperatures. He pointed out that it was remarkable that glaciers were
sometimes at lower elevations than snowless mountains, but fell back on the
well-worn explanation that this was caused by saltpetre in the earth where they were
formed. He noted that glaciers in the county of Skaftafell changed almost daily so
that men who had recently crossed bare plains found their tracks covered by ice on
their return, their original footprints still visible on either side of the new snout.
Horses that had fallen into deep holes in the glacier (moulins or crevasses) were
later found on the surface of the glacier itself, any holes they had fallen into now
uninterrupted, smooth ice (Horrebow 1758, pp. 3–4). A similar description had also
appeared in Saxo's *Gesta Danorum*.

But now it was the turn of Icelanders to undertake geographical expeditions on
behalf of the Royal Danish Academy of Sciences and Letters. The work begun by
Niels Horrebow was continued by the natural scientist and poet Eggert Ólafsson
(1726–1768) and the physician Bjarni Pálsson (1719–1779). They went on expeditions all around Iceland during the summers of the years 1750 to 1757 and with
their *Ferðabók* ('Travelogue') produced a magnificent overview of Iceland and its
people (Fig. 4.12). They were entrusted with collecting information about the
glaciers of Iceland and to record their locations, size and local conditions. Ólafsson
and Pálsson travelled very little in the central highlands, however, and thus did not

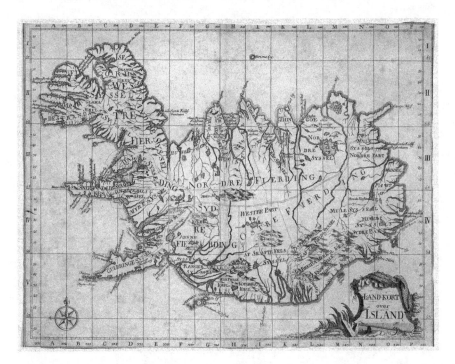

Fig. 4.11 The map of Iceland issued with the 1752 travelogue of Niels Horrebow (1712–1760), *The Natural History of Iceland*. NULI

enhance any knowledge or understanding of the origins and nature of glaciers. Ólafsson did most of the work, collating the data for the writing of the travelogue once Pálsson had been appointed Surgeon General of Iceland in 1760, and finally completed the book (in Danish) in 1766, two years before his death by drowning in Breiðafjörður. Jón Eiríksson (1728–1787; Fig. 4.13), a ministerial adviser, and the Norwegian historian Gerhard Schøning (1722–1780) had the book published in 1772 and included with it Knoff's map of Iceland (Ólafsson and Pálsson 1981; Eiríksson and Schøning 1981; Steindórsson 1981). There was a significant increase in place names from previous editions of the map, although much topographical detail discussed in the book was still missing, for indeed neither Ólafsson nor Pálsson had anything to do with the preparation of the map. The name Klofajökull (for Vatnajökull) appeared for the first time on a map. Langjökull, on the other hand was also used for Hofsjökull; people from eastern Iceland called Hofsjökull Langjökull, probably because they seemed as one when looking westward from the east. The travelogue finally appeared in an Icelandic translation by the botanist Steindór Steindórsson (1902–1997) in 1943.

It was noted in the travelogue that glaciers had become much larger in the 18th century than they had been during the years of settlement, that they were still expanding, and that new glaciers were being formed because of a worsening

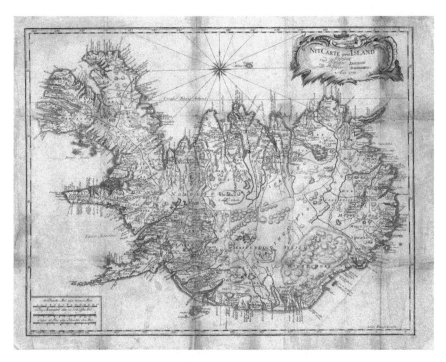

Fig. 4.12 The map of Iceland by Jón Eiríksson and Gerhard Schøning, published in Ólafsson and Pálsson's travelogue (1772). NULI

Fig. 4.13 Jón Eiríksson (1728–1787), ministerial adviser, helped bring about the publication of the travelogues of Eggert Ólafsson and Bjarni Palsson (1772) and Ólafur Olavius (1780). NMI

climate year after year, for example a new glacier had been seen to the north of Skarðsheiði in Borgarfjörður (Hornárdalsjökull), (Ólafsson and Pálsson 1981, Vol. 1, pp. 49–50). It was also noted that Drangajökull was expanding, although there were also reports that it sometimes retreated. They described the destruction and disruption of farms and their lands caused by glaciers, giving examples of small brooks that had become large glacial rivers, such as Tóftalækur turning into Almannafljót (now Hverfisfljót). Ólafsson also referred to jökulhlaups, which were considered to originate from marginal lakes, e.g. in the Jökulsá river on Sólheimasandur and in the Núpsvötn channels (probably from Grænalón) (Ólafsson and Pálsson 1981, Vol. 3, p. 143). Thirty-two glaciers were specifically named.

Ólafsson classified glaciers into three groups:

1. **Summit glaciers**, high mountains covered by firn where more snow collected than the sun could melt. This is the part of a glacier known as the accumulation zone today. These glaciers maintain themselves and even grow larger.
2. **Descending or creeping glaciers**, that have been pushed over or fallen from the high summit glaciers down onto the lowlands and been transformed from snow into ice. They survive because cold water and steam are borne by underground channels from the sea to the base of the glacier where they then freeze.
3. **Ground level glaciers**, that have come into existence on the lowlands in a similar way to creeping glaciers; a frozen brook is reinforced by seawater that arises from hidden, bottomless wells and then freezes to its base (Ólafsson and Pálsson 1981, Vol. 2, pp. 108–109).

Ólafsson believed that descending glaciers (now known as outlet glaciers) were unknown abroad and only existed in Iceland in its eastern and southern regions, examples being Skeiðarárjökull and Kötlujökull. This is the first time the word 'skriðjökull' or 'creeping glacier' occurs in a written work, for the concept is non-existent in other languages, and was probably even older in spoken Icelandic. Breiðamerkurjökull, on the other hand, was the only example of a ground level glacier in Iceland according to Ólafsson. It could not have descended onto the lowlands as Skeiðarárjökull had, though iceberg remnants from the eruption of Öræfajökull in 1362 might have provided the base for the glacier, which had then become enlarged because rivers had been dammed on their way to the sea and had frozen on the level ground. The connection between glaciers and the ocean was not only important in the maintenance of ground level glaciers, for Ólafsson believed large channels of seawater also reached from the ocean right up beneath the creeping glaciers of Mýrdalsjökull and Skeiðarárjökull, which spewed fire and had outburst floods and indeed extended down onto lowland only just above sea level itself. Like Vídalín before them, Ólafsson and Pálsson were tied to the academic theories they learned in Copenhagen and paid little heed to the opinion of the local people in the Skaftafell counties, although Vídalín did mention that glaciers crept down from the highlands in his thesis.

Exploration of Geitlandsjökull

Eggert Ólafsson and Bjarni Pálsson did not spend much time in the vicinity of glaciers, but their description of Geitlandsjökull is very interesting as it was a place well-known to Icelanders from oral traditions as having an allegedly geothermally heated, ice-free, grassy valley within its boundaries: Þórisdalur. The purpose of Ólafsson and Pálsson's visit to Geitlandsjökull in 1753, however, was not to search for hidden valleys and outlaws, but to seek enlightenment concerning the nature and origins of glaciers. Ólafsson was the first man to describe in detail crevasses and the ice towers covered in dirt (now called dirt cones) that he believed had been blown there by sand and dust storms from ice-free areas. In the summers, water had harrowed channels into the ice and the sand and dirt began to swirl upwards, eroding the tops of the ice chunks into sharp-pointed cones (Ólafsson and Pálsson 1981, Vol. 1, pp. 56–60). This explanation of the formation of pointed seracs is not far from the truth, though Ólafsson failed to grasp that the layers of sand prevented solar radiation from melting the ice further.

Ólafsson described pools of water on the surface of the glacier, 1–3 ft in diameter, as providing excellent drinking water, and concluded that it had seeped down into air bubbles and pores within the ice. He called these waterholes 'elf pools' because many people still believed that elves had created them. He also described a 20-m-high glacial moraine just a few steps away from the 3- to 5-m-high drop at the glacier's margin, from which it might be deduced that the glacier had been expanding. Gravel, stones and boulders were all mixed up in the moraine, which Ólafsson considered to have come from beneath the glacier, for it was not clear to him that the glacier had bulldozed this debris ahead of it, believing instead that water had spewed the rocks from beneath the ice and that the ice enclosing them had melted in front of the snout thus leaving a bare gap between the terminus and the moraine. Nor had it ever occurred to him that glaciers had covered a large part of Iceland in previous times. He did not see that erratic rocks, or 'Grettir's lifts,' indicated this, merely claiming that in earlier centuries men might have gathered to lift these rocks for sport and entertainment. Ólafsson later described the black, coned seracs and waterholes on the way up the northern part of Mýrdalsjökull (Ólafsson and Pálsson 1981, Vol. 2, p. 97). It should also be noted that in his academic thesis (for a bachelor's degree from the University of Copenhagen) he cites local farmers as claiming the pale colouring of the glacial rivers is caused by the sediment deposits they bear, though he himself believed it was the result of chalk (a species of rock not found in Iceland) being dissolved in the water. Ólafsson referred to Vídalín's theory that frost caused crevasses, but claimed that this was not enough, and involved perhaps meltwater collecting in streams on the glacier's surface and harrowing out channels that became deeper and more numerous as they flowed further down the glacier towards the snout. In support of his opinion, he noted that there were no crevasses on the summit of Geitlandsjökull, although he later described crevasses high up on Snæfellsjökull (Ólafsson and Pálsson 1981, Vol. 1, pp. 163–164; Fig. 4.14).

Fig. 4.14 Snæfellsjökull. A drawing from Eggert Ólafsson and Bjarni Pálsson's travelogue of 1772. NULI

Ólafsson and Pálsson were pioneers of glacial exploration in Iceland and helped removed much of the fear and superstition that many Icelanders held for the central highlands by climbing mountains and glaciers not far from human habitation and settlements: Hekla (1750), Geitlandsjökull (1753) and Snæfellsjökull (1754), the easternmost peak of which was then considered the highest mountain in Iceland (Fig. 4.15). The ghost of the guardian spirit Bárður Snæfellsáss did not prevent their ascent. The temperature at the summit was measured at −4.5 °C in the middle of the summer and they noted that there was no reason to assume the coldness of glaciers was due to particles of saltpetre. Ólafsson later described the crevasse-riddled southern arm stretching from Öræfajökull as looking similar to a huge pile of saltpetre, 'and that is perhaps the reason why some have believed that glaciers in Iceland were made from saltpetre.' (Ólafsson and Pálsson 1981, Vol. 2, p. 106). What also attracted their attention on Snæfelssjökull was that only a small part of the amount of water, which should be draining from such a large glacier, was actually visible, so they concluded the water must disappear into underground holes beneath it, from whence it flowed into the sea. They ascended Mýrdalsjökull at the end of August 1756 to explore the Kötlugjá ravine after the eruption that had begun there on 17 October 1755.

Yet another Icelander, Ólafur Ólafsson (Olavius, ca. 1741–1788; 1964–1965), made expeditions around Iceland (1775–1777) and wrote a *Ferðabók* ('Travelogue' 1780), which included Knoff's map of the country and information on farmsteads that had been abandoned due to the encroachment of Drangajökull in the western

Fig. 4.15 Part of a drawing of Snæfellsjökull by Pierre Ozanne. This was drawn at sea on the French expedition to Iceland in 1771–1772 and appeared in a book by Verdun de la Crenne (Paris 1778). NULI

fjords. The farm Fremrahorn in Reykjafjöður on the Hornstrandir coastline was enveloped by a glacier in 1706, and in 1741 the farm of Lón in Lónsvík bay (Lónhóll in Kaldalón) off the Ísafjarðardjúp fjord was destroyed, along with three other farms, by jökulhlaups from Drangajökull. He also reported that the summit of Gláma was mostly covered by a glacier. The book also contains a remarkable narrative of the upheavals in Öræfajökull in 1727, as recorded by Jón Þorláksson (1700–1790), the local parish priest (Olavius 1964–1965, Vol. 2, pp. 224–227; Eiríksson and Olavius 1964).

But curiosity about the central highlands was not limited to scientists and explorers. The outlaw and outcast Eyvindur Jónsson, Fjalla-Eyvindur (1714–ca. 1783), lived at Hveravellir to the east of Langjökull glacier, as well as in the Þjórsárver wetlands to the south of Hofsjökull, and in the Hvannalindir and Herðubreiðarlindir oases to the north of Vatnajökull. He was so familiar with the location of Grímsvötn that he should probably be included in any list of explorers of glaciers. About one and a half centuries after the journey of Jón Ketilsson from Hornafjörður, another person from Austur-Skaftafell County, Sigurður Þorsteinsson of Svínafell, in the Öræfi district, set off with another man at daybreak from Skaftafell in the spring of 1795 and reached the summit of an ice cap by mid-afternoon, from where they first saw a depression in the ice and then a glacial dome, to the north of which were black mountains, and in the far distance wilderness (Gunnarsson 1949, pp. 214–244).

4.6   The Pioneering Glaciologist Sveinn Pálsson

Before the 18th century drew to a close, Icelandic glaciology reached a high point with the writing of *Jöklarit* ('Treatise on Glaciers') by Sveinn Pálsson (1762–1840; Fig. 4.16). He used the research methods of Enlightenment natural scientists, researching phenomena in person, collecting facts, analysing and classifying data and then connecting the various pieces of information to present an overview. He was not directly involved in measuring or calculating work, however, indeed he lacked the individual tools and equipment to do so, but he spent much more time on his expeditions than the others who had travelled around the country before him. Pálsson was a prolific author, writing on such subjects as geology, botany, zoology and medicine, but his greatest contribution to scientific knowledge was his work on glaciology. Pálsson went on his scientific expeditions during the summers of 1791–1794, thanks to a grant from the Danish Academy of Natural Sciences, and from his diaries he produced his *Ferðabók* ('Travelogue'; Figs. 4.17 and 4.18) to which was attached his treatise on volcanoes and treatise on glaciers: *Draft of a Physical, Geographical, and Historical Description of Icelandic Ice Mountains on the Basis of a Journey to the Most Prominent of Them in 1792 to 1794.* Even though only 35 years had passed since Eggert Ólafsson and completed the travelogue of the expeditions made with his companion Bjarni Pálsson, Sveinn Pálsson added a great

Fig. 4.16 Sveinn Pálsson (1762–1840), physician and natural scientist, a pioneer in glaciology. NMI

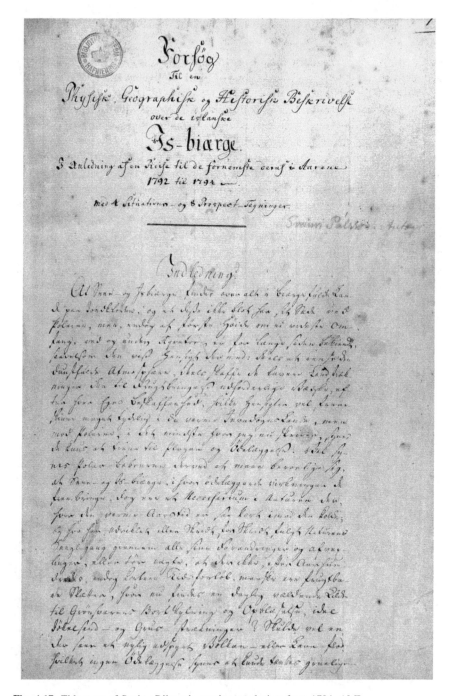

Fig. 4.17 Title-page of Sveinn Pálsson's treatise on glaciers from 1794. AMI

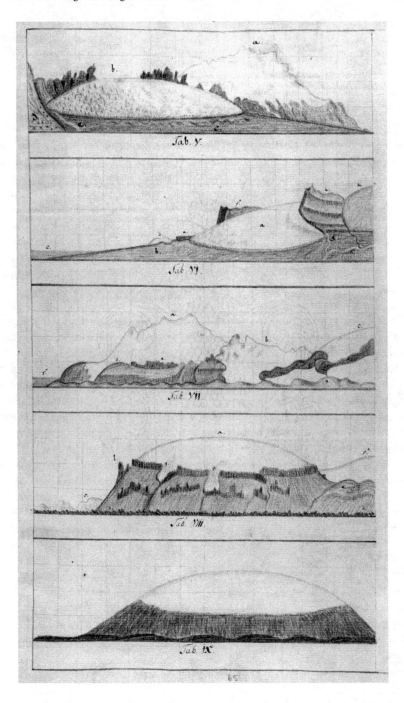

◄ **Fig. 4.18** Drawings from Sveinn Pálsson's 1794 treatise on glaciers. The *top* one shows a Öræfajökull and **b** Skeiðarárjökull as seen from Skeiðarársandur (**c**) near Lómagnúpur headland (**d**). The next one down is a view from the slopes of Skaftafell, west over Skeiðarárjökull (**a**), Skeiðarársandur (**b**), and Lómagnúpur (**f**). The third down is a view from Hornafjörður west towards Öræfajökull (**a**), Heinabergsjökull (**b**), and the glaciers of the Mýrar district (**c**) (see Chap. 8). The next to bottom picture is of Eiríksjökull (**a**) and Baldjökull (**f**). The *bottom* drawing is of Ok (Chap. 6). AMI

deal of important information to the knowledge of the glaciers of Iceland: their topography, classification, formation, and movements, and the interaction between glaciers and local communities in various locations. Sveinn Pálsson relied on the travelogue of Eggert Ólafsson and Bjarni Pálsson and 6 of 10 volumes of the works of Esaias Fleischer (1732–1804), his tutor at Copenhagen University. In the beginning he was much influenced by these predecessors, especially Fleischer, but as he progressed with his own writing he gradually came to a greater understanding, noticing things others had overlooked and describing them so well that others could easily understand them, and most of his conclusions have stood the test of time.

Sveinn Pálsson put forward the view that all glaciers had been formed in the same way: an annual layer of snow settles on top of another until the stack reaches a point where it cannot grow any larger because of summer melting. He disagreed with Eggert Ólafsson and Bjarni Pálsson's proposal of three different kinds of glaciers named after the causes of their formation (summit glaciers, creeping or descending glaciers and ground level glaciers; see Table 4.1) and also the idea that glaciers in Iceland were somehow different in nature to glaciers in other countries. Morphologically, however, two kinds of glacier could be categorised: summit glaciers and descending glaciers. He described the summit glaciers in similar terms as Ólafsson. They tower with bulging domes above other mountains, are covered in firn snow, so that they might even be called firn glaciers, and are formed due to the cold temperatures at such an elevation where precipitation falls as snow. Descending glaciers, on the other hand, he claims are formed from snow that has been transformed into ice and have crept down from the summit glaciers and thus might even be called ice glaciers to distinguish them from firn glaciers (Pálsson 1945, p. 431; Eyþórsson 1945). We now call ice glaciers ablation zones and firn

Table 4.1 Sveinn Pálsson's categorisation of glaciers (examples in brackets)

Summit glaciers or firn glaciers		
Domed glaciers	**Mountain glaciers**	**Peak glaciers**
(Mid and northern part of Klofajökull, Öræfajökull)	(Eastern part of Klofajökull, Drangajökull)	(Snæfellsjökull, Eyjafjallajökull)
Descending or ice glaciers		
Mountainside glaciers	**Creeping glaciers**	**Collapsing/falling glaciers**
(Hornárdalsjökull, to the north of Skarðsheiði)	(Skeiðarárjökull, Breiðamerkurjökull)	(Hoffellsjökull, Kvíárjökull)

glaciers accumulation zones. He did not believe that any glacier had fallen onto the lowlands, or that any glacier had been formed on level land next to the sea, although it was certainly possible that stretches of firn had broken off from summit glaciers and been thrust down onto the lowlands during volcanic eruptions. Summit glaciers might also be further categorised by their shapes and sizes into steep, mountain glaciers, large domed glaciers, and finally peak glaciers covering pointed mountain tops. Mountain glaciers might also be subcategorised into mountainside glaciers, creeping glaciers (now valley glaciers), and collapsing or falling glaciers (ice falls), (Pálsson 1945, pp. 430–432).

Sveinn Pálsson concluded correctly that the 'elf holes' that Ólafsson had described were formed because dark, small stones or pebbles absorb solar radiation better than the ice around it, which it melts, creating a hole in the ice which gradually deepens as meltwater runs into it. Sveinn Pálsson stated that glacial pyramids or dirt cones were made of ice covered with mud and gravel and not made totally of mud as had usually been asserted. He believed the mud and dirt had fallen from mountains and been transported from the valley head when a glacier scraped up soft gravel as it advanced, burying ice that had broken off from the glacier's edge; mud and dirt could also be transported by water spurting up through crevasses. Ólafsson, who maintained that the dirt had been blown by winds onto the glacier, on the other hand, was much nearer to explaining the formation of ice-cone seracs than Sveinn Pálsson.

Sveinn Pálsson agreed with Vídalín that crevasses might be formed by frost, based on its behaviour in topsoil, and pointed out correctly that glacial crevasses are linked to the glacial bed's incline and that they can form on the surface wherever rivers melt subglacial caverns on the glacier bed. He also stated his opinion that crevasses could heal and close up again. Sveinn Pálsson was also clearly aware that glaciers bulldoze moraines ahead of them and leave them behind when they withdraw, although Eggert Ólafsson believed it was water that had carried them forwards from beneath the glacier. Sveinn correctly explained the formation of kettle holes whereby blocks of ice had sunk and become partly buried in mud and gravel and then melted. He described dirty grey or milky-coloured glacial rivers in detail, so totally different to clear, blue groundwater rivers, as well as their forceful impact in the creation of ice on the beds of rivers with strong currents when supercooled water from the surface strikes the bottom. He pointed out that glacial rivers had their sources far inside the summit glaciers and although meltwater seeped down into the glacier from all over its surface, the rivers emerged from beneath it in only a few places (Pálsson 1945, pp. 446–447). Pálsson had thus noticed a conundrum which glaciologists are still grappling with today, some understanding of which only first emerging toward the end of the 20th century.

Earliest Descriptions of Surging Glaciers
Sveinn Pálsson's description of glacial surges are unique in the history of glaciology. He believed that rivers in the bed of a glacier eased its advance by flushing away gravel supports upon which the glacier rested so that it slid forwards

under its own weight, and that river channels also became dammed by ice and sediments so that water formed lakes beneath the glacier, raising it and facilitating its sliding forwards. And he continues: 'Ultimately it so happens that the front edges of the glacier can no longer withstand the pressure of the water.... A tremendous flood bursts forth, water pouring out through all the cracks in the glacier carrying with it a large number of small icebergs as well as an enormous quantity of mud, and this is what causes the occurrence of these unusual jökulhlaups in glacial rivers.... There are unusual changes to the surface of the glacier itself. The glacial dome subsides and the glacier's snout, which is completely shattered, is borne away in the floodwaters, which in turn causes the glacier to be lowered significantly and jerk backwards as it seems... Icelanders usually describe this whole process as the glacier 'running' [surging].' (Pálsson 1945, p. 454). Such surges, stated Pálsson, are mostly in the broad and almost level descending glaciers such as Breiðamerkurjökull and Skeiðarárjökull, and he later actually described the glacial surge of Breiðamerkurjökull in 1793–1794. Pálsson reports that in the summer of 1794 the glacier to the east of the Jökulsá had advanced a full 200 fathoms from the previous year. He describes the glacier's margin thus: 'In some places there were large crevasses and pointed blocks of ice, in other places the ice was sundered like a piece of filigree, and yet other places large blocks of ice had tumbled over each other, and even where the glacier's edge still seemed in one piece, it was swollen and bulging outwards in the centre like a turf wall that is sodden with water and about to collapse. In addition to this, there were constant rumbling sounds from within the glacier. Small streams were bursting out here and there from the crevasses.... The glacier is supposed to have advanced in one shudder, so to speak, during Whitsuntide of that same year, 1794, without any accompanying increase of flowing water to speak of.' (Pálsson 1945, pp. 477–478). Pálsson then states that, as far as is known, these glacial surges do not occur at regular intervals, though such behaviour has been attributed to Skeiðarárjökull, but are rather the result of irregular outburst floods in the glacial rivers and the continuous disruption caused by migrating waters at the base of the glacier.

It was while he was at Breiðamerkurjökull, that Sveinn Pálsson claimed Ólafsson had been incorrect in believing the glacier had been sustained by seawater from the glacier's bed, and he followed this up with one of his own most remarkable insights: 'I cannot resist interjecting here an idea that has taken hold of me, no matter how ridiculous it may seem, while I was observing the eastern part of Breiðamerkurjökull before it advanced. Everyone knows how brittle tar can be, if it is very pure, and that it has the same capability as viscous material, even when cold, of gradually, very slowly and almost invisibly, filling up all the inclines and hollows and settling horizontally into one mass. If a few lumps of tar are put into a tilted vessel, it will soon be apparent after a short while that the tar has not only accumulated in the lowest hollows of the vessel but that all the lumps have amalgamated into one. Might it not be possible, I thought, to imagine that there was a similar element of viscosity in ice, and if that this was factually proven, it could be considered a new and contributing cause to the formation of many descending and creeping glaciers and also the reason why glacial crevasses disappear over a short period of time. There is not as yet any hard evidence to support this, but nonetheless I could not but think of this idea once again as I looked out over the descending glacier to the east of Kvísker from the summit of Öræfajökull' (Pálsson 1945, p. 478). Later, on 11 August 1794, as he looked down from Öræfajökull, he wrote: 'I paid special attention to the previously mentioned descending glacier that has crept forward just to the east of Kvísker [Hrútárjökull]. Its surface seems to be full of semi-circular grooves that lie right across the glacier, especially high up by the main glacier, its arc-shaped bulges pointing downwards to the lowland, as if this descending glacier had flowed forwards half-melted or as a thick, viscous material as I suggested in the previous section?' (Pálsson 1945, p. 495).

Nonsensical Ideas About the Origins of Glaciers
Sveinn Pálsson distinguished between jökulhlaups that emerged from glacially-dammed lakes and those originating from the meltwater of a subglacial geothermal area or volcanic eruption. Concerning the causes of outburst floods not connected to eruptions, he states that it is possible that: 'a similar power as that which emerges in hot springs is also at work on the hidden sources of water or lakes beneath the glacier. The opinion of Eggert [Ólafsson] and Bjarni [Pálsson] that the ocean reaches directly underneath the base of the glacier seems to me quite unlikely and unnecessary.' He then adds: 'For wouldn't there sometimes be evidence of maritime particles being flushed from beneath such glaciers if they had rested directly on the ocean? But everyone must make up their own minds concerning such nonsense, as indeed I have.' (Pálsson 1945, p. 502).

Planimetric maps of four of Iceland's main glaciers accompanied Sveinn Pálsson's treatise as well as eight perspective illustrations (cross sections). The outlines of the glaciers' margins were not based on survey measurements, as Pálsson relied on older data including Knoff's map in Niel Horrebow's book (1752) and the map of Johann Baptist Homann (1663–1724) from 1761, but the glaciers were depicted with their proper shape, even though much was to be desired concerning the accuracy of their size. Pálsson himself ascended the summits of many of the highest volcanoes: Öræfajökull, Eyjafjallajökull and Snæfell, as well as Hekla and Skjaldbreið. There is much information and knowledge on volcanic eruptions in Pálsson's treatise, especially descriptions of eight eruptions of Katla, and later in his life he wrote about the Katla eruption of 1823.

A Wide-Ranging Explorer
Sveinn Pálsson's treatise was the best account of Icelandic glaciers for almost a century, until the works of Þorvaldur Thoroddsen appeared, and should have deservedly become the most important landmark publication on the history of glaciology in the 18th century. But this was not to be. The Danish Academy of Natural Sciences sent Pálsson's works to the Topographical Society of Norway, which unfortunately did nothing with them. Pálsson did at least manage to give a copy of his treatise on glaciers to the Scottish missionary Ebenezer Henderson

(1784–1858) in 1815, the influence of which can be seen in Henderson's travelogue in English (Henderson 1818, Chaps. 6 and 7). No foreign visitor had travelled as extensively in Iceland as Henderson, who journeyed around the country for 13 months in the years 1814 and 1815 and he wrote many interesting descriptions of glaciers. A map of Iceland was included in Henderson's travelogue in which the location of Langjökull is corrected from previous maps and Hofsjökull is also included from the maps of Sveinn Pálsson and Homann. The northern margin of Vatnajökull is more clearly defined than on previous maps, though it is positioned slightly too far south of its true location. Skaftárjökull was not divided from the main part of the glacier and the Tungnaá, Skaftá and Hverfisfljót rivers were all shown to have separate sources. The missionary had probably gleaned all this knowledge from Pálsson's work, because for a long period of time the sources of the Tungnaá, Skaftár and Hverfisfljót were believed to have been one and the same, and they were still depicted as such on Gunnlaugsson's map from 1848 (or with very narrow gaps between them).

As late as 1832, just eight years before he died, Pálsson tried once more, unsuccessfully, to have the Literary Society of Iceland publish his treatise (Ólafsdóttir 1975). After Pálsson's death, the naturalist Jónas Hallgrímsson (1807–1845) bought the manuscript of his travelogue from his relatives for the Literary Society, thus saving it from oblivion, though he was nonetheless criticised for doing so. Þorvaldur Thoroddsen (1855–1921) brought Pálsson's works out of obscurity by writing about and discussing them in his own scientific publications, and a part of Pálsson's treatise was first published in the *Year Book of the Norwegian Tourist Association* 1882–1884 at the instigation of Thoroddsen and the Norwegian geologist Amund Helland (1840–1918). This did not include the most important part of the treatise, however, nor any of its maps and cross sections. All of Pálsson's travelogue and its accompanying *Treatise on Glaciers* and *Treatise on Volcanoes* were finally published in their entirety in 1945, edited and translated into Icelandic by Jón Eyþórsson (1895–1968) and Pálmi Hannesson (1898–1956).

Sveinn Pálsson's fate as a natural scientist had given many of his successors much to think about. Certain individuals prevented the Danish Academy of Natural Sciences from giving him any further support for his research expeditions after he had reached the age of 33, and nor was he appointed to any teaching position. Lacking a certain sense of ambition, he failed to fight for his own interests and thus suffered a fate of constant struggle and poverty. He lived isolated and was unable to do much by himself through the difficult times of the harsh recession of the late 18th century after the large loss of life caused by the natural disasters of the Laki Haze and the period of the Napoleonic Wars in the Danish empire. He farmed the land and survived the winters by fishing at sea in order to provide for his family. He ended up in a poorly remunerated and arduous medical practice covering the Árnes, Rangárvellir and Skaftafell counties as well as the Vestmannaeyjar isles. Thus for

35 years he served as the only medical practitioner for a district that covered a huge part of southern Iceland, reaching from the Hellisheiði moor in the west to the Skeiðarársandur plain in the east, while he himself was stationed at Suður-Vík in the Mýrdalur valley. He often took care of the sick in Austur-Skaftafell County as well. He was unable to employ and enjoy his skills as a natural scientist, but he always maintained his interest in research and scholarship in this field, and this must have provided him with a diversion and some satisfaction and comfort throughout the many adversities of his long life. On Sveinn Pálsson's death, the poet Bjarni Thorarensen (1786–1841) composed the following in his memory:

A mind so free and fertile
you formed and applied
in castles in every cliff
and in cloudy bowers,
each farm a flowery centre
of Elysian fields,
as you sought to commune
with the souls of sages.

The arrows of fate e'er failed
to fell your spirit,
your mind and its mazes
had many ways of escape; –
debating with the departed
or dealing in elvish goods,
the norns had ne'er a chance
of ensnaring it at home. (Thorarensen 1935, p. 189)

4.7 Gunnlaugsson's Map of Iceland and Hallgrímsson's Unfinished 'Account of Iceland'

The surveying of the Icelandic coast by the Danes from 1801 to 1818 ended with the first maps to accurately delineate the Icelandic shoreline. Until the end of the 19th century, however, Icelanders knew very little of the central highlands and indeed most of the journeys onto glaciers were made only by explorers until well into the 20th century. Knowledge of the location and size of glaciers had increased tremendously through the work of Björn Gunnlaugsson (1788–1876; Fig. 4.19), who surveyed and mapped Iceland in the years 1831–1844, resulting in the 1848 Map of Iceland on a scale of 1:480,000 (1:960,000 in 1849). Gunnlaugsson completed the map for the Copenhagen branch of The Icelandic Literary Society, although in his original brief he was only supposed to map the inhabited areas, not the highlands (Figs. 4.20 and 4.21). He rarely left the usual mountain trails between

Fig. 4.19 Björn
Gunnlaugsson (1788–1876),
mathematician and surveyor,
who surveyed Iceland 1831–
1843. NMI

SIGFÚS EYMUNDARSON.

inhabited areas and in several places he relied on the knowledge and reports of
others. The main glaciers are very prominent on Gunnlaugsson's map and many of
the outlet glaciers are named, though their outlines are not accurately depicted and
the map seems slightly skewed and distorted in many places. Nonetheless,
Gunnlaugsson's map was a tremendous achievement in the light of the lack of
money, difficult working conditions, and miserable hardship he had to endure while
completing it, for such was his persistence that few others would have been able to
do it, according to Þorvaldur Thoroddsen, who said that: '… men took advantage of
his kindness and unpretentiousness, for it is the way of the world to starve such men
while they are alive but praise them when they are dead, for then there was no
longer any need to pay them a decent wage' (Thoroddsen 1892–1904, Vol. 3,
p. 322). Gunnlaugsson's official salary at the Reykjavik grammar school was later
increased, however, and he was shown other marks of respect and honour. The
Literary Society also received much credit for supporting this project, though there
were those who thought that only the government should sponsor such map making
and that the society should limit itself to the publication of good books.

   At the instigation of Jónas Hallgrímsson, the Literary Society began in 1838 to
collect parish and county reports from local pastors and sheriffs for an account of
Iceland that could be equated with the Map of Iceland (Gunnlaugsson 1848, 1849;
Guðmundsson 1839, p. 78; Hallgrímsson 1989, Vol. 3, pp. 125–202). With the

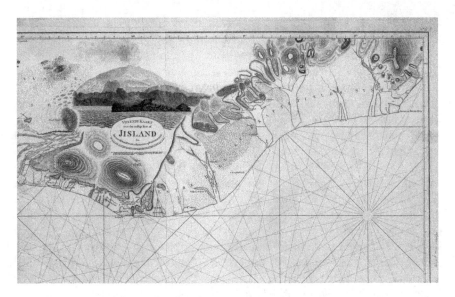

Fig. 4.20 Danish coastal map of the southern shoreline of Iceland from 1823. NULI

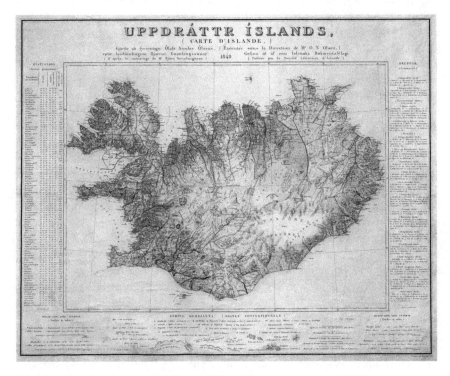

Fig. 4.21 Map of Iceland from 1849 compiled by Björn Gunnlaugsson. NULI

death of Hallgrímsson before the completion of his *Íslandslýsing* ('An Account of Iceland'), research into the country's natural history by Icelanders fell into decline until Thoroddsen revived its prestige in the late 19th century. Finally, in the beginning of the 21st century, the parish and country reports have now all been published, except for those of the Vestmannaeyjar isles, though they had proved valuable to both Gunnlaugsson and Thoroddsen, and not least the Danish scientist Kristian Kålund (1844–1919), who travelled around Iceland during the years 1872–1874 and wrote a large and important work in the years 1877–1880 on the historical topography of Iceland (Kålund 1984–1986).

In 1839–1840 the Danish government sent two Danish natural scientists to Iceland, Johannes Japetus Smith Steenstrup (1813–1897), an associate of Jónas Hallgrímsson, and Jørgen Christian Schythe (1814–1877). Their brief was not only to study the natural features of Iceland, with particular attention to finding exploitable minerals such as sulphur, but also to explore and increase general knowledge on the natural history and conditions of Iceland. They were to examine in particular if there was any truth in the assertion that the climate in Iceland had worsened since the age of settlement and whether outlet glaciers such as Breiðamerkurjökull had expanded, and indeed whether all glaciers descend into valleys down onto the lowlands, or just a few of them. Furthermore, they were to ascertain if descending glaciers flowed into valleys during volcanic eruptions and if they turned into water, mud and rocky lava during jökulhlaups because the volcanoes themselves were structured from alternative layers of ash, ice and lava (Thoroddsen 1892–1904, Vol. 4, pp. 21–23). In his report, Steenstrup compared the glacial moraines of Skeiðarárjökull and Sólheimajökull with boulders in Denmark, which geologists had not yet realised had been transported there by a glacier, so little was still known about glacial remnants at that time. Steenstrup was later a professor at Copenhagen University and one of the most educated contemporary natural scientists of Denmark. Almost three decades after Steenstrup's expedition with Hallgrímsson, the Swiss botanist Oswald Heer (1808–1883) analysed the plant fossils they had collected and proved beyond dispute that they were indeed the remains of ancient deciduous forests that had flourished in Iceland during the Miocene epoch more than 6 million years ago (the second to last period of the Tertiary when basalt rock was formed). It was thus quite clear that the climate had been much warmer in Iceland than in present times, with an average mid-summer temperature of 15–20 °C, similar to what is now common in the Mediterranean (Heer 1868). From the days when Eggert Ólafsson had been the first man to describe the plant fossils he had found at Brjánslækur on the Barðaströnd coast, reckoning them to be remnants of large forests from many centuries ago, scientists had debated as to whether they had been borne to Iceland as driftwood from warmer climes, or had been formed at the bottom of the ocean and later been uplifted from the sea.

As far as Jørgen Christian Schythe is concerned, he left with Pastor Sigurður Gunnarsson (1812–1878; 1949, 1950), a teacher and later a dean in eastern Iceland, during the great sea-ice summer of 1840, on what was the fourth journey, as is known of, along the Vatnajökulsvegur highland trail to the west of Jökulsá á Brú river that leads south through Grágæsadalur, crosses the Kreppa river just north of its confluence with the Kverká, and then stretches westward along the northern margins of Vatnajökull. They encountered tremendous snowstorms and only managed to reach the northern part of Jökuldalur with great difficulty. A number of horses died on this disastrous trip and it was clear that Gunnlaugsson had been very lucky with the weather the previous year and that this track would never be a national highway. Schythe then came to Iceland to do research on the eruption of Hekla in 1845 and he published an important work on this (Schythe 1847). A few years later, he settled in Chile, became a university teacher and novelist, then a regional governor, and later a manager of the Chilean National Bank until his death in 1877.

4.8 The Search for Traces of Ice-Age Glaciers in Iceland

Although Iceland had few natural scientists in the 19th century, the nation was not isolated from foreign studies and trends in natural sciences and benefited from the strengthening of scientific research in Europe, for many highly-qualified scientists came to Iceland, especially from Scandinavia and Germany (Fig. 4.22). Just before the middle of the 19th century, in 1846, two pioneers of natural sciences visited Iceland, the Germans Wolfgang Sartorius von Waltershausen (1809–1876), a geologist, and the chemist Robert Wilhelm Bunsen (1811–1899). Their guide was the Danish general and geologist Haagen Vilhelm Mathiesen (1812–1897). Waltershausen found many striations marks on rocks from ice over a wide area, both in the mountains and down by the sea, though he believed that they were not caused by outlet glaciers advancing, but by being formed by pack ice and frozen water on ocean cliffs while land was being uplifted from the sea. As evidence of this, he pointed out that he had found grooves heading in different directions on one and the same rock (Waltershausen 1847; Bunsen 1847, Thoroddsen 1892–1904, Vol. 4, p. 178; Kjerulf 1853). Although he referred to ideas concerning an Ice Age, which had been circulating for the previous ten years, as nonsense, he did not deny that more recent glaciers might also have left behind striations and moraines. He described Arnarfellsjökull (Múlajökull), its layers of dirt in the ice, and the triple folds of moraines at its margins. Waltershausen (1847) was one of the first geologists to analyse the Icelandic 'móberg' rock, calling it palagonite, and believed it had originated on the ocean bed. Mathiesen (1846, 1864) and Bunsen (1847) wrote accounts of ancient glacial moraines, erratic rocks, and glacial striation and leaned towards the theory of the Ice Age. Mathiesen was the first man to claim in print (1846) that Iceland had been covered largely or completely by a glacier and quoted as evidence striation marks that headed from the central highlands out to sea. This was 37 years after the English botanist William J. Hooker (1785–1865; 1811), who had been amazed by the erratic boulders he had found dispersed around a hill in the vicinity of Reykjavík, also wrote an important work on sea ice and ocean currents around Iceland (Mathiesen 1864).

◄ **Fig. 4.22**   Early explorers of glaciers in Iceland (*Left* to *right*, *top* to *bottom*). Ebenezer Henderson (1784–1858), Scottish priest and missionary, who wrote an important travelogue on Iceland. Jónas Hallgrímsson (1807–1845), Icelandic natural scientist and poet. Japetus Steenstrup (1813–1897), Danish natural scientist, travelling companion of Jónas Hallgrímsson. Jørgen Christian Schythe (1814–1877), Danish natural scientist, travelled around Iceland 1839–1840. Haagen V. Mathiesen (1812–1897), Danish geologist and general; visited Iceland in 1846. Wolfgang Sartorius von Waltershausen (1809–1876), German geologist; visited Iceland in 1846. Robert Wilhelm Bunsen (1811–1899), German chemist; visited Iceland in 1846. Theodor Kjerulf (1825–1888), Norwegian geologist; visited Iceland in 1850. Otto Martin Torell (1828–1900), Swedish geologist; travelled round Iceland in the summer of 1857. Carl Wilhelm Paijkull (1836–1869), Swedish geologist; compiled first geological map of Iceland in 1865. Amund Helland (1846–1918), Norwegian geologist; mediated the work of Sveinn Pálsson abroad; visited Iceland in 1881. Friedrich Konrad Keilhack (1858–1944); began taking part in geological research projects in Iceland in 1883 which continued for more than fifty years

Was Iceland Previously Covered by a Glacier?

A few years later, two pioneers of the theory of the Ice Age came to Iceland, the Norwegian Theodor Kjerulf (1825–1888) and the Swede Otto Martin Torell (1828–1900). Their view that a glacier had formerly covered their homelands in Scandinavia became stronger during their sojourn in Iceland as they travelled widely in the country, examining in detail moraines and other remnants of glaciers in their forefields, calculating the direction of grooves in the surface of bedrock, which they believed had been carved by an ice-age glacier; if the striations headed in different directions, then they had been formed during differing times of a glacial advance. At 25 years of age, Kjerulf came to Iceland in the summer of 1850 and was the first to measure the country's snowline, pointing out that it varied considerably according to region and climate. Kjerulf later became the director of the Norwegian Geological Survey in 1858. Torell travelled around Iceland for six months in the summer of 1857, and his observations at the southern margins of Vatnajökull and in the north along the Kjölur highland route proved very useful when he established that it had been ice-age glaciers from Scandinavia advancing south to northern Germany that had transported rocks and other glacial remnants there and not pack ice or the Great Flood. Torell measured the first glacial movements in Iceland on Svínafellsjökull in the Öræfi district, which edged forward 2.5 ells (1.5 m) in six days in the summer of 1857 (Gadde 1983, p. 82; Torell 1857). It was almost 80 years before a glacier's advance was again measured in Iceland, Hoffellsjökull moving around 2 m in 24 h during the Swedish-Icelandic expedition in the summer of 1936. Accompanying Torell was the physician Niels Ohlson Gadde (1834–1904), who wrote an entertaining travelogue on his visit to Iceland in 1857. Torell later became one of the leading natural scientists in Sweden, a professor of geology and zoology, director of the Swedish Geological Survey, and the pioneer of systematic scientific research on the North and South Poles. His expedition to Iceland thus marked the very beginning of his illustrious career. Another Swedish geologist, Carl Wilhelm Paijkull (1836–1869), examined the glacial remains and deposits on the outwash plains of Mýrdalsjökull and the southern sides of Vatnajökull in 1865, describing the formation of outwash plains from glacial sediment layers and drawing attention to jökulhlaups and the eruptions of Katla, of which little was then known abroad. Paijkull claimed that glacial striation of bedrock along glacially eroded valleys and fjords all indicated that Iceland had been covered by a glacier. He compiled the first draft of a geological map of Iceland, based on Gunnlaugsson's map from 1848, but it only covered a tenth of the country (Pajkull 1866, 1867).

The Norwegian geologist Amund Helland (1846–1918), who came to Iceland in 1881, stated that all its valleys, lakes, and fjords were glacially eroded. He wrote a great deal about glaciers, glacial outwash plains, and glacial rivers, describing jökulhlaups and volcanic eruptions and believed that beneath the southern coast lay fjords that had been filled up with glacial till and sediment loads from jökulhlaups. He was the first scientist to measure the water flow rates and sediment loads in Iceland's glacial rivers. Without even dismounting from his horse, Helland would dip bottles into the muddy rivers flowing from the southern verges of Vatnajökull and evaluate the sediment load from beneath the glacier and its erosive powers. He calculated that the glacier eroded from its bed 0.65 mm a year. The summer of 1881 was unusually cold and there was little water in the glacial rivers, for extensive measurements during the second half of the 20th century indicate that the glacier had eroded on average about 3 mm from its bed every year (Tómasson 1976, 1990; Björnsson 1979). Helland also believed the odour from the Jökulsá river on the Sólheimasandur plain was caused by subglacial geothermal activity and rejected the theory that it was due to the erosion of sulphur from bedrock. But Helland will also be remembered, along with Thoroddsen, for his role in getting a section of Sveinn Pálsson's treatise on glaciers into print in the 1882–1884 edition of the *Year Book of the Norwegian Tourist Association*. Helland (1882, 1883a, b) still considered Pálsson's work as innovatory, even though it was almost a century since his travelogue had been written.

4.9 Þorvaldur Thoroddsen's Overview of Iceland's Geography and Geology

And now it was the turn for an Icelandic explorer and geologist, Þorvaldur Thoroddsen, to present a collective overview of Iceland and its glaciers (Fig. 4.23). Every summer from 1881 to 1898, he systematically visited every part of Iceland and was the first natural scientist to dwell for weeks, sometimes even months, in the central highlands. No Icelander had previously visited so much of the country as Thoroddsen. In addition to this achievement, he also wrote more books than any other Icelandic natural scientist has done, right up until the present day. He continually published the results of his research as he travelled round the country, and the comprehensive edition of his work was published just after the turn of the 20th century. Thoroddsen increased the knowledge of Icelandic glaciers, even though he was not a theorist who advanced glaciology as such, for he described the prevalent

Fig. 4.23 Þorvaldur
Thoroddsen (1855–1921),
Icelandic geologist. NMI

climatic conditions and pointed out that the existence of very large glaciers in
Iceland was due to high levels of precipitation and cool summers. He measured and
compiled a contoured map of the snowline over a wide area of Iceland, drawing
attention to how variable it could be from year to year (Thoroddsen 1907–1911,
Vol. 2, pp. 1–67). He discussed the expansion of glaciers since the time of settlement and believed that there had been no substantial changes since then in their
subsequent size, or in the routes of communication alongside them; they were the
same as they had always been. He did, however, conclude that Drangajökull was
retreating. Thoroddsen produced the first geological map of the whole of Iceland (in
1901 on a scale of 1:600,000), though it was based on Gunnlaugsson's map from
1848. Thoroddsen's map depicted the outlines of all the main glaciers of Iceland,
though without contour lines. He did not survey mountains but calculated their
height from barometric measurements. He corrected the errors of Gunnlaugsson's
map, especially concerning the central highlands to the west and north of
Vatnajökull, though the northern margins of Vatnajökull are nonetheless still drawn
better on Sveinn Pálsson's map of 1794. When Thoroddsen began his research in
1881, only 25 outlet glaciers had ever been mentioned in print, but in his published
works Thoroddsen named a total of 130 glaciers, many of which had been named in
the unpublished work of Sveinn Pálsson, which Thoroddsen had read. Thoroddsen
also gave an account of subglacial volcanic eruptions and jökulhlaups, though he

did not describe the origins of the latter as being within glaciers. Nor did he actually ascend glaciers any more than his compatriots had done.

Thoroddsen's geological map depicted the direction of glacial striation in all parts of the country, altogether at 170 places, from which he concluded that a glacier from the last glacial period had crept all the way from the central highlands down to the shoreline, 1000 m thick above central Iceland, though above the western fjords there had been a special glacial dome hardly much more than 400–500 m thick (Thoroddsen 1892, 1901). A few nunataks had stood up through the ice shield such as Bláfjall (1225 m high), in the Mývatn district in the northeast, for he found no signs of glacial striation there. The geologist Sigurður Þórarinsson later came to the conclusion (1937) that the outermost regions of basalt bedrock in the eastern fjords, mid-northern Iceland, and in the western fjords had never been covered by a glacier (Þórarinsson 1937). Many other geologists have since been of the same opinion. But Thoroddsen (1892, 1892–1904, 1905/1906) believed that even though nunataks had stood above the ice sheet, all life had been destroyed in the last glacial period and once it had ended both fauna and flora had established themselves once again (the *tabula rasa* theory). The Swedish entomologist Carl H. Lindroth (1905–1979; 1931, 1965) later agreed with the view that nunataks had stood up from the ice sheet, but maintained that Iceland's insect life and various plants would have survived the glacial period on such ice-free environments. Living organisms would have been borne from such central areas and taken root elsewhere once the glaciers had melted. To this day, biologists are still presenting arguments and evidence both for and against these two hypotheses (Ægisdóttir and Þórhallsdóttir 2005). The most recent studies suggest that the present day plant species in Iceland do not have a long history, indeed that they have borne here after the last glaciation. Seeds could have been transported by birds, or even been blown by winds high in the atmosphere; it is also possible that plants or seeds floated to Iceland on icebergs or driftwood. The only forms of life that have been found in Iceland that have survived an ice age there are two species of freshwater crustaceans. These two species have never been found anywhere else in the world, and genetic research indicates that they have evolved in isolation in Iceland previous to its last glacial epoch.

4.10   Detection of Ancient Climate Changes in Marine and Terrestrial Sediments

We now turn to traces of ancient glaciers and climate change in layers of topsoil. As noted above, Heer's analysis of fossilised remains of vegetation in layers of lignite in basalt formations indicated that 3–14 million years ago the climate was much warmer than it is today. Fossils of marine animals on dry land also bear witness to changes in sea levels. Shells on land are mentioned in the *Book of Settlements* (Benediktsson 1968 (1), p. 71) and Eggert Ólafsson and Bjarni Pálsson found seashells on the Tjörnes peninsula high above the present sea level. Their accounts of layers of peat below the tideline also implied that sea had covered land since they had been formed (Ólafsson and Pálsson 1981, Vol. 1, pp. 7–8; Vol. 2 p. 201). Nonetheless, it was not until the Dane Otto Andreas Lowson Mørch (1828–1878; 1869) researched seashells from Hallbjarnarstaðakambur crest on Tjörnes in 1871, that it became clear that these marine fossils were from the end of the Tertiary period (Pliocene) when the sea inundated Tjörnes. The ice age was then beginning in Iceland. Helgi Pjeturss (1872–1949; 1906b) later pointed out that there were numerous layers with various kinds of shells. Guðmundur G. Bárðarson (1880–1933; 1906) later classified the layers of shells on Tjörnes by age and found that the sea had gradually grown colder as the layers were forming. He also discovered, by researching mollusc shell strata in the layers of soil around Húnaflói Bay, that the climate must have been warmer when they were formed than it is currently; a species of mollusc that lives in spiral shells (Purpura lapillus), could now no longer survive on the northern shores of Iceland due to the coldness of the sea. Geologists have since traced in detail changes in climate and sea levels from about 4–2.5 million years ago (the first part of the Pleistocene to the Pliocene) in deposit strata 500 m thick on Tjörnes. Marine fossils reveal that the sea was 2–5 °C warmer until the last glacial period began and till covered the sediment layers.

At first it was believed the glacial epoch had been one and continuous, but the 19th century ended with the discovery by the geologist Pjeturss (1900, 1901, 1902, 1903, 1904, 1905a, b, 1906a, 1908, 1909) that it had been interrupted by at least one interglacial period when glaciers had disappeared from most of Iceland. He found layers of till between stratas of lava; the lava had flowed during interstadial periods, while during cold periods they were covered by ground moraine. Pjeturss also found glacial moraines merged with palagonite formations and was the first to show, in 1899, that this was from the last glacial period, although it was still unclear how the palagonite rock had been formed. He was also the first scientist to find and draw attention to fossil layers on the Búlandshöfði headland on Snæfellsnes and to demonstrate that they had been formed alternately during interstadial and glacial periods. It was also clear to him that outlet glaciers descending southward from Vatnajökull (Breiðamerkurjökull and Hoffellsjökull) covered ocean formations from an interstadial period when the sea level was 50–60 m higher than it is now. Pjeturss described and analysed the sediment layers in the Fossvogur valley (now part of greater Reykjavík), where oceanic and glacial deposits lay one on top of the other in turn. They are now believed to have been formed at the end of the last glacial period, when the ice sheet margin retreated and the sea followed it inland (Hjartarson 1989; Hjartarson and Ingófsson 1988; Norðdahl 1991; Geirsdóttir and Eiríksson 1994; Norðdahl and Pétursson 2005). Guðmundur G. Bárðarson also described the changes in sea levels when a glacier from the last glacial period melted into the ocean, the resulting large increase of water in the sea then flooding far inland as the overburden pressure of the glacier had thrust down the bedrock. The land was then uplifted as the pressure of the ice was removed and for a while the tidelines of Iceland were much lower than they are currently.

It had long been assumed that the ice sheet from the last glacial period had gradually and uninterruptedly disappeared from Iceland completely. Pjeturss (1910) pointed out, however, that this had not been the case. Indeed, he had found glacial moraines on Langanes in the far northeast, the Skagi peninsula in the west, and in southern Iceland, which bore witness that in some places the ice sheet had been static or had even begun to advance again. Bárðarson (1921, 1923) also found glacial moraines in the inland heads of the Breiðafjörður and eastern Borgarfjörður fjords that revealed there had been a halt to the retreat of the ice caps of the last glacial period. Kjartansson (1939, 1943, 1945, 1960, 1962, 1964, 1965, 1966a, 1968, 1969) then demonstrated that the entire ice shield covering the central highlands had advanced, pushing up a terminal moraine across all the southern lowlands of Iceland from the eastern Efstadalsfjall all the way east to Vatnsdalsfjall above the Fljótshlíð slopes. The moraine lies athwart the Þjórsá at the Búði waterfall and thus Kjartansson named the moraine the Búðaröð line and called the cold period which had triggered the advance of the glacier about 10,000–9500 years ago the Búðaskeið (internationally: the Younger Dryas period). Þorleifur Einarsson (1931–1999; 1960, 1961, 1963, 1968; Geirsdóttir et al. 1997) then found signs of an earlier glacial advance around 12,200–11,900 years ago, which he named the Álftanesstig (internationally: the Elder Dryas period). Evidence has actually come to light that the ice sheet had come to a halt or advanced for a while even more often than during these two periods, though on the whole it had retracted. The ice finally disappeared so quickly from the central highlands about 8000 years ago that the huge Þjórsárhraun lava field had flowed unhindered for 150 km on its way from the Tungnaáröræfi wilderness to the southern shores of southern Iceland. Kjartansson (1964, 1966a) had discovered this while researching the expansion of the lava fields around Hekla and the river Tungnaá.

4.11   From Ice-Age Glacial Remnants to Modern Glaciers

Many German, Scandinavian, Danish and Austrian scientists continued doing research into the remnants of ice-age glaciers, and the formation of outwash plains from glaciofluvial deposits, which Torell, Helland, Thoroddsen and Pjeturss had begun. The real object of their studies was often far from the glaciers of Iceland, however. The Danes saw in Iceland similarities with the heaths of Jutland, the Austrians made comparisons with the Alps, and the Germans believed that through research into the central highlands of Iceland they could explore similar environments in northern Germany at the end of the last glacial period, for the main ice shield of Scandinavia had extended that far south. German scientists were thus mostly interested in the broad outlet glaciers of Langjökull and Hofsjökull and the northern margins of Vatnajökull, as they believed the climate in these locations would have been the most similar, especially concerning temperature and precipitation, to that of their homeland at that time. To the north of Vatnajökull there were southerly winds that had lost most of their precipitation and heat on their journey from the warm Gulf Stream across cold ice sheets. The steep flanks of the southern outlet glaciers of Vatnajökull, on the other hand, might well be likened to the environs of the Alps, where narrow outlet glaciers descend onto the flat lowlands of southern Germany. Attention was also paid to the outlet glaciers on the Tröllaskagi peninsula in northern Iceland and the glaciers of the western fjords, Drangajökull and Gláma, where the snowline in Iceland is at its lowest. The German geologists Konrad Keilhack (1858–1944; 1883, 1885, 1886, 1934), who first came to Iceland in 1883, and Walther von Knebel (1880–1907; 1905, 1906), who drowned, along with another man, in Lake Askja in 1907, traced and analysed signs of the last glacial period and the directions of glacial striations in the central highlands. Research into volcanoes drew German scientists at the beginning of the 20th century to Askja in the Ódáðahraun lava desert to the north of Vatnajökull, and this also resulted in a great increase in knowledge concerning the margins of Dyngjujökull and Brúarjökull and the Kverkfjöll mountains during the years 1910–1912. The Finns Leiviskä (2005) and Okko (1956) and the Swede Gunnar Hoppe (1953, 1968, 1982, 1995) described and analysed glacial remains in Iceland and compared them with those of their own countries. The main interest of all these geologists, however, was centred on land formations in the forefields of Iceland's ice sheets (see also Spethmann 1908, 1912; Backlund 1939; Todtmann 1952, 1953, 1955a, b, 1957, 1960; Schwarzback 1983; Kristjánsson 2001; Maizels 1991; Price 1969, 1971, 1982). It was not until the 1920s that specific research began to be carried out on the actual glaciers themselves: what were their main features and characteristics, how were they formed, how did they maintain themselves, how did they move, and how did they respond to climate changes. Indeed, it was this research into contemporary glaciers that was to provide the key to understanding glaciers of the past (Fig. 4.24).

Foreign Travellers Ascend Icelandic Glaciers
Very few men had been known to have climbed glaciers in Iceland by the beginning of the 19th century. Icelanders themselves had little interest in unnecessary trips up mountains and into the wilderness; indeed, daily life outdoors was already harsh enough in the farmsteads and villages of the country. Towards the end of the 19th century, however, there was an increase in the number of ascents of Icelandic glaciers, though this was mostly done for sport and pleasure. These climbers were nonetheless curious about their environment and described much of what they encountered of the local conditions, locations and shapes of mountains, land formations and oases, passable and impassable routes, the sources and outlets of the main glacial rivers that hindered their progress, glacial snouts and jagged nunataks. Of these adventurers and climbers, mention should be made of William Lord Watts (1851–1920), an English lawyer in his early twenties, an energetic and courageous traveller, who climbed the highest dome of Mýrdalsjökull with two Icelanders in 1874, and ascended Vatnajökull three times, successfully crossing it on the final trip (Fig. 4.25). Watts (1875, 1876) wrote two books about his journeys, and although these reveal no significant scientific gains from these expeditions on Vatnajökull, he did express his certainty that the glacier had a great influence on the climate of northern Iceland by shielding it from the moisture borne by southerly winds. Two

Fig. 4.24 Explorers of glaciers in Iceland (*Left* to *right*, *top* to *bottom*). Sigurður Gunnarsson (1812–1878), dean and travelling companion of Björn Gunnlaugsson and Jörgen Schythe. Páll Pálsson (1848–1912), 'Glacier Páll', from Hörgsland at Síða; teacher and travelling companion of Watts. Ögmundur Sigurðsson (1859–1937), headmaster and travelling companion of Þorvaldur Thoroddsen. William L. Watts (1851–1920), led the first expedition across Vatnajökull. Johan Peter Koch (1870–1928), surveyor, colonel in the Danish army, and explorer of glaciers. Daniel Bruun (1856–1931), archaeologist and captain in the Danish army. Niels Nielsen (1893–1981), Danish geographer, who worked for years in Iceland. Gunnar Hoppe (1914–2005), Swedish geographer and Icelandophile. Helgi Pjeturss (1872–1949), Icelandic geologist and philosopher. Guðmundur G. Bárðarson (1880–1933), Icelandic geologist. Guðmundur Kjartansson (1909–1972), Icelandic geologist. Þorleifur Einarsson (1931–1999), Icelandic geologist

Fig. 4.24 (continued)

years after he climbed Vatnajökull, Watts was presumed dead, but a later sighting and more recent evidence indicates that he lived and worked in California in later years, prospecting for oil until 1920 (Gíslason 2008).

Foreign scientists made a considerable number of journeys over the central highlands between Langjökull and Hofsjökull and along the northern margins of Vatnajökull early in the 20th century due to an interest in ancient mountain trails. In 1901 Daniel Bruun (1856–1931; 1902a, b, 1914, 1921–1927), then a captain in the Danish army, and Elías Jónsson (1863–1929), a farmer at Aðalból in Hrafnkelsdalur valley, travelled on horseback along the eastern verges of Brúarjökull to the south of Snæfell to investigate the credibility of stories of journeys across the glacier to Hornafjörður in previous centuries (Fig. 4.26). A year later, Bruun, along with the teacher Ögmundur Sigurðsson (1859–1937), searched for routes up onto the southern flanks of Vatnajökull as they rode to the inland heads of valleys by Skálafellsjökull and Heinabergsjökull while heading in the direction of the Breiðubunga dome, stacking cairns as they went for the proposed surveying expedition of the Danish General Staff. The English adventurer Frederick William Warbreck Howell (1857–1901) traversed Langjökull glacier with two other Britons in August 1899, the first known crossing of this glacier. Along with two Icelanders, he also climbed the highest peak of Öræfajökull, Hvannadalshnúkur, in 1891 (Fig. 4.27), they being the first three men to achieve this feat as far as is known (Guðmundsson 1999). With his two companions, Howell ascended the glacial plateau of Langjökull from Flosaskarð pass, and from the summit they could see above the clouds all the way to Snæfellsjökull in the far west, and to Vatnajökull in the east; they descended the Fögruhlíð slopes from Langjökull. The German Heinrich Erkes (1864–1932; 1911, 1914), who travelled in the central highlands for years, described the northern reaches of the Sprengisandur gravel plain and the highlands from Hofsjökull east to the northern margins of Vatnajökull. He shared Daniel Bruun's (1902a, b, 1914, 1921–1927) interest in old

Fig. 4.25 Drawing of Pálsfjall on Vatnajökull from *Across the Vatna Jökull or Scenes in Iceland* (1876) by William Lord Watts. NULI

Fig. 4.26 Riding across Brúarjökull, Kverkfjöll ahead. In *front* is Elías Jónsson, farmer at Aðalból in Hrafnkelsdalur and in the rear Daniel Bruun. Daniel Bruun, 1927. NULI

mountain trails and they both wrote interesting and informative books of their journeys in Iceland.

Crossing Vatnajökull from North to South
The first crossing of Vatnajökull in the 20th century was in the late summer of 1904, almost thirty years after Watts' journeys. This time two Scotsmen, Wigner (1905) and Muir (1905a, b; Muir and Wigner 1953), crossed the glacier from

(a)

APPROACHING THE SUMMIT OF THE ÖRALFA JÖKULL.

(b)

THE ACTUAL SUMMIT OF THE ÖRÆFA JÖKULL.

Fig. 4.27 Drawings of Hvannadalshnúkur, the peak of Öræfajökull, from Howell's travelogue of 1893; Howell was the first man to climb the peak in 1891. NULI

Fig. 4.28 Alfred L. Wegener (1880–1930), geologist, Vigfús Sigurðsson (1875–1950), carpenter, Lars Larsen (b. 1886), sea captain, and Johan Peter Koch (1870–1928), surveyor, in western Greenland, July 1914. RLC

Brúarjökull on the eastern side, heading to the nunataks of Esjufjöll at the top of Breiðamerkurjökull, before turning southwest to Grænalón, a journey on the glacier of 128 km in 13 days. Eight years were to pass after the Scotsmen's effort before Vatnajökull was traversed again, this time in 1912 by the Dane Johan Peter Koch (1870–1928), a colonel, who in the years 1903–1904 had been director of the surveying of the southern verges of Vatnajökull, which will be discussed below. Along with him was Alfred Lothar Wegener (1880–1930), a German meteorologist and geologist, a pioneer in supporting the theory of continental drift with evidence from his research expeditions (Figs. 4.28 and 4.29). Koch and Wegener first wanted to cross Vatnajökull from north to south to prepare themselves for their expedition to northern Greenland. Their attempt succeeded beyond all expectations. At the end of June they set off on horseback in extraordinarily beautiful weather with 14 horses and covered 65 km on an 18-hour journey from Brúarjökull, east of the Kreppa river, to Esjufjöll on Breiðamerkurjökull. Koch (1912) described the flora and fauna and the small glacially-dammed lakes of Esjufjöll. From there they headed east over the glacier to the edge of the Veðurárdalsfjöll and the Þverártindur pinnacle, which Koch knew well from his surveying work in 1903 and 1904 (Koch 1905–1906). Once this had been accomplished they rode back the same way they had come, but with a visit en route to the western Kverkfjöll to explore the geothermal areas there. Koch pointed out that the greatest height they attained while crossing Vatnajökull

Fig. 4.29 Koch and Wegener's expedition crossing Vatnajökull glacier in June 1912. RLC

was in the southern part of the glacier and that a ridge covered by ice extended southwest from the Kverkfjöll mountains.

Koch and his companions had no skis or sledges with them on this expedition. In the deep and soft snow of June, they led their horses and tramped a pathway themselves through the snow using Canadian snowshoes, the lighter men and horses going first. At journey's end, Koch believed there was no longer any reason to doubt the truth of ancient stories of journeys across Vatnajökull; in the much firmer snow of August they might have completed the same journey as they had made in just over half the time. The horses proved so successful, both to ride and as packhorses on the glacier, that Koch and Wegener took Icelandic horses with them on their trip to Greenland, 1912–1913, and with them Vigfús Sigurðsson (1875–1950; 1948), later nicknamed 'Greenland-farer,' who was to be in charge of them. The extraordinary lucky weather conditions on Vatnajökull, exceptionally calm and bright, created an inflated confidence in Icelandic horses, for most other expeditions onto the glacier had to endure far worse conditions.

There was now but a short time until the greatest discovery of all was about to be made. On the fourth expedition of modern times onto Vatnajökull, in the late summer of 1919, two Swedish students, Hakon Wadell (1895–1962) and Erik R. Ygberg (1896–1952), came upon the ancient Lake Grímsvötn in the centre of the glacier. Wadell and Ygberg had originally planned to climb Mýrdalsjökull, but had been deterred from this by the eruption of Katla the previous year, and so they had decided to ascend Vatnajökull instead. There was an amazing silence about their discovery until the Grímsvötn volcano dramatically drew attention to itself fifteen years later.

In the summer of 1926 there was the first all-Icelandic journey onto the glacier of the 20th century and, furthermore, this expedition was carried out simply for pleasure by young men who wanted to achieve something athletic and daring and to have something else to look back on other than the dreary mundanity of everyday life (Benediktsson 1977). All the men were from Hornafjörður: Helgi Guðmundsson (1904–1981) from Hoffell, Unnar Benediktsson (1894–1973) from Einholt, and Sigurbergur Árnason (1899–1983) from Svínafell, and on 15 July they set off onto the glacier on skis, with a bit of equipment on a sledge, inland from Hálsaheiði. They then crossed the glacier to Dyngjujökull, just west of the Kverkfjöll, where they descended to the west of Kistufellskriki in the evening of 18 July. From there they crossed the Ódáðahraun lava desert to Svartárkot farmstead in Bárðardalur in the northeast, returning to Hornafjörður by the exact same route. They accomplished this journey on home-made shoes and skis and slept in one fur sleeping-bag in a field tent, thus proving decisively that it was possible to traverse Vatnajökull with very simple gear (Fig. 4.30).

Exploring the Volcanic Eruption in Grímsvötn

Grímsvötn erupted in March 1934, and so for the first time there was an expedition onto Vatnajökull to explore and research the volcano sites and the sources of jökulhlaups onto Skeiðarársandur. Until this point in time, geoscientists had only studied the aftermath outside of glaciers for evidence of such previous volcanic eruptions; Icelanders had seen plumes of volcanic smoke rising into the skies before, had watched the flashes of lightning, heard the rumbling thunder and shrill whining of an eruption, but always at a distance from farmsteads or villages, sometimes hundreds of kilometres away. They had also been aware of ash falls carried by the winds to distant parts of the country and seen jökulhlaups trundle mud, rocks and icebergs over outwash plains and into the sea. But now, with the acquired experience and perseverance learned from exploring mountains and the central highlands, men were ready to take on visiting the very volcanoes themselves. It would be difficult, of course, to approach violent and turbulent volcanic plumes, not least if the crater were to be underwater or beneath a glacier, but this was exactly why there could be no hesitation or waiting until the eruption ended, for then the glacier would quickly cover up all traces of it, and it would be much more difficult to gain direct and fresh knowledge of what exactly had occurred. The expedition had to be planned so as to be able to respond to unknown or unexpected circumstances on the way: flood channels, gaping crevasses, the hauling of sledges through layers of ash, and tephra perhaps being hot enough to melt narrow gullies in the snow. Poisonous gases from the volcano might cover the glacier's surface, and the volcano itself might even erupt through new vents right beneath their feet. This expedition is described in greater detail in the chapter dealing with Grímsvötn, but right now we turn to what geologists were most curious about and hoped to find evidence of from these erupting volcanoes.

Fig. 4.30 Early 20th-century explorers of glaciers in Iceland (*Left* to *right*, *top* to *bottom*). Helgi Guðmundsson (1904–1981), from Hoffell in Hornafjörður. Unnar Benediktsson (1894–1973), from Einholt in the Mýrar district of Austur-Skaftafell County. Sigurbergur Árnason (1899–1983), of Svínafell in Hornafjörður. Steinþór Sigurðsson (1904–1947), Icelandic astronomer. Helgi Hermann Eiríksson (1890–1974), Icelandic mine engineer, headmaster of the Technical College in Reykjavík. Jón Eyþórsson (1895–1968), Icelandic meteorologist and founder of the Iceland Glaciological Society. Jóhannes Áskelsson (1902–1961), Icelandic geologist. Skarphéðinn Gíslason (1885–1974), from Vagnstaðir in the Suðursveit district. Emmy Mercedes Todtmann (1888–1973), German geologist

The Riddle of How Palagonite is Formed

Scientists hoped that the 1934 eruption of Grímsvötn would provide an answer to the question as to how palagonite rock was formed. This had been a riddle to geologists ever since von Walterhausen had first drawn attention to palagonite in the middle of the 19th century. Most geologists had come to the conclusion that palagonite was formed in eruptions beneath a glacier, the basalt magma super-cooling when coming into contact with ice or water, thus solidifying into palagonite instead of flowing as lava. Thus Niels Nielsen (1893–1981; 1937a, b; Nielsen and Noe-Nygaard 1936) and Arne Noe-Nygaard (1908–1991; 1940), later well-known professors of geology in Copenhagen, expected to find newly-formed palagonite rock in Grímsvötn. They were to be greatly disappointed, for the volcanic material emerging from the eruption looked completely different to palagonite so that the origin of the latter was still unclear. Furthermore, there was nothing to indicate that mountains of palagonite rock had been built up beneath glaciers, for in spite of the innumerable eruptions of Grímsvötn and Katla over the last few centuries that had pumped such abundant material into their respective glaciers, no palagonite mountain had arisen. Some geologists still leaned towards earlier ideas. Von Knebel (1880–1907; 1912) believed the palagonite mountains of the Kjölur interior and in the Ódáðahraun lava desert were the remains of a highland massif that had been sundered by water erosion in interstadial intervals between glacial periods, while Hans Reck (1886–1937; 1910) had the opinion that sections of the Earth's crust around the mountains had subsided along fault lines that were all around them. Hans Spethmann (1883–1957) leaned towards Knebel's erosion theory, but thought that the fault lines might also play a part in the formation of palagonite mountains. The Swiss Sonder (1938) actually believed the mountains had been thrust upwards while their surroundings had remained unchanged. Josef Keindl (1903–2007) agreed with the theory, so prevalent among geologists of the 1920s, that table mountains, or tuyas, were created by tectonic displacements along fault lines. This theory had not yet been scientifically substantiated, however, for indeed it was difficult to reach the ledges of fault lines at the foot of mountains which were often covered with scree and surrounded by historic lava flows. The Danish geographer Niels Nielsen also agreed with this tectonic displacement theory. No Dane either before or since has ever been so involved in researching the natural history of Iceland, as he worked in the country from 1924 to 1937. Having observed the effects of volcanic activity, erosion and the movement of tectonic plates on the landscape of the central highlands, Nielsen (1933) was the first geologist to put forward the opinion that Iceland was constantly sliding along the rift that lies right through the island. Nielsen was accompanied on his travels by various Icelanders including Pálmi Hannesson, who carried out botanical and biological research and was later headmaster of the Reykjavík grammar school, Sigurður Thoroddsen (1902–1983), an engineer, and Steinþór Sigurðsson (1904–1947), both of whom worked at surveying. Sigurðsson was later much involved in research in the central highlands, including Grímsvötn.

In the summer of 1941, the geologist Guðmundur Kjartansson (1909–1972; 1943, 1961, 1966b), began carrying out research on the Hlöðufell and Skriða mountains in Árnes County, and two years later published his hypothesis that palagonite mountains were piled up during explosive eruptions of basalt magma in vaults that had been melted upwards into glacial ice. On being supercooled in meltwater, the magma rock fractures into glass-like volcanic gravel, which accumulates and is restrained by water pressure and walls of ice, becoming hardened palagonite before the moulds are removed from outside the amalgamation of glass shards. If the volcanic material piling up manages to break through the surface of the glacier, the water would cease to stream toward the eruption vent, and lava would flow over the summit of the palagonite platform to form a dome of lava above it from crystalized basalt; an explosive eruption is transformed into a lava flow eruption and a table mountain or tuya arises (see also Bemmelen and Rutten 1955). Most scientists now accept Kjartansson's hypothesis on the formation of palagonite tuyas, and indeed this is how the island of Surtsey arose out of the sea in the offshore eruption of 1963–1964, and Kjartansson lived long enough to see volcanic material on its surface be transformed into hard palagonite rock. During the compilation of a map of Vatnajökull's subglacial topography in the 1970s (to be discussed later), it soon became clear that there were mountain ridges beneath the glacier that had been formed during subglacial eruptions and had never succeeded in breaking through the surface of the ice to allow their spines to bask in broad daylight. One of these ridges was formed in the eruption to the north of Grímsvötn in 1938, and this was further enlarged during the 1996 eruption.

4.12   Changes in Glaciers and Climate Since the Settlement of Iceland

The opinions of scholars and scientists on changes in the glaciers of Iceland since its settlement were for a long time varied. Bishop Oddur Einarsson reported glaciers were growing larger at the end of the 16th century, Þórður Vídalín stated that more glaciers were expanding than retreating at the end of the 17th century, and Árni Magnússon, Páll Vídalín, Ísleifur Einarsson and Eggert Ólafsson all clearly described the expansion of glaciers over vegetated land at the end of the 18th century as causing the abandonment of farms. Sveinn Pálsson also stated at the end of the 18th century that most people believed the glaciers were indeed advancing, but pointed out that they also sometimes advanced and then retreated, or that some receded whilst others advanced. He indicated that some glacial changes were caused by volcanic activity and some by climate change. The Scottish priest, Henderson, maintained early in the 19th century that the climate of Iceland had changed since the days of the settlement. From Jónas Hallgrímsson's writings it might be assumed that he did not think the climate had changed much since the settlement, but one might also suspect that he had known better and had tried to conceal this as he wanted to encourage and put heart into Iceland's farmers. Þorvaldur Thoroddsen later described alterations in glaciers but tried not to link them with climate change, opining that Iceland's climate had been more or less the same since the settlement with fluctuations between harsher and milder periods. He stated that although many believed that glaciers had expanded considerably since the settlement, he saw no reason in coming to this conclusion from perusing annals and written sources. Pjeturss (1916, 1959), on the other hand, provided strong evidence that glaciers were much larger now than they had been between the time of the settlement and the 13th century. The Frenchman Charles Rabot (1856–1944; 1894), however, believed he could see a connection between the climate in Iceland and the recorded advances of some of its glaciers at the end of the 19th century, most of the glacial data he cited coming from the works of Thoroddsen himself. The geologist Guðmundur G. Bárðarson (1880–1933) collected records that bore witness to considerable changes in glaciers since the settlement period, especially in inhabited areas along the southern margins of Vatnajökull. His conclusions, however, were not published until after his death in 1934. The most comprehensive overview of glacial changes since the end of the 16th century was produced by Sigurður Þórarinsson in 1943, in which he came to the conclusion that many glaciers had reached their most extensive size in the middle of the 18th century, although some had continued to grow larger until the middle of the 19th century. Since then, however, all glaciers had in general retreated and were now smaller than they had been around 1680. Around the same time, the mid-1940s, Þórarinsson (1940) also produced his overview of the size and retraction of glaciers all over the world along with the rising of global sea levels.

During the first quarter of the 20th century, it became indisputably evident that glaciers had shrunk in size since the turn of the century and that this suggested the centuries-long advance of glaciers was coming to an end. An international committee on glaciological research encouraged the regular measuring of glacial advances and retreats and the IMO, recently formed in 1926, issued work schedules that included annual measurements of glacial changes in connection with its climate research. In the spring of 1927, Guðmundur G. Bárðarson and Þorkell Þorkelsson (1876–1961), a physicist and director of the IMO, suggested that the Icelandic Society of Sciences become involved in raising funds for measuring the length of glaciers and the setting up of measuring equipment in front of their termini, so-called 'glacier markers' (Bárðarson 1934, p. 11). The Cultural Fund of Iceland later supported the research projects instigated by the engineer Helgi Hermann Eiríksson (1890–1974) and the meteorologist Jón Eyþórsson (1895–1968). In the summer of 1930, Eiríksson (1932) studied five glaciers in Hornafjörður and recorded how much they had receded from the time of the Danish General Staff map of 1903. During the same summer, Eyþórsson (1931) studied the outlet glaciers of Mýrdalsjökull and Eyjafjallajökull, an outlet glacier of Tindfjallajökull, four glacial tongues of Snæfellsjökull, and, in 1931, four outlet glaciers from Drangajökull. Eyþórsson also recorded the height of the firn line on glaciers and its connection with precipitation and summer temperatures. Since then the systematic measurement of glacial snouts in Iceland, and thus all glacial changes, has been carried out regularly, now under the direction of the Iceland Glaciological Society (Eyþórsson 1963; Rist 1967–1987; Sigurðsson 2005; Sigurðsson et al. 2007).

Beginning the Study of Weather and Mass Balance on Glaciers

Climate warming had become an irrefutable fact by the 1920s, when it had finally become possible to research scientifically the climate changing processes in the atmosphere, in the ocean, and on land. Glaciers retreated, but data on the size of glacial snouts was not sufficient in itself to link this process to climate change, it would also be necessary to examine the accumulation and melting of snow and to observe how climate affects glacial mass-balance, its expansion or shrinkage. It was in this manner that a fuller understanding of the origins of both historic and ancient glaciers was attained. The only available knowledge of weather conditions on glaciers came from the travelogue descriptions of glacier adventurers such as Watts, Wigner and Muir, Wadell and Ygberg, Roberts, and finally Nielsen, who all described important aspects of meteorological conditions on Vatnajökull: frequent storms and almost endless snowfalls in which up to 2 m of snow could fall within three weeks during the spring and summer. Meteorological research on glaciers and in the highlands still needed to answer questions concerning how constant the extreme cold was in this huge mass of snow and ice, and how long did the summer ablation period last?

During the International Polar Year of 1932–1933, Danish and Swiss scientists ran a weather observation post at a height of 825 m on Jökulháls ridge on the eastern flanks of Snæfellsjökull (Kristjánsson and Jónsson 1998). The Austrian Hans Slanar (1890–1955) measured temperatures and wind speeds and observed the summer climate on the central highlands near Gæsavötn to the north of Vatnajökull for a short period in July 1931.

The pioneer of investigating the 'economy' of glaciers—profit and loss of snow/ice—was the Swede Hans Jakob Konrad Wilhelmsson Ahlmann (1889–1974), who on many expeditions introduced the systematic research of the relationship between mass balance and weather conditions around the North Atlantic, and who came to Iceland in 1936 in order to research a glacier in a maritime climate with a large amount of precipitation and a great deal of summer ablation. Vatnajökull was chosen as it lay in the path of low depressions that crossed the Atlantic Ocean and was to a certain extent similar to northwestern Europe during part of the last glacial period. The Swedish-Icelandic expedition onto Vatnajökull of 1936–1938, the largest glacial research expedition and project in Iceland until the 1970s, had first been discussed by Hans Ahlmann and Jón Eyþórsson, the Icelandic meteorologist, in Jötunheim in Norway in 1924–1925, while they were working on weather observations on Mt. Fannaråken, at a height of 2000 m above sea level, and on glacial measurements on the nearby Styggedal glacier (Figs. 4.31, 4.32, 4.33 and 4.34). Ahlmann wanted to use Greenlandic sledge-dogs on Vatnajökull, for they could haul their own victuals (stockfish and liver paste) on low-slung toboggans or pulks, were tough and hardy, and could take care of themselves when snowed under in bad weather. Horses were ill-suited in the soft and slushy snow of early summer, needed a great deal of fodder, had to be shielded in bad weather, and

Fig. 4.31 Expedition leaders Hans Wilhelmsson Ahlmann (1889–1974), geographer, and Jón Eyþórsson (1895–1968), meteorologist, resting at the end of an expedition on Vatnajökull in 1936. Eyþórsson had begun surveying the position of glacial snouts in 1930, a practice still continued by volunteers to this day. He led the French-Icelandic expedition of 1951 that investigated the thickness of Vatnajökull using seismic reflection measurements. Eyþórsson established the Iceland Glaciological Society in 1951 and was its chairman and secretary until his death in 1968. He also did research on the history of climate and the encroachment of sea ice on the shores of Iceland. Carl Mannerfelt, 1936

they could gain no sustenance from eating snow. Dogs could travel faster than horses on a glacier, which was important when there could be sudden changes in the weather and groups had to be ready to set off when the right opportunity arose. But as with so many other expeditions on Vatnajökull, the Swedish-Icelandic one continually met with blizzards, relentlessly drifting snow, thick snowfalls, icing, and then finally torrential rain that resulted in members' clothing being wet through. The main camp they made during this horrendous weather they called Djöflaskarð, or Devil's Pass. The men slept on the pulks to protect themselves from the cold and dampness of the tent floor.

Along with the expedition leaders were two students from Stockholm University, Carl Otto Mannerfelt (1913–2009), who later wrote his doctoral thesis

Fig. 4.32 Jón Jónsson (1900–1941) from Laugar in Biskupstungur, policeman and glacier climber, and Skarphéðinn Gíslason (1895–1974), engineer and collector of glacial data. The photograph was taken in 1936. Carl Mannerfelt

Fig. 4.33 Carl Otto Mannerfelt (1913–2009), Swedish expert in land formation, took part in the Swedish-Icelandic expedition onto Vatnajökull in 1936

Fig. 4.34 Sledge-dogs on Vatnajökull in the summer of 1936. Carl Mannerfelt

on land formations resulting from the melting of ice-age glaciers in the mountains of Sweden and Norway (and who later became CEO of Esselte), and Sigurður Þórarinsson (1912–1983; Fig. 4.35). Also part of the team were the Swede Mac Lilliehöök (1912–1936), who knew how to run sledge-dogs, and Jón Jónsson (1900–1941) from Laugur, an extremely strong man. The members of the expedition lived to a certain extent on newly invented powdered and condensed foods, finding them quite edible. The accumulation of snow on the glacier was measured by digging trenches and drilling through the winter snow and from its thickness and density calculating how much water was bound in its layers, the so-called water equivalent. By fortuitous good fortune, a layer of ash from the 1934 Grímsvötn eruption had been englaciated and so it was easy to calculate the amount of snow that had fallen since the tephra had settled. Summer ablation was measured from yardsticks thrust down into the glacier. The expedition operated on the eastern side of Vatnajökull and calculated accurately the snowfall, ablation rates, movement and drainage of its outlet glacier, Hoffellsjökull, in Hornafjörður. Sigurður Þórarinsson

Fig. 4.35 Sigurður Þórarinsson (1912–1983), Icelandic geologist. The photograph was taken around 1960. Þórarinsson began glaciological research on Skeiðarársandur outwash plain after the Grímsvötn eruption of 1934 and took part in the Swedish-Icelandic Vatnajökull expeditions during the years 1936–1938. He investigated mass balances, ice flows, and the drainage of glaciers, as well as changes in glaciers in previous centuries, establishing the history of jökulhlaups from eruptions in Grímsvötn, Katla and Öræfajökull. His research into tephra layers has been utilised in evaluating glacial movements and the age of glacial ice. Þórarinsson was chairman of the Iceland Glaciological Society from 1969 until his death in 1983

continued these measurements over the summers of 1937 and 1938 with the assistance of local farmers, especially Guðmundur Jónsson (1875–1947) from Hoffell and Skarphéðinn Gíslason (1895–1974) from Vagnstaðir in the Suðursveit district. These research projects were carried out when climatic warming, which began in Iceland in the 1920s, was reaching a high point and this gave them additional value and importance. The results were published by Ahlmann and Þórarinsson in many scientific articles, and Ahlmann (1938) also wrote one of the best travelogues and commentaries on the Icelandic national way of life and character that has ever been written by a foreigner. He was a very competent author and an artistically gifted man, indeed he had studied art in Paris in 1913–1914, but for most of his working life he was an academic and he ended his career as the Swedish ambassador to Norway from 1950 to 1956.

Fig. 4.36 Guðmundur Jónasson (1909–1985), truck and coach driver with his snowmobile Gusi.
Árni Kristjánsson

Iceland Glaciological Society

In the middle of the 20th century, conditions for glaciological research were
significantly improved with the establishment of the Iceland Glaciological
Society (22 November 1950) and the opening up of a route through the
Tungnaöræfi wilderness onto Tungnaárjökull after the mountain truck driver
Guðmundur Jónasson and his associates had discovered a passable ford for
motorised vehicles, the so-called Hófsvað, in the summer of 1950 (Fig. 4.36).
Up until then all journeys onto Vatnajökull had been made from the
Ódáðahraun lava desert or from the lowlands to the south of the glacier,
especially from Kálfafell in the Fljótshverfi district. The formation of the
air-rescue team also gave a shot in the arm to glacier expeditions when the
Skymaster aircraft of Loftleiðir airlines, Geysir, crash-landed on Bárðarbunga
dome in the autumn of 1950. All the passengers and crew on board the
aircraft were rescued, but the wreck of the plane was still on the glacier. An
aircraft with ski landing gear from the NATO Iceland Defence Force, which

was sent to carry out the rescue operation, could not take off again after landing on the glacier at a height of 2000 m and so was left there to be covered by snow (Fig. 4.37). In April the following year, the plane was dug out of the snow and pulled southward down Skaftárjökull to Innri-Eyjar, south of the Laki craters, from where it was possible for the aircraft to take off and fly back to Reykjavík. Skiers and snowmobile drivers with experience of glacial conditions have since taken an active and important part in glaciological research in Iceland.

In its first year of operation, 1951, the IGS instigated the measuring of ice thickness on Vatnajökull with seismic soundings, and on most of its expeditions since 1976 the subglacial topography of Vatnajökull has been explored using radio echo soundings. Ever since its inception, members of the society have taken part in the measuring of snout locations over a wide area of Iceland. With its annual expeditions onto Vatnajökull since 1953, members have observed the battle between ice and fire in Grímsvötn, where the glacier has covered Iceland's most active volcano and where geothermal activity continuously melts ice, causing jökulhlaups. These outburst floods hindered the completion of the national highway or ring road around the country for decades, and indeed they still threaten the bridges on Skeiðarársandur. Members of the society have drilled into the glacier for ice cores and into Grímsfjall for geothermal power, raising a geothermally powered generator there, not to mention a sauna, in the centre of the glacier. The society now has 11 buildings: 9 cabins and two storage sheds. There are two cabins at Jökulheimar at the edge of Tungnaárjökull, one built in 1955 and the other in 1965. In addition there is a garage there from 1958, and a petrol storage facility built in 1963. There are two cabins on Grímsfjall, one from 1957 and the other raised in 1987 (plus storage shed in 1994), a cabin in the Kverkfjöll mountains (from 1977), another on Goðahryggur ridge (from 1979), one in the Esjufjöll mountains (from 1977), and another by Mt. Fjallkirkja (1228 m) on Langjökull, built in 1979. Finally, there is the society's oldest cabin, Breiðá, on Breiðarmerkursandur. The surface area of all these structures covers about 285 m^2, and by building these cabins the society has made expeditions, rescue work, and research on glaciers more accessible and considerably easier. The society also organises lectures on glaciers and publishes a journal, Jökull, its contents a mixture of society news and events along with both general and academic articles.

Fig. 4.37 Rescue aircraft on Bárðarbunga dome of Vatnajökull in 1950. Árni Kjartansson, 1951

4.13   Cartography of Iceland's Glaciers

The turn of the century in 1900 marked no particular watershed in the history of geological research in Iceland, but it was certainly a turning point in glacial cartography. The year that Þorvaldur Thoroddsen published his geological map of Iceland, and a century after the beginning of the second survey of Iceland's coastline (1801–1819), the DGS enthusiastically embarked on compiling a full map of Iceland that was to continue, with short intervals, until the establishment of the Icelandic republic in 1944 (Nørlund 1938, 1944; Böðvarsson 1996; Sigurðsson and Williams 2008). The surveying work was discontinued during World War I, and again between the years 1921 to 1929, after Iceland had gained Home Rule in 1918, but recommenced in 1930. In 1936 the directors of the Danish Geodetic Institute (which had replaced the DGS) appointed the Icelander Steinþór Sigurðsson, who had completed a major task in surveying the uninhabited hinterland of Iceland, to organise the completion of the surveying of Vatnajökull. Supplies were transported to the glacier on horses and the surveyors then continued on skis, hauling sledges with their supplies and equipment and dwelling in strongly-constructed tents. The DGI also had oblique aerial photographs of the glacier taken from aircraft in

1937–1938, but they were so white it was impossible to discern points on the glacial surface which could be used as a basis for the triangulation system of measurement. On the other hand, the actual shapes of the glaciers were discernible on these aerial photographs, and the contour lines on the General Staff's map were thus meant as guidelines rather than exact records of their elevations. The surveying begun by the DGS in 1901 came to an end in 1939, almost four decades after it had begun, and all the maps were fully processed and published before the end of World War II. The maps are for the most part accurate and revealed for the first time the exact expanse of the glaciers at the time of the maps' compilation (1903–1939). The elevation figures were incorrect, on the other hand, sometimes by a few dozen or even a hundred metres (Fig. 4.38).

Vertical aerial photographs taken from the fuselage of an aircraft helped improve cartography greatly. The United States Air Force took such aerial pictures of the greater part of Iceland from a height of 23,000 ft during the years 1945–1946 and the US Army Map Service completed the maps based on these photographs in the years 1948–1949. These maps seem to have been accurate regarding glaciers up to a

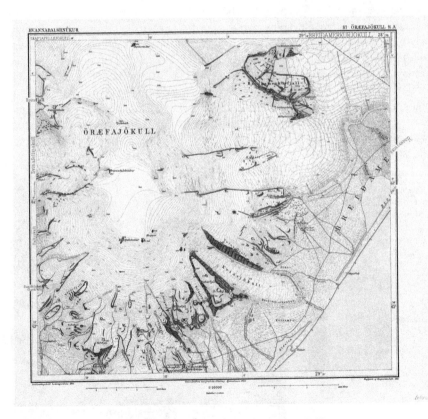

Fig. 4.38 Danish General Staff map of Öræfajökull from 1904. NLSI

height of 1000 m. The USAF took further aerial photographs between 1956 and 1961, only this time from a height of 15–18,000 ft, and in the years 1955 and 1956 a collaborative project involving the Americans, Icelanders and Danes began to establish a triangulation system for the whole country, including the central highlands and glaciers. The first year was one of the wettest summers in southern Iceland in living memory, with foul weather in the central highlands and tremendously cold and overwhelming blizzards on the glaciers that snowed in and almost buried the tents; weeks passed before there was good enough visibility for any surveying. Helicopters were of some use in these circumstances, but there were some mishaps even so, though not involving any injuries to personnel. The surveying projects were successfully completed in the summer of 1956, and many of the energetic assistants on Vatnajökull were members of the newly established IGS, including Guðmundur Jónasson (1909–1985), the mountain-truck driver, with his new Canadian Bombardier snowmobile.

Maps were also compiled of the glaciers' surfaces during a survey of subglacial topographies over the years 1980–2000, discussed in more detail below. Elevations were recorded on a great number of parallel sections at intervals of about 1 km and a map compiled by bridging the surfaces between these measuring lines. There were some gaps in the maps where it had been impossible to traverse the ice due to crevasses. At first the elevations were recorded with exact barometric pressure readings, but they were later calculated with the use of space satellites and the global positioning system, or GPS. A map of the surface of Hofsjökull was made in 1984, of the majority of Vatnajökull in the years 1980–2000, of Mýrdalsjökull in 1991 and Langjökull in 1997. There were also maps compiled by the US Army from aerial photographs taken in the period 1980–1987 (Fig. 4.39). They were very detailed and accurate of the whole of Mýrdalsjökull and over a wide area of other glaciers up to a height of 1000 m. The National Land Survey of Iceland and several national

Fig. 4.39 Aerial photograph of Múlajökull piedmont glacier, 14 August 1986. NLSI

employers' institutes such as the IRCA, the INEA, (previously the Electricity Supply Authority) and Landsvirkjun, the NPCI, have also made maps of individual glacial snouts, as have some foreign institutes, e.g. the University of Glasgow in Scotland. The company Loftmyndir ehf. ('Aerial Pictures') has also compiled maps of ablation zones of a good many glaciers through the automatic processing of digital aerial photographs. From 1972 onwards, distant scanning from satellites has been utilised in demarcating the margins of glaciers (Williams 1983, 1986; Williams and Þórarinsson 1973) and also, in more recent years, in the making of contoured maps (e.g. the western part of Vatnajökull and Langjökull from SPOT-satellites, in collaboration with the French, with an accuracy up to 1–2 m. Figures 4.40 and 4.41). In 1998 a very accurate contour-map of the western parts of Vatnajökull was made with radar measurements (SAR) from an aircraft (Magnússon et al. 2004, 2005). Maps of various parts of the Vatnajökull, Mýrdalsjökull, Hofsjökull, Langjökull and Drangajökull glaciers have also been revised from accurate GPS-calculations. An attempt is being made to revise all glacier maps every few years because glaciers are being constantly transformed in response to climate change, surges, volcanic activity, and because of subglacial geothermal heating. Even an evaluation of glacial change due to differing drainage patterns can be deduced from a comparison of maps from different periods of time. As a contribution to the International Polar Year of 2008–2009, the IMO and IESUI instigated a programme of airborne lidar mapping of all the glaciers of Iceland. This was to provide a basis for all future comparisons of changes in Icelandic ice masses and was completed in 2013 (Jóhannesson et al. 2013).

Fig. 4.40 Satellite image of Vatnajökull from a height of 700 km on 4 August 1999. Öræfajökull, Skeiðarárjökull, Síðujökull, Tungnaárjökull, Köldukvíslarjökull, Grímsvötn and the southern part of Bárðarbunga dome and the Kverkfjöll mountains are all clearly visible. Landsat7

Fig. 4.41 Satellite image from 16 August 2006 showing the four main glaciers of Iceland from a height of 700 km (Langjökull, Hofsjökull, Vatnajökull and Mýrdalsjökull). MODIS

4.14 Hidden Subglacial Topography Revealed

An important turning point in the history of subglacial topography occurred in the middle of the 20th century when, in the spring of 1951, two decades after British students had made a vain attempt to measure the thickness of the Vatnajökull ice cover, a French-Icelandic expedition travelled widely over Vatnajökull and measured the thickness of its ice in 33 locations using seismic reflection surveys (Eyþórsson 1951a, b, c, 1952; Joset and Holtzscherer 1954). Dynamite was buried a metre deep into the glacier's surface and detonated, and the time it took for the sound wave to travel to the glacier bed and back to the surface was recorded on a seismograph. It was then possible to calculate the depth of the glacier because the speed of the sound wave through ice is already known (about ten times the speed of sound in air). The expedition was directed by the meteorologist Jón Eyþórsson and the French geophysicist Alain Joset, and accompanying them were Stéphane Sanvelian, the driver of the snowmobile, Árni Stefánsson, a motor mechanic, who filmed the expedition, and Sigurjón Rist (1917–1994; Fig. 4.42), the pioneer of regular hydrologic surveys in Iceland from 1947 onwards, and he who measured the winter snowfall levels over the whole glacier. They ascended Breiðamerkurjökull and for just over a month they drove in two snowmobiles ('weasels') over a great part of Vatnajökull, marking a turning point in transportation methods on glaciers in

Fig. 4.42 Sigurjón Rist (1917–1994), hydrologist. He laid the groundwork for collecting hydrologic data in Iceland by keeping records of avalanches, conditions of glacial snouts, and measuring the thickness of Vatnajökull and Mýrdalsjökull (echo soundings). Rist had a world-wide reputation for his measuring of the flow rates of jökulhlaups, and he also measured the depths of all the lakes in Iceland. He was chairman of the Iceland Glaciological Society from 1983 until 1987

Iceland. The French left Iceland to continue measuring the thickness of glaciers in Greenland, where Alain Joset, together with the Danish engineer Jens Jarl, died while doing research on 4 August 1951. Research collaboration with the French continued in the summer of 1955 with a two-week project measuring ice thickness on Tungnaárjökull and Grímsvötn and later on the ice plateau of Mýrdalsjökull. Five years then passed until employees of the State Electricity Authority calculated from gravitational measurements the average thickness of the glacier at Grímsvötn and from there to the Kverkfjöll mountains on one of the spring expeditions of the IGS (Pálmason 1964; Sigurðsson 1970).

Research into the topography lying beneath glaciers threw new light onto their mode of existence. The Norwegian Theodor Kjerulf and Þorvaldur Thoroddsen had collected a great deal of data and knowledge on the snowline of Iceland's glaciers and Josef Keindl had previously noted that although Hofsjökull and Langjökull must be supported by mountains that rose above the central highland plateau, they nonetheless did not reach the elevation of the firn line. Hofsjökull and Langjökull had thus been formed in a colder climate than which now exists and had been maintained ever since by their own height. They were probably remnants from the last glacial period and if they were to disappear, e.g. as a result of volcanic eruptions, they would never reach their former size again in the current climate. During the French-Icelandic expedition on Vatnajökull, it came to light that over a wide area there was a 500–600 m thick ice cap above the highland plateau, which is 400–500 m lower than the firn lines of the glacier. If the glacier were to suddenly disappear, only about six small glaciers would come into being once more in the current climate on the peaks and domes that rise above the plateau: on Öræfajökull, Grímsfjall, the Háabunga, Breiðabunga and Bárðarbunga domes, and Kverkfjöll.

There was a turning point in investigating the thickness of glaciers in Iceland during the years 1976–1978, when geologists from the Science Institute of the University of Iceland began to employ a device that sent radio waves down to the glacial bed and successfully recorded their echoes reaching the surface of the ice. Such radio echo soundings had been utilised for about ten years on the frozen glaciers of Greenland and Antarctica, but had not until then been successful on temperate glaciers in which meltwater seeped through the ice. The cause of this was deemed to be that the radio waves became weakened as they travelled through thawing ice so that it was necessary to direct energy with a very narrow beam through the glacier. In order to do this the frequency had to be increased and the wavelengths shortened. But this still had no results by 1975 until it was suggested that the cause was not the weakening of the radio waves, but that their energy was irregularly dispersed when encountering water-filled cavities in the glacier. The echo from the glacial bed could not be discerned from other irregular echoes from within the glacier. The shortening of radio waves had thus in fact worsened matters, but when radio waves were made longer than the diameter of water channels and cavities, they reached down to a depth of 900–1000 m of a thick temperate glacier with little interference and the echo could be distinguished at the surface. The echo sounder sends out 5 MHz mono-pulses, 34 m long. A thousand pulses travel through the ice every second (at just over half the speed of light in a vacuum) and by measuring the time it takes for the wave to reach the glacier bed and return to the surface, the distance travelled through the ice can be measured and its thickness thus calculated. The echo sounding device is drawn by a snowmobile or snow-sledge across the glacier's surface and it views a continuous strip on the glacier's bed directly below the measuring line (200 m broad if the glacier is 600 m thick) and the ice thickness is recorded continuously on photographic film (Fig. 4.43). This was a revolution from when estimating ice depths could only be achieved by the time consuming method of the controlled explosions of dynamite at

Fig. 4.43 Cross section of radio echo sounding image that reveals a mountain 500 m high beneath Tungnaárjökull outlet glacier (Chap. 8)

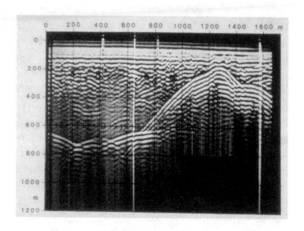

certain points on the glacier's surface. The first echo sounding trips were made according to a compass and a distance measuring apparatus, but the radio waves were later guided by a Loran device from stations at Hellissandur, Jan Mayen and the Faeroe Isles, and they are now guided by satellites in the Earth's orbit at a height of 1000 km (GPS). The height of the glacier was also calculated when driving along the measuring lines, and the height of the glacier's bed was found by deducting the thickness of the ice from the elevation of the glacier's surface. At first such heights were calculated from barometric measurements, but today's satellites can coordinate measurements with a margin of error of less than a metre. The measuring vehicle is driven along parallel lines, often with a distance of 500–1000 m in between, and by bridging the planes with calculations that come nearest to matching the measured values, a map can be compiled of both the glacier's thickness and the topography of its bed (Björnsson 1987, 1988, 2009; Sverrisson et al. 1980).

For the entire fourth quarter of the 20th century, radio echo soundings were made every April and May while snow covered the crevasses and before any serious melting began and swathes of slush formed on the glaciers. Thus a gradual knowledge was acquired, not only of the terrain beneath glaciers, of their mountains, valleys and land formations, but also of the water stored within the glaciers, the boundaries of water and ice divides of specific rivers in a glacier, and the source and channels of jökulhlaups from lakes, geothermal areas, and subglacial volcanic eruptions. Up until 1980, tents were used for housing researchers with varying conditions, but since then mobile cabins have been hauled around the glacier and this has improved productivity immensely.

Glaciological research has thus changed considerably from the days when men ascended glaciers on foot, harnessed to their supplies and equipment, which they pulled behind them or had them carried on horses or dog-drawn sledges, or since the days when Gnúpa-Bárður had all his livestock haul his baggage over the ice. Motorised vehicles increased efficiency, and they were succeeded by snowmobiles large and small, Bombardier caterpillar trucks, snow grooming machines, skidoos and finally off-road jeeps and even helicopters (Figs. 4.44 and 4.45); navigation equipment has also changed from compasses to Loran devices and satellite sensors. Specialised equipment revealed what the human eye could not see. The fitness and endurance of glacier researchers increased with improved clothing and camping equipment, wind- and waterproof synthetic materials replacing woollen clothing, fur and down sleeping-bags, and primus stoves for cooking gave way to gas-fired equipment. Fords across unbridged glacial rivers still need to be chosen carefully, for vehicles can become stuck in muddy areas, and there is still no real shelter on a glacier's plateau from storms and blizzards. The days are long gone, however, when most of researchers' time and energy could be wasted on struggling to survive in adverse weather conditions in leaky tents, only to be exhausted when the time came to begin scientific work.

Fig. 4.44 Snowmobiles haul mobile cabins across the glaciers. Finnur Pálsson

Fig. 4.45 Making a quick trip onto Langjökull in a helicopter to collect data. Eiríksjökull in the background. HB, October 2002

4.15 Main Emphasis of Glaciological Research at the Beginning of the 21st Century

From the 1960s onwards, a great deal of research has been directed at accumulating the basic facts and data concerning the glaciers of Iceland, how they maintain themselves through snow accumulation, how much they melt in the summer, the volume of water they discharge into glacial and groundwater rivers, and to what extent they move and change from year to year. Every spring and autumn, expeditions have gone up onto the glaciers to measure their mass balances and to examine their links with various meteorological features with the use of automatic weather monitoring stations. The transport of ice down to the ablation zone and to rivers has also been calculated from measuring the ice flow velocity of outlet glaciers.

It has been estimated that in the near future, if the climate continues to grow warmer in the North Atlantic, the mass balance of glaciers will be reduced, the utilisation of glacial and groundwater rivers and geothermal heat will change, and the sea level will rise; land will be uplifted near the glaciers once the overburden pressure is removed from the Earth's crust, and the erosion of land along the shoreline will increase, while the transportation of glaciofluvial deposits out to sea will decrease. The dispersal of precipitation will change, and consequently so will the areas and magnitude of vegetation. The effect on society and its inhabited locations will be enormous.

Glaciologists thus prepare models defining the current glaciers, their size and movements and connections with their mass balances and their recorded links to meteorological factors. All this information is then used to calculate the mass balance and response of glaciers and the water drainage from them in the coming years according to given scenarios related to probable climate changes. Water divides will change and the sources of rivers will migrate. If the forecasts prove correct, the glaciers will mostly disappear in the next 200 years, while the drainage from them might double during the years 1990 to 2050.

Research on the history of climate change and its effects on glaciers in Iceland are an important contribution to international scientific knowledge. In Iceland, in the middle of the North Atlantic, the oceanic and wind currents spawned in the tropical belt meet those of the polar region, so that the history of the climate, fauna and flora of the Quaternary and Tertiary periods are recorded in the various layers of soil and deposits of Iceland, both on land and under the sea. It is precisely in Iceland that extremely active glaciers have responded rapidly to climate change and moulded landscape and ocean beds through erosion and sediment deposits, making it an important research laboratory for studying climate change due to processes in the atmosphere and ocean before the pollution caused by mankind began to take effect. A knowlege and understanding of these processes is vital to be able to predict and evaluate climate change in the future.

Iceland's Glaciers as one Dynamic Research Laboratory

The data on glaciological changes in Iceland over the last millennium is substantial because of the close, centuries-long proximity of Icelanders to glaciers. Ice sheets in Iceland reveal the properties of temperate glaciers from the end of the last glacial period. The influence of global warming can also be monitored here, as it is anticipated that it will take place more rapidly in the northern regions than over the globe as a whole; this is partly due to diminishing snow cover and sea ice, which in turn reduces solar reflection from the Earth. Glaciers outside of the polar regions are thus the most sensitive ones to climate change. It is now believed that a lasting increase in temperature of about 1 °C will bring about a 15–20 % reduction in size of the glaciers outside the polar regions, and that their average mass balances will shrink about 0.8 m a year. Although they only cover around 4 % of the world's glacial areas, their shrinking has caused a third of the rising of world sea levels since the beginning of the 20th century by more or less 1–2.5 mm/year. Temperate glaciers are at freezing point so that meltwater on their surfaces does not freeze when it seeps into them but flows unhindered out to sea. Nowhere is there greater precipitation or ablation than on these glaciers and in the coming years it is estimated that they will shrink even more rapidly due to the influence of the so-called greenhouse effect, the latest models predicting that the global sea level might rise by around 50 cm by the year 2100.

When looking back over the history of glaciological research, it can be seen that Icelanders have made vital contributions in many of its fields. Up until 1800, there was no place on Earth where there was a greater understanding of the nature of glaciers than in Iceland. The oldest history and records of glacial movements are preserved in Iceland, and the first theories concerning how they did indeed move also emerged here. No better examples of jökulhlaups, glacial surges and the effects of subglacial volcanic eruptions can be found other than in Iceland. It was in this country that the mystery of the formation of palagonite rock was solved and important data provided in the search for traces of ice-age glaciers. The future history of Icelandic glaciers could well be a short one, but it will contain important lessons for the polar regions that will continue to enjoy a cohabitation with glaciers for some time to come.

References

Ægisdóttir, H., & Þórhallsdóttir, Þ. (2005). Theories on migration and history of the North-Atlantic flora: A review. *Jökull, 54*, 1–16.

Ahlmann, H. (1938). *Land of Ice and Fire*. Lewes, K. & Lewes, L. (Trans.) London: Kegan Paul, Trench & Trubner. Original Swedish version: *På skidor och till häst i Vatnajökulls rike*. Stockholm: Norstedt, 1936.

Anderson, S. (Trans.) (1997). *Bard's Saga* [Bárðarsaga Snæfellsás] In *The Complete Sagas of Icelanders*. Hreinsson, V. (ed.) Vol. 2 pp. 237–266. Reykjavík: Leifur Eiríksson.

Árnason, J. (1961). *Íslenzkar þjóðsögur og ævintýri*. 6 vols. In: Á. Böðvarsson, & B. Vilhjálmsson, (Eds.), Reykjavík: Þjóðsaga, Hólar.

Backlund, H. (1939). Islandprobleme. *Geologische Rundschau*, 30, 625–630.

Bárðarson, G. (1906). Purpura Lapillus L. I hævede Lag paa Nordkysten af Island og Mærker efter Klima- og Niveauforandringer ved Húnaflói, Nord-Island. *Videnskabelige Meddelelser fra Naturhistorisk Forening i Köbenhavn* (pp. 177–185).

Bárðarson, G. (1921). Fossile skalaflejringer ved Breiðafjörður í Vest-Island. *Geologiska Föreningens i Stockholm Förhandlingar*, 22, 323–380.

Bárðarson, G. (1923). Fornar sjávarminjar við Borgarfjörð og Hvalfjörð. *Rit Vísindafélags Íslendinga*. 1. Reykjavík.

Bárðarson, G. (1932). Jarðmyndanir á Snæfellsnesi. *Árbók Ferðafélags Íslands*. pp. 60–66.

Bárðarson, G. (1934). Islands Gletscher. Beitrage zur Kenntniss der Gletscherbewegungen und Schwankungen auf Grund alter Quellenschriften und neuesten Forschung. *Rit Vísindafélags Íslendinga, 16*. Reykjavík.

Bemmelen, R., & Rutten, M. (1955). *Table Mountains of Northern Iceland*. Leiden: E. J. Brill.

Benediktsson, G. (1977). Frá Hornafirði til Bárðardals yfir Vatnajökul. *Jökull*, 27, 100–108.

Benediktsson, J. (Ed.) (1968). *Íslendingabók. Landnámabók*. In *Íslenzk fornrit*. Vol. 1. Reykjavík: Hið íslenska fornritafélag.

Björnsson, H. (1979). Glaciers in Iceland. *Jökull, 29*, 74–80.

Björnsson, H. (1987). Könnun á jöklum með rafsegulbylgjum. In Þ. Sigfússon (Ed.), *Í hlutarins eðli* (pp. 279–292). Reykjavík: Menningarsjóður.

Björnsson, H. (1988). Hydrology of ice caps in volcanic regions. *Rit Vísindafélags Íslendinga, 45* (Reykjavík).

Björnsson, H. (2009). *Jöklar á Íslandi*. Reykjavík: Opna.

Böðvarsson, Á. (1996). *Landmælingar og kortagerð Dana á Íslandi. Upphaf Landmælinga Íslands*. Reykjavík: Landmælingar Íslands.

Bruun, D. (1902a). Sprengisandur og Egnen mellem Hofs- og Vatnajökull. Undersøgelser foretagne I Sommeren 1902. *Geografisk Tidsskrift, 16*, 23–46.

Bruun, D. (1902b). Ved Vatna Jökulls Nordrand. Studier af Nordboernes Kulturliv 4.1. *Geografisk Tidsskrift*. Copenhagen: Nordiske Forlag.

Bruun, D. (1914). Islænderferder til Hest over Vatna-Jökull i ældre Tider. *Geografisk Tidsskrift, 22*, 4–13.

Bruun, D. (1921–1927). *Turistruter paa Island. 5 vols*. 1: Reykjavik og kysten rundt; 2: Udflugter fra Reykjavik; 3: Gennem beboede egne; 4: Fjeldveje gennem Islands indre Højland; De øde Egne nord for Vatna-Jökull. Copenhagen: E. Jespersen & Gyldendal.

Bunsen, R. (1847). Beitrag zur Kenntniss des islandischen Tuffgebirges. *Annalen der Chemie und Pharmacie, 61*, 265–279.

Einarsson, O. (1971) [1585]. *Íslandslýsing - Qualiscunque descriptio Islandiae*. Pálsson, S. (trans.) Reykjavík: Bókaútgáfa Menningarsjóðs.

Einarsson, Þ. (1960). Geologie von Hellisheiði (SW-Island). *Sonderveröffentlichungen des Geologischen Institutes der Universität Köln. 5*.

Einarsson, Þ. (1961). Pollenanalytische Untersuchungen zur spät- und postglazialen Klimageschichte Islands. *Sonderveröffentlich-ungen des Geologischen Institutes der Universität Köln. 6*.

Einarsson, Þ. (1963). Pollen-analytical studies on the vegetation and climate history of Iceland in Late and Post-Glacial times. In Á. Löve & D. Löve, (Eds.), *North Atlantic Biota and their History* (pp. 355–365).

Einarsson, Þ. (1968). *Jarðfræði. Saga bergs and lands*. Reykjavík: Mál og Menning.

Eiríksson, H. (1932). Observations and measurements of some glaciers in Austur-Skaftafellssýsla in the summer 1930. *Rit Vísindafélags Íslendinga, 12* (Reykjavík).

Eiríksson, J. & Schøning, G. (1981) [1772]. Nyt Carte over Island. In E. Ólafsson & B. Pálsson, (Eds.), (1981) *Ferðabók Eggerts Ólafssonar og Bjarna Pálssonar um ferðir þeirra á Íslandi árin 1752–1757*. 2 vols. Steindórsson S. (trans.) Reykjavík: Örn og Örlygur. Inside cover.

Eiríksson, J. & Olavius, Ó. (1964–65) [1780]. Nyt Carte over Island. In Ó. Olavius (Ed.), *Ferðabók: landshagir í norðvestur-, norður- og norðaustursýslum Íslands 1775–1777*. 2 vols. Steindórsson, S. (trans.) Reykjavík: Bókfellsútgáfan. p. 144.

Erkes, H. (1911). Das isländische Hochland zwischen Hofsjökull und Vatnajökull. *Petermanns Geographische Mitteilungen. Part II., 57*, 140–143.

Erkes, H. (1914). Der Anteil der Deutschen an der Erforschung Inner Islands. *Die Erde, 12*, 225–228.

Espolín, J. (1825). *Íslands Árbœkur í Sögu Formi*. 4 vols. Copenhagen: Hið Íslenska Bókmenntafélag.

Eyþórsson, J. (1931). On the present position of the glaciers in Iceland. Some preliminary studies and investigations in the summer 1930. *Rit Vísindafélags Íslendinga, 10*.

Eyþórsson, J. (1945). Skýringar. In J. Eyþórsson, P. Hannesson, & S. Steinsdórsson, (Eds.), *Ferðabók Sveins Pálssonar. Dagbœkur og ritgerðir 1791–1797* (pp. 747–770). Reykjavík: Snælandsútgáfan.

Eyþórsson, J. (1951a). Fransk-íslenzki Vatnajökulsleiðangurinn, marz-apríl 1951. *Jökull, 1*, 10–14.

Eyþórsson, J. (1951b). Jöklamælingar 1950 og 1951. *Jökull, 1*, 16.

Eyþórsson, J. (1951c). Þykkt Vatnajökuls. *Jökull, 1*, 1–6.

Eyþórsson, J. (1952). Landið undir Vatnajökli. *Jökull, 2*, 1–4.

Eyþórsson, J. (1963). Variation of Iceland glaciers 1931–1960. *Jökull, 13*, 31–33.

Gadde, O. (1983). *Íslandsferð sumarið 1857*. Erlingsson, G. (trans.) Akranes: Hörpuútgáfan.

Geirsdóttir, Á., & Eiríksson, J. (1994). Sedimentary facies and environmental history of the late-glacial glaciomarine Fossvogur sediments in Reykjavík, Iceland. *Boreas, 23*, 164–176.

Geirsdóttir, Á., Harðardóttir, J., & Eiríksson, J. (1997). The depositional history of the Younger Dryas-Preboreal Búði moraines in south-central Iceland. *Arctic and Alpine Research, 29*(1), 13–23.

Gíslason, V. (2008). Personal communication by e-mail; material as yet unpublished.

Guðmundsson, J. (1936) [1650]. Áradalsóður. In *Huld: Safn alþýðlegra frœða íslenzkra*. Vol. 2. (2nd ed.) pp. 48–62.

Guðmundsson, S. (1999). *Þar sem landið rís hœst. Örœfajökull og Örœfasveit*. Reykjavík: Mál og menning.

Guðmundsson, Þ. (1839). Félagsins ástand og athafnir. *Skírnir, 13*, 72–83.

Gunnarsson, S. (1949) [1877]. Miðlandsöræfi Íslands. In J. Eyþórsson, & Hannesson P. (Eds.), *Hrakningar og heiðavegir*. Akureyri: Norðri. pp. 214–244.

Gunnarsson, S. (1950) [1876]. Um öræfi Íslands. In J. Eyþórsson, & P. Hannesson (Eds.), *Hrakningar og heiðavegir*. Vol. 2. Akureyri: Norðri. pp. 196–209.

Gunnlaugsson, B. (1835). Um fund Þórisdals. *Skírnir, 9*, 104–107.

Gunnlaugsson, B. (1836). Um Þórisdal. *Sunnanpósturinn, 2*(8), 113–126.

Gunnlaugsson, B. (1848). *Uppdráttr Íslands*. Scale 1:480,000. Reykjavík & Copenhagen: Hið Íslenska bókmenntafélagið.

Gunnlaugsson, B. (1849). *Uppdráttr Íslands*. Scale 1:960,000. Copenhagen: Hið Íslenska bókmenntafélagið.

Gunnlaugsson, B. (1953). Kannaður Þórisdalur. In J. Eyþórsson, & P. Hannesson, (Eds.), *Hrakningar og heiðavegir*. Vol. 3. pp. 137–148.

Hallgrímsson, J. (1989). *Ritverk Jónasar Hallgrímssonar [Works of Jónas Hallgrímsson]* 4 vols. Hannesson, H., Valsson, P. & Egilsson, Y. (Eds.). Reykjavík: Svart á hvítu.

Heer, O. (1868). *Flora fossilis arctica*. Zürich: F. Schulthess.

Helland, A. (1882). Om Islands Jøkler og om Jøkulelvenes Vandmængde og Slamgehalt. *Archiv for Mathematik og Naturvidenskab, 7*, 200–232.

Helland, A. (1883a). Om Vulkaner i og under Jøkler på Island og om Jøkulhlaup. *Nordisk Tidsskrift, 6*, 368–387.

Helland, A. (1883b). Islændingen Sveinn Pálssons Beskrivelser af islandske Vulkaner og Bræer; Meddelte af Amund Helland. *Den Norske Turistforenings Årbog for, 1882*, 19–79.

Henderson, E. (1818). *Iceland; or the Journal of a Residence in that Island during the Years 1814 and 1815*. Edinburgh: Oliphant, Waugh & Innes.

Hjartarson, Á. (1989). The ages of the Fossvogur layers and the Álftanes end-moraine, SW-Iceland. *Jökull, 39*, 21–31.

Hjartarson, Á., & Ingólfsson, Ó. (1988). Preboreal glaciation of Southern Iceland. *Jökull, 38*, 1–16.

Hooker, W. (1811). *Journal of a tour in Iceland in the summer of 1809*. London: Vernon, Hood & Sharpe.

Hoppe, G. (1953). Några iakttagelser vid islandska jöklar sommaren 1952. *Ymer, 73*, 241–265.

Hoppe, G. (1968). Grímsey and the maximum extent of the last glaciation of Iceland. *Geografiska Annaler, 50*, 16–24.

Hoppe, G. (1982). The extent of the last inland ice sheet of Iceland. *Jökull, 32*, 3–11.

Hoppe, G. (1995). Brúarjökull. *Glettingur, 5*(2), 38–41.

Horrebow, N. (1758). *The natural history of Iceland*. Anderson, J. (Trans.) London: A. Linde.

Jóhannesson, T. & Björnsson, H., et al. (2013) Icelandic Met Office Progress Report on airborne lidar mapping of the ice surfaces of Icelandic glacires and ice caps. Web. Accessed December 2013. http://www.vegagerdin.is/Vefur2.nsf/Files/Maelingar_yfirbord_isl_jokla_lidar-framv2013/$file/M%C3%A6lingar_yfirbor%C3%B0_%C3%ADsl_j%C3%B6kla_lidar-framv2013.pdf.

Joset, A., & Holtzscherer, J. (1954). Expedition Franco-Islandaise au Vatnajökull, mars-avril 1951. Resultats des sondages seismiques. *Jökull, 4*, 1–33.

Kålund, P. (1882). *Bidrag til en historisk-topografisk beskrivelse af Island*. Copenhagen: Gyldendal.

Kålund, P. (1984–1986). *Íslenzkir sögustaðir*. 4 vols. Matthíasson, H. (trans.) Reykjavík: Örn og Örlygur.

Keilhack, K. (1883). *Vergleichende Beobachtungen an isländischen Gletscher und nord-deutschen Diluvial-Ablagerungen*. Jahrbuch des preussischen Landesanstalt und Bergakadamie. Berlin. pp. 159–176.

Keilhack, K. (1885). *Reisebilder aus Island*. Reisewitz: Gera A.

Keilhack, K. (1886). Beiträge zur Geologie der Insel Island. *Zeitschrift der Deutschen Geologischen Gesellschaft, 38*, 376–449.

Keilhack, K. (1934). Riesiger Gletscherrückgang in Nordwest-Island von 1844–1915. *Zeitschrift für Gletscherkunde, 21*, 365–370.

Kjartansson, G. (1939). Stadier i Isens Tilbagerykning fra det sydvestislandske Lavland. Skuringmerker. En isdaemmet sö. Marine dannelser. Postglacial tektonik. *Meddelelser fra Dansk Geologisk Forening, 9*(4), 426–458.

Kjartansson, G. (1943). Yfirlit og jarðsaga. In G. Jónsson, (Ed.), *Árnesingasaga*. *Náttúrulýsing Árnessýslu*. Vol.1. Reykjavík: Árnesingafélagið í Reykjavík. pp. 1–249.

Kjartansson, G. (1945). Íslenzkar vatnsfallategundir. *Náttúrufræðingurinn, 15*(2), 113–126.

Kjartansson, G. (1960). *Jarðfræðikort af Íslandi*. Map 3. Suðvesturland. Scale: 1:250,000. Reykjavík: Náttúrugripasafn Íslands.

Kjartansson, G. (1961). Móbergsmyndunin. In V. Gíslason & S. Þórarinsson (Eds.), *Náttúra Íslands* (pp. 31–66). Almenna bókafélagið: Reykjavík.

Kjartansson, G. (1962). *Jarðfræðikort af Íslandi*. Map 6. Miðsuðurland. Scale: 1:250,000. Reykjavík: Náttúrugripasafn Íslands.

Kjartansson, G. (1964). Ísaldarlok og eldfjöll á Kili. *Náttúrufræðingurinn, 34*, 9–38.

Kjartansson, G. (1965). *Jarðfræðikort af Íslandi*. Map 5. Mið-Ísland. Scale: 1:250,000. Reykjavík: Náttúrugripasafn Íslands.

Kjartansson, G. (1966a). Nokkrar nýjar C^{14} aldursákvarðanir. *Náttúrufræðingurinn, 36*, 126–141.

Kjartansson, G. (1966b). Stapakenninginn og Surtsey. *Náttúrufræðingurinn, 36*, 1–34.

Kjartansson, G. (1968). *Jarðfræðikort af Íslandi*. Map 2. Mið-Vesturland. Scale: 1:250,000. Reykjavík: Náttúrugripasafn Íslands.

Kjartansson, G. (1969). *Jarðfræðikort af Íslandi*. Map 1. Norðvesturland. Scale: 1:250,000. Reykjavík: Náttúrugripasafn Íslands.

Kjerulf, T. (1853). *Bidrag til Islands geognostiske Fremstilling*. Naturvidenskaberne: Nyt Magasin f. 7.

Knebel, W. (1905). Der Nachweis verschiedener Eiszeiten in den Hochflachen des Inneren Islands. *Centralblatt für Mineralogie, Geologie und Paleontologie, 6*, 546–553.

Knebel, W. (1906). Zur Frage der diluvialen Vergletscherungen auf der Insel Island. *Centralblatt für Mineralogie, Geologie und Paleontologie, 8*, 232–237.

Knebel, W. (1912). *Island. Eine naturwissenschaftliche Studie*. Stuttgart: E. Schwitzerbart.

Koch, J. (1905–1906). Fra Generalstabens topografiske Avdelings Virksomhed paa Island. *Geografisk Tidsskrift, 18*, 1–14.

Koch, J. (1912). Den danske Ekspedition til Dronning Louises land og tværsover Nordgrønlands Indlandsis 1912–1913. Rejsen tværsover Island i Juni 1912. *Geografisk Tidsskrift, 21*, 257–264.

Kristjánsson, L. (2001). Historical notes on earth science research in Iceland by German expeditions, 1819–1970. In G. Schröder, (Ed.), Geschichte und Philosophie der Geophysik, IAGA IDC History Newsletter No. 42. pp. 138–154.

Kristjánsson, L., & Jónsson, T. (1998). Alþjóða-heimsksautaárin tvö og rannsóknarstöðin við Snæfellsjökul 1932–33. English summary: The two International Polar Years and the field research station at Snæefellsjökull, W-Iceland 1932–33. *Jökull, 46*, 35–47.

Larson, L. (Trans.) (1917). *The King's Mirror*. [13th Cent.] Library of Scandinavian Literature, vol. XV. New York: Twayne.

Leiviskä, I. (2005) *Ísland. Land frosts og funa*. Kjærnested, B. (trans.) Reykjavík: Hið íslenska bókmenntafélag.

Lindroth, C. (1931). Die Insektenfauna Islands und ihre Probleme. *Zoologiska bidrag fran Uppsala, 13*, 105–599.

Lindroth, C. (1965). *Skaftafell Iceland. A Living Glacial Refugium*. Copenhagen: Munksgaard.

Magnússon, Á. (1955) [ca. 1702]. *Chorographica Islandica*. In Ó. Lárusson (Ed.), *Safn til sögu Íslands*. Second series. Vol. 2. Reykjavík: Hið íslenska bókmenntafélag. pp. 1–120.

Magnússon, E., Björnsson, H., Pálsson, F. & Dall, J. (2004). Glaciological application of InSAR topography data of W-Vatnajökull aquired in 1998. *Jökull, 54*, 17–36.

Magnússon, E., Björnsson, H., Dall, J., & Pálsson, F. (2005). Volume changes of Vatnajökull ice cap, Iceland, due to surface mass balance, ice flow, and sub-glacial melting at geothermal areas. *Geophysical Research Letters, 32*(5), L05504.

Maizels, J. (1991). The origin and evolution of Holocene sandur deposits in areas of jökulhlaup drainage, Iceland. In J. Maizels & C. Caseldine (Eds.), *Environmental change in Iceland: Past and present* (pp. 267–279). Dordrecht: Kluwer Academic Publishers.

Mathiesen, H. (1846). Über die Entstehung des Monte Nuovo und die neueste Hekla-Eruption. *Neues Jahrb. f. Mineral. Geognosie*. pp. 586–595.

Mathiesen, H. (1864). Strömningernes Beliggenhed og Drivisens Forekomst I de nordlige Atlanterhav og tilgrænsede Polarhave. *Tidsskrift for Søvæsen*. 9. 12 pp.

Mørch, O. (1869). Faunaula molluscorum Islandiæ. Oversigt over Islands Bløddyr. *Videnskab. Medd. fra den naturhist. Foren i København for, 1868*, 185–227.

Muir, T. (1905a). The physical geography of Iceland. *Scottish Geographical Magazine, 31*, 254–257.

Muir, T. (1905b). Notes on the weather on the Vatnajökull during August and September 1904. *Journal of Scottish Meteorological Society, 13*, 33–37.

Muir, T., & Wigner, J. (1953). Gönguför yfir þveran Vatnajökul fyrir hálfri öld. *Vísir. Christmas issue*. pp. 7–8, 25–29.

Nielsen, N. (1933). Contributions to the physiography of Iceland with particular reference to the highlands west of Vatnajökull. *D. Kgl. Danske Vidensk. Selsk. Skrifter. Naturvidensk. og Mathem. Afd.* 9 Række. IV.5. pp. 183–286.

Nielsen, N. (1937a). *Vatnajökull. Kampen mellem Ild og Is*. Copenhagen: H. Hagerup.

Nielsen, N. (1937b). A volcano under an ice-cap. Vatnajökull, Iceland, 1934–36. *The Geographical Journal, 90*, 6–23.

Nielsen, N., & Noe-Nygaard, A. (1936). Om den islandske "Palagonitformation"s Oprindelse. En foreløbig Meddelelse. *Geogr. Tidsskrift., 39*, 89–122.

Noe-Nygaard, A. (1940). Sub-glacial volcanic activity in ancient and recent times. Studies in the Palagonite-System of Iceland 1. *Folia Geographica Danica*. 67 pp.

Norðdahl, H. (1991). Late Weichselian and Early Holocene deglaciation history of Iceland. *Jökull* , *40*, 27–50.

Norðdahl, H., & Pétursson H. (2005). Relative sea-level changes in Iceland: new aspects of the Weichselian deglaciation of Iceland. In C. Caseldine, A. Russell, J. Harðardóttir, & Ó. Knudsen (Eds.), *Iceland. Modern processes and past environments* (pp. 25–78). Amsterdam: Elsevier.

Nørlund, N. (1938). Denmark. *Photogrammetric Engineering, 4*(3), 119–120.

Nørlund, N. (1944). *Islands Kortlægning. En historisk Fremstilling.* Geodætisk Instituts Publikationer. 7. Copenhagen: Munksgaard.

Okko, V. (1956). Glacial drift in Iceland, its origin and morphology. *Acta Geographica, 15*, 1–133.

Ólafsdóttir, N. (1975). Úr bréfum Sveins læknis Pálssonar. *Árbók Landsbókasafns* (pp. 10–39).

Ólafsson, E., & Pálsson, B. (1981) *Ferðabók Eggerts Ólafssonar og Bjarna Pálssonar um ferðir þeirra á Íslandi árin 1752–1757.* Steindórsson S. (trans.) Reykjavík: Örn og Örlygur.

Olavius, Ó. (1964–1965). *Ferðabók: landshagir í norðvestur-, norður- og norðaustursýslum Íslands 1775–1777: ásamt ritgerðum Ole Henckels um brennisteinsnám og Christian Zieners um surtarbrand. Íslandskort frá 1780 eftir höfund og Jón Eiríksson.* 2 vols. Steindórsson, S. (Trans.) Reykjavík: Bókfellsútgáfan.

Paijkull, C. (1866). *En Sommar på Island. Reseskildring.* Stockholm: Albert Bonnier.

Paijkull, C. (1867). Bidrag till kännedomen om Islands bergsbyggnad. *Kgl. svenska Vetenskaps Akademiens Handlingar, 7*(1), 50.

Paijkull, C. (2014) [1868]. *A Summer in Iceland.* Barnard, R. (Trans.) Swedish original: *En Sommar på Island. Reseskildring.* 1866. Stockholm: Albert Bonnier. Cambridge: Cambridge University Press.

Pálmason, G. (1964). Gravity measurements in the Grímsvötn area. *Jökull, 14*, 61–66.

Pálsson, S. (1945) *Ferðabók Sveins Pálssonar. Dagbækur og ritgerðir 1791–1794.* In: Eyþórsson, J., Hannesson, P. & Steindórsson, S. (Trans. and Eds.). Reykjavík: Snælandsútgáfan.

Pjeturss, H. (1900). The glacial palagonite-formation of Iceland. *The Scottish Geographical Magazine, 16*, 265–293.

Pjeturss, H. (1901). Moræner i den islandske Palagonitformation. *Oversigt over Det Kgl. Danske Videnskabernes Selskabs Forhandlinger, 5*, 147–171.

Pjeturss, H. (1902). Fortsatte Bidrag til Kundskab om Islands "Glaciale Palagonit-Formation". *Geol. Fören. i Stockholm Förhandlinger, 24*, 367.

Pjeturss, H. (1903). On a shelly boulder-clay in the so-called Palagoniteformation of Iceland. *Quarterly Journal of the Geological Society, 59*, 356–361.

Pjeturss, H. (1904). Om nogle glaciale og interglaciale Vulkaner. *Oversigt over Det Kgl. Danske Videnskabernes Selskabs Forhandlinger.* pp. 211–267.

Pjeturss, H. (1905a). *Om Islands geologie.* Doctoral dissertation. University of Copenhagen.

Pjeturss, H. (1905b). Das Pleistocän Islands. Einige Bemerkungen zu den vorläufigen Mitteilungen Dr. W. v. Knebels. *Centralblatt f. Mineralogie, Geologie und Palaeontologie, 24*, 740–745.

Pjeturss, H. (1906a). Zur Forschungsgeschichte Islands. Einige Worte, durch die Entgegnung Dr. W. v. Knebels hervorgerufen. *Zentralblatt für Mineralogie, Geologie und Palaeontologie. 18.* pp. 566–68. (Also in: *Monatsberichten der Deutschen Geologischen Gesellschaft*, pp. 274–287).

Pjeturss, H. (1906b). Nokkur orð um loftslagsbreytingar á Íslandi og orsakir þeirra. *Andvari, 31*, 149–164.

Pjeturss, H. (1908). Einige Hauptzüge der Geologie und Morphologie Islands. *Zeitschrift der Gesellschaft für Erdkunde zu Berlin.* pp. 451–467.

Pjeturss, H. (1909). Über marines Interglazial in der Umgebung von Reykjavík. (Plus two maps). *Zeitschrift der Deutschen Geologischen Gesellschaft, 61*, 274–287.

Pjeturss, H. (1910). Island. *Handbuch der Regionalen Geologie, 4*, 1–22.

Pjeturss, H. (1959) [1916]. Jöklar á Íslandi í fornöld. *Skírnir.* 90. pp. 429–430. Rpt in *Ferðabók dr. Helga Pjeturrs.* 1959. Reykjavík: Bókfellsútgáfan. pp. 215–218.

Price, R. (1969). Moraines, sandur, kames and eskers near Breiðamerkurjökull, Iceland. *Transactions of the Institute of British Geographers, 46*, 17-43.

Price, R. (1971). The development and destruction of a sandur, Breiðamerkurjökull, Iceland. *Arctic and Alpine Research, 3(3)*, 225–237.

Price, R. (1982). Changes in the proglacial area of Breiðamerkurjökull, southern Iceland: 1890– 1980. *Jökull, 32*, 29–35.

Rabot, C. (1906). Les variations des glaciers de l'Islande méridionale de 1893-1894 á 1903-1904 d'apres la nouvelle carte d'Islande. *Zeitschrift für Gletscherkunde, 1*, 132–138.

Reck, H. (1910). Ein Beitrag zur Spaltenfrage der Vulkane. *Centralblatt f. Min, 6*, 166–169.

Rist, S. (1967–1987). Jöklabreytingar (Glacier variations in metres). *Jökull*, 17–37.

Scheuchzer, J. (1723). *Itinera per Helvetiae alpinas regiones facta annis 1702–1711.* Leyden: Peter van der Aa.

Schwarzbach, M. (1983). Deutsche Islandsforscher im 19. Jahrhundert - Begegnungen in der Gegenwart. *Jökull, 33*, 25–32.

Schythe, J. (1847). *Hekla og dens sidste Udbrud, den 2den September 1845.* Copenhagen.

Scudder, B. (Trans.) (2002). *Egil's Saga.* London: Penguin.

Scudder, B. (Trans.) (2005). *Grettir's Saga.* London: Penguin.

Sigurðsson, H. (1978). *Kortasaga Íslands frá lokum 16. aldar til 1848.* Reykjavík, Bókaútgáfa Menningarsjóðs og Þjóðvinafélagsins.

Sigurðsson, O. (2005). Variations of termini of glaciers in Iceland in recent centuries and their connection with climate, In C. Caseldine, A. Russell, J. Harðardóttir & Ó. Knudsen (Eds.), *Iceland. Modern processes and past environment* (pp. 241–255). Amsterdam: Elsevier.

Sigurðsson, O., Jónsson, T., & Jóhannesson, T. (2007). Relation between glacier-termini variations and summer temperature in Iceland since 1930. *Annals of Glaciology, 42*, 395–401.

Sigurðsson, O., & Williams, R. (2008). *Geographic names of Iceland's glaciers: historic and modern.* Reston, VA: U.S. Geological Survey.

Sigurðsson, S. (1970). Gravity survey on Western Vatnajökull. *Jökull, 20*, 38–44.

Sigurðsson, V. (1948). *Um þvert Grænland með kapt. J.P. Koch 1912–1913.* Reykjavík: Ársæll Árnason.

Sonder, R. (1938). Zur magmatischen und allgemeinen Tektonik von Island. *Schweiz. Mineralogische Mitteilung, 18*, 429–436.

Spethmann, H. (1908). Der Nordrand des islandischen Inlandeises Vatnajökull. *Zeitschrift für Gletscherkunde, 3*, 36–43.

Spethmann, H. (1912). Forschungen am Vatnajökull auf Island und Studien uber seine Bedeutung für die Vergletscherung Norddeutschlands. *Zeitschrift. d. Gesellschaft f. Erdkunde.* Berlin. pp. 414–433.

Steindórsson, S. (1981). *Íslenskir náttúrufræðingar.* Reykjavík: Menningarsjóður.

Sverrisson, M., Jóhannesson, Æ., & Björnsson, H. (1980). Radio-echo equipment for depth sounding of temperate glaciers. *Journal of Glaciology, 25*(93), 477–485.

Thorarensen, B. (1935) *Kvæði,* Vol. 1. Helgason, J. (ed.) Copenhagen: Hið íslenzka fræðafélag í Köbenhavn.

Thoroddsen, Þ. (1892). Islands Jøkler i Fortid og Nutid. *Geografisk Tidsskrift 11*, 111–146.

Thoroddsen, Þ. (1892–1904). *Landfræðissaga Íslands.* [Geological History of Iceland]. 4 vols. Copenhagen: Hið íslenska bókmenntafélag.

Thoroddsen, Þ. (1901). *Jarðfræðikort af Íslandi.* Map. Scale 1:600.000. Copenhagen: Carlsberg Fund.

Thoroddsen, Þ. (1905/06). Die Gletscher Islands. Island, Grundriss der Geographie und Geologie. *Petermanns Geographische Mitteilungen, 152*, 1–161; 153, pp. 162–358.

Thoroddsen, Þ. (1907–1911). *Lýsing Íslands.* 2 vols. Copenhagen: Hið íslenzka Bókmentafélag. (Reprinted 1932).

Todtmann, E. (1951/1952). Im Gletscherrückzugsgebiet des Vatna Jökull auf Island, 1951. *Neues Jahrbuch für Geologie und Paläontologie.* 1951 (11). pp. 335–341; 1952 (9). pp. 401–411.

Todtmann, E. (1953). Am Rand des Eyjabakkagletschers 1953. *Jökull, 3*, 34–37.

Todtmann, E. (1955a). Kringilsárrani, das Vorfeld des Brúarjökuls, am Nordland des Vatnajökull. *Jökull, 5,* 9–10.

Todtmann, E. (1955b). Übersicht über die Eisrandlagen in Kringilsárrani 1890–1955. *Jökull, 5,* 8–10.

Todtmann, E. (1957). Kringilsárrani, das Vorfeld des Brúarjökull, am Nordrand des Vatnajökull. *Neues Jahrbuch für Geologie und Paläontologie, 104,* 255–278.

Todtmann, E. (1960). Gletcherforschungen auf Island (Vatnajökull). *Universität Hamburg Abhandlungen aus dem Gebiet der Auslandkunde* (Naturwissenschaften). Bd. 65, Reihe C, Bd. 19.

Tómasson, H. (1976). The sediment load of Icelandic rivers. *Nordic Hydrological Conference, 5,* 1–16 (Reykjavík).

Tómasson, H. (1990). Glaciofluvial sediment transport and erosion. In Y. Gjessing, J. Hagen, K. Hassel, K. Sand, & B. Wold, (Eds.), *Arctic hydrology. Present and future tasks. Hydrology of Svalbard—Hydrological problems in cold climate.* Report 23. Oslo: Norwegian National Committee for Hydrology, pp. 27–36.

Torell, O. (1857). Bref om Island. *Öfversigt af Kgl. Vetenskaps-Akademiens förhandlingar.* 18. pp. 325–332. *Vatnajökull. Barátta elds og ísa.* Hannesson, P. (Trans.) Reykjavík: Mál og menning.

Vídalín, P. (1965a) [1754]. Afterword to German translation of *Jöklarit* by Vídalín Þ. [Abhandlung von den isländischen Eisbergen] *Hamburgisches Magazin* 2. pp. 9–27, and 5. pp. 197–218. Ásmundsson G. (Trans.) Reykjavík, Ferðafélag Íslands. pp. 39–44.

Vídalin, Þ. (1965b) [1695]. *Jöklarit [Thesis on Glaciers].* Ásmundsson G. (Trans.) Reykjavík, Ferðafélag Íslands. (Original Latin title: Dissertationcula de montibus Islandiæ Chrystallinis). pp. 19–37.

Waltershausen, W. (1847). *Physische-geographische Skizze von Island mit besondere Rücksicht auf vulkanische Erscheinungen* (p. 85). Göttinger Studien, Göttingen: Handenhoek und Ruprecht.

Watts, W. (1875). *Snioland or Iceland, its Jökulls and Fjalls.* London: Longman.

Watts, W. (1876). *Across the Vatna jökull or, Scenes in Iceland.* London: Longman.

Wigner, J. (1905). The Vatna Jökull traversed from North-East to South-West. *Alpine Journal, 22,* 436–448.

Williams, R. (1983). Satellite glaciology of Iceland. *Jökull, 33,* 3–12.

Williams, R. (1986). Glacier inventories of Iceland. Evaluation and use of sources of data. *Annals of Glaciology, 8,* 184–191.

Williams, R. & Þórarinsson, S. (1973). ERTS-1 image of the Vatnajökull area: General comments. *Jökull, 23,* 1–6.

Þórarinsson, S. (1937). The main geological and topographical features of Iceland. *Geografiska Annaler, 19,* 161–175.

Þórarinsson, S. (1940). Present glacier shrinkage and eustatic changes of sea-level. *Geografiska Annaler, 22,* 139–159.

Þórarinsson, S. (1965). Þórður Þorkelsson Vídalín og jöklarit hans. *Vídalín's Jöklarit* (pp. 11–16). Ferðafélag Íslands: Reykjavík.

Part II
The Glaciers of Iceland

Chapter 5
Glaciers of Southern Iceland

Mýrdalsjökull, Eyjafjallajökull, Tindfjallajökull, Torfajökull and Hekla

The finest leaf in our laurel band, Katla, greatest wonder of this land ....

(Ólafsson 1832, p. 200).

Abstract This chapter has a detailed description of the types, shapes, sizes, locations, and all relevant scientific data of the glaciers in southern Iceland, as well as their recorded local histories, folklore and legends: Mýrdalsjökull, Eyjafjallajökull, Tindfjallajökull, Torfajökull and Hekla. Three of them are well-known for their glacier-volcanic interaction during both historical and recent times. Beneath Mýrdalsjökull (540 km^2) is the Katla caldera, which has erupted 20 times since the settlement of Iceland and produced some of the most tremendous jökulhlaups on Earth, some of which are described here. Katla has also been the inspiration for many stories and legends in Icelandic culture. The last eruption of Eyjafjallajökull (70 km^2), in 2011, produced ash that caused problems for air traffic worldwide. Tindfjallajökull (11 km^2) covers a dormant volcano; its caldera was partly formed in a catastrophic event during the last glaciation. Torfajökull (10 km^2) is situated in one of the largest geothermal areas in Iceland, and from the 15th to 18th centuries was shrouded in myths concerning highland bandits. The volcano has not erupted since the last glaciation, but has shown some activity with injections of magma through fissures to the northeast. Hekla, the most famous volcano in Iceland, has small glaciers on its summit and northern ridge and has produced jökulhlaups during eruptions. It has erupted 20 times since 1104, most recently in 1991.

5.1 Mýrdalsjökull Ice Cap

No glacier poses a greater threat to Iceland than Mýrdalsjökull, (Figs. 5.1, 5.2, 5.3 and 5.4) for beneath it lies hidden the country's most infamous volcano, Katla, which has burst through its glacial cover twenty times since the settlement of Iceland, causing enormous, rampant jökulhlaups, that have swept over gravel plains, roads, bridges, farmsteads, and cultivated land (Figs. 5.5 and 5.6). Its

© Atlantis Press and the author(s) 2017
H. Björnsson, *The Glaciers of Iceland*, Atlantis Advances
in Quaternary Science 2, DOI 10.2991/978-94-6239-207-6_5

211

Fig. 5.1 Aerial view from above Kötlujökull westward over Mýrdalsjökull. In the distance are Eyjafjallajökull, Tindfjallajökull and Hekla. Ragnar Axelsson

volcanic eruptions, accompanied by deafening barrages of thunder and lightning, have destroyed human life and livestock in adjacent rural areas. The greatest volume of its volcanic ash fall is within the first twenty-four hours, but eruptions have lasted as long as two weeks and even up to four months (e.g. Þórarinsson 1975; Larsen 2000). The ash fall has been so thick that people could hardly see each other at arm's length, sails on ships 600 km east of the country have become pitch black, grazing lands in the Faroe Isles have been damaged, and the streets of Bergen in Norway have been covered in ash. Katla's jökulhlaups are among the largest meltwater floods on Earth, which, together with its extensive ash falls, makes it the most dangerous volcano in Iceland. As I write these words, Katla has been biding its time for a longer period than ever before since the end of the 11th century. The last time Katla erupted was in 1918 (Jóhannsson 1919; Sveinsson 1919; Hannesson 1934; Karlsson 1994; Tómasson 1996; Figs. 5.7, 5.8 and 5.9a, b).

Katla and Jökulhlaups Katla meltwater floods attain an enormous flow rate within a few hours (100–300,000 m³/s), many times the maximum of the Skeiðarársandur jökulhlaup in 1996, although Katla's main floods have lasted only for 10–12 h with a few frequent flood surges later. Shortly after the volcano's column of ash and fire bursts through the ice cap and reaches into the skies, a surge of water rushes from beneath the glacier's snout and through its surface, harrowing long ice gorges deep into the glacier. An hour later, the flood's frontal wave will have reached as far as 20 km on its way to the sea. An example of such a torrent can be found in the jökulhlaup of 11 May 1721, when the inhabitants of Höfðabrekka

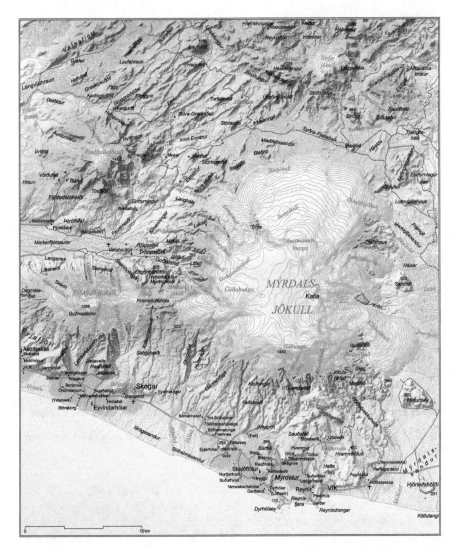

Fig. 5.2 Map of Eyjarfjallajökull and Mýrdalsjökull and their environs. HB 2009

farm had to ascend Hákollur (Háafell), to the north of the farm, to be able to see over the jumbled blocks of ice to the farmstead of Hjörleifshöfði, which was then at the foot of the headland. In 1625 the meltwater flood was said to be 15 fathoms (28 m) deep in the channels in the middle of Mýrdalssandur, and even ocean-going ships were able to float in hollows near the farmstead of Þykkvibær in the Álftaver district, 25 km from the glacier, indicating depths of at least 5 m. During the jökulhlaup of 1660, the waters rose as high as 150 m as they reached up to Klofagil gully on Hafursey (Fig. 5.2).

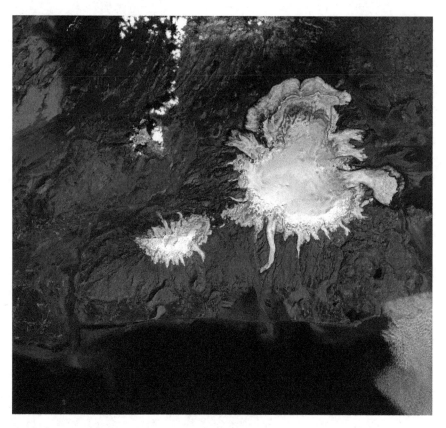

Fig. 5.3 Satellite picture of Mýrdalsjökull, Eyjafjallajökull and Tindfjallajökull taken from a height of 820 km on 14 August 2004. SPOT

Since the settlement of Iceland (ca. 870 A.D.), no less than 18 of the 20 erup-tions traced to Mýrdalsjökull have resulted in meltwater floods rushing southeast-ward onto Mýrdalsandur. All of these jökulhlaups have flooded southward, to the west of Hafursey and Hjörleifshöfði, and eastward, from beneath the centre of the glacier into Álftaver and the Skálm and Landbrotsá rivers, but the largest floods have also emerged in Kriki and from there into the Leirá and Hólmsá rivers. Over the last 4000 years, all of several dozen meltwater floods have flowed from beneath Sólheimajökull down onto Sólheimasandur and Skógasandur, but only twice since the settlement of Iceland. At least ten prehistoric jökulhlaups in the last 8000 years are believed to have flowed northwestward from beneath Entujökull, gouging out the precipitous sides of Markarfljótsgljúfur ravine and dispersing sediment over the Markarfljót river plains and Landeyjar region as far west as the river Þjórsá. There may also have been lesser jökulhlaups during volcanic eruptions from beneath the more westerly Merkurjökull, Krossárjökull and Tungnakvíslarjökull, which might have flowed into the Ljósá, Þröngá and Krossá rivers. Huge, dry channels, ravines

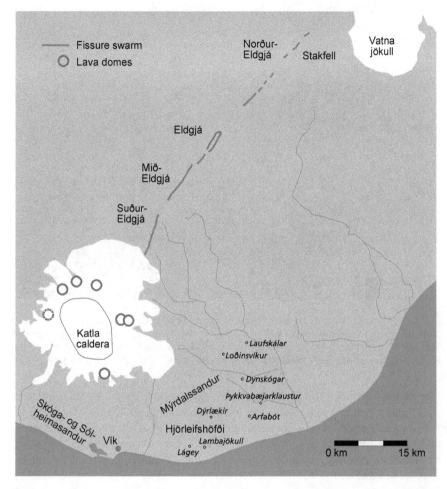

Fig. 5.4 A long chain of volcanic fissures stretches 80 km from the Katla caldera through the Eldgjá gorge almost to Vatnajökull. The Mýrdalssandur plain had vegetation during the time of Iceland's settlement. The farm names bear witness to a community destroyed by Katla's meltwater floods

and gullies, scoured lava fields, thick layers of sediment, and scattered boulders at some distance from present watercourses all bear witness to these meltwater floods.

Katla jökulhlaups have flooded all of the main part of Mýrdalssandur, flowing on both sides of Hafursey, surrounding Hjörleifshöfði (a 221-m-high mountain of palagonitic tuff), reaching as far east as the Skálm river and into Álftaver and Meðalland, and also northeast from Kriki into the Leirá and Hólmsá rivers. Traces reveal that meltwater floods have also emerged from the outlet glaciers to the north of Kötlujökull as far as Öldufellsjökull and down onto Mælifellssandur from the sections of the Eldgjá fissures covered by a glacier (Fig. 5.4). From contemporary descriptions of jökulhlaups, it can be ascertained that high, leading flood waves had

Fig. 5.5 Múlakvísl braided river on the western side of Mýrdalssandur. Water converges under the bridge from the gravelly plains before fanning out and harrowing its way through vegetated land. In the background, Mýrdalsjökull is hidden by clouds. The bridge was swept away by a jökulhlaup, but replaced by a temporary one within 96 hours, in July 2011. HB, August 1999

moved at a speed of 6–20 km/h, later increasing rapidly (20–60 km/h), the torrent transporting volcanic ash, mud, sand, sludge, basal till, and boulders torn up from beneath the glacier. The sludge becomes separated and settles, land rises, and old channels are filled with mud, but the floodwaters surge onwards. Icebergs break off from the glacier's margins and are borne onward in the currents of sludge right out into the sea and then float offshore, some reaching as far away as Reykjanes, the southwest peninsula of Iceland. Blocks of ice have run aground on the outwash plains, where they melt and create kettle holes. So much sediment is transported out to sea that the shoreline is extended for hundreds of metres after each eruption. After the 1660 flood, fishing boats could no longer be launched from Víkurklettar rocks and Skiphellir cave, and a dry, shingle beach reached as far south of Höfðabrekkuháls, where fishing boats had previously been moored in waters 20 fathoms deep (38 m). In 1721 an ice floe ran aground where there had once been waters 150 m deep. There are examples of jökulhlaups having created tsunamis out at sea, flood waves coming ashore at Þorlákshöfn and Grindavík, and whisking away boats, shacks and fish-oil barrels in the Vestmannaeyjar (Westman) islands. A sudden collapse of offshore sedimentary deposits piled up on shallow seabeds could also have caused an underwater landslide, resulting in coastal flood waves travelling quite a distance at an alarming speed.

Where Mýrdalssandur is today, there were once lush, cultivated and prosperous farmlands at the time of the settlement (Sveinsson 1947). Nearly forty known

Fig. 5.6 The barren Mýrdalssandur stretches 15 km from the grass-grown Hjörleifshöfði headland to the base of Kötlujökull. During Katla jökulhlaups there would be one constant sea of water between the headland and the glacier. Hafursey, with its good grazing pastures (to the right, 582 m), rises singly from the plain. HB, May 2004

farmsteads were there, and such place-names as Dynskógar, Laufskálar and Loðinsvíkur indicate the land's fertility. Soon after Iceland had accepted Christianity, there were four parishes in the area: Höfðahverfi, between Hjörleifshöfði and Hafursey, Álftaver, further east by the Kúðafljót river, Dynskógahverfi to its northwest, and Lágeyjarhverfi to its south and west. There had been very few Katla jökulhlaups down onto Mýrdalssandur in the centuries previous to the settlement of Iceland, but once the area had been settled and cultivated the farmers soon began to suffer from them, and by the end of the 15th century only Álftaver parish remained, as it still does, even though Katla jökulhlaups have often destroyed vegetation there and threatened farmsteads as far east as Meðalland. The floods flushed away pastures and grassy fields, livestock died in the sludge, and even woods were smothered in volcanic ash. The jökulhlaups were followed by destructive sand and dust storms. There has been no loss of human life in an eruption of Katla since lightning struck and killed two people at Skaftártunga in 1755, although the sheriff of Skaftafell County died with three others in a surging floodwave in the Landbrotsá in Álftaver almost two months after an eruption had ended in 1823.

Sediment deposits have extended Mýrdalssandur after every jökulhlaup (Figs. 5.10, 5.11 and 5.12). The Hjörleifshöfði headland is now a single-standing

Fig. 5.7 Katla erupts, bursting through a glacier 400 m thick. A cloud of volcanic ash and embers rises to a height of 10 km. Kjartan Guðmundsson, 1918

mountain 2.5 km from the sea at Kötlutangi spit, which had been created by deposits from a Katla jökulhlaup. Up until the 14th century, the sea had reached the headland where the settler Hjörleifur Hróðmarsson probably landed shortly after the year 870. The *Book of Settlements* (from the early 12th century) states: 'in those days there was a fjord stretching right up to the headland,' referring to Kerlingarfjörður to the west of Hjörleifshöfði (Pálsson and Edwards 1972, p. 20). After the last Katla jökulhlaup in 1918, Kötlutangi became the southernmost point of Iceland, having been extended about 3 km, so that the shoreline now existed where previously the ocean had been 20 fathoms deep. In some jökulhlaups, huge blocks of ice covered in sand were piled up on the plain at the edges of the floodwaters. The ice has long since disappeared, but traces of the blocks are still visible in the form of gravel ridges. One of them, called Höfðabrekkujökull, came into existence in 1755 to the east of the Kerlingardalsá, in front of Höfðabrekka, and was up to 17 m high, 1000 m long, and 600 m wide. Another is Austurjökull, to the east of Múlakvísl near Kötlujökull, and even further east is Lambajökull, between Hjörleifshöfði and Bólhraun. Árni Magnússon (1663–1730) described Lambajökull at the beginning of the 18th century as 'gravel mounds and not a glacier' (Magnússon 1955, p. 27). The title of

Fig. 5.8 In 1918 the volcanic cloud could be seen from Ægissíða at Holt as rising midway between Þríhyningur and Eyjarfjallajökull. This postcard removes any doubt as to the volcano's location. Anon

Fig. 5.9 (**a** and **b**) A Katla meltwater flood bears with it gigantic blocks of ice which are strewn across Mýrdalssandur; some of them are 20–40 m high. Kjartan Guðmundsson, 1918

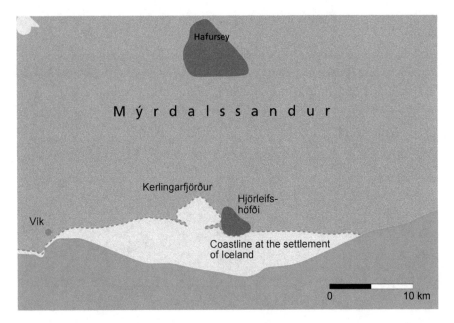

Fig. 5.10 When Iceland was first settled, the sea reached as far inland as Hjörleifshöfði. It is believed that the then existing Kerlingarfjörður was navigable by ship. Katla jökulhlaups have since filled in this fjord and extended the coastline by as much as 5 km

glacier had stuck because the ice had persisted in the piles for decades. In the jökulhlaup of 1755, the blocks of ice that had lain under the glacial till from 1721, resurfaced and were rotated once again. Contemporary accounts of Katla eruptions and jökulhlaups written by Magnússon (1907–1915), Salómonsson (1907–1915), Sigurðsson (1907–1915), Stefánsson (1907–1915), Þorleifsson and Guðmundsson (1907–1915), Guðmundsson (1907–1915), Steingrímsson (1907–1915) and Pálsson (1907–1915) were collected in *Safn til Sögu Íslands* (1907–1915) [Collected Writings on the History of Iceland] (see References).

5.1.1 *Mýrdalsjökull and Its Subglacial Volcanic System*

Mýrdalsjökull is Iceland's fourth largest glacier, covering around 540 km^2 of land. It is also the country's southernmost glacier and rises steeply from gravel plains only a short distance from the sea up to a height of 1200–1500 m. A huge amount of precipitation is accumulated on its ice cap as masses of moisture come ashore, rapidly ascend in height and then cool. At around 1300 m above sea level, there is a 60 km^2 glacial plateau, surrounded by elevations that rise a further 100–200 m above it. They are Hábunga (1497 m) to the south, Kötlukollar (1320 m) to the east, the nunatak Austmannsbunga (1377 m) to the north, and Goðabunga (1510 m)

Fig. 5.11 Fishing boats were launched from Skiphellir cave on Höfðabrekka up until 1660. This ancient boat-house is now 2 km from the sea. HB, July 1999

to the west, which is connected to the Eyjafjallajökull volcano through Fimmvörðuháls ridge. The proximity of a dozen ice cauldrons up to 30 m deep and a few hundred metres in diameter, all distributed around the glacial plateau, bear witness to constant subglacial geothermal activity. The main ice divides lie along the high verges of the elevations, and between them the outlet glaciers advance from the plateau down onto the lowlands: Sólheimajökull, southward onto Sólheimasandur between Hábunga and Goðabunga, Entujökull, northwestward between Goðabunga and Austmannsbunga, and Kötlujökull (also called Höfðabrekkujökull), eastward towards Mýrdalssandur. Kötlujökull is the largest of these glaciers, accumulating ice over the whole of its plateau and flowing more than 20 km on its way down onto Mýrdalssandur from its ice divide with Entujökull. Broad snouts descend eastward to a height of 200–400 m, and one large tract forms its northern part down to a height of 600–650 m. Sléttjökull and Botnjökull advance to the north of the glacial plateau, while Sandfellsjökull transports ice to the northeast of Kötlukollar. In the south, steep outlet glaciers descend along the narrow ravines of Jökulsárgil, Klifurgil and Vesturgil, while in the northwest Tungnakvíslarjökull, Krossárjökull and Merkurjökull advance to the tributary sources of the Markarfljót. There is a huge amount of meltwater discharge in the summer and many rivers flow from the glacier.

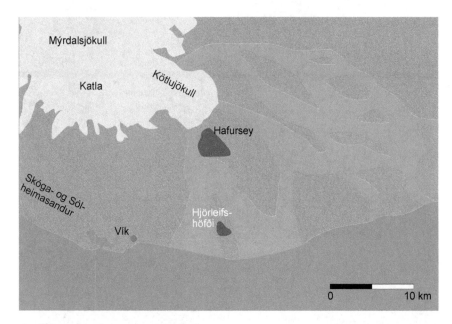

Fig. 5.12 The Katla jökulhlaup of 1918 emerged from beneath the glacier in three main arteries and spread widely over Mýrdalssandur as far east as Meðalland, 30 km from the glacier's margin. HB, July 2008

Mýrdalsjökull lies at the southern end of the eastern volcanic belt and covers one of Iceland's most active volcanoes. At its southern end is the Katla caldera, and stretching northeast from it, and a fair way towards Vatnajökull glacier, is an 80-km-long fissure swarm through the Eldgjá gorge. This volcanic system has been active for 200,000 years (Þórarinsson 1967; Robson 1957; Sæmundsson 1979, 1982; Jakobsson 1979a, b). Katla has been its centre and has had explosive eruptions that have often spewed into the air alkaline volcanic material, although there is evidence of about twenty pre-settlement eruptions that produced acidic tephra. At other places along this line of fissures, there have been eruptions with tremendous lava flows, such as from Eldgjá in 934 (which produced 19 km^3 of volcanic material). The Katla caldera could have been formed during the last glacial period, and the large strata of ash that can be found over a wide area of the North Atlantic are from this volcanic system and have been identified in ice cores from boreholes in Greenland as being from about 75,400 and 77,500 years ago. An explosive eruption that spewed out more than 10 km^3 of rhyolite tephra about 12,000 years ago (10,600 B.C.) could have caused the roof of the magma chamber to sink, enlarging the crater substantially, if not mostly creating it. Red-hot, glowing ash flooded down the southern sides of the volcano and formed the ignimbrite rock formations associated with Sólheimar and widely visible in the Mýrdalur valley. Ash was borne more than 1500 km away to Scandinavia and is widely found in sediment strata in the North Atlantic all the way from Greenland to Norway, as far south as latitude 45°N, and is named after Vedde, a short distance from Ålasund in Norway, where it was discovered in freshwater sediments; here in Iceland, however, it is known as the Skógar-sediment layer from a farm in Fnjóskadalur. In modern times there could have been between 150 and 200 eruptions along the whole volcanic system. It is estimated that the total volume of tephra from subglacial eruptions, especially from the caldera, is 30–35 km^3, and there was an additional 15 km^3 of lava flows, for the most part from the swarm of fissures. There are two suboceanic ridges, Kötluhryggir, stretching 70 nautical miles to the south of the edge of the country's continental shelf, which might also have been formed from volcanic material from Katla (Hammer 1984; Lacasse et al. 1995; Grönvold et al. 1995; Mortensen et al. 2005; Óladóttir et al. 2005, 2008).

The mountain massif of Mýrdalsjökull is from palagonitic tuff that has been formed mostly under the glacier during the second half of the last glacial period and which has been greatly rifted and eroded by ice and water. Around the caldera are rhyolite domes that have slowly been pushed up to the surface as slow-flowing lava: Entukollar, Austmannsbunga, Eystri- and Vestari-Kötlukollur, and Gvendarfell.

5.1.2 Katla: Stories and Legends

The first written sources on Mýrdalsjökull are in connection with volcanic eruptions and jökulhlaups. The oldest narratives are rather obscure, but in the light of modern knowledge it is usually possible to determine the probable course of events. The *Book of Settlements* states: 'When Lodmund was an old man, another sorcerer, Thrasi, was living at Skógar. It happened one morning that Thrasi saw a great flood of water, and by means of his witchcraft he directed the flood east to Solheimar. Lodmund's slave saw this and told him the sea was flooding the land from the north. Lodmund had gone blind by then, and he told his slave to get a basin and bring him a sample of what he called sea-water. When the slave came back, Lodmund said, "This doesn't seem like sea-water to me." He told the slave to lead him to the flood. "Put the point of my staff into the water," he said. There was a ferrule on the staff, and Lodmund held the staff with both hands and bit the ferrule. Then the flood began to turn westward back to Skogar. In this way, each of the sorcerers kept directing the flood away from his farm, until they met each other at a certain ravine. So then they came to an agreement that the river should flow where the distance to the sea was shortest. The river is now called Jokuls River, and forms the Quarter boundary.' [Sólheimasandur was created by that deluge. The Quarter boundary is there with the Jokuls River in the middle.] (Pálsson and Edwards 1972, p. 116; Nordal 1928). Sólheimasandur and Skógasandur, however, are now believed to have been formed by jökulhlaups prior to the settlement of Iceland.

The Saga of Ólafur Tryggvason relates the trials and tribulations of the Christian missionary Þangbrandur who came to Iceland in 997: 'On the day when Hall and Thangbrand rode from Kirby [Kirkjubær], the ground opened suddenly under Thangbrand and his horse sank into the cleft; he himself sprang off the horse's back, and escaped by God's aid and the help of his companions' (Sephton 1895, p. 345). *Kristnisaga* describes these events as follows: 'In the summer, Þangbrandr rode to the Althing with Hallr. But when they came to Skógahverfi, the heathens paid a man called Galdra-Héðinn to make the ground fall away beneath Þangbrandr. On the day they rode away from Kirkjubær from the home of Surtr, son of Ásbjörn, son of Ketill the foolish—all his forebears on the father's side were baptised—then Þangbrandr's horse fell down into the ground, but he jumped off its back and stood on the brink' (Grønlie 2006, pp. 41–42). Here an iceberg had been transported by a jökulhlaup and had thawed and Þangbrandur's horse had fallen into a kettle hole. There are those who think that 'The Prophecy of the Seeress' in the *The Poetic Edda*, believed to have been composed in Iceland in the late 10th century, contains a description of a Katla eruption:

The sun turns sable,
land sinks under sea,
banished from the sky
are clear bright stars.
Steam bursts forth
with flashes of fire,
burning hot flames
flicker at heaven itself. (Sigurðsson 1998, p. 16)

Up until the 17th century, the sources relating to Katla eruptions are unreliable. There are narratives concerning a Katla jökulhlaup late in the 9th century (894) which destroyed six farms and all the grassland between the Hólmsá and Hafursey. There is also mention of a meltwater flood accompanying an Eldgjá eruption (934) and the so-called Höfðá river floodburst (1179), which destroyed the Höfðahverfi district and filled up the ancient Kerlingarfjörður (Höfðá river is now called Múlakvísl). The Sturluhlaup flood (1311) destroyed the Lágeyjarhverfi area and transformed the landscape of Álftaver, filling Kúðafjörður with glacial till. Many drowned in this flood, except for the Sturla, after whom it is named, and who was supposed to have been saved, along with a child in a cradle, on an iceberg that drifted out to sea but which later ran aground once more. Geologists have not discovered layers of tephra from this time, however, and so Lágeyjarhverfi is not considered to have been destroyed until around 1500, by the same jökulhlaup that destroyed the farm of Dynskógar. Höfðahlaup flood (1416), as its name implies, struck mostly at Hjörleifshöfði.

The following folktale was current about Katla before the Reformation: 'Once upon a time, after Þykkvibær had become a monastery, an abbot lived there who employed a housekeeper called Katla; she was very moody and temperamental and she owned a pair of breeches that were of such a nature that whoever wore them could run indefinitely without becoming tired. Katla only used these breeches when absolutely necessary; many were afraid of her magical powers and temper, including the abbot himself. There was a shepherd at Þykkvibær called Barði who often had to suffer Katla's harsh rebukes if any of the sheep he was tending went missing. One day during the autumn, the abbot went to a banquet and his housekeeper went with him; Barði was supposed to have rounded up all the sheep before they returned. The shepherd could not find the sheep as quickly as he was supposed to, so he decided to borrow Katla's breeches, and then did all the running he needed to round up the sheep. When Katla returned home, she quickly became aware that Barði had borrowed her breeches and she secretly drowned him in a barrel of sour whey, which traditionally stood by the main door, and she left him there in the barrel. No one knew what had become of him, but as the winter progressed and the whey began to evaporate, folk began to hear her muttering: "Barði will soon appear." And then she began to realise that her evil deed would soon be discovered and that she would not go unpunished, so she put on her breeches, ran out of the monastery, heading northwest to the glacier, where she apparently threw herself headlong into it, for no one ever saw her again. And right after this there was a jökulhlaup from the glacier which headed towards the monastery and Álftaver. From this came the belief that her sorcery had caused the flood, and the gorge has been called Kötlugjá ('Katla's gorge') ever since; the area which the jökulhlaup had made desolate was called Kötlusandur' (Árnason 1961, Vol. 1, pp. 175–176).

5.1.3 History of Research on Katla

The first contemporary, written description of an eruption of Katla (1580) was published by Bishop Oddur Einarsson (1559–1630) in his *Íslandslýsing* ('Account of Iceland') in 1590, and since 1625 there are extant witnesses' narratives of all Katla's later eruptions. The actual location of the eruptions was seldom described, however, although it is evident that they have been in individual craters and small fissures far enough in the east of the volcano for jökulhlaups to have descended onto Mýrdalssandur. In 1625 a small volcanic aperture was sighted to the east of the main crater six days after the eruption began. In the convulsions of 1721, the glacier began to subside due to melting ice, and crags or a mountain came to light that had been covered by the glacier for more than a century. The boundaries on Mýrdalssandur had previously been calculated from these rocks and so boundary disputes arose once the crags became visible again. In 1755 the eruption was in two separate places at first, one of them being more to the western side of the caldera, to the north of Holt in Mýrdalur. The later eruptions of 1823 and 1918, and their locations, were described in great detail.

No one knows who first climbed Mýrdalsjökull, but Eggert Ólafsson and Bjarni Pálsson ascended it from the north, from Mælifellssandur, south of Strútur, at the end of August 1756 in order to explore Kötlugjá gorge after the eruption that had begun on 17 October 1755 and continued intermittently until the middle of 1756. They doggedly approached Katla even though it was hard to find their way due to a snowstorm. Newly formed piles of rough-hewn pumice stones and pebbles bore witness to continuing activity in the volcano. Later that autumn, on 6 October, they saw from the east 'the terrifying jaws' of Kötlugjá. 'A great valley is to be seen in the glacier where it starts to slope eastward. Above the gorge, black peaks rear up through the glacier, and at the sides are rows of crags down to the foot of the mountain. In front of the mouth of this valley can be seen how impossible and almost incomprehensible the powers of the natural forces of fire and ice can be when they work in tandem. Many cliffs have collapsed one on top of another there, and are torn apart with deep chasms and gaping precipices. These mountainous ruins can be most likened to a turbulent ocean, for the cliffs that have either collapsed or been tossed up anew lie as in waves, and the channels which the floodwaters have burst apart with flying rocks and remnants of icebergs are nonetheless visible in spite of all the upheavals that have taken place. As we looked out over this terrain from various good viewpoints around Höfðabrekka, it became clear to us that we would have to abandon completely our plan to walk on the southern side of the glacier' (Ólafsson and Pálsson 1981, Vol. 2, p. 142). After Pálsson and Ólafsson had visited Katla, a very proud Ólafsson composed his exaggerated eulogy to this legendary mountain:

Panegyric to Katla

> We've travelled far and wide o'er land
> Wetlands, wastelands, lava and sand
> Mountains, glaciers, and their streams
> Caves and cliffs, canyons and ravines,
> Our journey prospers by all ways and means.
> The finest leaf in our laurel band,
> Katla, greatest wonder of this land,
> A blooming rose with crimson rays,
> Snowy headdress worn in eerie haze,
> Wise men of the world sing her praise.
> Let us arise, this mount to ascend,
> Our forefathers' fortune will attend
> And blessings on our joy reflect!
> Fair nymphs our sanctum will protect
> As we render Katla all due respect. (Ólafsson 1832, pp. 200–201)

Sveinn Pálsson ascended the glacier in 1793, and in describing the crater bowl at its summit it became clear to him that Kötlugjá had been filled up over the previous four decades with an accumulation of snow and the influx of ice. There had clearly been an enormous increase in understanding since the time when Eggert Ólafsson had maintained that only the sea could fill up Katla's maw. In the summer of 1874, fourteen years after an eruption, the Englishman William Lord Watts (1851–1920), along with two other Icelanders, ascended the glacier's plateau and described a hollow in the shape of a horse-shoe full of ice (Watts 1875). In just over six weeks after the 1823 eruption, Pastor Jón Austmann (1787–1858; 1907–1915), the parish priest of Þykkvibæjarklaustur, along with four other men, went up onto

Sandfellsjökull to Kötlugjá gorge, but the way to the eruption sites was impassable due to crevasses. These travelling companions had almost been the first men ever to see Katla with their own eyes. Jón described the eruption site in the southeastern part of the glacier from a northeastern dip in the highest edge of the peak now called Austmannsbunga. A deep horse-shoe shaped ice gorge stretched from northeast to southwest, curving to the southeast at Hábunga peak from where there was a trench in the direction of Kötlujökull. Björn Gunnlaugsson depicted the Kötlugjá as extending in a northwest to southeast direction on his map from 1848. With the eruption in 1860, there is an account of one volcanic outlet on the eastern side of the glacier, but on the fourth day of the eruption a small jökulhlaup flowed onto Sólheimasandur, which could indicate that there had been some subglacial melting in the west of the glacier, although this jökulhlaup could also have originated from a lake at the edge of Sólheimajökull.

There has been much debate about the exact location of the eruptions in 1918. Gísli Sveinsson (1919), the sheriff of Skaftafell County, maintained its source had been 1500–2000 m to the north of the highest edge of Hábunga, in a fissure 0.8–1.0 km long in a north-south direction. According to the description of Jóhannsson (1919), a teacher in Vík, the eruption came from two craters, one to the east, one to the west, the gap between them clearly discernible from Vík when the volcanic smoke was not too thick. It seems there were a number of vents at both places. After taking bearings from the east, it was later reckoned that the volcanoes had been further north, just southwest of Vestari-Kötlukollur, in the same area that two ice cauldrons later sank in elevation in June 1955 and a sudden jökulhlaup flowed from beneath Kötlujökull, sweeping away bridges over the Múlakvísl and Skálm rivers; one of the cauldrons was 80 m deep and 1050 m in diameter, and the other about 15 m deep and 700 m broad (Þórarinsson 1957; Rist 1967b). Seismographic data supported the hypothesis that there had been a small eruption under the southernmost cauldron. A basal map of the glacier reveals a 60–150 m high row of knolls at the rim of the crater with a diameter of 300–500 m beneath a glacier 300–400 m thick. However, pictures taken by a photographer from Vík, Kjartan Guðmundsson (1885–1950), on 23 June 1919, from the top of Hábunga in the direction of the Kötlukollur knolls, reveal that the locations of the eruptions were not at the cauldrons mentioned above. The picture of the column of smoke, taken from Ægissíða at Holt (Fig. 5.7), confirms that the eruption's source was just to the north of the highest edge of Hábunga, in the same area as the 1823 eruption, at a height of about 800 m and beneath 400 m of glacial ice. This is in accordance with Gisli Sveinsson's description, and where Þorvaldur Thoroddsen marked the location of Katla on his geological map of 1901.

Early maps of Iceland had already depicted Eyjafjallajökull (Austurjökull) as a landmark for ships sailing to Iceland (Fig. 4.4). Eyjafjallajökull, Mýrdalsjökull and Sólheimajökull are all marked on the Iceland map of the Bishop of Hólar, Guðbrandur Þorláksson (1542–1627), from 1570 (published 1590; Fig. 4.5). On Sveinn Pálsson's map from 1795, Eyjafjallajökull and Mýrdalsjökull are connected by the perennial snow cover of Fimmvörðuháls ridge, the so-called Lágjökull ('low glacier') while he named what is now Eyjafjallajökull as either Hájökull ('high glacier'), or Guðnasteinn (Fig. 5.13). Extending from it northward were unnamed outlet glaciers (now called Gígjökull and Steinsholtsjökull). Pálsson placed the name Mýrdalsjökull on the most easterly part of the ice cap and depicted Kötlugjá as crossing directly above the outlet glacier to its southeast. The northwestern part of Mýrdalsjökull he called Botnjökull and its northernmost point Emstrujökull, where Entujökull, Merkurjökull, Goðalandsjökull, Krossárjökull and Tungnakvíslarjökull are situated today. He depicted Sólheimajökull, but used its name for a much larger area than just the outlet glacier. The lake that Árni Magnússon (1663–1730) had shown in his sketch map as being in front of the snout had disappeared (Fig. 4.9). Pálsson's glacier map from 1795 showed Torfajökull and Tindfjallajökull, although he did not delineate their margins. Torfajökull, which had various locations on different maps, was now finally marked in the correct place. The cluster of glaciers to which Pálsson had given the one name, Eyjafjallajökull, was divided up by Björn Gunnlaugsson (1788–1876) on his map from 1849 (Fig. 4.21) into Eyjafjallajökull furthest west, as is the case today, while the eastern part is separated into three: Mýrdalsjökull southernmost, nearest to Mýrdalur, Goðalandsjökull to the north and west, and Merkurjökull northernmost. He depicted Kötlugjá gorge in a southeast to northwest direction on the eastern side of Mýrdalsjökull. On his geological map from 1901, however, Thoroddsen depicted Katla with a single dot on the southeastern part of the ice cap, marking the spot where Katla had erupted in 1823, and again in 1918, but otherwise depicted the glaciers as on Gunnlaugsson's map.

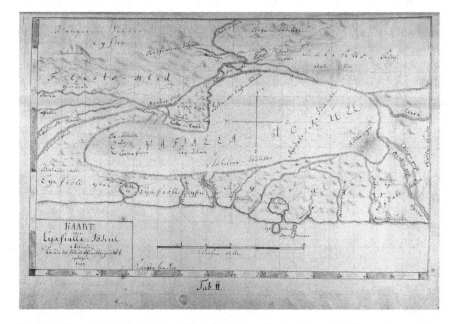

Fig. 5.13 Map of Eyjafjallajökull and Mýrdalsjökull from Sveinn Pálsson's 1795 treatise on glaciers

Fig. 5.14 The evolution of Mýrdalsjökull. The series of pictures shows the results of numerical models illustrating the formation of Mýrdalsjökull. Before a glacier covered Katla, it probably erupted through a lake in its caldera

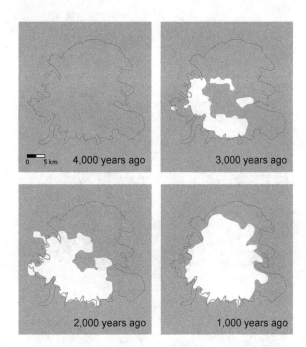

The first accurate maps of the southeastern part of Mýrdalsjökull were made by the Danish General Staff during the years 1904 to 1907, and of the main glacier by the Danish Geodetic Institute between 1937 and 1938 (on a scale of 1:100,000). The contour lines were drawn according to surveys along the edges of the glacier and from oblique aerial photographs, although these show the form of the glacier's surface rather than its elevation. In later editions of these maps, the position of the glacier's margins was based on aerial photographs. The first contoured map of the glacial plateau was made by the astronomer Steinþór Sigurðsson in 1943 (Nørlund 1944; Sigurðsson 1978; Björnsson 1977; Björnsson et al. 2000; Mackintosh et al. 2000). Cartographers of the U.S. Army then later drew up a map based on aerial photographs from the years 1945–1946 (on a scale of 1:50,000). They could only trace contour lines on the dirty glacial snouts (a few km from their margins), but above the firn line they were drawn according to the Danish maps. The maps of the surface depicted outlet glaciers and the main outlines of the ice cap with a depression in the glacier's centre surrounded by icy domes.

Only rough estimates have been applied to creating an image of the formation of Mýrdalsjökull during the Holocene (Fig. 5.14), in comparison with the more detailed data used for the image of Langjökull (Chap. 6). The same applies for computer-generated images of the development of Mýrdalsjökull (Fig. 5.15), which is based on current predictions of climate change in Iceland (see Figs. 3.17 and 3.18).

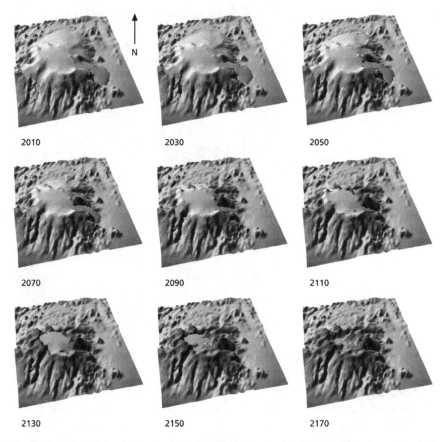

Fig. 5.15 Projected development of Mýrdalsjökull into the late 22nd century (assuming A1B scenario of IPCC 2007, see Chap. 3.9). The size of Mýrdalsjökull until 2170, projections based on predicted climate conditions in Iceland

5.1.4   The Subglacial Topography and Geology of Mýrdalsjökull

A survey of the thickness of the Mýrdalsjökull ice cap and its subglacial landscape began in the summer of 1955 with seismic reflection measurements at nine locations on the glacier's dome. Measurements revealed a blanket of ice 300–400 m thick (Rist 1967a). In 1977 the ice thickness was measured with radio echo soundings in a few cross sections on the glacier, which revealed an awesome subglacial terrain. In the centre of Mýrdalsjökull the ice was 500–600 m thick and this confirmed that it covered a large caldera. Satellite pictures (ERTS Landsat) of the surface landscape had also previously given a strong indication that the glacier was hiding a vast crater.

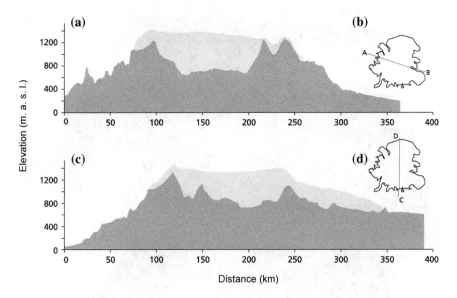

Fig. 5.16 Cross section of Mýrdalsjökull displaying Katla's caldera. The glacier rises to a height of 1300 m over a wide area and is up to 700 m thick

In the spring of 1991 an expedition from the Science Institute of the University of Iceland engaged in mapping both the surface and bedrock topography of the glacier. The surface elevations were calculated with great accuracy using barometric altimetry. The glacier was traversed in a great number of lines while its thickness was also being measured through radio echo soundings. Mýrdalsjökull is 740 m thick at its greatest density at the source of Kötlujökull in the northern part of the caldera, where there is an area of 12 km^2 area covered by ice more than 600 m thick (Fig. 5.16). Sólheimajökull transports as much as 500–600 m thick ice from the ridge between Hábunga and Goðabunga, which stretches for 1–2 km on the inside of the caldera's rim. The ice is only 150–200 m thick above the edges of Hábunga and Goðabunga. Outside of the caldera, it is up to 450 m in thickness above the continuation of Eldgjá gorge under the glacier. The thickness of the main part of Sléttjökull is 200–300 m, as is that of Sólheimajökull. On average, Mýrdalsjökull is 230 m thick and its total volume is about 140 km^3. The area within the caldera's rim covers 100 km^2 and its volume of ice 45 km^3, of which nearly a quarter is below an elevation of 950 m. The distribution of area and volume compared to elevations reveals that only about 20 % of Mýrdalsjökull's glacial bed is above an elevation of over 1000 m, as opposed to 55 % of its surface.

Let's now remove the ice cover from Mýrdalsjökull (Figs. 5.17, 5.18, 5.19, 5.20 and 5.21); what we now see before us is the second largest volcano in Iceland (after Hofsjökull volcano) with a volume of 380 km^3 over an area 300 km^2 in size. The lower reaches of the volcano are believed to be from basalt, although rhyolite can be seen in the nunataks that extrude through the ice covering the caldera's rims. The

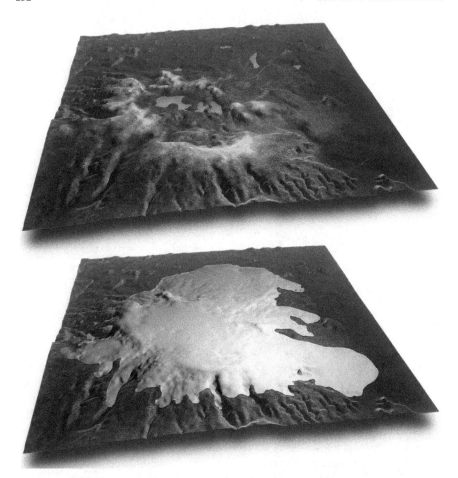

Fig. 5.17 Computer-generated images of Mýrdalsjökull and its subglacial base. Mýrdalsjökull is about 540 km² in size and covers the volcano Katla, which has a base area of 300 km². The mountain rises up to the caldera rim that encloses an elliptical-shaped area of 110 km². The caldera is 650–750 m deep within these rims, through which three outlet glaciers have gouged deep passes: Kötlujökull to the east (on the *right*), Sólheimajökull to the south, and Entujökull to the northwest. Just over half of the glacier's surface reaches a height of over 1000 m, but only a fifth of its bed does so. If Mýrdalsjökull were to disappear, a lake might form in the caldera basin which would be larger than Lake Askja in the Dyngjufjöll mountains (Fig. 8.108). Vegetation would reach a height of 600–700 m, but snow would still cover the highest mountains for most of the time

mountain rises to a height of 1300–1380 m at these rims, but within them is a 650–750 m deep caldera which reaches down to an elevation of 650 m above sea level. The caldera's rims enclose an elliptical area of 110 km² with a 14 km longer axis in a southeast to northwest direction, and 9 km shorter axis in a southwest to northeast direction. A few single-standing mountains have been formed from volcanic eruptions in the caldera's basin, which is lower and flatter in the north than in the

Fig. 5.18 Outlet glaciers descend steep ravines from the southern rim of Katla's caldera. From *left* to *right*: Klofningar, Gvendarfell, Jökulshöfuð, and Huldufjöll. In the distance, Kötlujökull descends onto Mýrdalssandur. HB, July 2008

Fig. 5.19 A view west over Hábunga on Mýrdalsjökull. Gvendarfell is in the *centre* of the photograph. To its west descends Hafursjökull, and the glacier associated with Thoroddsen is to its east. Lake Gæsavatn can be seen *left forefront*. In the distance are Eyjafjallajökull and the Tindfjöll mountains. HB, July 1999

Fig. 5.20 When the sun is low, the shadows sharpen the edges of the Katla caldera by Hábunga (on the *left*) and the ice cauldrons above subglacial geothermal areas. Kötlujökull glistens as it descends, rugged and crevassed, through the caldera's main pass. HB

south. Just to the east of Goðabunga, and to the north of Hábunga, are ridges and single pinnacles that reach above a height of 1100 m and around them are depressions that are below 750 m. There is a ridge about 5 km long to the east of Goðabunga and another nearly 3 km north-northwest from the eastern part of Hábunga. The eruption in 1755 could have originated in a fissure a few kilometres long running out east from Goðabunga and reaching far into the drainage basin of Kötlujökull, which is why the meltwater had flooded eastward onto Mýrdalssandur. This was the greatest eruption in Mýrdalsjökull since the settlement of Iceland and it discharged 1.5 km³ of tephra. The volcanic vents near the beginning of Sólheimajökull could explain the unusual behaviour of this outlet glacier, which was similar to a sudden surge. Eggert Ólafsson wrote in his *Ferðabók* ('Travelogue') in 1772 that, while the eruption was taking place, the glacier had 'moved in waves, rising and falling, until it finally billowed out so much that it was now twice the size it had been previously' (Ólafsson and Pálsson 1981, Vol. 2, p. 94). Furthermore, even Eyjafjallajökull glacier was lowered by the volume of melting ice so that ice-free peaks and black crags between them rose up out of the glacier to become visible for the first time in living memory. Ólafsson claimed that there had been three contiguous volcanic vents, but the sheriff at Holt, Jón Sigurðsson, counted as many as five craters about two months after the eruption had begun, and said that by its end three columns of flames were visible.

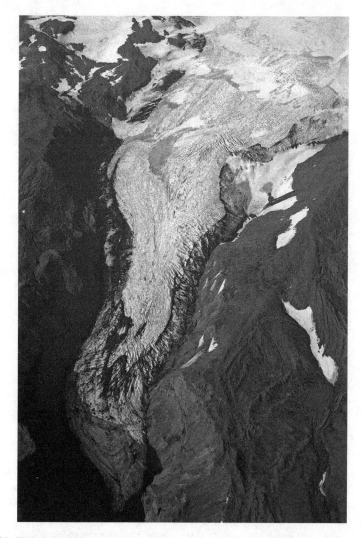

Fig. 5.21 Klifurárjökull falls steeply into a narrow ravine, south of Hábunga. HB, July 2009

About 3 km to the north of Hábunga there is a single ridge in an east-west direction parallel to the edge of the caldera. About 2 km inside the eastern edge of the caldera is a row of peaks in a north-northwest direction which might be recent craters. The eruptions in 1823 and 1918 could have originated in a 2–3 km long ridge north-northwest from the eastern edge of Hábunga. According to the diary of Pastor Jón Austmann, the volcanic smoke in 1823 was seen from Vík above the crest of the highest crags of Hatta. In 1860 a column of smoke was sighted a little to the east of due north of Vík and could also be seen above the western edge of Hatta. There is now a 300–400 m thick glacier over the volcanic vents from 1918, which

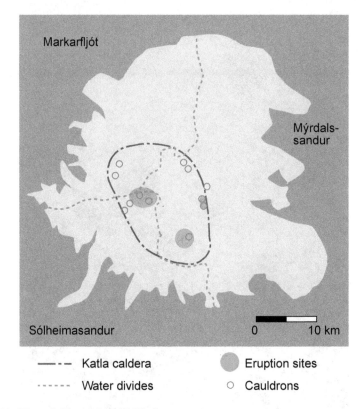

Fig. 5.22 Water divides on Mýrdalsjökull

are about 900 m above sea level. The glacier is so thick in these parts that the gorges and crag belts, which had previously been described near the volcanic vents, had not been solid bedrock but drainage and runoff channels on the surface of the glacier covered in black ash. Kötlugjá was thus an ice gorge that covered the volcanic vents and was gouged out lengthwise by running meltwater.

Three outlet glaciers have gouged deep passes through the caldera's rim through which jökulhlaups have flowed during eruptions beneath the glacier. Outburst floods could also have gouged out two narrow ravines northward from Austmannsbunga, one heading due north and the other to the northeast. Of the three glacially abraded passes in the caldera rim, the lowest is at a height of 750 m between Hábunga and Kötlukollar and leads southeast to Kötlujökull. Sólheimajökull descends southwestward through a 1050 m high pass between Hábunga and Goðabunga. A pass in the northwest, probably 1000–1100 m high at its lowest, leads onto Entujökull. These main passes might be named Kötluskarð pass, Sólheimaskarð pass and Entuskarð pass. They are all water-gouged channels from outburst floods caused by volcanic eruptions within the caldera.

The conditions on Mýrdalsjökull are such that water does not accumulate at the base of the caldera, but is forced out through its passes by the immense pressure of overlying ice (Fig. 5.22). Meltwater flows from Kötlujökull, from the whole eastern part of the caldera, an area of 60 km^2, onto Mýrdalssandur, and from an area of 20 km^2 down onto Sólheimasandur and Skógasandur, and from another area of 20 km^2 down from Entujökull into the Markarfljót. Outside of the caldera, the water divide lies between Mýrdalssandur and the Markarfljót down the centre of Sléttjökull. In terms of the whole of the glacier, water is discharged from an area of approximately 310 km^2 onto Mýrdalssandur, from an area of 110 km^2 onto Sólheimasandur and Skógasandur (Figs. 5.23, 5.24, 5.25, 5.26 and 5.27), and from an area of 170 km^2 into the Markarfljót (Figs. 5.28 and 5.29).

The ice cauldrons on the surface of the glacier are just within the caldera rim, from which water trickles down through vertical cracks in the rock seeking geothermal heat from shallow magma chambers in the Earth's crust. Meltwater, however, accumulates in small lakes at the base of the glacier beneath the cauldrons. The pressure of the ice at the base of the cauldrons closes drainage channels and so the lakes grow in size until the water is thrust out of them in small outbursts. Such jökulhlaups have flowed into the Múlakvísl from beneath ice cauldrons to the west of Kötlukollar and into Fremri-Emstruá from beneath cauldrons to the east of Goðabunga. An odour of sulphur from the Jökulsá on Sólheimasandur points to geothermally heated water flowing almost constantly from beneath cauldrons on the ridge between Goðabunga and Hábunga.

The main channel for outburst floods beneath Kötlujökull is on the other side of the Kötluskarð pass; it is about 1 km broad and 150 m deep and over a 5-km-long section it descends 150 m, before continuing into the Múlakvísl. At its lowest, the channel bed is 50 m below the land in front of the glacier's snout. The origins of the main channel are in the narrowest part of the glacier to the north of Huldufjöll mountains. In the largest jökulhlaups, the floodwaters reach as far as Kriki and descend eastward into the Leirá (this happened in 1918, 1755 and 1721). There is also a gigantic channel in the glacial bed high up under the centre of Entujökull, which is up to 300 m deep and 1 km broad, and flows over a 4-km-long section towards the beginnings of the Fremri-Emstruá. The glacial bed has been gouged down to sea level beneath Sólheimajökull, and is 100 m below the land at the glacier's snout. This is the lowest elevation point beneath Mýrdalsjökull.

Sandfellsjökull is divided from Sléttjökull by a ridge northeast from Austmannsbunga to Öldufellsjökull. An approximately 1.5-km-broad, V-shaped canyon with sheer walls 200–250 m high, runs parallel with it in a northwesterly direction just to the west of Austmannsbunga. The canyon is a continuation of the Eldgjá gorge which erupted in 934 A.D. and could later have been cut off by outburst floods. A few striking ridges lead away from the main volcano, probably made of palagonitic tuff, and a swarm of craters have piled up around the fissure vents. There are many single-standing peaks beneath Sléttjökull and Botnjökull, revealing that recent eruptions have thrown up landforms quicker than glacial ice has managed to erode them.

Fig. 5.23 Sólheimajökull descends steeply onto the plain named after it. Over the last 4000 years, dozens of meltwater floods have surged from beneath this outlet glacier during eruptions beneath Mýrdalsjökull, though only twice since the settlement of Iceland. HB, September 1997

Should the glacier disappear, a lake might form in the caldera basin with a drainage outlet over the lowest pass at an elevation of 750 m. The lake would be about 15 km^2 in size and almost 0.6 km^3 in volume; it would be rather larger than Lake Askja, but have only half its depth. In earlier times the passes might have been less abraded by ice and water and an eruption of Katla during an interglacial period would have burst directly upwards from a much bigger caldera lake and spewed tephra far and wide. Below the 800 m contour line, there might have been room for a lake 25 km^2 in size. On the ice-free areas beyond the lakes, on the other hand, eruptions would have resulted in lava fields from lava flows.

Fig. 5.24 River Jökulsá flowing innocently through Sólheimasandur in the summer of 1999. A few days earlier there had been a sudden jökulhlaup, which tore away strips of vegetated land. HB, 1999

Eruption of Katla in 1580

... the volcano was clearly visible and was so enormous it spewed huge rocks out into the ocean. And quite amazingly, loud booming and banging sounds, like the rumbling of canon fire, were heard in far distant parts of the country, i.e. in the north and west of the land, though those who lived near to the mountain were unaware of them. (Einarsson 1971, p. 43)

Oddur Einarsson's description in his *Íslandslýsing* ('Account of Iceland,' 1585) contains the first contemporary record of an eruption of Katla, and also the first mention in the history of volcanoes of a rumbling sound heard in faraway parts of the country but not near the volcanoes themselves. The explanation of this phenomenon is that when cold air lies low over the land, sound can be carried by higher and warmer layers of air to where echoes from them are once more audible at ground level.

Fig. 5.25 Ground moraine, a mixture of boulders, rocks, pebbles, gravel and clay, bulldozed onto Sólheimasandur by the glacier. HB, May 2004

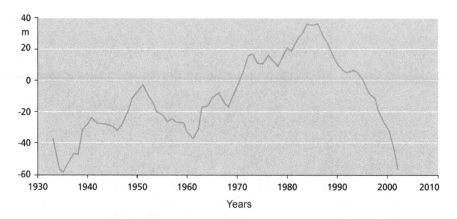

Fig. 5.26 Annual changes in the position of Sólheimajökull's snout 1933–2007. Sólheimajökull is a very sensitive monitor of climate change. Its snout is constantly active and has advanced and retreated in accordance with varying meteorological conditions

Fig. 5.27 Vegetation takes root in rugged ground moraine as Sólheimajökull retreats. HB, 2002

Fig. 5.28 A prehistoric catastrophic jökulhlaup after an eruption in Mýrdalsjökull gouged out the 120-m-deep Markarfljótsgljúfur canyon. HB, 2003

Fig. 5.29 A jökulhlaup over the Markarfljót river plains about 1200 years ago was so sudden that tree trunks remained vertical when buried, even though the flow of sludge destroyed their upper parts. The remains of the wood at Drumbabót were revealed by wind erosion removing their gravel cover. Eyjafjallajökull and Mýrdalsjökull look peaceful in the background, but proved to be very volatile in 2010. HB, 2008

Eruption of Katla in 1625

> There was so much fire, loud noises and bangs in the air above us and on the earth around us, and the fire glowed so much on our hats and other clothing, that it seemed we were all in one flame or surrounded by glowing coals. But it did not have such a colour as our natural fire doth have, but rather a phosphorescent light of fresh fish, or St. Elmo's fire, or Jack o' Lantern light as some doth call it. It was said that fire or bursts of flame covered and flowed along the earth just so as in the sky, so that everything seemed as one ball of fire in the moment it arrived … the same fire, which from the said darkness emerged, was sometimes so bright that it was shinier than any worldly fire or light can be (apart from the sun), sometimes red, and sometimes all the colours of the rainbow. And immediately after the bursts of lightning came accompanying violent cracks of thunder, crashes and bangs …
> (Magnússon 1907–1915, Vol. 4, pp. 205–206)

This is the description of Þorsteinn Magnússon, sheriff of Þykkvibær in Álftaver, dated 15 August 1625. This is the first eye-witness account of an eruption of Katla, and one of the most important descriptions of a volcanic eruption to be recorded since Pliny the Younger described the eruption of Vesuvius which destroyed Pompei in 79 A.D. There was a tremendous ash fall on farming communities to the east of Mýrdalsjökull, as far east as Hornafjörður, and it was impossible for livestock to graze as far east as Skaftafell in the Öræfi district. As winter progressed the layer of ash became diluted from rain and constant winds and, in all, eighteen farms were abandoned in Skaftártunga, two at Síða, and three in Meðalland. The woods in the area were also smothered by ash and could never be utilised again.

Katla Jökulhlaup 1660

On Tuesday, which was the 6th Nov., the same glacial stream flowed onwards all day in one continuous flood, as far as to the aforementioned Múli, to the east of the farmstead of Höfðabrekka, its northern limit being at Kaplagarðar, and then as far as one might see to the east from the high mountains, and so all over the broad gravel plains there was not a single hillock that stood visible ... On Wednesday ... a small brook flowed to the east and in front of the farm at Höfðabrekka, a small part of which splattered into the graveyard so that a tint of the glacier's colours can be seen there, when all was over ... But at daylight on the 8th Nov., folk awakened to the sound or running water, which was such a severe torrent rushing on both sides of the church, bursting the church's outer wall with its first onrush, each flood wave also dashing through the farmstead, covering its wooden pillars with glacial clay, the waters up to mid-thigh, and we grabbing all loose things afloat that we could, though this was not without danger while the flood was in full spate; the rapid current then began to slacken, abating back to the farm as some kind of whirlpool from the west of the home field slope, each rebound from the mountain bringing such filth and sand to the farm that everything was flattened, though the highest points of the muddy plains were in front of and to the east of the farm, the glacier's continuous accumulation of sediments had smothered most of the church ... After 9th Nov. there was good weather here, no waters to be seen, so that the folk wished to return to their farmsteads from which they had fled up onto the higher hillsides, where they had erected covering for themselves during all these commotions, when there was heard an incomparable crashing sound of floodwater and everything in its approach shook and trembled, as it advanced more powerfully and hastily than the previous flood had ever done, and with greater harmful products and effects, for it so swept through that there was hardly anything left of the church that had already been covered in clay, and it went right directly through the farmstead, leaving not a single stone standing, as if there had never been any walls, a house or buildings there.' (Salómonsson 1907–1915, Vol. 4, pp. 217–218)

This is the account of Pastor Jón Salómonsson (d. 1697), dated 12 November 1660 at the Kerlingardalur farmstead. The farmstead buildings and church were rebuilt at a height of 140 m on the side of Höfðabrekka. There was a farm there until 1964, but the church was destroyed in a violent storm in 1924. There was hardly any ash fall, but the meltwater floods were voluminous and persisted for many days, flowing only from the very base of the glacial tongue, never breaking through to its surface.

Katla Jökulhlaup 1721

Niels Horrebow's account, from around 1750:

'After several warnings by shocks of an earthquake, it first disgorged fire which melted down the ice. A most rapid torrent of water ensued, bearing away with it an incredible quantity of sand and earth, and destroying all the ground it went over by washing away all the mould. The intire [sic] current rushed with the same violence into the sea and filled it up like a hill, to near three miles distance from the shore. It since gradually declined to its present condition, appearing not much above the surface of the water. Between this mountain and the sea there is a rock called Haver

Ey [Hafursey], to the top of which the two travellers retired. Though the inundation overspread all the adjoining parts to the height of several fathoms, and destroyed a deal of fine ground and grass, they notwithstanding, about a day and a half later, pursued their journey across the country that had been overflowed, and were able to give the best account of this frightful sight, which they beheld without any danger from the top of the rock Haver Ey. This mountain stands in an extensive sandy plain, called Middals Sand [Mýrdalssandur].

Many years before, the same misfortune happened to this place and destroyed the valley, where there was good grass. The houses that then stood thereon were intirely consumed, but it received no additional hurt this time, as being ruined before. The prodigious quantity of sand, stones, and earth, carried into the sea, may be ascertained from what still remains to be seen, as also from the account given of the island of Westman, which lies 72 [sic] miles out at sea, where the sea all of a sudden rose with a violent motion to such an uncommon height, that it was with the greatest difficulty, the fishermen saved their vessels from being tossed ashore, and washed over by the waves. Such a violent agitation of the sea, and at such a distance, sufficiently proves what an excessive quantity of sand, &c. must have been poured with the water into the sea.' (Horrebow 1758, pp. 12–13)

The eruption began 11 May and continued until the autumn and was one of Katla's biggest eruptions, the total volume of fallen ash estimated at 1 km^3. The jökulhlaup also accelerated unusually quickly and the maximum flow was very powerful. Indeed, the flood did not emerge from the leading edge of the glacier, as other previous meltwater floods were said to have done (e.g. 1660), but burst forth from the middle of the glacier, shattering and flowing over it. Mýrdalsjökull had increased in size significantly in the decades prior to this large eruption and the resulting upheaval was stupendous. The glacier sank in elevation during the eruption due to melting ice, and rock belts came into view that had been covered by the glacier for more than a century; they had been boundary landmarks in earlier times and thus were the source of new boundary disputes.

The farmstead at Hjörleifshöfði, which had been on the plain for centuries, was finally destroyed. The farm was deserted for thirty years until a new farmhouse was built up on the former headland itself, and the farm remained functional until 1936. The story goes that it was finally abandoned because it was haunted, ghosts succeeding in doing what Katla had failed to do since the settlement of Iceland.

Eruption of Katla in 1755
The largest tephra eruption of Katla lasted for 120 days, from 17 October 1755 until 13 February 1756. The total volume of new tephra has been estimated as being 1.5 km^3, twice the amount of the volcanic material in 1918. The eruption was from at least five vents in a fissure, and three columns

of fire were later seen in the Kötlugjá gorge. 'Sunday the 19th, around the time to get up, a terrible cloud could still be seen billowing up from the gorge with a hasty speed with smoke, fire, and mist, and a thick sandy cloud and frequent flashes of lightning, each cloud rising from the gorge in 2 places almost due north from my homestead, namely Holt in Mýrdalur valley, which is nearly two Danish miles from Kötlugjá...' (Account of Jón Sigurðsson, sheriff of Vestur-Skaftafell County, 19 December 1755 in Sigurðsson 1907–1915). The bailiff in Svínadalur, Jón Þorvarðarson, and his maid were struck and killed by lightning. Many horses were also killed, but no one has died in a Katla eruption since then. There were some men crossing Mýrdalssandur on the first day of the eruption who had not felt the earthquake but had heard the rumbling in the glacier. Three men travelling westward became aware of imminent danger when they were just east of Hafursey. They turned around and rode as fast as their horses could go. Their packhorses were all loaded with timber, which was all lost. To the south of Lóðinsvík they managed to save themselves by climbing high enough before the jökulhlaup came rushing past bearing ice floes as high as houses. They remained there for the night in a continual ash fall occasionally shattered by streaks of lightning. They escaped the following day once the floodwaters had abated.

Eyjafjallajökull has been lowered considerably because of the fiery upheavals in Kötlugjá gorge, and now two small, ice-free pinnacles stand up out of it with a pitch-black belt of crags between them that no one has seen in living memory. The glacier is at a distance of 5 miles from Kötlugjá.' From the travelogue of Eggert Ólafsson and Bjarni Pálsson (1772). (Ólafsson and Pálsson 1981, Vol. 2, p. 89)

Eruptions in Eyjafjallajökull and Katla, 1821–1823

Just before 6 p.m. (18:00 h) on 26 June, earth tremors were felt in Vík in Mýrdalur, which continued until about 9 p.m. (21:00 h), after which a cloud of smoke was also seen from Vík. A little later, two young shepherd boys on Höttutindur peak saw a glacial surge which seemed to them to roll slowly past Hafursey.

Near the middle of the night ... the glacial flood began with some small blocks of ice and glacial clay to the east of Víkurfjall, filling up the coastal lagoon almost up to the home field, south under the gravel ridge and west into the Víkurá river from whence it debouched into the sea, just to the east of Reynisfjall.' (Account of Sveinn Pálsson, 1824) (Pálsson 1907–1915, Vol. 4, p. 269)

The eruption was not powerful enough to break up large parts of the glacier and the jökulhlaup bore no icebergs with it, unlike meltwater floods from previous centuries, which then bore a great deal of gravel and glacial till with them, filling in the local trout pools. Mýrdalssandur was flattened and evened out and afterwards it was easier to cross than before. The eruption lasted for four weeks with occasional small jökulhlaups, one of which drowned the sheriff of Skaftafell County and three others in Landbrotsá. This flood was unusual in that water continued to reach as far as Álftaver for a long time. The inhabitants of Álftaver, Skaftártunga and Meðalland journeyed with their packhorses to the north of Eyjafjallajökull, over Mælifellssandur, and those from the Síða farmsteads did their autumn trading at Djúpivogur on the eastern coast. The Eyjafjallajökull eruption started on 19 December 1821 and continued until 29 December.

While this was going on, I saw to the north of Hafursey on Mýrdalssandur, a flood cascading from the lower reaches of the glacier, which seemed to spread out over the glacier and move higher up it, and I toyed with the idea that it was caused by water collecting inside the glacier, which was welling up more and more and spreading out.

Description of Reverend Jón Jónsson Austmann, priest at Mýrar in Álftaver. (Austmann 1907–1915, Vol. 4, p. 252)

Eruption of Katla in 1860

So much gravel was flushed into the sea between Hjörleifshöfði and Höfðabrekka that it reached as far as where it used to be 15 fathoms deep … The gravel between Höfðabrekka and Kerlingardalsá stretched a mile further out to sea than it did previously. Damage from the jökulhlaup was less than feared, but at Höfðabrekka all the lowland and newly sprouted vegetation were buried under gravel. On 22 May there was a great deal of activity in the glacier and there was a strong odour of sulphur in the air. Four men who were taken by surprise by the flood managed to escape it by getting onto Hjörleifshöfði. They had to make do there for a fortnight until it was possible to take them off in a boat from Vík.

Narrative of Hákonarson (1860, pp. 61–62, 67), pastor of Reynir and Höfðabrekka.

Eruption of Katla in 1918

'The waters that flooded over the western part of Mýrdalssandur came from beneath the base of the glacier along a gorge all of 2 km long, 500 m broad and 250 m deep, which had been formed between the mountain and the glacier up by Vatnsrásarhöfði head (on Vatnsrásarfjall, 486 m), where the glacier had previously been joined to it high up the mountainside. The waters now burst from beneath the glacier as if they emerged from the gravel, in great spouts as black as tar. The higher reaches of the glacier seemed whole for the most part. East towards Hafursey, on the other hand, the outlet glacier's margin was all riven, crevassed and shattered. A bit further to the east, the eastern jökulhlaup poured down from the surface of the glacier through another gorge.' (Markússon 1952; Fig. 5.30).

Fig. 5.30 At Laufskálar on Mýrdalssandur, there are a large number of cairns, which travellers have helped stack in the belief that it will guarantee them a safe journey across the plain. In the background is Mýrdalsjökull, where Katla builds up its strength and bides its time

5.1.5 Waiting for Katla

Earthquakes have been felt in Vík in Mýrdalur valley from between one to eight hours prior to any eruption melting its way through Mýrdalsjökull glacier and belching plumes of smoke into the sky. Almost the moment this happens, jökulhlaups have burst out from beneath the Kötlujökull outlet. With modern seismographs it is hoped that volcanic activity will be spotted earlier and that volcanic vents and sites of probable outburst floods will be speedily located. Any eruption beneath the glacier will almost certainly have dramatic and unpredictable consequences, though it is still considered probable that meltwater will run through already existing channels; an outburst flood from the 60 km^2 catchment area of Kötlujökull, within the caldera, would spill onto Mýrdalssandur, as 18 of the 20 jökulhlaups recorded since the settlement of Iceland have done; there will also be outburst floods from the 20 km^2 catchment area onto Sólheimasandur and Skógasandur, and from an equally large catchment area from Entujökull into the Markarfljót. It is hard to predict the routes and velocity of these floods. Some jökulhlaups have started at great speed, bursting out of the centre of the glacier, breaking it up and flooding over it, as indeed happened in the 1721 eruption. Other floods begin more slowly, as did the outburst in 1660, and emerge from under the roots of the glacial tongue, but tearing it apart nonetheless, breaking off and tumbling onwards chunks of ice. The outburst flood of 1918 emerged from beneath the glacier at the outlet of the Múlakvísl after flowing through a glacial ravine almost 2 km long and 500 m broad, but further to the east a jökulhlaup flooded out higher up on the surface of the glacier through a crevasse or gorge.

The size of an outburst flood is usually determined by the length of the volcanic vent and the thickness of the glacier which the volcano is melting above its crater. A great deal also depends on how much water has been accumulated before the outburst begins. Eruptions are most powerful at the start, when water probably cannot drain away from the volcanoes as quickly as it is created by melting due to the resistance of the ice tunnels. Water could thus accumulate for a while near the volcanic vents until it is bursting with sufficient power to break open the glacier.

It is now almost a century since Katla last stirred and Mýrdalsjökull is virtually in the intensive care of geologists. The Icelandic Meteorological Office (IMO) and the IES at the University of Iceland monitor earthquakes and tectonic tremors caused by movements of magma at the root of the volcano, and the hydrology section of the IMO constantly gauges water levels and conductivity in the Jökulsá on Sólheimasandur, and in the Markarfljót, Hólmsá and Múlakvísl rivers. With such surveillance, any increase in geothermal heating should be quickly noted. The IES, with the help of the ICAA, also regularly checks for any changes in the elevation levels of the glacier using radar from aircraft.

With reflection seismology, a 2-km-thick magma chamber has been detected under Katla's caldera, 4–5 km in diameter. Its ceiling is 1–2 km beneath the glacier's bed. Earthquakes up to 3.5 on the Richter scale are frequent and originate either at a depth of 3–5 km near the magma chamber or outside the caldera at Goðabunga (Tryggvason 1960; Björnsson and Einarsson 1974, 1981; Einarsson and Björnsson 1979; Guðmundsson et al. 1994; Sturkell et al. 2003a, b). The earthquakes and land uplift (3 cm in 2004), in an area which is 5–10 km^2 in size, indicate that magma is collecting at the roots of the volcano which is not very deep within the Earth's crust. With increasing pressure from the magma, the volcano starts to expand, bedrock over the chamber begins to fracture, and the upward movement of geothermal water and gases increases. Ice cauldrons above the rim of Katla's caldera have sunk lower and glacial meltwater escaped in short bursts. If the magma pressure reaches the limits of the Earth's crust's resistance, the volcano can erupt. On the 18 July 1999, there was a sudden meltwater burst of 20,000,000 m^3 of water from beneath Sólheimajökull, and earthquakes indicated that magma had rapidly ascended to the base of the glacier and caused a new cauldron, all of 60 m deep, on the glacier ridge between Hábunga and Goðabunga. Simultaneously, all the other present cauldrons in Mýrdalsjökull have become deeper. The land beneath Eyjafjallajökull became uplifted, a vent formed on Fimmvörðuháls ridge, and there was a notable increase of carbon dioxide emissions from beneath Gígjökull. All of which ended with an eruption in 2010 (see below).

The increasing accumulation of magma under Katla is considered an indication that it is preparing itself for action. It has often been noted that impending volcanic action has always been announced a few months, even a whole year, before an eruption of Katla, through changes to the surface of Mýrdalsjökull. When viewed from adjacent settlements, it has appeared to swell up in some areas and sink in others so that black peaks are seen peeping up through the glacier. The river flows have changed, their channels have been moved, the stench from the glacier has increased, there has been some minor flooding, and the glacier's runoff streams have even dried up. After the eruption of 1755, Eggert Ólafsson and Bjarni Pálsson pointed out that: 'This last eruption came completely without any warning, although it so happened that during the summer prior to the eruption, two rivers burst out from beneath it and flowed out onto Mýrdalssandur and hindered any travel there. It never occurred to anyone, however, that an eruption was in the offing, even though this was, and has always been, a definite sign of a rising heat in the bowels of the mountain and thus melting ice more common than usual' (Ólafsson and Pálsson 1981, Vol. 2, p. 89). On events preceding the eruption of 26 June 1823, Sveinn Pálsson, then a physician in Vík in Mýrdalur, wrote: 'Late in the summer of 1822, more and more people noticed that the whole eastern part of Mýrdalsjökull, around and descending from Kötlugjá gorge, seemed to be lower, so that several rocky outcrops which had never been seen before had now appeared. This was obviously connected with some of the runoff waters on Mýrdalssandur drying up all that spring. Nonetheless, right up until the jökulhlaup, for example in the Múlakvísl and Eyjará rivers, which emerge from beneath the western part of Kötlufalljökull, where there have often been meltwater floods, there had otherwise been no sign of glacial discoloration of the water, except for just a little in the Leirá' (Pálsson 1907-1915, Vol. 4, p. 267). Gísli Sveinsson, sheriff in Vík in Mýrdalur, writes in his report on the eruption of 1918: 'Finally, it should be mentioned that quite a while before the eruption people thought they noticed that Mýrdalsjökull billowed up, growing higher in the east (as seen e.g. from Álftaver), but sinking further to the west (over Outer-Mýrdalur), so that black peaks actually came into view that had previously been covered by the glacier' (Sveinsson 1919, p. 59). Páll Sveinsson, a grammar school teacher, states that: 'People still believe there had been signs of an impending eruption in that Múlakvísl was virtually dry all summer, and so the water would have been accumulating in the glacier. And finally it should be stated that there was an unusual stench from the glacier on the eastern part of Mýrdalssandur as the summer passed, both east and west of Loðinsvík, and probably west of Hafursey too' (Sveinsson 1930, p. 93). Kjartan L. Markússon, a farmer at Suður-Hvammur in Mýrdalur, says: 'I do believe there was good reason during the summer of 1918 to expect an eruption if people had been more attentive and had realised there was something wrong with Mýrdalsjökull, because there was a change from the usual that summer in that there was hardly any water running from the outlet glaciers above and to the west of Hafursey. The huge volume of water that had flowed from out of the glacier every summer since 1860, either over the outwash plain to the east of Hjörleifshöfði, or in the Múlakvísl channels, was scarcely visible during the summer of 1918. There was only a small brook with glacial water that flowed onto the plain, but it had so little water it never even reached the sea. In my opinion, the water's usual outlet channel from the glacier had been closed by ice and the waters had accumulated within or beneath it' (Markússon 1952, p. 11).

All these alleged precursors of a Katla eruption were recalled and recorded after the event and could have often taken place without Katla having erupted. What has been described here, however, could be the result of a geothermal area moving or increasing its strength. This is a recognised phenomenon of volcanoes; the geothermal temperature increased six months before the eruption of Askja in 1961. With increased melting, the surface of the glacier sinks and basal movement could increase so that ice advances and swells out. If geothermal heating increases enough in the whole caldera to form a cauldron on the glacier's surface above its centre, meltwater could accumulate at its base, because of the pressure of the ice around the cauldron, and become trapped there. During an eruption, the glacial dam is burst and the water rushes out. An eruption of Katla could be so powerful that a melt-water flood would be very rapid, even if no considerable amount of water had accumulated in the caldera before the eruption began.

The DCEPM, the IRCA, and the institutes of power distribution, are all on constant alert because of Katla. The national highway across the Mýrdalssandur, Sólheimasandur and Markarfljót plains is especially vulnerable to meltwater floods, which will be full of sediment and debris, and even icebergs from collapsed ice dams, and rush through continuously changing channels. The only alternative route, from spring until autumn, would be the Landmannaleið route (the northern Fjallabak mountain track) through the central highlands, passable in 4WD vehicles and trucks. This was indeed utilised as a detour for hundreds of otherwise marooned tourists and travellers when the bridge across the Múlakvísl was swept away by a jökulhlaup in July 2011. A temporary bridge was constructed within a remarkable 96 hours.

The health authorities have issued warnings that volcanic ash from eruptions can be very harmful to both men and beasts. A very fine ash cloud can be borne very quickly by the winds all over Iceland, the grains of which can irritate the nasal and breathing passages, reach the lungs and, through the bloodstream, the whole body. Toxic elements from the ash clouds can also settle on vegetation and infiltrate drinking water, thereby gaining entrance to digestive systems and causing inflam-mation, bloody diarrhoea and vomiting. Too much fluoride damages the dental enamel of grazing animals and causes a shortage of calcium, the fluoride binding the calcium into compounds which are difficult to dissolve, resulting in a poisoning called narcosis in sheep and cows and lock-jaw in horses (especially in mares about to give birth). The long-term toxic effects of fluorosis include the uneven wearing down of teeth and jaw bones. The toxic gases from a Katla eruption are not as destructive as those from volcanoes without glaciers, such as Hekla and the Laki craters, because with a subglacial eruption dissolvable material in the ash is rinsed out by the waters of meltwater floods. Eruptions in Eyjafjallajökull can, however, cause serious air pollution, and even air traffic disruption on a world-wide scale, as the eruption in 2010 demonstrated. On the other hand, lightning in ash clouds from Katla would be much more dangerous than in former eruptions, as nowadays it could disable communication systems and electrical power supplies.

5.2   Eyjafjallajökull

The stratovolcano Eyjafjallajökull rises 1660 m high at the southern rim of the central highlands and bears an 70 km^2 ice cap above an elevation of 1000 m (Figs. 5.31, 5.32, 5.33, 5.34, 5.35 and 5.36). The glacier towers over southern Iceland's coastal lowlands and was from the very earliest of times a well-known landmark to seafarers to Iceland. Its name is indeed derived from a view of them from the Vestmannaeyjar islands (i.e. 'the islands' mountain glacier'). The sea reached the wave-scoured cliffs of the Eyjafjöll mountains right up until the last glacial period about 11,000 years ago, but today the Skógafoss, Seljalandsfoss and Gljúfurárfoss waterfalls cascade from their cliffs about 5 km from the ocean. The foothills are from palagonitic tuff, basalt and conglomerate rock, which has mostly been piled up in the latter part of the ice age. Craggy mountains, capes and headlands jut outwards and gorges have been gouged out by the power of water. The mountainous regions have been greatly eroded, but there are copses still hidden in valley rifts and many place-names and stories bear witness to ancient woods. At the southern foot of the glacier there is the grassy Eyjafjallasveit district between the Markarfljót and Jökulsá rivers on Sólheimasandur with thriving farmsteads on the lowlands up into the foothills, for indeed the area has a mild climate apart from occasional gales.

At the top of the volcanic cone is an ice-filled crater or small caldera about 2.5 km in diameter, surrounded by palagonite peaks projecting upwards from the circular rim of the crater: Hámundur (1666 m), Guðnasteinn and Goðasteinn, and Innri- and Fremri-Skoltur. A folk tale has it that the local inhabitants had hidden all their heathen idols on Goðasteinn ('rock of the gods') when Christianity was established as the legal religion of Iceland. Guðni was said to be a giant and national hero in the same mode as Bárður of Snæfellsás, or else a slave who, after having tried to murder his master, had fled up onto the glacier where he was slain.

EYAFIALLA IOKUL, from HLIDERENDE.

Fig. 5.31 Eyjafjallajökull viewed from Hlíðarendi at Fljótshlíð. A picture by H. Holland (Mackenzie in 1811). NULI

Udsigt fra Fljótshlíð over Markarfljót mod Eyjafjalla-Jökull og Þórsmörk.
(D. B. teg.)

Fig. 5.32 A view over the Markarfljót braided rivers to Eyjafjallajökull and Mýrdalsjökull from Fljótshlíð, northwest of the glaciers. Drawing by Daniel Bruun, 1921

Fig. 5.33 Eyjafjallajökull with Heimaey in the foreground. Eruption in Eyjafjallajökull in 1823. Erin Bruun. RLC

The glacier is thickest, 260 m, over the eastern part of the caldera. Sveinn Pálsson ascended Eyjafjallajökull on 16 August 1793 and was the first person to describe the crater's bowl. He hewed the letter P and the year 1793 in the rock face of the

Fig. 5.34 Painting by Ina von Grumbkow (1908). NULI

Fig. 5.35 Dusk over glacier and ocean. The nunataks rise higher from Eyjafjallajökull with every passing year. HB, September, 1997

Fig. 5.36 The rivers Fremri- and Innri-Emstruá flow into the Markarfljót. Steinsholtsjökull and Gígjökull descend northward from Eyjafjallajökull onto the lowlands. Þórsmörk is on the other side of Almenningar on the left. HB, July 1999

only ice-free crag, Guðnasteinn, and noted it was made of *móberg*, or what was later to be called palagonite.

Fimmvörðuháls is a 1000–1100 m high ridge connecting the Eyjafjallajökull and Mýrdalsjökull glaciers. A continuous ice sheet and perennial snow covered the highest part of Fimmvörðuháls until near the end of the 19th century. There was previously an ancient mountain trail along which the farmers around Eyjafjall used to drive their sheep up into the summer pastures at Goðaland, and this has been a popular and much-used walking path.

In the current climate, the snowline is on average at about 1100 m high in the southern half of Iceland and thus glaciers settle on mountains that reach that height. Fimmvörðuháls rises to a height of 1132 m and is on the borderline of being able to sustain a glacier in the present climate, so that the existence and size of a glacier in this location provides a sensitive monitor of climate change. After the settlement of Iceland and until near the end of the 12th century, it is reckoned that circumstances had been similar to what they are now, but since then temperatures have dropped, the snowline has become lower, and a continuous ice cover formed between Eyjafjallajökull and Mýrdalsjökull. Glacial tongues had slowly forced their way into the ridge, from both east and west, but the gap between them had been closed due to accumulating snow piling up on the high ridge itself. In his treatise on glaciers at the end of the 18th century, Sveinn Pálsson actually called this continuous tract of ice Eyjafjallajökull, and the glacial ridge Lágjökull ('low glacier') which connected the western part of Hájökull ('high glacier,' now Eyjafjallajökull) to Mýrdalsjökull in the east. Fimmvörðuháls was covered by a glacier for all of the 19th and well into the first half of the 20th century. When a group of mountaineers built a hut at 1080 m on the ridge in 1940, 'the glacier had retreated considerably from this area, for it had been covered in ice at the turn of the century, except for the highest mounts, and so Fimmvörðuháls now protruded from the ice ...' (Einarsson 1960, p. 72). Thus Fimmvörðuháls ridge arose out of the ice like a nunatak in the interglacial period of the 20th century.

Outlet Glaciers and Glacial Rivers of Eyjarfjallajökull About 8 to 10 small snouts descend from the glacial ridge, southward into a gully ending at an elevation of 800–900 m and from which flows the water that in the lowlands runs in the Írá and Holtsá rivers, and to the east of Holtsós in the Svaðbælisá; from Kaldaklifsgil gully the glacier descends to a height of 600 m and discharges water into the Kaldaklifsá. To the north of the ice cap, 5–6 glacial snouts come to an end high up in the mountainside, although the lowest one descends Hvannárgil gully, from which the Hvanná flows into the Krossá to the west of Útigangshöfði. Two of Eyjafjallajökull's largest outlet glaciers also descend on the north side, Gígjökull, from which Jökulsá emerges, and Steinsholtsjökull, from beneath which runs the river Steinsholtsá. All of these tributaries flow north into the Markarfljót, which then runs west of the Eyjafjöll mountains to the sea. The name Markarfljót is probably drawn from the name of Þórsmörk, though in the *Book of Settlements* only the river name Fljót is mentioned.

Gígjökull (also called Falljökull or Skriðjökull) descends steeply, with many crevasses, through a pass from the crater bowl down to the lowland. Eystri- and Vestari-Skoltur cliffs stand covered in ice on both sides of the pass. High moraines, about 1 km from the glacier, mark its terminus at the end of the 19th century. The glacier still reached and filled the area within these moraines around 1930, but a decade later a lake appeared in front of the snout that has steadily increased in size ever since. Ice climbers like to practice their skills on the steep and rough surface of Gígjökull (Figs. 5.36 and 5.37).

Fig. 5.37   The glacial lake at the snout of Gígjökull was a popular stopping place on the way to Þórsmörk. It was filled up with sediments during the eruption of Eyjafjallajökull in 2010. HB, May 2004

In the autumn of 1944, an American B-17 bomber crash-landed on Eyjafjallajökull just west of Gígjökull. The crew made their way down off the glacier across the Markarfljót to Fljótsdalur. The wreckage from the plane has long since begun to emerge from the glacier. In the spring of 1952, an American Grumman Albatross rescue plane crashed on Eyjafjallajökull. Only one body was found at the crash site, and although four others survived the wreck, they later died of exposure on the glacier. Twelve years later, a body and the finger of another man, bearing a wedding-ring, were found, and in the summer of 1966 the bodies of the other three crew members were discovered in Gígjökull. Wreckage from the plane has since emerged. In the autumn of 1975, an American plane with a married couple on board crashed on Eyjafjallajökull.

To the northeast of the eastern part of the ice cap, a steep and crevassed outlet glacier named Steinsholtsjökull descends to the lowlands. It ends in a lake, Steinsholtslón, which formed in the middle of last century. The sides between the outlet glaciers are punctuated by deep ravines, between which protrude five headlands including Fremstihaus at Gígjökull and Innstihaus at Steinsholtsjökull, a ca. 400-m-high precipice. In 1967 a large section of Innstihaus (15,000,000 m³, 40,000,000 tonnes) crashed onto the glacial snout, causing shattered icebergs and huge boulders to surge into the lake and along the river valley in a flood wave of water, ice and rock (Kjartansson 1967a, b). The huge chunk fell so quickly that the resulting air pressure rushed ahead of it to ease its flow along the channel, the waters reaching a height of 75 m above the valley basin (Fig. 5.38). Gigantic boulders were strewn on the plains alongside the Steinsholtsá, while water and icebergs were borne down past the Markarfljót bridge ninety minutes later and into the sea at the Markarfljót estuary, about 40 km from the starting point. The water flow was believed to reach a volume and velocity of 2100 m³/s. The boulders are clearly visible today when driving into Þórsmörk, the largest being about 80 m³ in size and weighing 200 tonnes. According to the seismograph at Kirkjubæjarklaustur, 75 km away, the rock head fell on 15 January at 13:47:55. A deep fracture had formed in the mountain a fair distance inward from the rock face and it had gradually been split apart by a freeze-thaw process until the rock face suddenly collapsed.

Fig. 5.38 There used to be a lake at the snout of Steinsholtsjökull, but it disappeared when huge sections of bedrock crashed down onto the glacier in January 1967 and shunted it forwards, sending water, ice floes and rocks cascading down onto the Markarfljót river plain. HB, May 2004

Rivers from Eyjafjallajökull run through very flat terrain when they reach the lowlands and so often flood their banks in heavy rainfalls, as many of them bear large muddy deposits, which fill up their watercourses and force the rivers to change channels. The Írá, Holtsá and Kaldaklifsá have often damaged meadows under the southern slopes of Eyjafjöll and the Markarfljót plains have long been without vegetation. Rivulets from the Fljót also frequently flowed east along the southern flanks of Eyjafjöll into the Írá and Holtsá and all the way to Holtssós estuary, damaging land. The *Jarðabók* ('Land Register') of Árni Magnússon and Páll Vídalín states that in 1645 the Markarfljót had broken out eastward along Eyjafjöll, and Þórður Þorláksson's map of Iceland from 1668 actually depicts the mouth of this river to the south of Eyjafjöll at the Holtsós estuary. On Knoff's maps of 1733 and 1734, the river ran due south, but in the 18th century it harrowed its way westward into the Þverá and destroyed the meadows of the Fljótshlíð farms and flooded the Landeyjar area to the southwest. This is how the river Fljót has been sweeping all before it for thousands of years, and in the 250-m-thick sediment deposits of the Markarfljót outwash plain, layers of glacial till are interspersed with layers of earth mixed with remnants of vegetation and tree trunks (Haraldsson 1981; Smith and Haraldsson 2005).

In Jón Árnason's collection of folktales, there is a story of a monster in the form of a huge skate fish that a certain Jón, a wizard, had enlarged through magic and cast into the Þverá to protect the farmers of Fljótshlíð from the encroaching glacial waters, first and foremost from the Markarfljót. All the glacial waters then flowed into the eastern braided rivers of the Markarfljót and did a lot of damage along the coastal areas under Eyjafjöll. There are many accounts extant of those who have seen the monster in the river, an old, enlarged and dried skate, which had become a 12 to 15-m-long colossus with three or four heads silently peeping up out of the rushing waters. Ögmundur of Aurasel, on the banks of the Þverá, also became famous later in the 19th century for halting encroaching floodwaters from the Markarfljót, apparently through magic too, or so many believed (Óla 1928).

Farmers tried to protect themselves from the river Fljót by building various kinds of dykes on their lands, but in 1907 a 1-km-long levee was raised on the eastern side of the Markarfljót, from out of Seljalandsmúli, to prevent the Fljót from running along the southern flanks of Eyjafjöll. There have been many other levees since then and they have all coped well with restraining the Fljót. In 1934 the river was bridged, and at 242 m long it was for a time the longest concrete bridge in Iceland. In 1978 a bridge was built over the Markarfljót at Emstrur, northeast of the glacier, where previously there had only been a roped cable-car, thus opening up a route out of Fljótshlíð into the central highlands and onto the Fjallabak mountain route. The most recent bridge over the Markarfljót was completed in 1992, about 5 km south of the 1934 bridge.

5.2.1 Volcanic Eruptions Beneath Eyjafjallajökull

There have now been four eruptions under Eyjafjallajökull since the settlement. There were explosive eruptions in the main crater at intervals during the years 1821–1823, beginning on the evening of 19 December 1821, when flashes of fire could be seen above the glacier, and in the morning light a column of steam arose that soon became a column of ash. A tremendous rumbling and groaning could be heard underground, but there was no earthquake. A meltwater flood burst out from beneath Gígjökull and inundated the whole of the Markarfljót plain between Langanes and the parts of Fljótshlíð further inland, and it filled Þverá's channel right up to the farms in Fljótshlíð. The jökulhlaup increased in volume slowly enough, however, to enable the rescuing of livestock. The water levels also rose in Holtsá at the southern foot of the Eyjafjöll. The floods caused some damage to the land and floodwaters continued to burst into the Markarfljót until the following spring. The jökulhlaup left behind chunks of glacier on the Jökulsá plain that took two years to melt. The eruption abated in the beginning of the year 1822, and drew little attention to itself until 26 June, when the glacier began once again to belch fire and ash even more vehemently than before, and from up to seven vents. There was ash fall all over southern Iceland and as far west as Reykjavík. There were few flames visible by the end of July, although it continued to belch tephra until just before the end of 1822. A white cloud of steam emerged from it in 1823, however, and on 12 October that same year an eruption of Katla began. There had been considerable ash fall in the nearby districts, though mostly on the moors above the settlements. This ash fall ruined grazing lands and fluoride poisoning caused swelling joints and lameness in livestock. The eruption thus caused a great deal of damage in Fljótshlíð and under the southern flanks of Eyjafjöll. A large cauldron had formed in the glacier in the main crater.

In the annals of Björn Jónsson at Skarðsá, an entry for 1612 states: 'Eyjafjallajökull burst forth eastward all the way to the sea. Flames soared up; it could be seen everywhere in the north of the country' (Jónsson 1922, p. 200). This may have been a sighting of the Katla eruption which had begun on 12 October that year, though it is believed that an eruption in the top crater of Eyjafjallajökull preceded the one in Katla. There is a connection between the volcanic centres of Eyjafjallajökull and the Katla system through a volcanic fissure which runs from west to east along the Fimmvörðuháls ridge. The third eruption is believed to have happened early in the 10th century (ca. 920) at Skerin, which is a roughly 4-km-long sinter ridge that now rises up out of the glacier on a volcanic fissure that stretches northwest from the top crater. The eruption reached the base of the glacier and a meltwater flood descended northward into Langanes.

There are signs on Ystaskálaheiði and Miðskálaheiði of a prehistoric jökulhlaup from beneath the western margin of the glacier that had flowed down onto the lowlands into the Írá, Miðskálaá and Holtsá rivers and could have reached as far east as the Holtsós estuary, or even gone directly west through Tröllagil gully and along the Gljúfurá into the Markarfljót.

The history of Eyjafjallajökull eruptions has still not been fully researched, but it is clear that it is one of the few stratovolcanoes in Iceland that has erupted often in historical times. Most of its volcanic fissures are in an east-west direction although a few radiate outwards from the ridge to the west of the volcano. Some of them extend under the glacier and have caused jökulhlaups, most of which flow to the west and north. There are signs of a fresh fissure vent in its western side, while all the young volcanoes on Fimmvörðuháls (considered to be from the last glacial period or the beginning of historical times) with their volcanic ridges, crater mounds and lava tongues, are to the east of the main part of the mountain. On the high ridge to its north are volcanic centres. Many of the craters are half full of snow and there are small lakes in some of them. The lava there is from a kind of rock called ankaramite, which is common in the Eyjafjöll. The youngest remnants of eruptions are hardly eroded and are visible because they have not been covered by a thick glacial ice for very long. A few low-temperature geothermal areas are around Eyjafjallajökull and many travellers are familiar with the heated swimming pool at Seljavellir.

The Infamous Eruption of 2010 In the spring of 2010 an eruption took place in Eyjafjallajökull that became well-known all over the world because it severely disrupted international air traffic by spewing volcanic ash into the atmosphere over the North Atlantic. The origins of this infamous event can be traced back to the 1990s, as there had been signs of magma injections into the roots of Eyjafjallajökull that had increased geothermal activity beneath the glacier. Precise geodetic measurements during the winter of 1999–2000 revealed that the southern part of Eyjafjallajökull had been uplifted, probably because of this shifting of magma at a depth of a few kilometres, as had similarly happened in 1994 (Sturkell et al. 2003a; Pedersen and Sigmundsson 2004, 2006). In the autumn of 2000 there was a sudden surge of water in the Lambafellsá, which had its origins on Svaðbælisheiði above Seljavellir. A new river flowed from beneath the glacier just to the west of the source of the Laugará and a chemical analysis of its flow revealed the presence of geothermal waters. There had also been a faint sulphurous odour emanating from crevasses at the caldera's rim, and this was the first time that geothermal heat had been found at the top of the glacier. An ice cauldron also became evident in the glacier on Fimmvörðuháls ridge as well as an increase in carbon dioxide emissions from Gígjökull.

All this evidence implied an increasing influx of magma into the roots of the volcano and that it could be preparing itself for an eruption. During the summer of 2009, seismic activity and uplift of the volcano increased considerably and by mid-March the location of earthquakes was rising from a depth of 10 km to about 5 km, indicating that magma was ascending beneath the volcano. Suddenly, on 20 March, the magma was ejected laterally from the main vertical vent of the volcano and emerged as a lava flow in a 500-m-long fissure that opened on an ice-free area on Fimmvörðuháls, about 2–3 km from the centre of the caldera. An eruption had been expected for several years, but instruments did not detect the beginning of this outbreak. About 400 people were rapidly evacuated from the neighbouring districts to save them from a possible jökulhlaup, but they were allowed to return to their homes 24 h later. Several lava fountains rose 100–150 m up into the sky, and lava streams moved northward from the fissure, producing a lava field of 2 km². An 80-m-high cone was piled up above the main crater. The eruption seemed to have stopped by the end of the first week in April, but seismic activity continued, and on14 April a 2-km-long fissure opened inside the top crater of Eyjafjallajökull. The volcanic tremors were intensive and tremendous explosions took place in what were, in effect, three huge sinkholes or ice vents, which had been created by meltwater penetrating right through the 200-m-thick glacier.

This time, 800 inhabitants were evacuated from the threatened neighbouring districts of local farmsteads and communities, which had been developed since the last eruption 187 years ago. A jökulhlaup rushed from beneath Gígjökull glacier, pouringing through its crevasses, inundating the proglacial lake Gígjökulslón with sediments loads, before careening across the fluvial plain of Markarfljót, heading toward the bridge across the river. The peak discharge there was about 2000 m³/s. The bridge was saved by quick-thinking workers, who bulldozed the raised causeway on either side of the bridge to allow the flood to bypass it and flow out to sea. Some meltwater also drained southward from the volcano and caused damage to land in the Eyjafjöll district. The erupting magma was injected into water and exploded into finely grained ash; indeed, it was pulverised into even still smaller grains because of the gaseous content of the magma. The ash fall was so dense that local visibility was hardly more than 1 m and people had to use masks to protect themselves against air pollution. The thick ash cover damaged vegetation permanently in the neighbouring highlands, but ultimately turned out to be excellent fertilizer for the lowland pastures, where the thinner cover was washed away by rain.

The ash-plume rose to the top of the atmosphere and was transported by westerly winds over the European continent. This caused immense disruption to air traffic, and about 100,000 flights were cancelled. The 2010 eruption of Eyjafjallajökull was thus a grave warning as to the potential impact of future eruptions of Icelandic volcanoes on international air traffic.

5.3 Tindfjallajökull

Tindfjallajökull is a 10 km² ice cap on the plateau of Tindfjall, inland from Fljótshlíð (Fig. 5.39). The glacier covers the middle and northern part of an almost circular caldera (6–7 km in diameter, 35 km²) within the main volcanic centre (ca. 10 km long and 7 km broad) named after the mountains which draw their name from the sharp peaks which tower upwards from the rim of the caldera. To its north and east, outlet glaciers descend steeply down the outer flanks of the caldera. The highest peaks, Ýmir (1462 m) and Ýma (1448 m), are free from ice but have steep icefalls on their sides. Ýma was a troll woman, and Ýmir a giant, who came into existence from a mixture of frost and flames and from whom Odin and his brothers created heaven and earth. The lower peaks are Saxi (1308 m), Hornklofi (1237 m) and Tindur (1251 m), and northernmost is the red-coloured Sindri (1289 m). The Tindfjöll mountains are clearly visible from the lowlands of southern Iceland, but the glacier is best viewed from the Fjallabak mountain track. There are three mountain huts at the southwestern foot of the glacier (Einarsson 1960).

In previous times very few travellers reached Tindfjallajökull and its location was not well known. Eggert Ólafsson and Bjarni Pálsson incorrectly stated that it was northwest of Torfajökull and northeast of Hekla. When Sveinn Pálsson was crossing the area in 1793, the weather was so cold that there was no runoff water in the river Hvítmaga and so he gave little credence to the idea there was a glacier in the Tindfjöll mountains.

Six small outlet glaciers fall steeply from the glacial ridge at a height of 1100–1200 m, down to a height of 900–800 m. The Eystri-Rangá receives its water from two westerly snouts through the Valá, which flows along the Austurdalur, and the Blesá originates in a small glacial tongue a little further north. Both rivers tumble through a deep ravine into the Eystri-Rangá. Þórólfsá runs from the southwestern part of the glacier through Jökuldalur, between Bláfell and Tindfjöll and then flows between Þórólfsfell and Fljótsdalur into the Markarfljót. The Eystri-Botná has its origins in the largest outlet glacier from the glacial sheet, to the east of Lifrarfjöll, and the Vestri-Botná bears water from the southernmost part of the glacier. These rivers (also called Innri- and Fremri-Botná) descend through a ravine into Gilsá and from there enter the Markarfljót to the east of Þórólfsfell. There was previously a lot

Fig. 5.39 There are still glaciers within the Tindfjöll caldera, but rising highest are the ice-free pinnacles around its rim. HB, 2008

of geothermal heat between the Botná rivers, but the area has now mostly gone cold, although it is slightly warmer in Hitagil gully. Finally, the Hvítmaga discharges glacial water into the Markarfljót to the north of Grænafjall.

The main volcano has been built up over the last three glacial periods and its two interglacial interludes. The caldera within it, however, was partly formed in a catastrophic event 54,000 years ago when the whole southeastern part of the Tindfjöll volcanic cone exploded. The subsequent amalgamation of hot, acidic tephra and mud engulfed southeastern Þórsmörk in clouds of flame before becoming petrified as grey or golden ignimbrite rock, which can be widely seen at the base of palagonite mountains. This violent upheaval was in the middle of the last glacial period and testifies that the Þórsmörk area was not covered by a glacier at that time. It is estimated that the layer of sediment was the equivalent of 4 km^3 of solid rock, but when it first fell it would have been ca. 8–12 km^3 in volume. This was the same kind of explosion that occurred in Mount St. Helens in Washington State in America in 1980, but was a much more powerful one. Nothing is known of any eruption of the Tindfjöll volcanic system in modern historical times, but the most recent evidence of volcanic activity could be a 10–15 thousand-year-old lava flow with a spattering of glacial ice. Nonetheless, the volcano is still classified as active.

5.4 Torfajökull

There are heavy snowfalls in the southern part of the central highlands and the snow takes a long time to melt. It is here that Torfajökull is hidden, a ca. 10-km^2 ice cap in a mountain range after which it is named. As well as this glacier and another, Kaldaklofsjökull (<2 km^2), there are intermittent glacial snow patches in depressions between the peaks and craggy ridges in the Hrafntinnusker (1128 m) and Reykjafjöll (1165 m) mountains (Fig. 5.40).

Small tributaries flow from Torfajökull in all directions. Jökulgilskvísl runs north into the Tungnaá, and in the east waters flow into Syðri-Ófæra and from there into the Skaftá, while the glacier's southeastern end provides the highest sources of the Hólmsá. From the southwestern parts of the glacier there is a confluence of tributaries in Kaldaklof which then flow into the Markarfljót.

The glacier can hardly be seen from the highland road, but it is easy to climb and has marvellous views, for nowhere in Iceland can a more variable and colourful landscape be found. Its peaks are sharply pointed, either vertical or skewed, rounded or jagged, and there are gullies, gorges and ravines carved out by glaciers and rivers and eroded by frost and wind. Rhyolite mountains have little vegetation on their steep inclines and screes of grey, yellow, pink, red, blue, and black, gleaming obsidian lava, while mountains of palagonitic tuff are arrayed in moss, white glaciers and snow.

Torfajökull is situated in one of the largest geothermal areas in Iceland, which covers an area 140 km^2 and reaches from 600 m up to 1100 m above sea level.

Fig. 5.40 Looking south toward Torfajökull over Jökulgil canyon. Snævarr Guðmundsson, 2007

Almost all of the geothermal heat is within the huge, ancient caldera and stretches to the north of Landmannalaugar and as far west as Rauðfossafjöll to the east of Hekla. Hot water flows from beneath the Laugahraun lava field to Landmannalaugar and there is a further bathing spot at Strútslaugur. There are sulphur- and mud pools, hot water and steam springs, pools and carbon dioxide vents. The most hot spring activity is around Hrafntinnusker, where fumaroles and brooks with warm water have melted caves into the ice. Ice caves like these exist nowhere else in Iceland, except in the Kerlingarfjöll, and only in these two locations are active sulphur springs and active glaciers to be found in a rhyolitic area in Iceland.

The volcanic system named after Torfajökull glacier is a 50-km-long, 30-km-broad accumulation of sediment layers of volcanic rock from the glacial and interglacial periods from the last 700–800 thousand years. In the middle of the central volcano is by far the largest caldera in Iceland (18 km long and 13 km broad), which was probably formed late in the ice age but is now eroded and partially filled with volcanic material. The volcano is unique in Iceland because it discharges virtually nothing but rhyolite, and has created the country's largest rhyolitic area, 400 km^2 in size. Eruptions have usually been beneath the ice-age glacier so that rhyolite bluffs can be found here, e.g. Laufafell (1164 m). Moreover, the only obsidian lava fields in Iceland are to be found at the rim of the volcano's caldera and are believed to have come from four eruptions after the glaciers had retreated in recent historical times, perhaps even after the settlement of Iceland. The best known obsidian lava area in this area is Hrafntinnusker, which is probably between 8000 and 8700 years old.

Torfajökull volcano has not actually erupted since the last glacial period ended, but it has shown signs of activity at least 10 to 15 times with injections of magma from the main volcanoes to the northeast. This is where the Veiðivatnareinar volcanic fissures intersect with the Torfajökull system and then connect it to the fissure channel of the Bárðarbunga system which reaches as far as Dyngjufjöll, a distance of 80–190 km, Iceland's longest volcanic system. The last time a fissure opened in the Torfajökull system was in 1477 when magma was borne all the way from Bárðarbunga and caused one of the biggest eruptions since the settlement, the Veiðivötn fires. The Laugahraun lava field came into being at Landmannalaugar at the southernmost point of the volcanic chain, while Námshraun was formed to the south of Frostastaðavatn. The lake Ljótipollur was also created around the same time. In 871 there were some tremendous upheavals in the Veiðivötn trench, which created Vatnaöldur and Hrafntinnuhraun. If the same time scale between eruptions and rifting episodes is maintained in this area, further upheavals might be expected there before the end of this century.

Torfajökull is first depicted on Knoff's map from 1733. Eggert Ólafsson and Bjarni Pálsson reached it in 1756: 'It is one of the most amazing places in Iceland, like no other place, for boiling water comes up out of the glacier there ... When we came to the glacier's edge we came across a valley from out of which flowed a river, and a great hot spring was very noisily boiling there, lots of steam rising from it. Steam reached upwards from smaller hot springs here and there, but a short distance from the big spring a small peak jutted out of the glacier, and to its west was the second-largest hot spring which seemed to come out of a ridge of white gravel. From the top of Torfajökull there is an expansive view over the surrounding highlands and glaciers ... But because round-ups often go badly here and many sheep are lost, there is a popular belief that outlaws hide out in these mountains' (Pálsson 1945, p. 254). Columns of smoke rising into the skies can only have strengthened this belief in outlaw settlements at Torfajökull.

Sveinn Pálsson explored the area in 1793 and states: 'It is not a regular glacier any more than Tindfjallajökull is, but rather many and variously shaped mountains, with rounded summits joined together in a circle ...' (Ólafsson and Pálsson 1981, Vol. 2, pp. 95–96). The Danish geologist Schythe wrote about the area in 1847 and Þorvaldur Thoroddsen passed through it in 1888.

Torfajökull as Sanctuary During the Plague

The name Torfajökull is first found recorded in Árni Magnússon's *Chorographica Islandica* from 1702–1714 (Magnússon 1955, p. 26). 'Torfajökull, from the Fiskivötn lakes southward, is separate from all other glaciers. Torfahlaup, Torfavötn [Torfavatn], are all named after an outlaw called Torfi.' Torfi Jónsson, the sheriff of Klofi at Land (d. ca. 1504) fled from the last plague, which lasted from 1494 to 1495, 'with all his wretched band of followers up onto the Landmannarétt sheep grazing pastures ... and they came upon a broad and beautiful valley that seemed to lie alongside a glacier from east to west and with no other pass but that through which they had entered and from out of which flowed rivulets. For as far as they could see around this valley there was nothing else but the glacier and the clear sky. But where the glacier ended, the lower slopes were covered in woods right down to the lowlands, and where the woodlands ended, the plains began, as grassy as they were beautiful ... There is no mention of how long Torfi remained up at the glacier but it has since been named after him and called Torfajökull. It is said that when Torfi began to think of moving back from the glacier to the settlements, some of his household families were reluctant to leave, and he had given in to them and let them have the homestead he left behind. Since then, and until very recently, it has been said as the gospel truth that there were outlaws up by Torfajökull, and travellers who journeyed along the mountain trail from Rangárvellir in southern Iceland east to Skaftafell County, to the south of Torfajökull, thought they could see traces of smoke from the glacier in a northerly wind, as if timber was being burned. It was widely believed that these outlaws were the reason why sheep went missing from the round-ups, a not uncommon event. But a few years ago it was established that something other than outlaws caused the poor return of sheep in round-ups in Torfajökull, for men from Land got round to exploring Jökulgil, reaching far enough into the canyon to see that the valley was full of glaciers and quite uninhabitable, and thus totally incongruent with the stories they had heard of what it was supposed to have been like during the days of Torfi.' (Árnason 1961, Vol. 2, p. 136)

Torfajökull has also been associated with another Torfi, the foster-son of a farmer at Keldá at Rangárvellir, who ran away with his foster-father's daughter and jumped over Torfahlaup with her. Torfajökull and Eiríksjökull are the only two glaciers in Iceland that are named after persons.

5.5 Hekla

Hekla (1491 m) is known for its mantle of snow and not as a glacier (Figs. 5.41 and 5.42). Sheets of firn snow and small glaciers have long been in view, especially to the northwest on the volcanic ridge (Þórarinsson 1944, 1968; Kjartansson 1945). Eggert Ólafsson and Bjarni Pálsson waded knee-deep through freshly fallen snow when they became, as far as is known, the very first men to climb Hekla on 20 June 1750. 'The highest part of the mountain was covered in ice and snow. Nonetheless, it isn't a glacier because the snow usually melts in the summer except for the bits that lie in rocky gullies and depressions, as in other mountains that are not glacial mountains' (Ólafsson and Pálsson 1981, Vol. 2, p. 160). Sveinn Pálsson writes that Hekla was a cone-shaped, snow-capped mountain just like Snæfell. He ascended its summit on 27 August 1793 in a windy snowstorm, and once again in bright weather four years later.

Hekla is mentioned in the context of this book because of the jökulhlaups caused by its eruptions. Such meltwater floods are mentioned in written sources from the 18th century onwards, but they could also have occurred previously, though they have not been recorded as they probably caused little damage. Hannes Finnsson (1739–1796), the Bishop of Skálholt, describes the jökulhlaup in the Ytri-Rangá during the eruption of 1766 and states that they frequently accompany Hekla eruptions. He believed pumice had piled up in the river damming it, the river then bursting through this hindrance; he does not mention volcanic fires as having melted ice on the mountain.

The next eruption was on 2 September 1845, and in an important publication on this, the Danish geologist Schythe (1847) states that although the jökulhlaup in the Ytri-Rangá that day might have been caused by the river being dammed, it might also be assumed to be the result of the eruption melting snow and ice. Someone had informed him that the current was slowing down in the Ytri-Rangá about midday,

EYAFIALLA IOKUL, MOUNT HEKLA, & the RIVER ELVAS, from the Westward.

Fig. 5.41 Fairly large glaciers can be seen on the flanks of Hekla (on the *left*) in a painting from 1810. Mackenzie, 1811. NULI

Fig. 5.42 Glaciers on Hekla diminished rapidly and were covered with ash during the 20th century due to frequent volcanic activity and a gradually warming climate. HB, July 1999

but in the afternoon it was still impassable. Later that evening it was passable on horseback over the usual fords. Nonetheless, the river was so hot that it was impossible to hold one's hand in it and this indicated that glowing embers had rained down on the water, which had become heated. Hundreds of trout had been ejected onto the banks half boiled. Thoroddsen, in his history of Icelandic volcanoes (1925, p. 143), agreed with Schythe that the causes of the jökulhlaups had been the two mentioned above.

On the other hand, there is also extant a very informative description by Oddur Erlendsson (1818–1855), a farmer at Þúfa in the Landsveit district, dated 1848, published in Markús Loftsson's book on volcanic eruptions in Iceland: 'There then came such a flood of water in the Ytri-Rangá, originating from a short distance to the west of Hekla, under the western flanks of Næfurholtsfjöll, that it flooded its banks and was unfordable by horse because it was so hot; it bore with it so much glacial mud that it was light blue in colour, although the spring water remains clear. All the trout in the river were killed and about two hundred dead fish washed up around a few farmsteads in the central area along the Rangá, and there was a rotting smell from the heat in the water. The river abated in the evening and became passable ... people think this flood was the result of the glacier melting on the mountain, because it had become very large after so many years. People saw that the water that flowed into the Rangá had cascaded northward out of the mountain, then ran to the north of Sauðafell, and from there into an old meltwater channel and then into the eastern [outer] Rangá river basin, that lies to the west of the mountain.

The water had gouged a passage through the gravel and rocks. People also saw that these waters had flooded westward over the plain and west lower down from the eastern [outer] Rangá basin' (Erlendsson 1999, pp. 102–103). Erlendsson makes no mention of the river having been dammed by pumice deposits and Guðmundur Kjartansson has pointed out that this might well be ruled out because the wind had been blowing in a south-southeasterly direction.

Kjartansson (1951) researched the jökulhlaup from the Hekla eruption of 1947 in great detail. In the beginning, the flood cascaded down both sides of the Hekla ridge, though most of it northward before it became joined once more at Litla-Hekla and flowed in one current along Norður-Bjallar and Sauðafell before entering the sources of the Ytri-Rangá in the Rangá river basin, about 11 km to the north of Hekla's peak. It is believed that the flood had reached a flow rate of 900 m^3/s, double the average flow of the Þjórsá, just 2 km from the base of Hekla. The black tracks of the mudflows could be seen on Hekla's hard crust of snow from quite a distance for the first twenty minutes of the eruption. The flood wave then rushed southwestward through the farming communities at a speed of 8–9 km/h. It was first noted in the populated areas between the farms of Galtalækur and Næfurholt, before continuing at 5–6 km/h past one farm after another (Hólar, Svínhagi, Bolholt, Þingskálar) and then past the hamlet of Hella, about 60 km from the foot of Hekla, and finally 70 km all the way out to sea. The floodwaters increased for 1–2 h up to its high mark, after it had first been noticed, but there are no reports as to how slowly it abated. The highest velocity and volume of the flood was about 120 m^3/s, three times the average flow. Pumice floated in the water and the floods dispersed it along the river banks. A few dead trout were found on dry land.

From the top of Hekla the volcano spewed out superheated steam and red-hot lumps of lava which were strewn over the snow cover, mixing with and melting it quickly. To the north of the ridge, water flooded over the hard-packed firn down onto three outlet glaciers. There was little resistance as more and more meltwater was produced by the melting ice on the glaciers' surfaces. Glowing lumps of lava fell into the water heating it even more; the chunks of lava cooling as the water became warmer. As long as they remained hot and resilient, however, they did not disintegrate but were borne by the flow, so that there was lukewarm water down in the settlements, even though its route there passed through many layers of winter snow at the end of March. During the eruption of September 1845, there had been little snow down the sides of the glaciers and the perennial snow and water were much hotter in the central areas along the river Rangá. The streams of lava that flowed out of Hekla, when the mountain's ridge was ripped open, melted little snow, on the other hand, and had the meltwater flowed beneath the glaciers it would not have retained any heat.

The water flow covered the glaciers with mud and slurry and in some places gouged deep gullies into the ice. At the foot of the mountain, blocks of ice mixed with pumice became stranded, creating cone-shaped pumice holes when they melted. The century-old hummocks from the 1845 eruption were mostly swept away by the jökulhlaup in 1947. The water that flowed southeastward from off the Hekla ridge, on the other hand, seeped into the lava because there were no glaciers there and the mud slides came to a halt a short distance from the foot of the mountain.

The total volume of floodwater from the sides of Hekla was estimated at 3,000,000 m³, of which 2,500,000 m³ flowed down Norður-Bjallar, and of that almost 2,000,000 reached as far as the hamlet of Hella and onwards into the sea. About a third of all the floodwater eventually seeped into the lava or remained as pools in channels. Most of it had been melted from the mantle of snow on top of Hekla, but a few jets of steam were also released in the eruption which then became condensed and added to the flow of water.

A few jökulhlaups have flowed into the Ytri-Rangá during Hekla eruptions, which have been at 10-year intervals since 1970, and although the small glaciers and constant snow cover have been much reduced, it can be assumed that a wintertime eruption would still cause meltwater floods.

References

Árnason, J. (1961) [1862–64] *Íslenzkar þjóðsögur og ævintýri*. 6 vols. In Á. Böðvarsson & B. Vilhjálmsson (Eds.), Reykjavík: Þjóðsaga.

Austmann, J. (1907–1915). Skýrslur um Kötlugos. In Þ. Thoroddsen (Ed.), *Safn til sögu Íslands* IV (pp. 186–294). Copenhagen and Reykjavík: Hið íslenska bókmenntafélag.

Björnsson, H. (1977). Könnun á jöklum með rafsegulbylgjum. *Náttúrufræðingurinn, 47*(3–4), 184–194.

Björnsson, S., & Einarsson, P. (1974). Seismicity of Iceland. In L. Kristjánsson (Ed.), *Geodynamics of Iceland and the North Atlantic Area* (pp. 225–239). Dordrecht: Springer.

Björnsson, S., & Einarsson, P. (1981). Jarðskjálftar—"Jörðin skalf og ípraði af ótta." In *Náttúra Íslands* (2nd ed., pp. 121–155). Reykjavík: Almenna bókfélagið.

Björnsson, H., Pálsson, F., & Guðmundsson, M. (2000). Surface and bedrock topography of Mýrdalsjökull, Iceland: The Katla caldera, recent eruption sites and routes of jökulhlaups. *Jökull, 49*, 29–46.

Einarsson, G. (1960). Suðurjöklar. *Árbók Ferðafélags Íslands*.

Einarsson, O. (1971) [1585]. *Íslandslýsing - Qualiscunque descriptio Islandiae*. Pálsson, S. (trans.). Reykjavík: Bókaútgáfa Menningarsjóðs.

Einarsson, P., & Björnsson, S. (1979). Earthquakes in Iceland. *Jökull, 29*, 37–43.

Erlendsson, J. (1999) [1848] Katla eruption. In M. Loftsson (Ed.), *Rit um jarðelda á Íslandi* (pp. 101–119). Þorlákshöfn: Halla Kjartansdóttir.

Eyþórsson, J. (1945). Um Kötlugjá and Mýrdalsjökul. *Náttúrufræðingurinn, 15*(4), 145–174.

Grønlie, S. (Trans.) (2006). *The book of the Icelanders: The story of the conversion*. London: Viking Society for Northern Research.

Grönvold, K., Óskarsson, N., Johnsen, S., Clausen, H., Hammer, C., Bond, G., et al. (1995). Ash layers from Iceland in the Greenland GRIP ice core correlated with oceanic and land sediments. *Earth and Planetary Science Letters, 135*, 149–155.

Guðmundsson, J. (1907–1915). Sannferðug undirrétting um Kötlugjár jökulhlaup og þess verkanir. In Þ. Thoroddsen (Ed.), *Safn til sögu Íslands* IV (pp. 247–251). Copenhagen and Reykjavík: Hið íslenska bókmenntafélag.

Guðmundsson, Ó., Brandsdótttir, B., Menke, W., & Sigvaldsson, G. (1994). The crustal magma chamber of the Katla volcano in South Iceland revealed by seismic undershooting. *Geophysical Journal International, 119*, 277–296.

Hákonarson, M. (1860). Kötluhlaup á Sólheimasandi. *Íslendingur*. July 19, *1*(8), 61–62 and July 26, *1*(9), 67.

Hammer, C. (1984). Traces of Icelandic eruptions in the Greenland ice sheet. *Jökull, 3*, 51–65.

Hannesson, P. (1934). Kötlugosið síðasta. *Náttúrufræðingurinn, 4*(1), 1–4.

Haraldsson, H. (1981). The Markarfljót sandur area, Southern Iceland. Sedimentological, petrographical and stratigraphical studies. *Striae. 15*, 1–65 (Doctoral dissertation. Uppsala University, Sweden).

Horrebow, N. (1758). *The natural history of Iceland*. Anderson, J. (Trans.). London: A. Linde.

Jakobsson, S. (1979a). Petrology of recent basalts of the Eastern Volcanic Zone, Iceland. *Acta Naturalia Islandica*. 26. Reykjavík: Náttúrufræðistofnun Íslands.

Jakobsson, S. (1979b). Outline of the petrology of Iceland. *Jökull, 29*, 57–73.

Jóhannsson, G. (1919). *Kötlugosið 1918*. Reykjavík: Bókaverslun Ársæls Árnasonar.

Jónsson, B. (1922). [1612] Skarðsárannáll 1400–1640. *Annálar 1400–1800* (Vol. 1, pp. 49–272). Reykjavík: Hið íslenska bókmenntafélag.

Karlsson, Þ. (1994). Kötluhlaup 1918 – vangaveltur um eðli hlaupsins og hámarksreynsli. In *Kötlustefna, February 26, 1994* (pp. 10–12). Reykjavík: Jarðfræðifélag Íslands.

Kjartansson, G. (1945). Hekla. *Árbók Ferðafélags Íslands*.

Kjartansson, G. (1951). The Eruption of Hekla 1947–1948. Water flood and mud flows. *Rit Vísindafélags Íslendinga, 2*(4), 1–51.

Kjartansson, G. (1967a). The changing level of Hagavatn and glacial recession in this century. *Jökull, 17*, 263–279.

Kjartansson, G. (1967b). The Steinsholtshlaup, Central-South Iceland on January 15th, 1967. *Jökull, 17*, 249–262.

Lacasse, C., Sigurðsson, H., Jóhannesson, H., Paterne, M., & Carey, S. (1995). Source of Ash Zone 1 in the North Atlantic. *Bulletin of Volcanology, 57*, 18–32.

Larsen, G. (2000). Holocene eruptions within the Katla volcanic system, Iceland: Notes on characteristics and environmental impact. *Jökull, 49*, 1–28.

Mackintosh, A., Dugmore, A., & Jacobsen, F. (2000). Ice-thickness measurements on Sólheimajökull, southern Iceland and their relevance to its recent behaviour. *Jökull, 48*, 9–16.

Magnússon, Þ. (1907–1915) [1625]. Relatio Þorsteins Magnússonar um jöklabrunann fyrir austan 1625. In Þ. Thoroddsen (Ed.), *Safn til sögu Íslands* IV (pp. 200–215). Copenhagen and Reykjavík: Hið íslenska bókmenntafélagið.

Magnússon, Á. (1955). *Chorographica Islandica*. [ca. 1702]. In Ó. Lárusson (Ed.), *Safn til sögu Íslands* (2nd series, Vol. 2). Reykjavík: Hið íslenska bókmenntafélag.

Markússon, K. (1952). Hvenær gýs Katla næst? *Morgunblaðið, 39*(249), 11.

Mortensen, A., Bigler, M., Grönvold, K., Steffensen, J., & Johnsen, S. (2005). Volcanic ash layers from the last glacial termination in the NGRIP ice core. *Journal of Quaternary Science, 20*(3), 209–219.

Nordal, S. (1928). Þangbrandur á Mýrdalssandi. *Festskrift til Finnur Jónsson, 29. maj 1928*. Copenhagen: Levin & Munksgaard.

Nørlund, N. (1944) *Islands Kortlægning. En historisk Fremstilling*. Geodætisk Instituts Publikationer. 7. Copenhagen: Munksgaard.

Óla, Á. (1928). Skatan í Þverá: sögur um Ögmund í Auraseli. *Lesbók Morgunblaðs* (pp. 9–11), 15 January.

Ólafsson, E., & Pálsson, B. (1981). *Ferðabók Eggerts Ólafssonar og Bjarna Pálssonar um ferðir þeirra á Íslandi árin 1752–1757*. Steindórsson S. (Trans.). Reykjavík: Örn og Örlygur.

Óladóttir, B., Larsen, A., Þórðarson, G., & Sigmarsson, O. (2005). The Katla volcano S-Iceland. Holocene tephra stratigraphy and eruption frequency. *Jökull, 55*, 53–74.

Óladóttir, B., Larsen, A., Þórðarson, G., & Sigmarsson, O. (2008). Katla volcano, Iceland: Magma composition, dynamics and eruption frequency as depicted by the Holocene tephra layer record. *Bulletin of Volcanology, 70*, 473–493.

Ólafsson, E. (1832). *Kvæði Eggerts Olafssonar, útgefin eptir þeim beztu handritum er feingizt gátu*. Copenhagen.

Pálsson, S. (1883) [1795]. Forsög til en Physisk, Geographisk og Historisk Beskrifelse over de islandske Isbiærge (1792–1794). *Den norske Turistforenings Árbok* 1882–1883. Kristiania. 116 pp.

Pálsson, S. (1907–1915) [1826]. Lýsing á Kötlugjárgosinu 1823. In Þ. Thoroddsen (Ed.), *Safn til sögu Íslands* IV (pp. 264–294). Copenhagen and Reykjavík: Hið íslenska bókmenntafélag.

Pálsson, S. (1945). *Ferðabók Sveins Pálssonar. Dagbækur og ritgerðir 1791–1794.* In J. Eyþórsson, P. Hannesson & S. Steindórsson (Trans. and Eds.). Reykjavík: Snælandsútgáfan. (Thesis on glaciers: pp. 423–552).

Pálsson, S. (2004) [1795]. Williams, R and Sigurðsson, O. (trans.) *Draft of a physical, geographical, and historical description of Icelandic Ice Mountains on the basis of a journey to the most prominent of them in 1792–1794.* Reykjavík, Hið íslenzka bókmenntafélag.

Pálsson, H., & Edwards, P. (Trans.) (1972). *The Book of Settlements* [12th-13th cent.]. Winnipeg: University of Manitoba Press.

Pedersen, R., & Sigmundsson, F. (2004). InSAR based sill model links spatially offset areas of deformation and seismicity for the 1994 unrest episode at Eyjafjallajökull volcano, Iceland. *Geophysical Research Letters, 31,* L14610. doi:10.1029/2004GL020368.

Pedersen, R., & Sigmundsson, F. (2006). Temporal development of the 1999 intrusive episode in the Eyjafjallajökull volcano, Iceland, derived from InSAR images. *Bulletin of volcanology, 68,* 377–393.

Rist, S. (1967a). The thickness of the ice cover of Mýrdalsjökull, Southern Iceland. *Jökull, 17,* 237–242.

Rist, S. (1967b). Jökulhlaups from the ice cover of Mýrdalsjökull on June 25, 1955 and January 20, 1956. *Jökull, 17,* 243–248.

Robson, G. (1957). *The Volcanic Geology of Vestur-Skaftafellssýsla.* Doctoral dissertation. University of Durham.

Sæmundsson, K. (1979). Outline of the geology of Iceland. *Jökull, 29,* 7–28.

Sæmundsson, K. (1982). Öskjur á virkum eldfjallasvæðum á Íslandi. In H. Þórarinsdóttir, Ó. Óskarsson, S. Steinþórsson, & Þ. Einarsson (Eds.), *Eldur er í norðri* (pp. 221–239). Reykjavík: Sögufélag.

Salómonsson, J. (1907–1915). Skrif síra Jóns Salómossonar um hlaupið úr Mýrdalsjökli, Anno 1660. In Þ. Thoroddsen (Ed.), *Safn til Sögu Íslands* IV (pp. 216–219). Copenhagen & Reykjavík: Hið íslenzka bókmenntafélag.

Schythe, J. (1847). *Hekla og dens sidste Udbrud, den 2den September 1845.* Copenhagen.

Sephton, J. (Trans.) (1895). *The Saga of King Olaf Tryggwason* [sic]. London: David Nott.

Sigurðsson, J. (1907–1915). Relatio Jóns Sigurðssonar sýslumanns um Kötlugosið 1755. In Þ. Thoroddsen (Ed.), *Safn til sögu Íslands* IV (pp. 235–247). Copenhagen and Reykjavík: Hið íslenska bókmenntafélag.

Sigurðsson, H. (1978). *Kortasaga Íslands frá lokum 16. aldar til 1848.* Reykjavík: Bókaútgáfa Menningarsjóðs og Þjóðvinafélagsins.

Sigurðsson, G. (Ed.). (1998). *Eddukvæði.* Reykjavík: Mál og Menning.

Smith, K., & Haraldsson, H. (2005). A late Holocene jökulhlaup, Markarfljót, Iceland: Nature and impacts. *Jökull, 55,* 75–86.

Stefánsson, S. (1907–1915). Viðbætir Sigurðar Stefánssonar sýslumanns um hlaupin 1660 og 1721. In Þ. Thoroddsen, (Ed.), *Safn til sögu Íslands* IV (pp. 230–233). Copenhagen and Reykjavík: Hið íslenska bókmenntafélag.

Steingrímsson, J. (1907–1915) [1788]. Um Kötlugjá. In Þ. Thoroddsen (Ed.), *Safn til sögu Íslands* IV (pp. 190–199). Copenhagen and Reykjavík: Hið íslenska bókmenntafélag.

Sturkell, E., Einarsson, P., Sigmundsson, F., Geirsson, H., Ólafsson, H., Ólafsdóttir, R., et al. (2003a). Þrýstingur vex undir Kötlu. *Náttúrufræðingurinn, 71*(3–4), 80–86.

Sturkell, E., Sigmundsson, F., & Einarsson, P. (2003b). Recent unrest and magma movements at Eyjafjallajökull and Katla volcanoes, Iceland. *Journal of Geophysical Research, 108*(B8), 2369. doi:10.1029/2001JB000917.

Sveinsson, G. (1919). *Kötlugosið 1918 og afleiðingar þess.* Reykjavík: Stjórnarráð Íslands.

Sveinsson, P. (1930). Kötluför, 2. September 1919. In B. Björnsson (Ed.), *Vestur-Skaftafellssýsla og íbúar hennar* (pp. 67–73). (Also in *Jökull, 42*, pp. 89–93, 1992).

Sveinsson, E. (1947). Byggð á Mýrdalssandi. *Skírnir, 121*, 185–210. Från Mýrdalur. *Scripta Islandica, 14*, 42–49.

Thoroddsen, Þ. (1925). *Die Geschichte der isländischen vulkane: nach einem hinterlassenen Manuskript*. Copenhagen: Høst & Søn.

Tómasson, H. (1996). The jökulhlaup from Katla in 1918. *Annals of Glaciology, 22*, 249–254.

Tryggvason, E. (1960). Earthquakes, jökulhlaups and subglacial eruptions. *Jökull, 10*, 18–22.

Watts, W. (1875). *Snioland or Iceland, its Jökulls and Fjalls*. London: Longman.

Þórarinsson, S. (1944). *Tefrokronologiska studier på Island. Þjórsárdalur och dess förodelse*. Copenhagen. Munksgaard. (Also published in *Geografiska Annaler*, 1944).

Þórarinsson, S. (1957). The jökulhlaup from the Katla area in 1955 compared with other jökulhlaups in Iceland. *Jökull, 7*, 21–25.

Þórarinsson, S. (1967). Hekla and Katla. The share of acid and intermediate lava and tephra in the volcanic products through the geological history of Iceland. Iceland and Mid-ocean Ridges. In S. Björnsson (Ed.), *Rit Vísindafélags Íslendinga* (Vol. 38, pp. 190–197).

Þórarinsson, S. (1968). *Heklueldar*. Reykjavík: Sögufélagið.

Þórarinsson, S. (1975). Katla og annáll Kötlugosa. *Árbók Ferðafélags Íslands* (pp. 125–149).

Þorleifsson Þ., & Guðmundsson, E. (1907–1915). Skýrsla um Kötluhlaupið 1721. In Þ. Thoroddsen (Ed.), *Safn til sögu Íslands* IV (pp. 222–224). Copenhagen and Reykjavík: Hið Íslenzka bókenntafélagið.

Chapter 6
Glaciers of the Central Highlands

Langjökull, Þórisjökull, Eiríksjökull, Hrútfell and Ok, Hofsjökull, Kerlingarfjöll and Tungnafellsjökull

Ice-cold Eiríksjökull hath heard it all before.

(Hallgrímsson 1957, p. 126).

Abstract This chapter has a detailed description of the types, shapes, sizes, locations, and all the relevant scientific data of the glaciers in the central highlands of Iceland, as well as some of their recorded local histories, folklore and legends. Langjökull (870 km^2), is the second-largest ice cap in Iceland and Hofsjökull (830 km^2), the third largest. Smaller glaciers include Eiríksjökull (21 km^2), Þórisjökull (25 km^2), Hrútfell (<10 km^2), Ok (<1 km^2) and Tungnafellsjökull (33 km^2), as well as the small valley glaciers in Kerlingarfjöll (<3 km^2). All of these glaciers figure prominently in early descriptions of journeys across Iceland through the central highlands. The western branch of the Icelandic volcanic zone lies beneath these glaciers, but no subglacial eruptions have occurred since the settlement of Iceland. Langjökull covers hyaloclastic ridges, tuyas and a lava dome, but Hofsjökull covers a huge caldera. Langjökull has a bed of porous lava and supplies large amounts of meltwater into the groundwater systems of Lake Þingvallavatn. Valuable data on the subglacial topography and mass balances of Langjökull and Hofsjökull have been collected, as well as vital information on glacial surges, especially of Langjökull into the proglacial lakes of Hagavatn and Hvítárvatn. The Holocene birth and evolution of Langjökull is also revealed through numerical modelling. Assuming a rise of temperature of 2.0 °C per decade, the glaciers of the central highlands will probably disappear within 100–150 years.

6.1 Langjökull

In the minds of Icelanders, Langjökull has most often been associated with ancient sagas, folk tales, and poetry about trolls and outlaws. The eponymous hero of one of Iceland's most famous sagas, Grettir the Strong, spent some of his time as an exiled outlaw around Geitlandsjökull, living for one winter in the grassy Þórisdalur valley, enclosed by glaciers, where innumerable sheep roamed and the giant Þórir ruled (Scudder 2005, Ch. 61). This valley was later described by Guðmundsson

© Atlantis Press and the author(s) 2017

H. Björnsson, *The Glaciers of Iceland*, Atlantis Advances
in Quaternary Science 2, DOI 10.2991/978-94-6239-207-6_6

(1936) in his 'Ode to Áradalur' as a fairy-tale like setting for outlaws, and one of the very first expeditions into the central highlands after the dark Middle Ages was made there in 1664 with the aim of making Christian converts of its supposed inhabitants. Outlaws also sojourned just to the west of Langjökull in the Surtshellir cave, and to its east, at Hveravellir, was the lair of Iceland's other well-known outlaw, Eyvindur, who lived there with his wife Halla. Although there are three Þjófadalurs (thieves' valleys) and one Þjófakrókur (thieves' corner) in the vicinity of Langjökull, this long and quiescent ice cap posed no real threat, indeed it gained a fairy-tale aura connected to a belief in a better world up in the mountains (Gunnlaugsson 1835, 1836; Sveinsson 1862; Grímsson 1944–1948; Sigurðsson 1997).

Langjökull is the second-largest glacier in Iceland with a surface area of 870 km^2 and a volume of 207 km^3 of ice (Figs. 6.1, 6.2, 6.3, 6.4 and 6.5). Its name is drawn from its length, as it is 55 km long from its snout at Hagavatn all the way to its northeastern terminus in Hundadalur. It rises above its local terrain like a large

Fig. 6.1 Lake Hvítárvatn and Langjökull in winter. The large mountain jutting into the lake is Skriðufell tuya, straddled by Suðurjökull on the left, and Norðurjökull on the right. The lowland in the foreground to the right is Hvítárnes, and rising up from it on the other side of Hvítárvatn is Karlsdráttur, where there is abundant vegetation, including birch trees (Fig. 3.6). There are sources which suggest that people lived here in the early centuries of the settlement of Iceland. Hvítárvatn is about 420 m above sea level. Eiríksjökull is visible in the distance

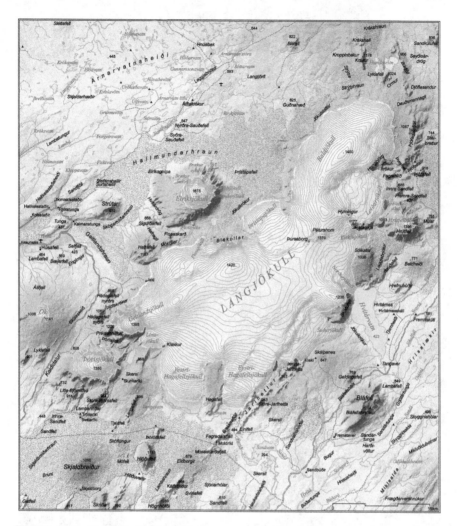

Fig. 6.2 Map of Langjökull and environs

whale on the route southwest over the central highlands, its back turned to
Borgarfjörður, its belly facing the Kjölur highland route, its tail rising high in the
northeast. Its broad front section is called by one name, Bláfellsjökull, and its
northeastern snout Baldjökull or Balljökull, but in the southwest the outlet glacier
Geitlandsjökull arises like the head of an animal. The glacier is broadest, at 30 km,
from the Jarlhettur pinnacles in the southeast to Hafrafell (1164 m) in the northwest, and narrowest, 13–15 km wide, on the saddle from Jökulkrókur, opposite
Eiríksjökull (1672 m), eastward to the shield volcano Leggjabrjótur, which also
marks the division of the ice cap between the two main parts of the glacier,
Bláfellsjökull and Baldjökull. The main ice divide on Langjökull lies from

Fig. 6.3 Satellite picture of Langjökull, Eiríksjökull and Ok. The photograph was taken from a height of 870 km on 17 August 2004. The snowline on Langjökull is clearly visible. SPOT

Geitlandsjökull (1395 m) northeastward along a high ridge over two summits, one 1385 m high and the other 1420 m high. The divide then turns eastward along the centre of the ridge and descends to a height of about 1150 m, to the southwest of the Þursaborg nunatak (1290 m), before veering northeastward again up onto the high ridge of Baldjökull, first reaching a glacial dome at an elevation of 1370 m and then finally rising up to Langjökull's highest summit at 1450 m, after which it slopes downwards to the northeast.

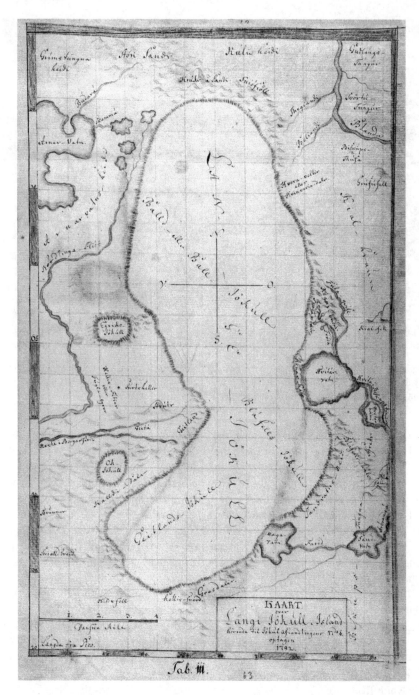

Fig. 6.4 The name Langjökull first appeared on a map by Sveinn Pálsson in 1792. Pálsson also correctly depicted the locations of the glaciers Eiríksjökull and Ok, and the lakes Hagavatn and Hvítárvatn. AMI

Fig. 6.5 View southeast over the crevassed Eystri-Helgafellsjökull. The Jarlhettur pinnacles (hyaloclastite ridges) extend in a row right across the centre of the photograph; beyond them is a glint of sunshine on Sandvatn. HB, September 1999

Langjökull's outlet glaciers extend like flippers from both sides of the high ridge, and runoff water drains westward into the Hvítá in Borgarfjörður, northward via the Seyðisá to the Blanda, and southward into the Hvítá in Árnes County, as well as into the Tungufljót and Brúará rivers. The southern and eastern flanks of the glacier are the most magnificent with their soaring mountain pinnacles surrounded by ice, and barren wastelands, glacial lakes and vegetated areas spread out below them. To the south of the ice divide, the western and eastern branches of Hagafellsjökull descend to an altitude below 500 m on both sides of Hagafell (917 m), which penetrates northward 8 km into the glacier, disappearing beneath it at a height of 900 m.

The two branches of Hagafellsjökull are the only outlet glaciers that have had their surges recorded (Björnsson et al. 2003; Théodórsson 1980). The eastern glacier has often calved into Hagavatn, the westernmost source of the Hvítá in Árnes County and into which a fifth of Langjökull discharges its runoff meltwater (196 km^2). The river Farið flows from the glacier's eastern corner over Nýjafoss waterfall into a narrow palagonite ravine along the eastern end of Brekknafjöll into Sandvatn; from there, it runs into the Hvítá, which then becomes the Ölfusá once it has been joined by the Sog tributary. Most of the water from Sandvatn used to flow through the Árbrandsá (or Ásbrandsá) into the Tungufljót and then reached the

Hvítá at Skálholt. A very memorable event occurred near the end of the 19th century when the Tungufljót first became a glacial lake as the glacier extended into Hagavatn. The SCSI, however, has now dammed Sandvatn so that a major flood would be necessary for glacial runoff waters to reach the Tungufljót (though this did actually occur as recently as 1999). This area is rather barren, though there is some vegetation between the outlet glaciers and rocks on the southern slopes of Hagafell, believed to have been named after the farm Hagi in the Grímsnes district, whose farmer drove his sheep there for pasture around 1800. The ITA built a cabin at Hagavatn by Einifell, just a short distance from the glacial snout, in 1942.

The outlet glacier Suðurjökull flows eastward from the high ice cap, south of the Skriðufell tuya at 1210 m, and up to Skálpanesdyngja shield volcano (874 m), to the southeast of which is the rocky expanse of Skálpanes along the shores of Hvítárvatn. Late in the 18th century, it was possible to walk on dry land between the glacier and Hvítárvatn, but around 1800 the glacier crept into the lake and closed off this route (Lárusson 1970). Glaciers advanced throughout the 19th century due to the cooling of the Little Ice Age. Three small glacial tongues even extended to the very edge of Skriðufell at this time, but they have since disappeared. A broad snout from Suðurjökull stretched all the way into Hvítárvatn until well into the mid-20th century, but today it has withdrawn 2 km from it. Norðurjökull still calves into the lake to the north of Skriðufell. It is so hemmed in by mountains, however, that even during the time when the Little Ice Age was flourishing, at the end of the 19th century, it was unable to spread out in the same way as its southern counterpart, Suðurjökull.

6.1.1   Karlsdráttur and Kirkjuból

Just to the east of Norðurjökull, at the northern end of Hvítárvatn, is a special oasis of vegetation around the springs at the southern end of Leggjabrjótur called Karlsdráttur, the remnants of moors covering the highlands some six to seven centuries ago (Fig. 3.6). The ruins of a fishing hut are said to be there, left by a man from Skálholt who once fished the lake with nets. Norðurjökull begins on the ridge between Bláfellsjökull and Balljökull, and from whence Kirkjujökull flows to the north of Leggjabrjótur. This latter glacier is demarcated in the north by a mountain spur west of Hrútafell (1396 m) that rises to its highest at Fjallkirkja ('church mountain'; 1230 m), from which the glacier draws its name. The IGS built its hut Kirkjuból there in 1979 at a height of 1180 m. The nunataks Hyrningur (1330 m), Péturshorn (1350 m) and Þursaborg (1.290 m) rise up through the highest edges of Kirkjujökull. The river Fróðá flows from Kirkjujökull between Leggjabrjótur and Baldheiði (778 m) to the north of Hvítárvatn (422 m). Further north, between Fjallkirkja and Jökulkrókur, is a broad bowl-shaped glacier called Leiðarjökull ('guiding glacier'), so-named because there is a route across it towards Hveravellir in the central highlands. The Jökulkrókur corner cuts into Baldjökull, to the northwest of the Þjófadalur and Fögruhlíð valleys, and ends at the precipitous

Fig. 6.6 The ITA's first lodge was erected in the Hvítárnes wetlands in 1930. One of its bunk beds is said to be haunted. Skriðufell tuya, or table mountain, is in the background, straddled by Suðurjökull and Norðurjökull

Hengibjörg cliffs. The steep, nameless glacier to the north of these cliffs, which descends in the direction of Rauðkollur peak, is the northernmost source of the Fúlakvísl at an altitude of nearly 950 m in Hundadalur. Fúlakvísl flows to the east of the Hrútfell and Baldheiði mountains and fans out in many braided streams before running southward through the grassy wetlands of Hvítárnes into Hvítárvatn. The Hvítá in Árnes County thus receives runoff meltwater from just over half of Langjökull (53 %, 440 km²), spanning an area from Hagavatn through Hvítárvatn, northward into Jökulkrókur to the northwest of Hrútfell. This area's ice reservoir is about 60 % of the entire volume of the glacier, 125 km³. In addition, the Hvítá also receives runoff water via Jökulfall river from 70 km² of Hofsjökull glacier. The ITA built its first cabin in the Hvítárnes wetlands in 1930, and trips there increased greatly in number once the Hvítá had been bridged at Hvítárvatn in 1935 (Figs. 6.6, 6.7, 6.8 and 6.9). Scientists and travellers on the Kjölur highland route have long been drawn to this lake, one of the most impressive and beautiful places of Iceland's interior, with its mixture of rugged outlet glaciers, steep mountainside screes, luxuriant wetlands, and multifarious birdlife, all crowned by a majestic horizon of mountain peaks (Keilhack 1886; Howell 1893; Bisiker 1902; Wunder 1910a, b; Ólafsson 1930; Keindl 1930, 1932; Kemel 1930; Ötting 1930; Nørvang 1937; Matthíasson 1961, 1980; Sigurðsson 2001).

6.1.2   Langjökull as the Source of Groundwater Springs

The meltwater that flows from Langjökull to northern Iceland comes from an 18 km² strip of the northernmost part of the Kráksjökull, named after Krákur (1188 m), the highest mountain on Stórisandur between Arnarvatnsheiði and the Kjölur highland route. The runoff water flows into the spring-fed Seyðisá and then later into the river Blanda. A third of Langjökull's runoff water, from the whole of the glacier's western side (33.5 %, 290 km², 50 km³), is borne into the Hvítá in

Fig. 6.7 The *top* photograph (**a**) was taken by Frederick Howell in Hvítárnes on 5 August 1899 (Ponzi 2004), the one *below* (**b**) a century later, in 1999, from the same place and a similar viewpoint. HB

Fig. 6.8 The drawing at the *top* was made by Bruun in 1889, when he made a halt in Hvítárnes on the Kjölur highland route. The horses and men in the foreground were inserted later. The drawing *below* is of southern Langjökull and the hyaloclastite ridge of Jarlhettur. NULI

Fig. 6.9 Hvítárgljúfur canyon below the Gullfoss waterfall was created by a catastrophic flood when a glacier-dammed lake on Kjölur emptied within 2–3 weeks about 9500 years ago. The total volume of the lake has been estimated as being 25 km³ and that the maximum flow rate reached 200,000 m³/s (Kjartansson 1964a; Tómasson 1993). HB, July 1999

Borgarfjörður (Fig. 6.10). A fair amount of meltwater flows into the Jökulstallar and northward to the Fljót rivers from the steep, nameless western flanks of Baldjökull. The water largely disappears into the porous lava of Hallmundarhraun and reappears in springs far from the glacier, accumulating in the main branch of the Norðlingafljót before merging with the Hvítá just below the northernmost farms in Mýrar County. Meltwater trickles into the gravel-strewn lava of Jökulkrókur, at a height of 600 m, from the only valley glacier reaching out westward from Langjökull, the 35-km-long Þrístapajökull, which faces the palagonite Þrístapafell (695 m). Further south, the steep Flosajökull descends to a height of 900 m in Flosaskarð pass, from where the tributary rivulets of the Hvítá run to the north of Hafrafell (1165 m). The English teacher Frederick W. W. Howell (1857–1901), along with two compatriots and two other Icelanders, ascended Langjökull from Flosakarð in the beginning of August 1899. As far as is known, they were the first men ever to walk lengthwise across the glacier, descending into Fagrahlíð valley in the northeast (Ponzi 2004). The Svartá flows from the northern side of

Fig. 6.10 The clear water cascading out of the lava in the Hraunfoss falls in Borgarfjöður is partly from Langjökull. HB

Geitlandsjökull through the Geitlandshraun lava field south of Hafrafell. The Hvítá receives its glacial colouring much further south from the most southern reaches of the Geitá, which has its source from a rivulet out of a valley between Prestahnúkur and the northernmost corner of Þórisjökull. Summer trips up onto the glacier for tourists from Húsafell begin on the fairly smooth flanks of Geitlandsjökull, which stretch down into Geitlandshraun. A glacial tongue extends northwestward to the southern Hádegisfell (1069 m) and steep icefalls tumble down from the southern flank of Geitlandsjökull to the foot of Prestahnúkur and into Þórisdalur. Lónsjökull, named after a lake without a drainage channel in front of its snout, descends between Geitlandsjökull and the nunatak Klakkur (1050 m), to the west of Hagafellsjökull. From the eastern side of this glacier, and from the western part of Vestri-Hagafellsjökull, glacial meltwater flows as groundwater streams into Lake Þingvallavatn, from a catchment area of 100 km^2 (Árnason 1976; Sigurðsson 1990; Sigurðsson and Sigbjarnason 2002; Björnsson 2002, 2011). Together with water from an additional 20 km^2 drainage basin in the southern and eastern parts of Þórisjökull, this means that Lake Þingvallavatn receives runoff meltwater from a catchment area of about 120 km^2.

A Confusion of Place Names, Outlaws, and Explorers

The name Langjökull first appeared on Sveinn Pálsson's map from 1792 (Fig. 6.4), where it is used as a collective name for a glacier that had previously been given several individual names from different directions: from Kaldidalur, the southwestern part was called Geitlandsjökull, while from the Þingvallasveit district the part of the glacier west of Hagavatn was called Skjaldbreiðarjökull, and from north of the Jarlhettur on the Kjölur highland route it was called Bláfellsjökull, and from the northwest from Hallmundarhraun lava field it was known as Balljökull. On the maps of Gerhard Schøning and Jón Eiríksson, which accompanied the travelogues of Ólafsson and Pálsson (1772, 1787, 1981) and Olavius (1780, 1964–1965), Langjökull was another name for Hofsjökull (Fig. 4.12). Sveinn depicted Eiríksjökull as an individual glacier, whereas previously it had either been considered as a part of Langjökull or its name had been used for both glaciers. On Knoff's map (1734), Eiríksjökull reached right up to Hvítárvatn (Fig. 4.10). The map of Guðbrandur Þorláksson (1590; Fig. 4.5) depicts Geitlandsjökull and Baldjökull. In ancient times, Balljökull might well have been a name for Eiríksjökull, for indeed no other glacier bears a more rounded ice cap (cf. the meaning of the word 'ball' in both German and English). Such was the confusion over nomenclature before the days of Sveinn Pálsson, however, that on Homann's map from 1761, Baldjökull was just a small glacier to the west of Kaldidalur in the region of Ok (1170 m). Gunnlaugsson (1835, 1836) named the glaciers as Sveinn Pálsson had done, but their size and delineation were more accurate; glaciers were breaking off into the Hvítárvatn and Hagavatn lakes and Hrútfell was joined to the main glacier (Fig. 4.21). Moreover, he also depicted as ice-free a passage from the east through Jökulkrókur into Þórisdalur, for the previous year, in 1834, Gunnlaugsson had seen from Bláfell that a valley ran through and divided the Geitlandsjökull part of the glacier. He could not resist investigating it, and so with another man he ascended Langjökull and drew a map of this valley, measuring its position from the nunatak Klakkur. Gunnlaugsson returned to the valley in mid-July of 1835, now with seven other men, determined to explore it thoroughly. He did not see a single blade of grass there, only ice and ground moraine. The leading figures of the Icelandic Literary Society thought this expedition a complete waste of time, 'all to prove a small dale should be completely bare of grass, indeed! Yes, I can well believe no moss can be found there!' wrote one of them (Thoroddsen 1902, Vol. 3 p. 310) Gunnlaugsson's curiosity over this shadowy valley and his descriptions of the landscape there met with widespread criticism. Right into the 1860s, both Gunnlaugsson and his fellow explorer, Sigurður Gunnarsson, were involved in defending their report against those, including priests, who still believed that outlaws inhabited remote nooks and crannies of Iceland's central highlands. It was widely believed that Gunnlaugsson and Gunnarsson had met such outlaws, but had been made to swear an oath to remain silent on the subject.

Þorvaldur Thoroddsen's map of Langjökull was for the most part similar to Gunnlaugsson's, although Thoroddsen depicts the outlet glaciers Flosajökull and Þrístapajökull, which Gunnlaugsson did not, and Þórisdalur valley was almost entirely encircled by the glacier (at least it was depicted so on a map published with an article by Thoroddsen in 1899). Ten years later, in 1909, the German Wunder (1912) and his companions, including some Icelanders, explored Þórisdalur and confirmed that it was possible to gain access to it from both the west and the east without crossing any glacier. They entered the valley westward from the direction of Hagavatn, and then ascended Langjökull northeastward from Þórisdalur at Vestari-Hagafellsjökull, and having crossed the glacier eastward, they descended

from the Skriðufell tuya to the west of Hvítárvatn. The glacier to the south of
Þórisdalur was thus clearly not a part of Langjökull and ought to bear a name of its
own, Þórisdalsjökull. Later, in 1918, this was shortened to Þórisjökull (1350 m) and
this name has prevailed, though there are those who would like to use the name
southern Geitlandsjökull. It should be noted, however, that on an earlier map from
1900, accompanying the travelogue of the Englishman Bisiker (1902), the glaciers
to the north and south of Þórisdalur were clearly depicted as separate entities.

6.1.3   The Subglacial Terrain of Langjökull

Langjökull glacier is at the northernmost point of a volcanic chain that has been
active for 6–7 million years and which connects the two main volcanic centres of
Prestahnúkur and Hengill (near Lake Þingvallavatn, e.g. Sæmundsson 1979;
Jakobsson 1979). The dolerite shield volcanoes Lyngdalsheiði (409 m) and
Baldheiði (771 m) came into existence during the last interglacial period 110–
130 000 years ago, when the region was ice-free; Ok (1170 m) and Skálpanes
(847 m) are much older. Once a glacial period began again, palagonite ridges were
piled up beneath the glacier and palagonite tuyas emerged where mountains rose up
through the ice and lava flowed onto their summits. The most well-known of these
ridges are the Kálfstindar peaks to the north of Lyngdalsheiði, the Jarlhettur pinnacles to the northeast of Hlöðufell, and the tuyas of Þórisjökull (1350 m),
Eiríksjökull (1675 m), Hrútfell (1396 m), Kjalfell (1008 m), Skriðufell (1235 m),
Bláfell (1204 m), Hagafell (920) and Hlöðufell (1186 m). At the end of the glacial
period about 10 000 years ago, the lava domes of Leggjabrjótur (1010 m) and
Hagafell were formed and stretched far beneath the surface of Langjökull, lava
flowing over the palagonite cliffs of Jökulstallar from craters now believed to be
beneath the glacier. At the northern terminus of Langjökull, just to the northeast of
Jökulstallar, there is a small lobe of lava believed to be from the same period. Later,
the shield volcanoes Skjaldbreiður (1060 m, about 9000 years old), Kjalhraun
(7800 years old), and the older Geitlandshraun (the younger lava field is 8900 years
old) were formed (Jakobsson 1979, Sæmundsson 1982; Sinton et al. 2005). From a
similar time period, the lava field between Langjökull and the northern side of
Kjalhraun was formed and is believed to have emerged from craters beneath the
glacier. It might be around 5000 years since there was a small eruption at the
summit of Hagafell, though it is not quite so long ago that the lava dome
Lambahraun was formed to the south of Langjökull (about 4000 years ago) as well
as the Krákshraun lava field to the north of it (4500 years ago). The only eruption in
this region after the settlement of Iceland was early in the 10th century when the
lava creating Hallmundarhraun flowed from three main craters near the glacier's
northwestern snout, the southernmost one of which may now be covered by the
glacier. The lava field reaches round the snout of Þrístapajökull, northward along

the western side of Baldjökull up to Jökulstallar, though the lava further north along
this western perimeter is from the last glacial period.

Langjökull's subglacial topography is in many ways similar to its surroundings,
its bedrock being formed by volcanic eruptions (Figs. 6.11, 6.12, 6.13, 6.14 and

Fig. 6.11 Computer-generated image of Langjökull and its glacial bed. Langjökull is the second
largest glacier in Iceland with an area of 870 km². The glacier is supported by many tuyas and
shield volcanoes, but it also covers a crater and a palagonite ridge. Only 10 % of the mountain tops
rise above the snowline in the current climate. To the east of Langjökull's high ridge is a
35-km-long valley that might be called Langidalur. If the glacier were to disappear, a 15-km-long
lake might form in the centre of this valley, up to 100 m in depth. There would be vegetation up to
a height of 600–700 m, but snow would remain for long periods of time on the highest mountains

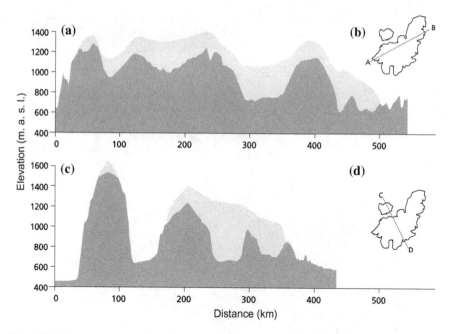

Fig. 6.12 Cross sections of Langjökull

6.15; Björnsson 2002, 2009, 2011). Langjökull might be covering lava, but volcanic centres beneath it have not been active since the settlement of Iceland. The mountains beneath the glacier rise over a wide area from its bed at a height of 700 m, but valleys and craters have harrowed themselves deeper into this plane. The glacier's bed is highest beneath the ice divide along the range heading northeastward from Þórisjökull to Baldjökull. On the ridge's high point the glacier is 150–250 m thick. At the southernmost point, Geitlandsjökull (1390 m) is resting on a 1250 m high tuya. Its flat summit has a diameter of around 5 km and it has sheer cliffs on all sides. To the south and west, the glacial bed rises up from a height of 800 m and to the northwest it descends to an altitude of 950 m to the foot of a shield volcano that ascends at an angle of 6° (1/10) to a 1230-m-high crater top that is about 10 km from the highest part of Mt. Geitlandstapi. The base of the lava dome is 850 m high and it has a diameter of 10 km and its total volume is about 10 km^3; this is 2/3 of the volume of the largest shield volcanoes in Iceland, Skjaldbreiður (1060 m) and Trölladyngja (1486 m), which are to the north of Vatnajökull and estimated to have a volume of 15 km^3 as they ascend 600–700 m above their local terrain. The Geitlandshraun lava field came from the crater of Geitlandsdyngja about 8900 years ago, around 2 km to the west of the present-day snout, down to a height of 700 m. The existence of this shield volcano is thus one piece of evidence that the ice-age glacier had by then disappeared from Iceland. The northeastern side of the lava dome only descends to a height of 1000 m because it is connected to a giant tuya massif with a 1200 m high summit that might be called

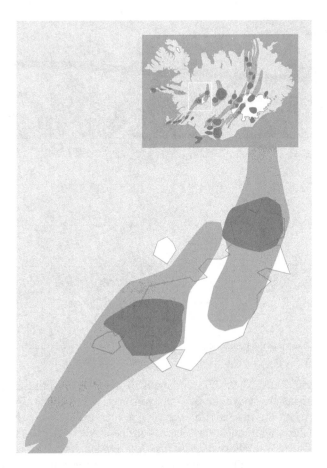

Fig. 6.13 Main volcanoes and fissure swarms beneath Langjökull

Flosastapi. Over its high dome lies a 200-m-thick glacier. The tuya extends beyond the glacial margins at Flosaskarð pass and 10 km southeastward beneath the glacier, and a further 10 km or more from the base of the lava dome to the northeast side up from Þrístapajökull. This tuya is lower but significantly larger in itself than Eiríksjökull, which has long been considered the biggest mountain in western Iceland. They both rise up from a base at a height of about 700 m, though the lava dome is probably higher under Eiríksjökull than Langjökull. The belts of Flosastapi's sheer cliffs face the Flosaskarð pass, Þrístapajökull and the southeast, and its convex ridge looks southwest to the Geitlandsdyngja lava dome. The Geitlandsstapi and Flosastapi massifs along with Geitlandsdyngja all form the highland bed beneath the southern section of Langjökull where the glacier reaches its highest elevation. From glacial striations on rocks it can be seen that, before this mountainous highland was formed, the glacier crept westward from the central

Fig. 6.14 The water divides of Langjökull. Most of the runoff water from Langjökull flows into glacial rivers in Borgarfjörður and Árnes County

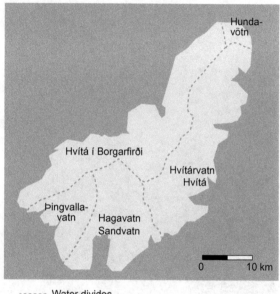

------ Water divides

highlands of Iceland, but after the tuyas arose during the last glacial period, the striation marks now head southward (Sæmundsson 2008).

In the foothills of the central highlands, to the east of the high ridge of Langjökull, is a 35-km-long valley which might thus be called Langidalur. At its southernmost end it is at a height of 500 m beneath the snout of Vestari-Hagafellsjökull, but it then curves towards the northeast, rising to a height of 750 m where the sheer cliffs of Hagafell tower on its right hand side and the convex base of Geitlandsdyngja rises to its west. Further north, the valley basin descends for 12 km to below 700 m until it rises again over a small balk to the northeast of Flosastapi tuya before descending to the snout of Þrístapajökull and opening out onto Jökulkrókur corner. At its lowest, the valley has an altitude of 600 m to the east of Flosastapi, but at no point along its length does it have an effect on the glacier's surface form. If the glacier were to disappear, a 15-km-long lake, up to 100 m in depth, might form in the centre of the valley below an elevation of 700 m, its runoff water possibly flowing to the northwest, down to Jökulkrókur. Its total surface area would be 30 km², similar to the present Hvítárvatn, but its volume would be almost double at 1.5 km³. This Langavatn would be dominated by the high, sheer precipices of Flosastapi, which would be reflected in its waters in the morning sunlight.

Langjökull's outlet glaciers gradually and evenly thin out as they descend westward from its high ridge to their snouts, for the ice is thickest to the east of the ridge, over the centre of Langidalur, to the southeast of Flosastapi (400–680 m) and on the ridge between the southern and northern part of Langjökull. The Hagafell glaciers, the two longest outlet glaciers from Langjökull (22 km), are only 200–

Fig. 6.15 The Eystri-Hagafellsjökull surged into Hagavatn in 1999. Geitlandsjökull is furthest away to the left, but in the lower right the river Farið flows toward the Hvítá. HB, September 1999

250 m thick when they descend from the high ridge, but 5 km further east they become 600 m thick as they progress southward over Langidalur, finally thinning out at their snouts. The eastern snouts of Langjökull extend over ridges, hollows, and peaks and the thickness of the ice is variable. To the west of Fjallkirkjan is a 1170 m high peak which is covered by ice 190 m thick. The Skriðufell tuya extends almost 10 km beneath the glacier and its 1100 m-high summit is covered by ice 200 m thick. The shield volcano Leggjabrjótur does not extend far beneath the glacier and the crater Sólkatla is about 1.5 km outside of it. The hollows on both sides of Skriðufell descend to a height of around 600 m, which is 100–200 m lower than the nearest terrain, so that two lakes would probably form there were the glacier to disappear. There would be a short barrier between the northernmost of these lakes and the Langavatn lake, and if there were a ravine between them (that has not been detected by radio echo soundings), the runoff water from both of them might flow into Hvítárvatn. Norðurjökull harrows itself down into a depression before descending, crevasse-ridden, 250 m over the cliff edges between Skriðufell and Leggjabrjótur, and then thickening again as it veers southeastward down to its terminus near Hvítárvatn. The Suðurjökull similarly advances, 400 m thick, over a

hollow southwest of Skriðufell, growing thinner as it ascends from it, but thickening again at the very end of the snout. Suðurjökull does not seem to descend from any barrier, but its narrow base suggests it crosses hard bedrock once it has crept up out of the hollow. The channel might originally have held craters, but they have now all been abraded by valley glaciers. The highest part of Hagafell extends almost 3 km under the glacier at a height of 1080 m, with an ice cover only 100 m thick. A line of palagonite mountaintops then rise from this in the direction of Leggjabrjótur and Hrútfell; there is a 950-m summit at the easternmost end of Eystri-Hagafellsjökull (beneath 150 m of ice) and another lower summit about 8 km northwest of Skálpanesdyngja (847 m). Further south, the Jarlhettur palagonite ridge disappears beneath the glacier in a more northerly direction towards Skriðufell. Beneath Eystri-Hagafellsjökull is a straight line of pinnacles, from southwest to northeast, parallel to Jarlhettur, and covering them are Jökulborgir, five uneven glacial hills.

Beneath the whole of Baldjökull is a very large massif, 30 km long and about 15 km broad with a flat summit at a height of around 1260 m. The ice above it is 190 m thick. To the east of this mountainous mass, and beneath the 400-m-thick Leiðarjökull, is a caldera 8–9 km in diameter, its rim demarcated by the summit of Fjallkirkja and the Hengibjörg cliffs. On its western side, cliff walls tower 400 m above its base, while on its eastern side there is a bowl about 200 m deep and in its centre rises a 100-m-high dome. The whole of Hrútfell (1396 m) would fit into this caldera. To the south of Oddnýjarhnjúkur pinnacle, there is now a cold former geothermal area from the post-glacial period at the perimeters of the caldera, and rhyolite can be found in Fögruhlíð near Jökulkrókur.

The Glacier Spilled into Hvítárvatn around 1800

There is a letter extant from the guardians of three churches in Biskupstunga district, dated 22 April 1844, answering an application 'as to whether we would not on our part wish to give farmers permission to begin grazing their sheep in the summer on the pastures to the north of Vötn belonging to the churches of Torfastaður, Bræðratunga, Haukadalur and Skálholt, as has previously been granted before the glacier had spilled into Hvítárvatn lake and thus come between them and Vötn.' It then later states: 'The northern sheep drives have once again become very sparse due to sheep numbers being reduced by the mange, but did not cease until—as has been reiterated—the glacier advanced into Hvítárvatn around the turn of the century. (Lárusson 1970). (Figs. 6.1, 6.7 and 6.8).

Enter a Glacier, Exit a Jökulhlaup

Lake Hagavatn, at the southern end of Langjökull, is believed to have been formed after lava flowed from the shield volcano Lambahraun about 3600 years ago and dammed the corrie between the Hagafell and Brekknafjöll mountains. The runoff water from this lake ran through the lowest pass in the Brekknafjöll and through the river Farið to Sandvatn. There were significant changes in the size of Hagavatn once

Eystri-Hagafellsjökull advanced in the 17th century and managed to dam the drainage channel through the pass where the Leynifossgljúfur ravine is today. The water level rose and the lake grew in size until the glacial dam eventually burst; jökulhlaups have been recorded as occurring in 1708, 1884, 1902, 1929, and 1939. Whenever the ice-dam gave way, the water would burst through, harrowing a narrow, deep channel into the Brekknafjöll's soft palagonite. The old channel was either broadened or deepened, or else a new and lower one was formed, leaving the older one dry. Once the flood was over, the water did not reach its former level, even though the glacial dam 'healed' and was whole again. According to Árni Magnússon and Páll Vídalín's land register (1981), the jökulhlaup of 1708 flooded meadows of the Vatnsleysa farm in the Biskupstunga district. Gunnlaugsson (1835) and Thoroddsen (1899) described the flow of water from Hagavatn as constant, though the lake is believed to have been the source of a jökulhlaup in 1884. Near the end of the 19th century, both of the Hagafell outlet glaciers had crept (or possibly surged) into Hagavatn, which then lay south of Hagafell with a water level 30 m higher than it is today, indeed Eystri-Hagafellsjökull then filled its current location. The glaciers then began to retreat at the end of the 19th century and in 1902 the lake waters burst out from under a glacial tongue, bearing 45,000,000 m^3 of water and lowering its water level about 4 or 5 m. There was a constant flow of water out of Hagavatn from then on until 1909, when the drainage channels were dammed once again and remained so for the next 20 years. In 1909 Wunder (1910a, b) arrived at this larger lake Hagavatn, into which both the Hagafell glaciers calved, and there were large icebergs floating on its surface. Early in 1929, when Keindl (1930) was travelling in the area, Hagavatn had become 4 km broad and 4–5 km long and the glacier had retreated just over 600 m since the turn of the century (625 m since 1902). It was impossible to reach Hagafell over dry land as Thoroddsen had done in 1883. On 16 August 1929, a large jökulhlaup burst through the rugged Leynifossgljúfur ravine in the Brekknafjöll to the west of the headland Stemmi and surged down the Farið into Sandvatn. There was a river from Sandvatn into the Hvítá, but the major part of the flood careened into the Tungufljót in Haukadalur, tearing down the bridge between Gullfoss and Geysir, 22 km from its origins, and inundating the meadows of Biskupstungur district, swirling away hundreds of cartloads of hay and leaving behind fields knee-deep in glaciofluvial mud. A levee now prevents water flowing from Sandvatn into the Tungufljót. The outburst flood of 1929 lowered the water level in Hagavatn around 6–7 m so that water could no longer reach westward in front of Hagafell where it had become possible to walk across a ledge along its shore after 1890. Right until near the end of the 1930s, the Eystri-Hagafellsjökull reached right across to Fagradalsfjall and dammed the drainage channel until a flood burst through the southeastern corner of Hagavatn and into Nýjafoss, and it still drains into this waterfall today. The water level then sank about 10 m and Hagavatn's surface area was reduced by half. The jökulhlaup caused considerable damage, both to bridges and river banks, and the flood wave itself reached populated areas along the Hvítá; indeed its high flood marks can still be seen on the bridge over the Ölfusá, in Selfoss, to this day (Ólafsson 1930; Wright 1935; Kjartansson 1938; Þórarinsson 1939; Green 1952; Sigbjarnarson 1967).

Langjökull began to retreat from its furthest extension around 1890 and by 1960 Eystri-Hagafellsjökull had withdrawn from Hagavatn (Figs. 6.16, 6.17, 6.18, 6.19 and 6.20). Around 1973 it had been shortened by just over 4 km and was 2 km away from the lake. In 1974, however, it surged again for 1200 m and was only 750 m short of re-entering the lake. Moreover, it eventually succeeded in doing so in 1980 when it advanced a further 900 m, causing a flood to burst through the river Farið, harrowing its channel 7–8 m deeper, the water level in Hagavatn sinking accordingly by the same amount; the lake now has a third of the surface area it had at the beginning of the 20th century. The pedestrian bridge over Farið, below the Stemmi pinnacle (521 m), was washed away in the floodwaters of 1980 and again in 1999, though it was quickly replaced on both occasions. After this, the glacier retreated once more, although it surged again briefly in 1998. The ITA erected a cabin by Einifell in 1942, just to the east of the Farið. All drinking and utility water has to be transported there. There are many enjoyable walking trails along the Jarlhettur pinnacles up to Langjökull and to Helgafell.

6.1.4   The Origins of Langjökull

A map of the topography of Langjökull's bed reveals that only 10 % of the mountain tops reach higher than the snowline in the current climate (Björnsson 2002, 2011). If the glacier were to disappear, only minor glaciers would settle on the highest mountains above a height of 1100–1200 m. Langjökull had thus come into existence when it was much colder than it is today and a large enough accumulation zone had been formed to sustain it. The glacier's survival since it was

Fig. 6.16 The surge of Hagafellsjökull into Lake Hagavatn in 1999 caused the river Farið to flood, sweeping away the ITA's footbridge, which was quickly replaced. Flooding had swept it away once before, in 1980. HB, July 1999

Fig. 6.17 The terminus of Eystri-Hagafellsjökull, near the Jarlhettur pinnacles, looked ragged and ugly during its surge in 1999, runoff water seeping out from beneath its entire snout

Fig. 6.18 The bedraggled Hagafellsjökull after its surge in 1999. HB, 1999

Fig. 6.19 Glacial surges from Langjökull. The only surges of Langjökull to have been recorded with certainty are those of Hagafellsjökull (Björnsson et al. 2003)

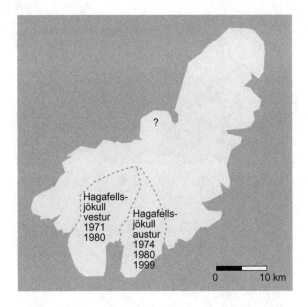

formed has been due to its own height, even though the climate today is warmer than at its creation.

Once the ice-age glacier had released its grip on Iceland, ice probably disappeared from the mountains beneath Langjökull so that there was no large, complete and continuous glacier there from about 9000 to 6000 years ago, when the annual temperature was as much as 3 °C higher than it is now. (Geirsdóttir et al. 2009; Axford et al. 2007, 2009; Flowers et al. 2007, 2008). There was no glacier between the Geitlandsstapi and Flosastapi tuyas, and the Geitlandsdyngja shield volcano was formed there almost 9000 years ago, its existence an important piece of evidence proving how rapidly the ice-age glacier disappeared from Iceland after the last glacial period.

With a cooling climate and increasing precipitation about 4000–5000 years ago, however, glaciers began to expand again on the highest peaks (Fig. 6.20). At first the glaciers covered the tuyas and shield volcano up from Flosaskarð pass. For a long time these glaciers were separate, so that the runoff water flowed from the southern glacier to the northwest of Langidalur to Þrístapafell, while only meltwater from the northern glacier ran into Hvítárvatn through the Fróðá region. But by the time of the settlement, the glaciers had been expanding for around 2000 years and had coalesced into one continuous Langjökull. It has remained a complete glacier ever since, though it had reached its largest expansion by the end of the 19th century after five centuries of a continual cold period. Numerical models based on temperature and precipitation data indicate that around 1600 the glacier had covered an area of about 900 km^2 (190 km^3) and by 1800 about 1000 km^2. It remained

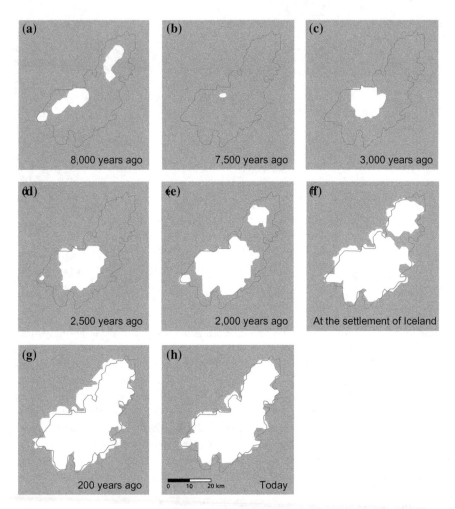

Fig. 6.20 Formation of Langjökull. The series of pictures shows the results of numerical models illustrating the formation of Langjökull glacier (Flowers et al. 2008)

more or less that size for the whole century, growing to 220 km³ by around 1840, but returning to its previous size by 1890. This picture is supported by evidence from travellers' observations, land formations, glacial moraines, and research on sediment layers in Hvítárvatn (Black et al. 2004; Geirsdóttir et al. 2009; Flowers et al. 2007, 2008). Of the sediments that have sunk to the bed of Hvítárvatn over the last 5700 years, about ¾ of them have done so during the last 400 years.

6.1.5 The Future of Langjökull

Langjökull has shrunk a great deal in the warm period of the 20th century. The shrinking was slow after the beginning of the century, but gathered pace from 1930 to 1960, the warmest interlude in Iceland's history. The glacial snouts retreated, many of them becoming 2 to 3 km shorter than when at their furthest extensions (Pálsson et al. 2012). The surface area of Langjökull has diminished from 1070 km^2 to 870 km^2, or about 18 %, from when the DGI maps were compiled around 1940. From 1900 to 1970, the glacier lost 40 km^3 or 16 % of its current mass volume, after which there was a respite to its retreat until the 1980s. During some of these years it might actually have grown slightly larger, but over the last ten years the shrinkage has rapidly increased once again. Since annual measurements began in 1996, Langjökull's accumulation zone has been only 10–40 % of the glacier's surface, and the firn line 200–300 m higher than in an average year. With each 100 m rise of the firn line, the mean mass balance of the glacier has shrunk by around 0.75 m per annum. These have been meagre years for Langjökull, winter precipitation below average (1000–2000 mm) and summer melting considerable (2000–4000 mm). The glacier's mass balance has been negative, being reduced from 2000 to 800 mm a year. The total loss of the mass balance between 1996 and 2004 was 10.8 m when spread equally over the whole glacier, which is equal to 5.3 % of its volume; this is a result of warm winters with little snowfall being followed by even warmer summers. If this continues much longer, Langjökull will not survive many more generations. If the climate is about 2 °C warmer in the coming years, the firn line might rise by as much as 300 m, reaching a height above 1400 m, meaning that there would hardly be any accumulation zone left on Langjökull and it would disappear by the end of this century (Fig. 6.21; Jóhannesson et al. 2006b; Guðmundsson et al. 2009a, b). Compared to more recent predictions concerning global warming over the next few decades (a continuous warming of 0.18 °C per decade until 2100 and then by 0.25 °C), Langjökull would mostly keep its shape until 2050, but its outlet glaciers would be shorter and nunataks would rise higher above its surface. The runoff meltwater would continually increase until a maximum is reached of 2.8 m more per year than today, after which the glacier would retract rapidly and its meltwater runoff begin to decrease. Around 2075 it would split into two, for there would be an ice-free way from Hvítárvatn in the south over to Jökulkrókur, to where Þrístapajökull now descends. At the next turn of the century, Baldjökull would be a very small ice cap, and Flosastapi tuya would have risen up through the ice, although there would still be a glacier along the basin of Langidalur. Ice would continue to descend onto it from Geitlandsdyngja, but Hagafell would have become totally free of ice. By about 2115 there would only be stagnant glacial ice remaining in Langidalur and, with no accumulation zone, this could disappear in the following 15–20 years. The only runoff water from the area would then be from precipitation.

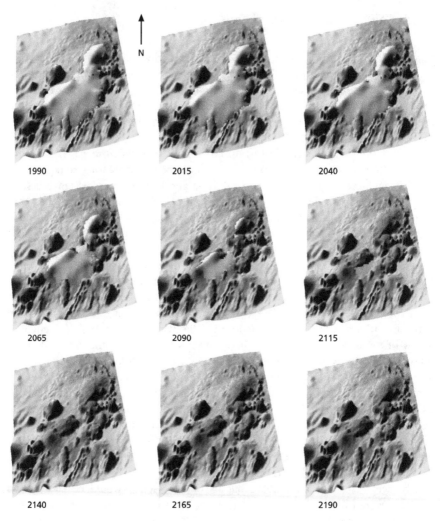

Fig. 6.21 The demise of Langjökull. Computer-generated images of the diminishing size of Langjökull over the next two centuries, based on current predictions of climate change in Iceland. It is estimated that the glacier could disappear within the next one hundred years (assuming A1B scenario of IPCC 2007)

6.2   Þórisjökull

Þórisjökull (1360 m high, surface area 25 km^2) is situated on the flat summit of a palagonite tuya that is 10 km long from west to east, and 6 km broad from north to south (Figs. 6.22, 6.23 and 6.24). In the south and east, the glacier reaches the edges of a belt of cliffs, but to its north a few unnamed outlet glaciers descend to a height below 800 m. The northwestern outlet glaciers of Þórisjökull discharge

Fig. 6.22 Satellite picture of Þórisjökull and the southern tip of Geitlandsjökull. Þórisdalur lies between them. Picture taken 22 August 2003. SPOT

water into the Geitá, but in the east, meltwater from the glacier flows into Þórisdalur. There is a gap of at least 1.7 km between Þórisjökull and Langjökull.

Until the beginning of the 20th century, the southern part of Langjökull and Þórisjökull were considered to be one complete glacier called either Geitlandsjökull or the two Suðurjökulls. A distinction was made between northern and southern Geitlandsjökull, and it is the southern part that is now called Þórisjökull. Þórisdalur was believed to be enclosed by Langjökull until 1910 when the German Ludwig

Fig. 6.23 View south toward Þórisjökull and Kaldidalur (*right*). The Geitá flows into the northern part of Hvítá. Between Geitlandsjökull (to the *left*) and Þórisjökull lies hidden Þórisdalur, and to its west Prestahnjúkur pinnacle (1226 m). Þjófakrókur ('thieves' corner') is between the Syðra- and Nyrðra-Hádegisfell mountains (1086 and 865 m), the northern one nearest in the photograph. HB

Wunder revealed (1912) that the southern glacier, which he called Þórisdalsjökull, was not in fact part of Langjökull. The shortened name of Þórisjökull came from Björn Ólafsson, the president of the ITA, who visited the area in 1918.

6.3   Eiríksjökull

Eiríksjökull glacier (21 km^2) is one of the most beautifully-shaped mountains of Iceland and has the highest summit in the western part of the country (Fig. 6.25). The glacier's high and smooth, convex ice cap rests on a plinth of the high cliffs of a palagonite tuya topped by a dolerite lava dome believed to have been formed in one of the largest volcanic eruptions of the late Quaternary period. The mountain's base covers an area of approximately 40 km^2 and its mass volume is 40–50 km^3. The glacier is reputedly 1675 m above sea level, rising just over 1000 m from its foothills in Flosaskarð pass or Jökulkrókur at sea level, and almost 1300 m above the lava on its western side. The ice cap reaches down to a height of 1300–1400 m, but its thickness has never been measured. There are no outlet glaciers on its southern and western flanks, but there are two prominent, semi-curvilinear terminal moraines, from a glacial bulge, to the west and southwest and 2 km apart, where a glacier had recently existed. A few outlet glaciers descend from the north and

Fig. 6.24 View southwest over Þórisjökull. In the background are Botnssúlur (*left*) and Skarðsheiði (*right*). To the right is Þórisdalur, and in the centre foreground there is a lake in the Fjallsauga crater

Fig. 6.25 Eiríksjökull is the largest single-standing mountain in western Iceland, proudly boasting a beautiful, domed ice cap. Aerial view westward over Flosaskarð pass to the glacier, from an aircraft above Langjökull. HB, 2007

northeastern glacial ledge, the largest being called Stórijökull by Hannesson (1958), though Thoroddsen had named it Klofajökull because the glacier had gouged a deep cleft into the mountain. Klofajökull was formed when three icefalls merged through a gap at the edge of the mountain. The firn line is probably at a height of 1200 m. Spurs of ice with a thin covering of ground moraine stretch out into the surrounding lowland and are called Jökulrani, which is about 2.5 km broad and extends the glacier in this coated form about 1 km from its outlets. To the east of Stórijökull is an outlet glacier called Þorvaldsjökull, and then further east another called Ögmundarjökull, of which little now remains visible (these glaciers were named thus by Pálmi Hannesson in 1923 in honour of Þorvaldur Thoroddsen and his associate Ögmundur Sigurðsson). The rocky pillar Eiríksnípa juts out from the northwestern corner of Stallur in the outlet glaciers to the north of Eiríksjökull. This Eiríkur was one of the outlaws from the Surtshellir caves, the Hellismen, who fled up this rock face on one leg to avoid being captured by a posse from the settlements. To the east of Eiríksnípa there were two outlet glaciers that have now retreated back up to the mountain edge, but because they trickled down from the main glacier in the shape of woollen long johns laid out to dry, Thoroddsen named them Eystri- and Vestri-Brók in 1898. In 1956 the eastern 'leg' still reached below the edge of the mountain, but not the western one. Glacial rivers do not flow from Eiríksjökull, but it is, however, a source of spring water.

6.4   Hrútfell

The ice cap of Hrútfellsjökull glacier (1396 m high, surface area 6 km^2) covers the top of the tuya Hrútfell (or Hrútafell), which rises steeply from its mountainous base at a height of about 600 m (Figs. 6.26 and 6.27). Its summit has cliffs on all sides, and there are five outlet glaciers that descend from it, three of them named. Its meltwater runoff reaches Hvítárvatn. The domed ice cap has also been called Regnbúðarjökull. On a satellite picture taken on 14 August 2004, the accumulation zone (white) is visibly less than a fifth of the glacier. With such a negative mass balance, Hrútfellsjökull will shrink so rapidly that the tuya could be glacier-free within a few decades.

6.5   Ok

On the northern side of the dolerite shield volcano of Ok (1190 m), to the west of Kaldidalur, is a small glacial patch (less than 1 km^2) that reaches down to a height of 950 m (Figs. 6.28 and 6.29). This glacier has diminished in size rapidly during the 20th century and is now completely below the snowline. Eggert Ólafsson and Bjarni Pálsson mentioned Ok in their travelogue (1981), but without stating whether it had a glacier or not, though Pálsson (1945) depicted Okjökull on his map of

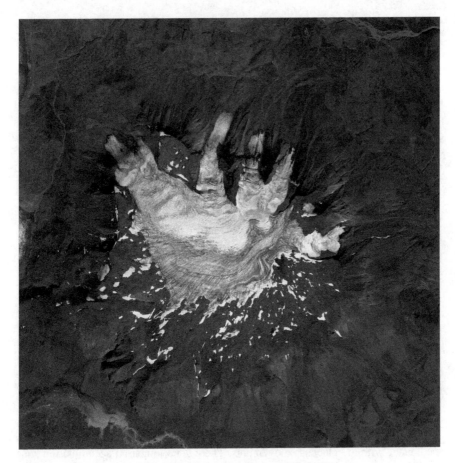

Fig. 6.26   Satellite picture of Hrútfell taken on 14 August 2003. SPOT

Langjökull from 1792. Thoroddsen described the glacier as a regularly-shaped, snow-white mound, like an eggshell, although further down its flanks it was succeeded by scattered patches of snow which gave the mountain the appearance of a spotted leopard. No individual outlet glaciers crept down from the glacier and all its meltwater runoff seeped into the dolerite lava. The ice cap is depicted as being about 38 km^2 on Thoroddsen's geological map from 1901, but on the map of the Danish General Staff from 1910, it was only around 15 km^2, and only 5 km^2 on the map of 1945; the glacial patch had a surface area of just over 3 km^2 on an aerial photograph from 1978, just less than 2 km^2 on a satellite picture from 2004, and it has been reduced even more since then, the crater at the top of the lava dome now glacier-free.

Once upon a time, the glacier had crept forwards in all directions from the top of Ok. Pure winter snow settled every year on the sand that had been blown onto its surface the previous summer. But instead of the layers of sand being one on top of

Fig. 6.27 Aerial view over the Innri- and Fremri-Sandfell mountains (888 and 927 m), looking south toward Hrútfell. Between Hrútfell and Langjökull is Sanddalur (on the *right*). Snævarr Guðmundsson, 2002

the other, they were shunted downwards and became lined up in concentric rows in accordance with the glacier's movements. There is no longer an accumulation zone on Ok and so every year a thin sliver is sliced off from the surface of the entire glacier, revealing the layers of sand like rings in a tree trunk. The mini-glacier seems to be lying over a small summit dome with layers of sand encircling its centre.

6.6 Hofsjökull

Hofsjökull is the third largest glacier of Iceland after Vatnajökull and Langjökull, but until a short time ago it had been the least visited or explored of these large glaciers. Hofsjökull is almost in the dead centre of Iceland and highland routes pass it on both sides, the Sprengisandur route to its east, and the Kjölur and Eyfirðingavegur routes to its west (Figs. 6.30, 6.31 and 6.32). Many large rivers flow from this glacier: Jökulfallið or Jökulkvísl runs southward through Kjölur into the Hvítá just below Hvítárvatn; the Blanda, on its western side, and the two northern Jökulsá rivers flow north into the Héraðsvötn lakes in Skagafjörður; the

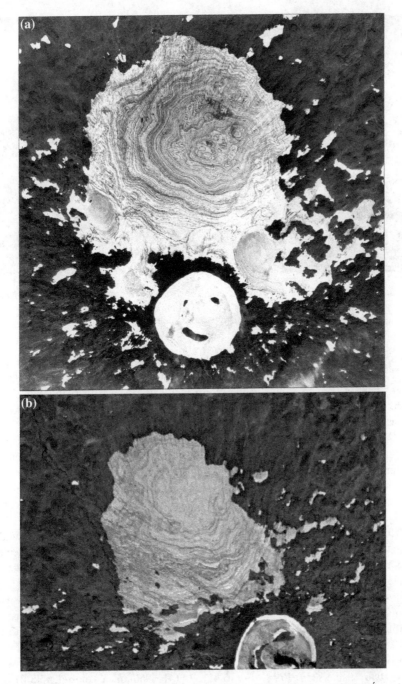

Fig. 6.28 The top picture of Ok (**a**) is taken from an aircraft in September 1978 (LMÍ), and the lower picture (**b**) from a satellite in August 2004. SPOT

Fig. 6.29 The postglacial dolerite shield volcano Ok is to the west of Kaldidalur. The glacier has a surface area of <1 km^2 and reaches a height of 1140 m. The crater itself is now glacier-free, but there is a small glacial patch on the northern edge of the mountain

Þjórsá tributaries flow east and south (Figs. 6.33 and 6.34). The Þjórsárver wetlands, the largest continuous oasis of vegetation in the Icelandic central highlands, owes much of their existence to the runoff meltwater from Hofsjökull (Figs. 6.35, 6.36, 6.37, 6.38, 6.39 and 6.40; Þórhallsdóttir 1988). Apart from Þjórsárver, Arnarfellsbrekka slope to the south, and the Álftabrekkur slopes to the northwest, Hofsjökull is otherwise surrounded by barren gravel plains, glacial till, and lava. The inhabitants of southern Iceland used to call it Arnarfellsjökull after Arnarfell Major, but the people in the north called it Hofsjökull after the farm of Hof in western Skagafjörður, and this is the name which has prevailed. For a long time the glacier was confused with Langjökull, especially in eastern Iceland, and Hofsjökull's outlet glaciers were not even named.

This confusion is notable in the cartographic history of Hofsjökull. Guðbrandur Þorláksson's map of 1590 (Fig. 4.5) depicts it as Arnarfellsjökull with rivers flowing northwestward from it to Skagafjörður, although it also seems to discharge water into the Hvítá in Árnes County and the river Þjórsá. On Knoff's map of 1734 (Fig. 4.10), and later Horrebow's of 1752 and Homann's of 1761, which were compiled from Knoff's original, the glacier is named Hofsjökull, but on the maps of Gerhard Schøning and Jón Eiríksson (Fig. 4.12), accompanying Eggert Ólafsson and Bjarni Pálsson's travelogue of 1772, and the map of Ólafur Olavius in 1780, it is called 'Hofs-Iökull al. Lange-Iökull'. Sveinn Pálsson made a distinction between these two glaciers in 1794 (Fig. 6.41), but Hofsjökull (Arnarfellsjökull) still remained an oval dome without outlet glaciers or place names on his map. On Björn Gunnlaugsson's map (Fig. 4.21), the outlines of Hofsjökull became more correct,

Fig. 6.30 Múlajökull creeps down from the high dome of Hofsjökull, held in on both sides by Hjartafell (*left*) and Kerfjall (*right*). Once it reaches the lowland it splays out, forming Iceland's most beautiful outlet glacier. Ahead of Múlajökull are small lakes in the arc-shaped glacial moraines. To the west of Hjartafell is Nauthagajökull, from which flows the Fremri-Múlakvísl on the far left. Arnarfellskvísl can be seen emerging from beneath the eastern margin of Múlajökull on the other side of the picture. There used to be an ice-dammed lake in the cleft between Kerfjall and Múlajökull, but there is no water there now. The southern and eastern flanks of Kerfjall have plenty of vegetation. Furthest east, there is a glimpse of the lower mountainside of Arnarfell Major. HB

but the outlet glaciers were still without names except for Blágnípujökull in the southwest, to the east of the peak Blágnípa, north of the Kerlingafjöll mountains. Gunnlaugsson never visited the northern side of Hofsjökull.

6.6.1 Explorations of Hofsjökull

In the nineteenth century, the Dane Jørgen Christian Schythe (1814–1877; 1840) passed Hofsjökull on a very calamitous journey he made along the Sprengisandur highland route in the summer of 1840, though he did initially have at least one good night's stay at the foot of Arnarfell. Iceland's national poet, Jónas Hallgrímsson, composed the following poem on Schythe's rather self-centred narrative about his journey:

(a)

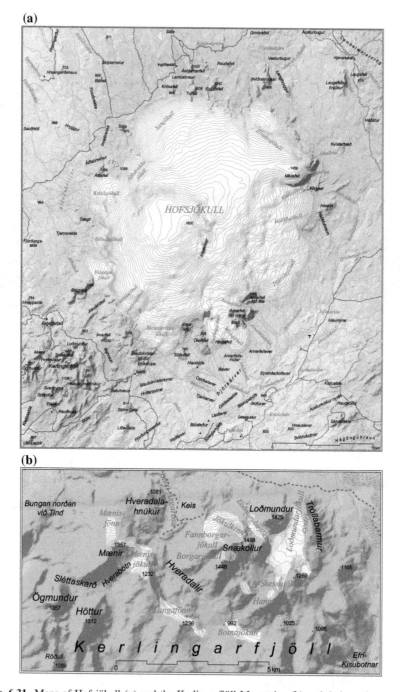

(b)

Fig. 6.31 Maps of Hofsjökull (**a**) and the Kerlingarfjöll Mountains (**b**) and their environs

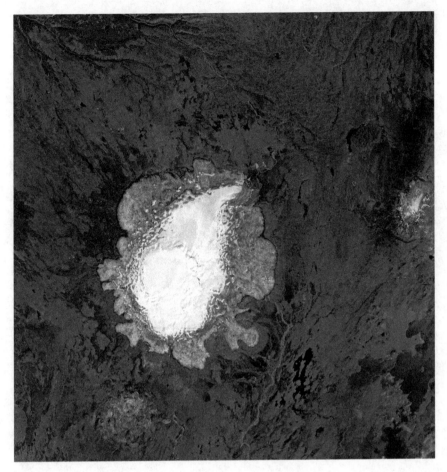

Fig. 6.32   Satellite image of Hofsjökull taken from a height of 820 km on 14 August 2004. SPOT

Under the icy eaves of Arnarfell
Human habitats far away
– I swear I am telling the truth –
Slept some Danes there yesterday
But e'er they arose next morning,
The fires were lit quick and fast
Victuals must be boiled and basted
For breakfast's a vital repast
Few, if any, know much at all
Of mountains and glaciers so vast
They should be telling us facts instead
About Arnarfellsjökull and its past. (Hallgrímsson 1957, p.127)

Fig. 6.33 Water divides of Hofsjökull. Most of the meltwater from the glacier flows into the rivers Þjórsá and Blanda

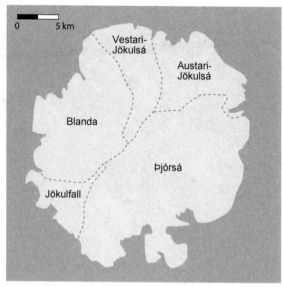

------ Water divide

Fig. 6.34 Surges from Hofsjökull have been especially notable in Múlajökull and Þjórsájökull

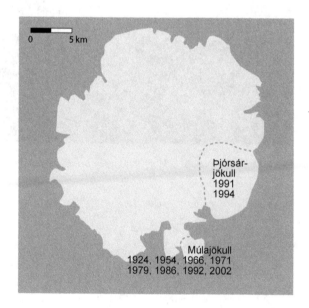

Other scientists who visited the southern margins of Hofsjökull along the Sprengisandur highland route in the 19th century included the Germans Bunsen (1847) and Wolfgang Sartorius von Waltershausen (1847), the Dane Bruun (1902, 1921–1927), and the Icelander Þorvaldur Thoroddsen, who described in exact detail the mountains at the glacier's margin and the locations of the rivers that emerged from beneath it. The German Erkes (1911, 1914) explored the highlands to the east

Fig. 6.35 The springs in the southernmost part of the Þúfuver wetlands. Hofsjökull, on the horizon to the north, is a vital mainstay of the Þjórsárver wetlands, providing them with both water and shelter. The outlet glaciers and mountains in view are, from west to east: Blautukvíslarjökull, Ólafsfell, Nauthagajökull, Hjartafell, Múlajökull, and standing out just above it is first Kerfjall and then Arnarfell Major, to the east of which can be seen Þjórsárjökull. The dark mound to the east of Arnarfell is Biskupsþúfa, and travellers used its shelter as a stopping place on the ancient Sprengisandur highland route. The edge of the rift lower down on the hill, with woolly willow, indicates that it once had much more vegetation. The yellow moss by the springs has been torn up by pink-footed geese searching for roots. HB, July 1999

Fig. 6.36 Dwarf fireweed and garden angelica blooming in front of Múlajökull in the Þjórsárver wetlands. Þóra Ellen Þórhallsdóttir, 2007

of Hofsjökull in the beginning of the 20th century, as indeed did the Swiss Hermann Stoll. In August 1911, the Germans Wunder (1912) and Eugen Weissenberg rode to the Kerlingarfjöll mountains, and from Blágnípa they ascended the dome of Hofsjökull on horseback, the first men to do so, as far as is known.

Fig. 6.37 There is permafrost in the Þjósárver wetlands and palsa hillocks are formed from ice lenses and frozen soil during frost heaving. Wetland vegetation disappears and dry-land plants gradually begin to take root. Palsa hillocks in the Þjósárver are usually up to about a metre high, but can be dozens of metres long. The picture is taken in Múlaver, while in the background Nyrðri- and Syðri-Háganga mountains are most prominent. Þóra Ellen Þórhallsdóttir, July 2008

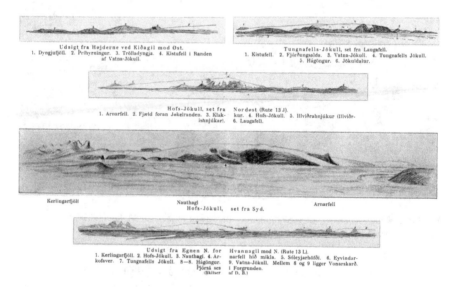

Fig. 6.38 Drawings of the central highlands of Iceland by Daniel Bruun from the year 1925. NULI

EYVINDARKOFAVER, WITH ARNARFELLSJÖKULL IN THE DISTANCE.

[To face page 60.

Fig. 6.39 Drawing of Hofsjökull, as seen from Eyvindarkofaver, taken from John Coles' book *Summer Travelling in Iceland* (1882). NULI

From there they continued westward via Álftabrekkujökull, to the north of the Gægir nunatak, to Strýtur on the Kjölur route, and then on to Hveravellir. Wunder sketched a contour map of Hofsjökull, stating the summit was 1635 m high and that its ablation zone reached a height of 1300 m on its western side. In August 1930, the Austrian Keindl (1903–2007; 1932), with his associate Josef Gitner, crossed Hofsjökull from southwest to northeast. Keindl drew a rough outline of Hofsjökull's contours and claimed its ablation zone reached a height of 1000 m, pointing out that, although the glacier was supported by mountains rising up from the highlands around it, not all of them reached as high as the firn line. Hofsjökull and Langjökull would have been created in a colder climate than the present one, and the glaciers' existence is now sustained by their own elevation; they were probably remnants from the last glacial period and should they disappear, e.g. due to volcanic eruptions, glaciers of a similar size would probably not form there again in current conditions. In 1951 the grammar school student Magnús Hallgrímsson (b. 1932), along with a British physician, became the first Icelander to ascend the glacial plateau of Hofsjökull, and he ascended it again in 1956, only this time with a group of colleagues. They climbed up onto the glacier from the southern side of Blágnípujökull, crossed the glacial dome northeastward, and descended by Laugafell; they were on their way north to Eyjafjörður from Biskupstungur in the south.

Fig. 6.40 View westward from Hofsjökull. Múlajökull is nearest with its arc-shaped moraines ahead of it. Múlajökull and Nauthagajökull used to meet up in front of Hjartafell. The marginal moraines they left behind are clearly visible on the picture, like a badly-fitting cape around the mountain's shoulders. To the west of Nauthagajökull is Ólafsfell, a long but not high mountain, and on its other side the valley indent Jökulkriki. Nautalda mound is the single terminal moraine on its forefield. Söðulfell stands out against Blautukvíslarjökull, and beyond it, in the distance, are the Kerlingafjöll mountains. HB

The Danish geographer Niels Nielsen explored the central highlands of Iceland extensively, and in 1924 he followed the same trail that von Waltershausen had taken in 1846. Nielsen described three glacial moraines in front of Arnarfellsmúlar and the tracks left by the one that had extended furthest. The southern terminus of Múlajökull had thus advanced about 800 m from the time of von Waltershausen's visit, for the moraine was only 200 m from the glacier when Nielsen was there, whereas it had been 1000 m away only eight decades previously. Nielsen did much pioneering work that is still referred to today (1927, 1933), although having listened to a lecture on one of his research expeditions in Iceland, the poet and satirist Jón Helgason was overcome by the same scepticism and mockery as his predecessor Jónas Hallgrímsson. Helgason presented Nielsen's lecture as follows:

I sailed o'er the sea to Iceland
Where frost melts from cliff and cave
And travelled widely its lava fields,
Its grasslands and gravel plains.

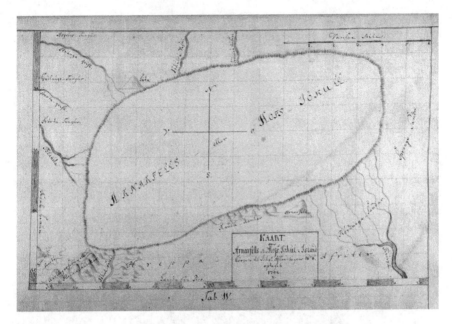

Fig. 6.41 Sveinn Pálsson's map of Arnarfells- or Hofsjökull from his 1795 treatise on glaciers. The name Hofsjökull is now the accepted one and is drawn from the farmstead Hof in Vesturdalur, inland from Skagafjörður

I rode my horse all by myself,
Straight-backed and, oh, so brave,
Though these servants of mine,
Who attend me here this night,
Had led it by the reins
Though the trail was tiresome
And storms splashed my attire
My findings were by no means few
And worth stating all the more:
Under the western side of Hofsjökull
A small crag came into view
– And as far as I'm aware,
It's ne'er been discovered before. (Helgason 1986, p. 215)

Regular research on Hofsjökull began with measuring the exact location of the snouts of the Blágnípujökull, Nauthagajökull and Múlajökull outlet glaciers in 1932, and that of Sátujökull in 1950. The glacier's mass balance has been calculated every year since 1988 (Björnsson et al. 2013, and accompanying references). The DGI had the first geodetic maps of the glacier compiled in 1937, after which the U.S. Army Map Service compiled a map from aerial photographs from 1945–1946. On satellite pictures (ERTS-1) from 1973, the outline of the glacier's surface indicated that a caldera lay hidden beneath it. An exact map of Hofsjökull was only finally obtained from altimetric data in 1983, while its subglacial topography was simultaneously being explored through radio echo soundings (Björnsson 1986, 1988, 1991).

6.6.2 Hofsjökull at the Beginning of the 21st Century

Hofsjökull currently has a surface area of around 830 km^2 and a volume of just under 200 km^3 (Figs. 6.31, 6.42, 6.43, 6.44 and 6.45). The average thickness of the ice cover is 225 m. The glacier is domed, like a circular shield of ice, 30–40 km in diameter and 1800 m high, surrounded by individual mountains, between which its

Fig. 6.42 Computer-generated images of Hofsjökull and its subglacial topography. Hofsjökull is 830 km^2 in size. It is convex in shape, like a circular shield, and it covers the largest volcano in Iceland. Beneath the glacier lies hidden a 650-m-deep caldera, full to the brim with ice. The subglacial terrain is at its lowest height at Arnarfell beneath the beautifully-shaped glacial tongue of Múlajökull. There would be lakes in this depression and within the caldera should the glacier disappear. Vegetation would then reach as high as 600–700 m, though snow would always be present on the highest mountains

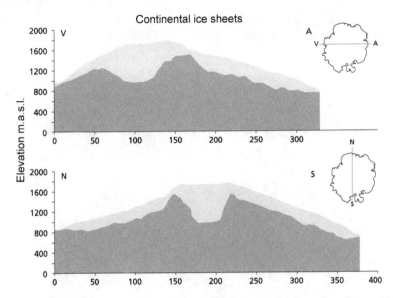

Fig. 6.43 Cross sections of caldera beneath Hofsjökull. The ice in the centre of the caldera is about 750 m thick

outlet glaciers force their way, descending steeply and with many crevasses in all directions (Björnsson 1986, 2009). The main ice divides are on a ridge of ice from its highest elevation northeastward to the palagonite tuya of Miklafell (1456 m; Fig. 6.46), which appears to have vertical cliffs when seen from the Sprengisandur highland route. Hofsjökull's largest individual outlet glacier, the 19-km-long Þjórsárjökull extends southeastward between Miklafell and Arnarfell Minor (1143 m), its terminus descending to a height of 650 m. A few ice-free peaks of rhyolite cliffs, Hásteinar (1630 m), reach 100 m above the cloak of ice high up on Þjórsárjökull's summit. Further down, half way to the snout and due north of Arnarfell, are Lágsteinar cliffs (name given during a radio echo sounding expedition in 1983). There is an indent in the terminus alongside a palagonite ridge to the east of Nyrðri-Hásteinar, and a little further north, just southwest of Háalda (938 m), Háölduhraun, believed to be 8000 years old (e.g. Kaldal and Víkingsson 1990), appears from beneath the glacier in a kilometre-broad swathe of lava that stretches 15 km southward along the glacial margin. The source of the Þjórsá is in the Bergvatnskvísl tributary in the northern reaches of the Sprengisandur central plain, but about 20 km further southwest it flows into Háöldukvísl, the most northerly river outlet from Hofsjökull. Once this converges with the Fjórðungakvísl from Tungnafellsjökull, the river then takes the name Þjórsá. The outlet glacier Rótarjökull advances between the rhyolite Arnarfell Minor (1140 m) to the north, and Arnarfell Major (1137 m), further south. These two mountains are almost

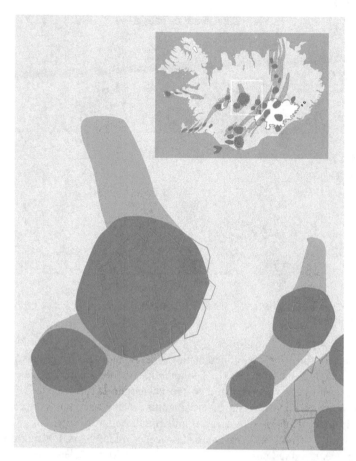

Fig. 6.44 Main volcanoes and fissure swarms beneath glaciers in the central highlands of Iceland. The diagram depicts the locations of the main central volcanoes in Hofsjökull, the Kerlingafjöll mountains and Tungnafellsjökull

exactly the same height, but the northern one is less noticeable and has very little vegetation.

Múlajökull, the largest outlet glacier from the southern flanks of Hofsjökull, descends in a 7-km-broad snout from a height of 1800 m before rapidly squeezing its way, with tremendous abrasive force, through a 2-km-broad cleft between durable rhyolite mountains to the west of Arnafell Major. It then veers southeastward between Jökulbrekka slope and Hjartafell and fans out as a piedmont glacier down to a height of 610 m. There are stories of a settlement of outlaws in a valley behind Arnafell Major, which could not actually be seen until stumbled upon; two brothers were supposed to have lived in the wild there (Briem 1959). There is a glacial lake on each side of Múlajökull, and jökulhlaups have flooded southward from both of them. To the west of the glacier is Ólafslón, north of Ólafsfell

Fig. 6.45 The eastern side of Hofsjökull. The rocks jutting out on the glacier's crest are Hásteinar and they demarcate the outer rim of the Hofsjökull caldera. The large outlet glacier on the right is Þjórsárjökull, from which flow the tributaries of the Þjórsá into the Þjósárver wetlands. To the left of Þjósárjökull is Arnarfell Minor, very steep but clad with moss high up its sides. Then comes the very narrow Rótarjökull and Arnarfell Major. Just in front of the main part of this mountain, a lower, single peak stands out, Arnarhaus. Continuous vegetation reaches as high as 1000 m on Arnarfell's flanks and is also rapidly taking root on the flood plains below to the south and east. HB, August 1996

(926 m), while the lake Jökulker is to the south of the foot of Arnarfell, south of Kerfjall. It used to be full of water and icebergs, but is now empty for long periods of time. Three principal rivers flow from Múlajökull: Arnarfellskvísl runs from beneath its northern margin, Fremri-Múlakvísl flows to the south of Hjartafell, and Innri-Múlakvísl emerges from its centre. The southern slope of Arnarfellsbrekka rises above the glacier, its lower reaches covered with grass, woolly and tea-leaved willow bushes, large angelica plants, northern bilberries, wood crane's bill, and golden root (altogether 97 species of flowering plants). Vegetation disappears above an elevation of 800 m and is replaced by moss covered screes, and then higher up at the top by bare screes. The many rows of curvilinear moraines, caused by previous advances of this outlet glacier into its forefield, are called Arnarfellsmúlar, from which Múlajökull gets its name. In between these hills are innumerable small lakes looking like pearls on a necklace. The glacier's snout has now retreated 2 km from

Fig. 6.46 Miklafell comes into view when crossing the Sprengisandur highland route northward, with Þjórsárjökull on the left. HB

its furthest terminal moraine. The moraines nearest the terminus have little vegetation, but huge angelica plants grow further away from it, as well as golden root bushes and broad swathes of a large number of other flowering plants. It is believed that the outlaw Fjalla-Eyvindur had a hovel here in these hills, which was destroyed in 1762 by men from Árnes County.

Nauthagajökull creeps south between Ólafsfell (926 m) and Hjartafell down to Nauthagi meadow, so named in 1847 when two bulls, which had been lost during a household move between northern and southern Iceland, were found grazing there. It also has warm springs. From the centre of Nauthagajökull flows the Miklakvísl, or great river, a name it lives up to. Blautukvíslarjökull creeps forwards to the west of Ólafsfell and Jökulkriki, the Blautakvísl flowing from it, though it also discharges water into Miklakvísl through streams in Jökulkriki to the east of Söðulfell (765 m), where lava also disappears beneath the glacier. The Blautakvísl flows further to the west through a large ravine down to the Blautukvíslareyrar gravel banks by Eiríksnípa (789 m), and to the northwest of these yet another lava field extends beneath the glacier's margin along a section 4 km long. The braided rivulets of the Blautakvísl are often difficult to cross due to quicksands and constantly changing channels, the water sometimes even flowing east into the Miklakvísl. The spring-water river Hnífá emerges from its basin just to the west of the Blautukvíslareyrar, and bears groundwater from beneath Illahraun lava field, between Hofsjökull and the Kerlingarfjöll, although glacial water from the Miklakvísl sometimes flows into and blends with it. The nunatak Tanni ('Tooth', 1564 m) can be glimpsed high up on Blautukvíslarjökull; it was named after the snowmobile bulldozer with spiked caterpillar tracks, which was used during the radio echo sounding expedition on the glacier in 1983 (Björnsson 1986).

The Þjórsá receives meltwater from almost half of Hofsjökull, its catchment area reaching from Blautakvísl to Háöldakvísl and up onto the eastern side of the glacial plateau. On both banks of the Þjórsá are the Þjórsárver wetlands (150 km^2), the largest isolated oasis of fertile grasslands and marshes in the central highlands of Iceland, and probably the vegetated area in the country that is as near as possible to being the most pristine and undisturbed by man. The Þjórsárver can thank these glacial meltwaters for their existence. To the west of the Þjórsá, the groundwater-fed river Hnífá, the glacial rivers Blautakvísl and Miklakvísl, and the two branches of the Múlakvísl and the Arnarfellskvísl, all flow through the wetlands.

In the western reaches of the Þjórsárver are the Hnífárver wetlands, bounded in the west by Fjórðungssandur, then by Tjarnaver as far as to Blautakvísl, and then by Oddkelsver between the Blautakvísl and Miklakvísl rivers. Allegedly, there was once an outlaw in these parts called Oddkell, who had lain with his sister and had eight children by her, and that the Bishop of Skálholt's servants had killed him when they stopped at Tjarnaver while travelling on an episcopal visitation. Another version has it that Oddkell had attacked some travellers, who had killed and buried him in a gravel moraine south of the Þjórsá, and this was why the place was named after him. Further south, at the confluence of the Þjórsá and Blautakvísl, is Sóleyjarhöfði headland. To the east of Miklakvísl, south of Nauthagi, are the Illaver wetlands which are difficult to traverse as they are so marshy. Next are the Múlaver wetlands, and then further inland the Arnarfellsver wetlands. Arnarfellsalda hill lies opposite the Arnarfellsmúlar moraines and to its west are the ruins of a shack by a small lake. Two labourers from Laugardælar, in the south near Selfoss, Gunnlaugur Bergsson and Jón Jónsson, fled there in the autumn of 1848 to escape justice after being found guilty of theft, but they were soon found and sentenced to 8 and 10 years hard labour. Jón Jónsson later became a Mormon in America and returned to Iceland as a missionary for his new faith. To the north of the hill are bridle paths, believed to be the remains of the ancient Arnarfellsleið track, which had been used when the route east over the Þjórsá at Sóleyjarhöfði ford was impassable. The Hnífá was then forded in Hnífárver, the track continuing through the Nauthagi meadows, along the Múlar moraines to Arnarfell, then along the glacier's margin and northward over the numerous Þjórsá tributaries, one by one, north of Háumýrar, the highest grassland of Holtamannaafrétt sheep pastures (630 m), and long since considered the beginning of the Sprengisandur highland plain. To the east of the Þjórsá are the Þúfuver and Eyvindarver wetlands, named after the outlaw Fjalla-Eyvindur, who had his best dwelling place in the highlands of the Þjórsárver.

Water flows southwestward from Hofsjökull's glacial tongues reaching toward Þverfell (1032 m), Brattalda moraine and Blágnípa pinnacle (1068 m), into the Jökulkvísl or Jökulfall rivers, all from a catchment area of about 50 km^2. Blágnípujökull is divided by the tuya Blágnípa north of Jökulkriki. The Jökulfall emerges from the glacier to the south of Blágnípa, and is joined by streams emerging from beneath Illahraun lava field and Loðmundarjökull in the Kerlingarfjöll mountains as it continues on its way south to the Hvítá in Árnes County. Illahraun was formed by lava from a row of craters near Brattalda and is very rough and difficult to traverse. Lava that originated from volcanoes on the northern side of Blágnípa is now covered by glacial till. To the east of Blágnípujökull is a passable route northeastward up onto the summit of Hofsjökull.

Blöndujökull creeps westward, north of Blágnípujökull, from the high dome of Hofsjökull. Further north are Kvíslajökull and then Álftabrekkujökull, all discharging water into the river Blanda. Blanda also receives glacial meltwater from the western part of Sátujökull, or in all from a catchment area of about 220 km^2 of the glacier's surface area. The borders of the catchment area supplying the Blanda reaches high up onto the glacier from the southern sides of Blöndujökull by Fjórðungsalda, just to the north of Blágnípa, then circles the high dome of Hofsjökull and the western part of Sátujökull, from where the Strangakvísl flows. The source of Svartakvísl and Blöndukvísl is further south, and the river Blanda receives glacial runoff water from Langjökull through the Seyðisá. The Gægir nunatak, as the name implies, peeks out from Kvíslajökull a short distance from its terminus. The vegetated slopes of Álftabrekkur to the west of the palagonite ridge of Álftafell (1084 m), at Kvíslajökull's northern margin, is a stopping place with grazing meadows for horses on journeys through the central highlands on the Eyfirðingavegur route, which ran just to the north of Sáta (965 m), a single-standing mountain of palagonitic tuff.

Vestari-Jökulsá receives its runoff meltwater from the more easterly part of Sátujökull to the west of Krókafell (996 m) and from its terminus to the north of Tvífell (1006 m) right up to the top of the glacier, a catchment area totalling about 90 km^2. The river emerges from beneath the glacier in two main branches on each side of Jökultunga, but also takes water from the Fossá and Bleikálukvísl to the east of Tvífell. Vestari-Jökulsá then flows north through a ravine into Hofsdalur. The lava flows from Lambahraun, which probably originated from the volcanoes beneath the glacier, enclose the palagonite mountains of Tvífell and Ásbjarnarfell (1025 m). Ingólfsskáli, the cabin of the Touring Association of Skagafjörður, is situated to the west of Ásbjarnarfell, from where there is an easily passable route up onto Sátujökull and all the way up to the dome of Hofsjökull. The tributary sources of the Austari-Jökulsá are to the west of the Illviðrahnjúkahraun lava field, with branches on either side of Illviðrahnjúkur (988 m), a rather bare mountain of palagonitic tuff, and in the Langakvísl tributaries to the south of Langihryggur ridge (891 m), though the southernmost source of the river can be found in the Jökulkvísl from beneath Klakksjökull to the south of Miklafell; altogether, this is a catchment area of 120 km^2. Klakkur (1010 m) rises 2 km to the east of Klakksjökull's snout, its thin pinnacle reaching 300 m above its surroundings. Austari-Jökulsá is a more voluminous river than Vestari-Jökulsá, though neither of them receive water from Hofsjökull's glacial plateau. Their combined catchment area is about 210 km^2. Austari-Jökulsá flows north into Austurdalur and joins the Vestari-Jökulsá further south of the Tunguháls farm in Tungusveit district, where Austurdalur and Vesturdalur converge, and from there on the river is called Héraðsvötn. This completes the circular tour of Hofsjökull's outlet glaciers and glacial rivers.

6.6.3   The Subglacial Terrain of Hofsjökull

Beneath Hofsjökull is the country's largest volcano, the midpoint of volcanic activity in the central highlands of Iceland during the Quaternary period. The volcano towers in the middle of a 90-km chain connected to Iceland's main volcanic zone from Reykjanes peninsula in the west, through Langjökull, and then east and north into Þingeyjar County (Fig. 3.12). The Hofsjökull zone includes both the Kerlingarfjöll mountains and Hofsjökull as it stretches northward with fissure swarms into Skagafjörður through the Hofsafrétt and over Eyvindarstaðaheiði, and southwestward from the Hofsjökull volcano to Kerlingafjöll and on to Hreppar and even as far as Skeið in the southern lowlands. The swarms contain a large low temperature geothermal belt. Due to very slow tectonic movements along this rift zone, the main volcano has managed to accumulate volcanic material for long enough in one place to become the largest volcano in Iceland in terms of volume. The basalt lava fields at the outlet termini to the north, east and south, all show that the volcano has been active in the most recent post-glacial period, although there have been no known eruptions after the settlement of Iceland. Volcanic vents linked to the lava fields on the glacier's margins are not immediately visible, although 2 km to the south of the snout, a single-standing, 150-m-high and 1-km-broad volcanic cone arises from Tvífell, which could be the source of the Lambahraun lava field. There are quite a number of palagonite mountains and a few tuyas around Hofsjökull, and rhyolite can be seen at the glacier's margins and in its nunataks.

Beneath Hofsjökull lies hidden a 650-m-deep caldera, full to the brim of ice and with its lowest opening to the west (Figs. 6.42 and 6.43). The bottom of the caldera is at a height of 980 m, at the same elevation as the edge of the glacier. The caldera's rim is highest on the eastern side, at 1650 m, where the Hásteinar peaks of schistose rhyolite (1630 m) rise up from beneath the ice, indicating the mountain was not covered by a glacier when lava was flowing there. The nunataks Gægir and Tanni (1564 m) are peaks outside of the caldera's rim, but single Lágsteinar peaks rise up from a ridge in a southeastward direction. Geologically speaking, the caldera is young, indeed probably from the last glacial period, and eruptions have not managed to fill entirely it with volcanic material. At a height of 1000 m, the volcano's base area is about 200 km^2. The ice in the centre of the caldera is around 750 m thick.

There is a pass in the western verge of the caldera's rim, the lowest height of which is unknown, though it could be at an elevation as low as 1200 m. This would be high enough for a lake to form there, similar to Lake Askja in the Ódáðahraun lava field, if the glacier were to disappear (Fig. 8.108). Blöndujökull creeps westward through this pass, for the most part almost 200 m thick as it descends. Meltwater does not seem to accumulate in the caldera's basin, but drains westward from it instead. Other outlet glaciers that transport ice from the caldera (Sátujökull, Þjórsárjökull and Múlajökull) are only 100–250 m thick where they flow over its edges, but they gradually become thicker as they descend from the high dome until they sprawl outwards, becoming thinner at their snouts.

Two ridges over 1100 m high extend southward from Hofsjökull's caldera. One marks the western boundary of Blautukvíslarjökull, to the south of which are lava fields which continue beneath its margins. The other ridge is about 6 km further east, the peak Tanni rising up from it, and stretches south towards a lava field in Jökulkriki to the east of Blautukvíslarjökull's terminus. Beneath the centre of Blautukvíslarjökull is a narrow, 250-m-deep ravine that might have been gouged out in a tremendous jökulhlaup during an eruption on the southern side of the main volcano and which continues as Blautukvíslargljúfur ravine when it appears from beneath its snout. A large ridge extends northeastward from the caldera in the direction of Illviðrahnjúkar peaks, and the sources of the Illviðrahnjúkahraun lava field are probably at its northernmost point. There is another high ridge, at just over 1300 m, extending southwestward from Miklafell that is parallel to the ridge coming northeastward from the caldera. There is a trough-shaped valley between these two ridges, closed at both ends, and its base elevation is as low as 900 m; this is where Hofsjökull's ice cap is at its thickest outside the caldera, at 580 m. Without a glacier, there would be a lake there. Hofsjökull glacier is thus supported by the rims of the caldera, Miklafell, and the extending ridges from both of them.

About two thirds of the glacier's bed is above 1000 m in height, but only a ninth of it is above a height of 1300 m. The terrain beneath Hofsjökull is at its lowest elevation furthest south, beneath the beautifully-shaped glacial tongue of Múlajökull alongside Arnarfell Major. A 550-m-thick outlet glacier has harrowed out a narrow valley in the cleft between Hjartafell and Jökulbrekka down to an elevation of 500 m above sea level. Beneath its tongue, furthest down, its basin reaches as low as 450 m, its bedrock 150 m lower than the forefield at the glacier's snout. A beautiful lake might form there, should the glacier disappear. The glacial snouts of Þjórsárjökull, southeast of Lágsteinar, and Blautukvíslarjökull, below the ravine to the south of Tanni, have gouged out 100-m-deep trenches to below a height of 600 m. Water erosion has had a major part in this, possibly through a jökulhlaup due to a subglacial volcanic eruption. The two outlet glaciers, to the west and north of Hofsjökull, on the other hand, only descend to a height of 800 m at their lowest. Finally, it should be pointed out that the indent in the centre of Þjórsárjökull, to the south of the source of Háölduhraun lava field, is caused by a palagonite ridge that continues northwestward into and beneath the glacier.

6.6.4   Origins and Future of Hofsjökull

Under current climatic conditions, only 10 % of the bed of Hofsjökull, and 40 % of its surface, would be above the snowline in an average year (Björnsson 1986, 1988). This is the accumulation zone on which it still survives. Should the glacier disappear, there would only be a few ice carapaces hunching around the caldera's rim above a height of 1200 m, above the Tanni and Hásteinar nunataks and on the northern edges of the volcano. Glacial patches might also be established on Miklafell. If the climate becomes 2–3 °C warmer, the firn lines would rise to a

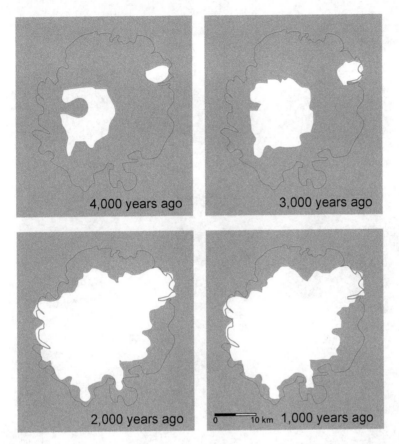

Fig. 6.47 Formation of Hofsjökull. The series of pictures shows the results of numerical models outlining the formation of Hofsjökull

height of 1400–1500 m and hardly any snow would accumulate on the mountains. This is probably how all the ice completely disappeared from Hofsjökull volcano in the warm period about 9000 to 6000 years ago, when the mean annual temperature was all of 3 °C higher than it is now. With a cooling climate and increased precipitation about 4000–5000 years ago, the glacier succeeded once again in forming on the outer sides of the caldera's rim and on the ridge leading from it to Miklafell. Outlet glaciers extended from this ice sheet in all directions, and avalanches fell into the caldera, gradually filling it. Hofsjökull was probably formed more rapidly than Langjökull, because its supporting terrain is at a higher elevation (Figs. 6.47 and 6.48).

Hofsjökull has shrunk so much in the warm period of the 20th century, that many of its termini are now 2 to 3 km from their previous furthest extensions. Since the DGI maps were compiled in 1937–1938 and up to the year 1973, the surface area of Hofsjökull has been reduced from 996 km^2 to 830 km^2 (a shrinkage of

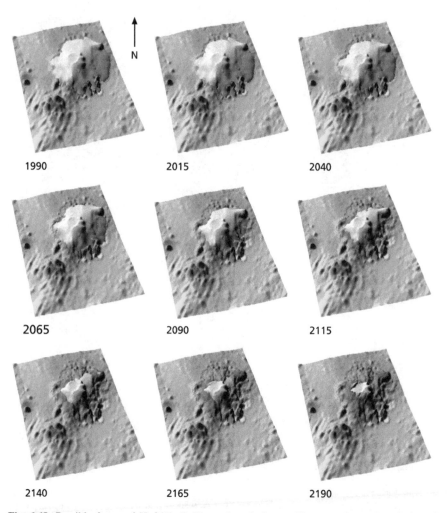

N

| 1990 | 2015 | 2040 |

| 2065 | 2090 | 2115 |

| 2140 | 2165 | 2190 |

Fig. 6.48 Possible future of Hofsjökull. The series of pictures illustrates the projected size of Hofsjökull until the end of the 22nd century, based on current predictions of climate change in Iceland (assuming A1B scenario of IPCC 2007)

16 %), and today the glacier is 830 km^2 or about 83 % of what it was seventy years ago. Since annual measuring began in 1988, the average mass balance of the whole glacier has only been positive on four occasions (1988/89 and 1991 to 1994), but in all other instances it has been negative. On average, Hofsjökull's surface lost about 3–4 m of water equivalent over a 17-year period from 1988 to 2005. All its outlet glaciers have retreated since 1995, and its total surface area has been reduced by about 3.5 % since 1986.

It has been estimated that Hofsjökull's mean annual mass average shrinks by 0.6 m, in water equivalent, for every degree Celcius of increased global warming.

In accordance with current projections of warming in the coming years, of about 0.18 °C per decade until the year 2100 and then onwards of 0.25 °C, the lowest part of Múlajökull should have disappeared by around 2050 and been replaced by a lake to the north of the Múlar hills. There should also be a lake in front of Blautukvíslarjökull, which by then would only be a narrow glacial tongue above the 5-km-long basin of the ravine. Hofsjökull would be by then only about 70 % of its current size in square kilometres, and the outlines of the caldera's rims and the steep outlet glaciers creeping from them would be clearly discernible. By 2100, Hofsjökull's surface area would only be half of what it is today (and 40 % of its volume), Múlajökull would have completely disappeared, and the ravine finally emerged from beneath Blautukvíslarjökull. The outlet glaciers descending from the caldera's rims near Hásteinar peaks would be as ragged as the eastern slopes of Öræfajökull are today. The view from Sprengisandur highland plain would be similar to that from the Jökulsárlón lagoon westward over Breiðamerkursandur. Around the year 2150, Hofsjökull's surface area would be a fifth of what it is now and all ice would have disappeared from Miklafell, though a glacier would still extend over the caldera's rim and even down its southern and eastern flanks. Its largest outlet glacier would then be above the trough between the caldera and Miklafell. It would also, by then, have been 20 years since the final remnants of Langjökull had disappeared. Hofsjökull would survive the 22nd century, but by its end there would only be ice within the caldera itself, and even that would disappear entirely soon thereafter. The runoff meltwater from Hofsjökull would increase by about 1.2 m a year, averaged over the whole glacier's surface according to its current dimensions, and would reach a maximum about 2060. By 2100 it would be 0.4 m per annum more than it is now, and would return to its current meltwater levels by around the year 2160 (Jóhannesson 1997; Jóhannesson et al. 1995a, b, 2006a, b; Aðalgeirsdóttir et al. 2006; Guðmundsson et al. 2009b; Björnsson and Pálsson 2008).

6.7 Kerlingarfjöll Mountains

The Kerlingarfjöll mountains (140 km^2) are the volcanic centre on the rift zone named after Hofsjökull. They rise sharply from a plateau 600–700 m high at the water divides of the Hvítá and Þjórsá rivers, and about ten of their peaks reach over a height of 1100 m (Fig. 6.49). The mountains are mostly composed of rhyolite, have been eroded by wind and water, have sparse vegetation, and are very colourful. There was tremendous volcanic activity among these volcanoes during the warm and glacial periods of the last ice age, but none since the end of the last glacial period. The eruptions piled up huge rhyolite mounds that are now the largest peaks of the Kerlingarfjöll. Some of these volcanic eruptions broke through the ice and formed tuyas, two of which are Höttur and Loðmundur. The Hveradalir valleys, which open up onto the Árskarðsá and Kisa rivers, cut across these mountains from the northwest to southeast, and the mountains to the west and east are thus

Fig. 6.49 Glaciers on the sides of the highest rhyolite summits of the Kerlingafjöll mountains are retreating rapidly. This used to be a popular skiing area during the summer, but with an increasingly warmer climate, skis have been replaced by hiking boots, for the area is indeed renowned for its natural beauty, colourful terrain, and geothermal heat. Snævarr Guðmundsson, 2000

appropriately named Vesturfjöll and Austurfjöll. The highest peaks in the Austurfjöll are Fannborg (1458 m), Loðmundur (1432 m) and Snækollur (1488 m), while in the Vesturfjöll the next-highest are Mænir (1340 m) and Ögmundur (1357 m).

Two calderas were formed in the Kerlingarfjöll during subsidence in a massive rhyolite eruption in the last glacial period. The eastern caldera is the larger one, with an area of about 17 km^2, while the western one has an area of about 14 km^2. Within them is a very warm geothermal area covering around 7 km^2. In the westernmost part of the main caldera are the Neðri-Hveradalir valleys, which are divided into Vesturdalur, Miðdalur or Fannardalur and Austurdalur, while in the southeast there are the Efri-Hveradalir valleys, where the geothermal area is hottest, especially on the flanks of Snækollur. The third geothermal area (1.6 km^2) is in Hverabotn (in Kerlingardalur), connected to the western caldera in a basin to the south below Mænir, where there is also the source of the Kerlingará. Volcanic material has filled in the calderas while geothermal activity has transformed them, gouging out large gullies and a ravine that are called valleys in that area. These calderas were covered by ice during the glacial period and contained a lake similar to the one now in Grímsvötn. The thick sediment layers found there can be traced back to jökulhlaups from the Hveradalir valleys at the end of the last glacial period.

There are a great number of hot pools in this area, as well as constantly boiling puddles, powerful, bubbling mud pools, fumaroles, and sulphur springs, all with colourful surroundings due to the local saline and sulphuric fallout. Temperatures as high as 150 °C have been recorded in some of these fumaroles and the mud pools change location from year to year. First, the hot steam rises through a vent and boils sediment layers so that they decompose and form mud pools. The mud then blocks up the vent so that the boiling steam must search for a new opening, and thus new mud pools appear while others disappear (Áskelsson 1946; Grönvold 1972).

There are about ten glaciers on the edges of the eastern caldera of the Kerlingarfjöll, all at an elevation of around 1200–1300 m, the lowest snout descending to a height of 800 m. The total surface area of these glaciers is about 3 km^2. The largest glaciers are Loðmundarjökull Eystri, Jökulkinn, Mænisjökull, Langafönn and Botnajökull. Snow patches could be found in valleys and gullies on the sides of the highest mountains until quite recently, but they have now shrunk rapidly due to increased warming. Where snow patches, ice and fumaroles meet, ice caves are formed, and such caves also used to exist in the Kerlingarfjöll, but not anymore. The name of these mountains is drawn from the troll Kerling, who was petrified as a 25-m-high, dark pillar of palagonitic tuff in the middle of a pale, rhyolite scree to the south of Kerlingartindur peak in the Vesturfjöll mountains.

In previous centuries, it was often believed that the Kerlingarfjöll were inhabited by outlaws and so they were rarely visited, indeed the area had little vegetation and was consequently not utilised for grazing until after the middle of the 19th century. The Hvítá was bridged in 1933, a road reached Kerlingarfjöll in 1936, and in 1937–1938 the ITA built a cabin at Ásgarður, its second after the one at Hvítárnes. The Fannborg tourist information centre is now in Ásgarður and hiking routes in the area have become popular in recent years. The road for vehicles extends from the Kjölur highland route, but there is also a track to the north of Loðmundur that runs southeast along the Illahraun lava field to the mountain cabin of Setrið and then onwards through Fjórðungssandur to the Þjórsá by Norðlingaalda. Should a geothermal power station ever be built in the Kerlingarfjöll, this would mean the coming of drilling platforms, man-made reservoirs, and electricity pylons, all of which would completely change the area's landscape.

6.8 Tungnafellsjökull

Tungnafellsjökull is a domed glacier about 10 km long and 5 km broad (covering ca. 33 km^2) to the west of Vonarskarð pass and southwest of the palagonite Tungnafell (1387 m), from which it draws its name (Fig. 6.50; Sæmundsson, 1982). The glacier covers most of one of the two calderas of the main volcano,

Fig. 6.50 View northeast across Tungnafellsjökull, which covers most of an ancient caldera. The Sprengisandur highland route lies to the west of the glacier. Mt. Herðubreið can be discerned in the distance. HB, 2008

which has not been active since the settlement of Iceland, though there are volcanic fissures to the northeast of the glacier from the post-glacial period. Tungnafellsjökull's main outlet glaciers are Fremri- and Innri-Hagajökull, which descend northwestward from the glacier's dome. The highest point of Tungnafellsjökull is in the southeast at Háhyrna (1530 m). The subglacial landscape of Tungnafellsjökull has not yet been investigated. There is a circular caldera just to the southeast of the glacier, and rhyolite lava can be found within it. The first geologist to explore the area was Reck (1911) in 1908, followed by Hermann Stoll, who spent some time there in 1911. The ITA has built a hut at the mouth of the Nýidalur, or Jökuldalur, to the south of the glacier.

References

Aðalgeirsdóttir, G., Jóhannesson, T., Björnsson, H., Pálsson, F., & Sigurðsson, O. (2006). The response of Hofsjökull and southern Vatnajökull, Iceland, to climate change. *Journal of Geophysical Research, 111*, F03001. doi:10.1029/2005JF000388.

Árnason, B. (1976) *Groundwater systems in Iceland, traced by deuterium. Rit Vísindafélags Íslendinga, 42*. Reykjavík.

Áskelsson, J. (1946). A contribution to the geology of Kerlingarfjöll. *Acta Naturalia Islandica, 1* (2), 15 pp.

Axford, Y., Miller, G., Geirsdóttir, Á., & Langdon, P. (2007). Holocene temperature history of northern Iceland inferred from subfossil midges. *Quaternary Science Reviews, 26*, 3344–3358.

Axford, Y., Miller, G., Geirsdóttir, Á., & Langdon, P. (2009). Climate of the Little Ice Age and the past 2000 years in northeast Iceland inferred from chironomids and other lake sediment proxies. *Journal of Paleolimnology, 41,* 7–24.

Bisiker, W. (1902). *Across Iceland.* London: Edward Arnold.

Björnsson, H. (1986). Surface and bedrock topography of ice caps in Iceland mapped by radio echo soundings. *Annals of Glaciology, 8,* 11–18.

Björnsson, H. (1988). Hydrology of ice caps in volcanic regions. *Rit Vísindafélags Íslendinga, 45.* Reykjavík.

Björnsson, H. (1991). Hofsjökull: landslag, ísforði og vatnasvæði. *Náttúrufræðingurinn, 60*(3), 113–126.

Björnsson, H. (2002). Langjökull: forðabúr Þingvallavatns og Hengilsins. In P. Jónasson & P. Hersteinsson (Eds.), *Þingvallavatn: undraheimur í mótun* (pp. 136–143). Reykjavík: Mál og Menning.

Björnsson, H. (2009). *Jöklar á Íslandi.* Reykjavík: Opna.

Björnsson, H. (2011). Langjökull: The storehouse of Lake Thingvallavatn and the Hengill area. In P. Jónasson & P. Hersteinsson (Eds.), *Thingvallavatn: A Unique World evolving* (pp. 140–145). Reykjavík: Opna.

Björnsson, H., & Pálsson, F. (2008). Icelandic glaciers. *Jökull, 58,* 365–386.

Björnsson, H., Pálsson, F., Sigurðsson, O., & Flowers, G. (2003). Surges of glaciers in Iceland. *Annals of Glaciology, 36,* 82–90.

Björnsson, H., Pálsson, F., Guðmundsson, E., Magnússon, S., Aðalgeirsdóttir, G., Jóhannesson, T., et al. (2013). Contribution of Icelandic ice caps to sea level rise: Trends and variability since the Little Ice Age. *Geophysical Research Letters, 40*(1–5), 1546–1550. doi:10.1002/grl.50278.

Black, J., Miller, G., Geirsdóttir, Á., Manley, W., & Björnsson, H. (2004). Sediment thickness and Holocene erosion rates from a seismic survey of Hvítárvatn, central Iceland. *Jökull, 54,* 37–56.

Briem, Ó. (1959). *Útilegumenn og auðar tóttir.* Reykjavík: Menningarsjóður.

Bunsen, R. (1847). Beitrag zur Kenntniss des islandischen Tuffgebirges. *Annalen der Chemie und Pharmacie, 61,* 265–279.

Bruun, D. (1902). Sprengisandur og Egnen mellem Hofs- og Vatnajökull. Undersøgelser foretagne I Sommeren 1902. *Geografisk Tidsskrift, 16,* 23–46.

Bruun, D. (1921–1927). *Turistruter paa Island,* 5 Vols. 1: Reykjavik og kysten rundt; 2: Udflugter fra Reykjavik; 3: Gennem beboede egne; 4: Fjeldveje gennem Islands indre Højland; De øde Egne nord for Vatna-Jökull. Copenhagen: E. Jespersen & Gyldendal.

Coles, J. (1882). *Summer travelling in Iceland.* London: John Murray.

Erkes, H. (1911). Das isländische Hochland zwischen Hofsjökull und Vatnajökull. *Petermanns Geographische Mitteilungen, 57*(2), 140–143.

Erkes, H. (1914). Der Anteil der Deutschen an der Erforschung Inner Islands. *Die Erde, 12,* 225–228.

Flowers, G., Björnsson, H., Geirsdóttir, Á., Black, J., & Clarke, G. (2007). Glacier fluctuation and inferred climatology of Langjökull ice cap through the Little Ice Age. *Quaternary Science Reviews, 22,* 2337–2353.

Flowers, G., Björnsson, H., Geirsdóttir, Á., Miller, G., Black, J., & Clarke, G. (2008). Holocene climate conditions and glacier variations in central Iceland from physical modeling and empirical evidence. *Quaternary Science Reviews, 27,* 797–813.

Geirsdóttir, Á., Miller, G., Axford, Y., & Ólafsdóttir, S. (2009). Holocene and latest Pleistocene climate and glacier fluctuations in Iceland. *Journal of Quaternary Science Review.*

Green, R. (1952). Sedimentary sequence in the Hagavatn basin. *Jökull, 2,* 10–16.

Grímsson, H. (1944–1948). Sagan af því, hversu Þórisdalur er fundinn. *Blanda, 8,* 333–355.

Grönvold, K. (1972). *Structural and petrochemical studies in the Kerlingafjöll region, central Iceland.* Doctoral dissertation, University of Oxford.

Guðmundsson, J. (1936) [1650] Áradalsóður. In *Huld: Safn alþýðlegra fræða íslenzkra* (Vol. 2, 2nd ed., pp. 48–62).

Guðmundsson, S., Björnsson, H., Pálsson, F., & Haraldsson, H. (2009a). Energy balance and degree-day models of summer ablation on the Langjökull ice cap, SW Iceland. *Jökull, 59*, 1–18.

Guðmundsson, S., Björnsson, H., Aðalgeirsdóttir, G., Jóhannesson, T., Pálsson, F., & Sigurðsson, O. (2009b). Similarities and differences in the response of two ice caps in Iceland to climate warming. *Hydrology Research, 40*(5), 495–502.

Gunnlaugsson, B. (1835). Um fund Þórisdals. *Skírnir, 9*, 104–107.

Gunnlaugsson, B. (1836). Um Þórisdal. *Sunnanpósturinn., 2*(8), 113–126.

Hallgrímsson, J. (1957). *Kvæði og sögur*. Reykjavík: Mál og Menning.

Hannesson, P. (1958). *Frá óbyggðum—ferðasögur og landlýsingar*. Reykjavík: Bókaútgáfa Menningarsjóðs.

Helgason, J. (1986). *Kvæðabók*. Reykjavík: Mál og Menning.

Howell, F. (1893). *Iceland pictures*. London: The Religious Tract Society.

IPCC. (2007). *Climate change: The science of climate change*. Cambridge, UK: FCambridge University Press. Website: www.ipcc.ch

Jakobsson, S. (1979). Outline of the petrology of Iceland. *Jökull, 29*, 57–73.

Jóhannesson, T. (1997). The response of two Icelandic glaciers to climatic warming computed with a degree-day glacier mass balance model coupled to a dynamic glacier model. *Journal of Glaciology, 43*(143), 321–327.

Jóhannesson, T., Sigurðsson, O., Laumann, T., & Kennett, M. (1995a). Degree-day glacier mass balance modelling with applications to glaciers in Iceland, Norway and Greenland. *Journal of Glaciology, 41*(138), 345–358.

Jóhannesson, T., Sigurðsson, O., Laumann, T., & Kennett, M. (1995b). Degree-day glacier mass balance modelling with applications to glaciers in Iceland, Norway and Greenland. *Journal of Glaciology., 41*(138), 345–358.

Jóhannesson, T., Aðalgeirsdóttir, G., Ahlstrøm, A., Andreassen, L., Björnsson, H., de Woul, M., et al. (2006a). The impact of climate change on glaciers and glacial runoff in the Nordic countries. In S. Árnadóttir (Ed.), *The European Conference of Impacts of Climate Change on Renewable Energy Sources, Reykjavík, Iceland, June 5–6* (pp. 31–38). Reykjavík: NEA.

Jóhannesson, T., Aðalgeirsdóttir, G., Björnsson, H., Bøggild, C., Elvehøy, H., Guðmundsson, Sv, et al. (2006b). Mass balance modeling of the Vatnajökull, Hofsjökull and Langjökull ice caps. In S. Árnadóttir (Ed.), *The European Conference of Impacts of Climate Change on Renewable Energy Sources, Reykjavík, Iceland, June 5–6* (pp. 39–42). Reykjavík: NEA.

Kaldal, I., & Víkingsson, S. (1990). Early Holocene deglaciation in central Iceland. *Jökull, 40*, 51–66.

Keilhack, K. (1886). Beiträge zur Geologie der Insel Island. *Zeitschrift der Deutschen Geologischen Gesellschaft., 38*, 376–449.

Keindl, J. (1930). Beobachtungen auf einer Studienreise nach Island. Sommer 1929. *Mitteilungen der Österreichischen Geographischen Gesellschaft., 73*, 164–174.

Keindl, J. (1932). Untersuchungen über den Hofs- und Langjökull in Island. *Zeitschrift für Gletscherkunde., 20*, 1–18.

Kemel, F. (1930). Geologische Beobachtungen in der Gegend des Hvítárvatn in Island. *Neue Jahrbuch, 19*(1), 267–275.

Kjartansson, G. (1938). Um nokkur jökullón og jökulhlaup í Harðangri og á Íslandi. *Náttúrufræðingurinn, 8*, 113–124.

Kjartansson, G. (1964). Ísaldarlok og eldfjöll á Kili. *Náttúrufræðingurinn, 34*, 9–38.

Lárusson, M. (1970). Hvenær lokaðist leiðin norður? *Saga, 8*, 264–267.

Magnússon, Á., & Vídalín, P. (1981). [1702–1712]. In Melsteð, B. and Jónsson, F. (Eds.), *Jarðabók Árna Magnússonar og Páls Vídalíns* (13 Vols, 2nd edn.). Reykjavík: Hið íslenska fræðafélag. Originally published in Copenhagen, 1913–21.

Matthíasson, H. (1961). Árnessýsla. *Árbók Ferðafélags Íslands*.

Matthíasson, H. (1980). Langjökulsleiðir. *Árbók Ferðafélags Íslands*.

Nielsen, N. (1927). Der Vulkanismus am Hvítárvatn und Hofsjökull auf Island. *Meddelelser fra Dansk Geologisk Forening, 7*(2), 102–128.

Nielsen, N. (1933). Contributions to the physiography of Iceland with particular reference to the highlands west of Vatnajökull. *D. Kgl. Danske Vidensk. Selsk. Skrifter. Naturvidensk. og Mathem. Afd.* 9 Række, IV.5, 183–286.

Nørvang, A. (1937). Dødisbælterne ved Langjökulls Sydöstrand. *Meddelelser fra Dansk Geologisk Forening, 9*(2), 186–198.

Ólafsson, B. (1930). Der Durchbruch des Hagavatn auf Island. *Petermanns Geographische Mitteilungen*, 76 pp.

Ólafsson, E., & Pálsson, B. (1978, 1981). *Ferðabók Eggerts Ólafssonar og Bjarna Pálssonar um ferðir þeirra á Íslandi árin 1752–1757*. (S. Steindórsson, Trans.) Reykjavík: Örn og Örlygur.

Olavius, Ó. (1787). *Oeconomisk Reise igiennem de nordvestlige, nordlige, og nordøstlige Kanter af Island*. Copenhagen: Gyldendal.

Olavius, Ó. (1964–1965). *Ferðabók: landshagir í norðvestur-, norður- og norðaustursýslum Íslands 1775–1777: ásamt ritgerðum Ole Henckels um brennisteinsnám og Christian Zieners um surtarbrand. Íslandskort frá 1780 eftir höfund og Jón Eiríksson*, 2 Vols (S. Steindórsson, Trans.). Reykjavík: Bókfellsútgáfan.

Ötting, W. (1930). Beobachtungen am Rande des Hofsjökull und Langjökull in Zentralisland. *Zeitschrift für Gletscherkunde, 18*, 43–51.

Pálsson, F., Guðmundsson, S., Björnsson, H., Berthier, E., Magnússon, E., Guðmundsson, S., et al. (2012). Mass balance of Langjökull ice cap, Iceland, from ∼1890 to 2009, deduced from old maps, satellite images and in situ mass balance measurements. *Jökull, 62*, 81–96.

Pálsson, S. (1945). *Ferðabók Sveins Pálssonar. Dagbækur og ritgerðir 1791–1794*. (J. Eyþórsson, P. Hannesson & S. Steindórsson, Trans. & Eds.). Reykjavík: Snælandsútgáfan (Thesis on glaciers: pp. 423–552).

Ponzi, F. (2004). *Ísland Howells*. Mosfellsbær: Brennholtsútgáfan.

Þórarinsson, S. (1939). The ice dammed lakes of Iceland with particular reference to their values as indicators of glacier oscillations. *Geografiska Annaler, 21*, 216–242.

Þórhallsdóttir, Þ. (1988). Þjórsárver. *Árbók Ferðafélags Íslands* (pp. 83–115).

Reck, H. (1911). Glazialgeologische Studien über die rezenten und diluvialen Gletscherbebiete Islands. *Zeitschr. für Gletscherkunde., 5*, 241–297.

Sæmundsson, K. (1979). Outline of the geology of Iceland. *Jökull, 29*, 7–28.

Sæmundsson, K. (1982). Öskjur á virkum eldfjallasvæðum á Íslandi. In H. Þórarinsdóttir, Ó. Óskarsson, S. Steinþórsson, & Þ. Einarsson (Eds.), *Eldur er í norðri* (pp. 221–239). Reykjavík: Sögufélag.

Sæmundsson, K. (2008). Personal communication.

Schythe, J. (1840). En Fjeldreise i Island i Sommeren 1840. *Naturhistorisk Tidsskrift, 4*, 331–394, 499–500.

Scudder, B. (Trans.) (2005). *Grettir's Saga*. London: Penguin.

Sigbjarnarson, G. (1967). The changing level of Hagavatn and glacial recession in this century. *Jökull, 17*, 263–279.

Sigurðsson, E. (1997). Þórisdalur og ferð prestanna 1664. *Fræðslurit Ferðafélag Íslands, 5*, 40 pp.

Sigurðsson, F. (1990). Groundwater from glacial areas in Iceland. *Jökull, 40*, 119–146.

Sigurðsson, F., & Sigbjarnarson, G. (2002). Grunnvatnið til Þingvallavatns. In P. Jónasson & P. Hersteinsson (Eds.), *Þingvallavatn: undraheimur í mótun* (pp. 120–135). Reykjavík: Mál og Menning.

Sigurðsson, O. (1998). Glacier variations in Iceland 1930–1995. From the database of the Iceland Glaciological Society. *Jökull, 45*, 3–26.

Sigurðsson, O. (2001). Jöklar á Kili. Kjölur og Kjalverðir. *Árbók Ferðafélags Íslands*, 184–223.

Sinton, J., Grönvold, K., & Sæmundsson, K. (2005). Postglacial eruptive history of the Western Volcanic Zone, Iceland. *Geochemistry, Geophysics, Geosystems, 6*, Q12009. doi:10.1029/2005GC001021.

Sveinsson, B. (Ed.) (1862, October 21). Um Þórisdal og frá ferð þeirra sjera Helga Grímssonar og sjera Bjarnar Stefánssonar þangað, árið 1664. *Íslendingur, 3*(11), 81–88.

Theódórsson, T. (1980). Hagafellsjöklar taka á rás. *Jökull, 30*, 75–77.

Thoroddsen, Þ. (1899). *Höjlandet ved Langjökull paa Island*. Copenhagen: Det kongelige danske geografiske selskab.

Thoroddsen, Þ. (1902). Landfræðisaga Íslands. Kaupmannahöfn. Hið íslenska bókmenntafélag.

Thoroddsen, Þ. (1905/06). Die Gletscher Islands. Island, Grundriss der Geographie und Geologie. *Petermanns Geographische Mitteilungen, 152*, 1–161; *153*, 162–358.

Tómasson, H. (1993). Jökulstífluð vötn á Kili og hamfarahlaup í Hvítá í Árnessýslu. *Náttúrufræðingurinn, 62*(1–2), 77–98.

Waltershausen, W. (1847). *Physische-geographische Skizze von Island mit besondere Rücksicht auf vulkanische Erscheinungen*. Göttingen.

Wright, J. (1935). The Hagavatn Gorge. *The Geographical Journal, 86*(3), 218–234.

Wunder, L. (1910a). Beobachtungen am Langjökull und im Thorisdalur auf Island. *Petermanns Geographische Mitteilungen, 56*(2), 123–126.

Wunder, L. (1910b). Gletschertouren in Island. *Zeitschrift der deutschen und österreichischen Alpenvereins, 41*, 46–58.

Wunder, L. (1912). *Beträge zur kenntnis des Kerlingarfjöllgebirges, des Hofsjökulls und Hochlandes zwischen Hofs- und Langjökull in Island*. Leipzig and Berlin: B. G. Teubner.

Chapter 7
Glaciers of Northern and Western Iceland

Small Glaciers Between Skagafjörður, Eyjafjörður and Skjálfandi Bay; Drangajökull and Snæfellsjökull

Pastor Jón: '*If one looks at the glacier long enough, words cease to have any meaning on God's earth.*'
(Laxness, *Under the Glacier*, 1972, p. 126).

Abstract This chapter has a detailed description of the types, shapes, sizes, locations, and all the relevant scientific data of the glaciers in northern and western Iceland, as well as some of their recorded local histories, folklore and legends. Many small glaciers are found in the mountain ranges between the northern fjords of Skagafjörður, Eyjafjörður, and Skjálfandi Bay (100–250 in number and 40–100 km^2 in area depending on how they are counted). They are rapidly disappearing due to a warmer climate. Earlier traditional travelling routes across these glaciers into valley heads are also described. A true ice cap, Drangajökull, along with its outlet glaciers (145 km^2), is located in the northwestern fjords, where the snowline is lowest in Iceland, at 700 m above sea level. In previous centuries, it was easier to travel between fjords over this ice cap rather than to make the same journey by sea. Historical records of the glacier date back to the early 17th century when its advance destroyed farms. Surges are also common. The stratovolcano Snæfellsjökull (10 km^2), is covered by an ice cap and is one of Iceland's best known and popular destinations in earlier travelogues, as it towers majestically at the end of Snæfellsnes peninsula. Indeed it became the setting of a best-selling French novel in the 19th century. Current climate predictions, however, mean that it might totally disappear within a few decades.

7.1 Glaciers Between Skagafjörður and Eyjafjörður

The mountain range between Skagafjörður and Eyjafjörður is the largest section of highland that extends from the centre of mainland Iceland and is also the country's largest area of basalt rock (Figs. 7.1, 7.2, 7.3, 7.4). The mountain range is 85 km from north to south and 35 km from west to east. Many of its summits reach over 1300 m above sea level. The three highest areas are in the centre of the range, on the water divide between Skagafjörður and Eyjafjörður between Skíðadalur and the outer

© Atlantis Press and the author(s) 2017

H. Björnsson, *The Glaciers of Iceland*, Atlantis Advances in Quaternary Science 2, DOI 10.2991/978-94-6239-207-6_7

Fig. 7.1 View south and west across the range of mountains between Skagafjörður and Eyjafjörður, the highest region of basalt rock in Iceland. Ice caps from the last glacial period gouged deep valleys into the mountains and there are still over a hundred small glaciers to be found there. The largest is Tungnahryggsjökull and across its easternmost part lay the Tungnahryggsleið trail from the head of Hörgádalur, along Barkádalur and over into Austurdalur (Kolbeinsdalur) in Skagafjörður. Above is a view over the western part of Tungnahryggsjökull. The Tungnahryggur ridge is *bottom left* on the photograph, and to its *right* can be seen the head of Vesturdalur (Kolbeinsdalur). The pointed peak of Grasárdalshnúkur (1268 m) is high up on the *right hand side*. Far in the distance, just *left of centre*, is Mælifellshnúkur peak (1138 m) in Skagafjörður. Ágúst Guðmundsson, 1996

Fig. 7.2 Glaciers creep down Hjaltadalur (*left*) and Grjótadalur (*right*), with Hafrafell (1311 m) in between the valleys. On the right is the farm of Reykir in Hjaltadalur, inland from Skagafjörður. HB

reaches of Svarfaðardalur, the mountains between Hörgárdalur and Þorvaldsdalur, and finally those to the east of Öxnadalur, where the highest summit of northern Iceland can be found: Kerling (1536 m). The mountain massifs are dissected by deep valleys on both sides and outlet glaciers advanced down the main ones to the

Fig. 7.3 Map of mountains and glaciers between Eyjafjörður and Skagafjörður

shoreline during the last glacial period, extending northeastward from Svarfaðardalur and Hörgárdalur, northward from Þorvaldsdalur into Eyjafjörður, and northwestward from Hjaltadalur and Kolbeinsdalur into Skagafjörður. Rough and jagged ridges and mountains rose up between these ice-age glaciers, and they are still there today, although their lower slopes and foothills are now vegetated and less precipitous. The striations in their rock faces, visible over a wide area, reveal how high up the valley sides the glaciers reached, often up to 1000 m in the innermost valleys, though to lower heights nearer the sea as the valleys opened out (Thoroddsen 1892;

Fig. 7.4 The northern part of the mountain ranges between Eyjafjörður and Skagafjörður, from the villages of Siglufjörður (on the *left*) to Dalvík (on the *right*) at the northernmost mouth of Svarfaðardalur. Outlet glaciers from the last glacial period advanced along the valleys while the mountain ridges rose up between them from the ice. There are still glaciers on the highest mountains as can be seen on this satellite image taken in the summer of 2007. The snowline here is from 1000 m down to 800 m, and individual glaciers descend to a height of 600 m

1892–1904, 1907–1911; Þórarinsson 1937; Steindórsson 1949; Einarsson 1968; Kaldal 1978; Víkingsson 1978; Sigbjarnarson 1983; Caseldine 1983).

Today there are only glaciers in the remoter valleys of this mountain range, though they are certainly numerous, even if they are small. They are reckoned to number between 100 and 250, with a total surface area covering 40–100 km², depending on how they are classified (Björnsson 1979, 1991). They all zealously work at gnawing out corries and basins at valley heads, breaking down cliffs and mountain stacks, and transporting chunks of bedrock and gravel into glacial rivers. There are very visible end moraines from the middle of the 18th century in many of the glaciers' forefields. Most of the valley glaciers of northern Iceland had by then advanced the furthest since the country had been settled and had destroyed earlier moraines while doing so. Although the cold period had continued into the 20th century, precipitation in northern Iceland had not been sufficient to maintain the expansion of glaciers, as had occurred in southern Iceland. Today there are now sparsely vegetated moraine screes in front of glaciers that have now been retreating for almost a century. Rock glaciers are also common in this area (e.g. Kugelmann 1989; Häberle 1991; Stötter and Wastl 1999).

7.1.1   Glaciers Formed in Valley Heads

Although the mountains between Skagafjörður and Eyjafjörður are high, they have no large glaciers, for there is insufficient precipitation and the summers are too warm. On the summits of the highest mountains, listed earlier, the precipitation is estimated to reach, on average, 2000–3000 mm a year. Certain individual mountains may attract more precipitation, but snow has disappeared from most of the mountain tops by the end of summer, having been blown away by winter storms or melted in the summer warmth. The snow that does accumulate in the valley heads, however, can contain many times the usual amount of precipitation because of additional gains from spindrift off mountain edges and from frequent avalanches from cliffs. The summer ablation does not succeed in melting all the snow in the corries and ravines over a wide area, especially if snow patches face northward and are mostly sheltered from sunlight. On the Tröllaskagi peninsula there are numerous corrie and cirque glaciers high up in hanging valleys, and glaciers also fill up the heads of remoter valleys. Most of these glaciers are above 1000 m in height and reach up mountainsides until they are too steep for snow to settle on them (Fig. 7.5, 7.6, 7.7, 7.8). In some places, a snow-covered pass links glaciers between valleys. Indeed, there are few real valley glaciers remaining on Tröllaskagi; they have retreated from the lower valleys higher up into inland valley heads and corries. The valley glacier that stretches furthest in central northern Iceland is Gljúfurárjökull (2 km²), enclosed by mountains that reach a height of 1200–1300 m. It descends to an elevation of about 580 m above sea level and can be seen from the village of

Fig. 7.5 View south over the eastern part of Tungnahryggsjökull, which creeps down into Austurdalur. From the *left* can be seen Eiríkshnjúkur (1305 m), Hólamannahnjúkur (1406 m), and Tungnahryggur ridge. Further away, in the centre, is Barkárdalsjökull

Fig. 7.6 View northwest over Barkárdalsjökull. The glacier's accumulation zone is restricted to the base of the cliffs. The Hólamannavegur pathway went up onto Barkárdalsjökull, through Hólamannaskarð pass and down along Tungnahryggsjökull (on the *right*) into Vesturdalur and then on to the bishopric of Hólar in Hjaltadalur. Drangey island in Skagafjörður can also be seen (famous site of the outlaw Grettir the Strong's last stand), and across the fjord is Tindastóll (989 m). Ágúst Guðmundsson

Fig. 7.7 Jagged mountain edges separate the cirque glaciers. View south along Tungnahryggur ridge as it divides Tungnahryggsjökull into two parts. To the south of the ridge is Hólmannahnjúkur (1406 m). Árni Tryggvason

Fig. 7.8 Lambárdalsjökull descends northward from Mt. Kerling (1538 m) into Eyjafjörður. The lowest part of the glacier is a large heap of ground moraine, from beneath which flows the Lambá. Ágúst Guðmundsson

Dalvík with Mt. Blekkill towering above it at just over a height of 1260 m. During the years 1976–1983, Gljúfurárjökull actually advanced about 150 m. Its ice attains a maximum thickness of about 120 m.

There is a tremendous fluctuation in glacial mass balances on the Tröllaskagi peninsula from year to year. After some cold summers, there is snow covering all of the glaciers into the autumn (e.g. the sea-ice years of 1965 and 1968), and almost the whole glacier is one accumulation zone, but during other summers almost all the winter snowfall melts. There are even examples of the accumulated snow of many years disappearing from glaciers in some warm summers, such as the one in 1984. All the glaciers respond very rapidly to fluctuations in the climate and they have often advanced over the last 200 years (around 1810, 1840–1850, 1870–1880, around 1890, 1920, around 1940 and finally during the years 1970–1980). In the warm period during the first part of the 20th century, the glaciers retreated from the low valleys up into the higher valley heads and corries until their expenditure (ablation) became so little that income was sufficient for them to survive. In the latter half of the 20th century, the accumulation zone consisted of over a half and up to two thirds of glaciers in corries and heads of valleys, but this was not sufficient for them to be able to advance along the valleys once again. Conditions for the survival of even glaciers in shadowy nooks are becoming increasingly difficult due to the warmer climate of the last few years (Björnsson 1971, 1972).

7.1.2   Location of Northern Glaciers

Precipitation is at its maximum where the mountains are highest between Svarfaðardalur and Skíðadalur, Hörgárdalur, Hjaltadalur and Kolbeinsdalur, and only a little less so between Öxnadalur and Eyjafjarðardalur. The area with the third highest amount of precipitation is in the lower mountains to the north of Svarfaðardalur, near the ocean. The dominating direction of the wind bearing precipitation and the shape of the mountain summits determine how and where snow accumulates. Spindrift increases snow accumulation more often in corries and cirques below summits that are large and flat, for example, than in those below summits which have jagged ridges and pinnacles. Glaciers often survive longest, and far from general view, in basins that face north.

The snowline on Tröllaskagi peninsula is lower to the north because it grows colder towards the sea and there is increased precipitation. Mountains south of Öxnadalur need to be 200 m higher than to the north of the further reaches of Svarfaðardalur to be able to form glaciers. There are kinks in the snowline contours, however. It becomes lower on flanks facing north and then becomes higher in the northern mountains of valleys that face the sun when at its highest in the sky. Even though the northern mountains of Svarfaðardalur reach as high as 1200 m, they only have snow patches because most of its valley heads face south. Snow accumulation from spindrift also explains the existence of perennial snow and small glaciers below the flat summits of mountains south of Öxnadalur that do not reach a height of 1200 m (Eyþórsson 1956).

7.1.3 Mountain Trails

The glaciers were far from local farmsteads, but were best known by overland travellers between Eyjafjörður and Skagafjörður. Altogether, there are written sources for 12 ancient mountain trails over glaciers into valley heads, the majority of them to and from the old bishopric at Hólar. Many of these trails crossed glaciers and were hardly passable in the summer due to open crevasses and the slipperiness of the ice at that time of year. Some of them were impassable on horseback because of crags. From Eyjafjörður there was a track up Hörgárdalur into Barkárdalur, from where there were two ways to continue: the Tungnahryggsleið route went up the very steep but easily passable Tungnahryggur ridge, north across Tungnahryggsjökull and then down into Austurdalur. The Hólamannavegur trail, which was impassable for horses, went through the deep Hólamannaskarð pass (at a height of about 1210 m) and then across a crevassed glacier and over difficult crags into Vesturdalur and then down into Kolbeinsdalur.

From Svarfaðardalur there were two main routes out of the inner side valley of Skíðadalur. One of them went into Þverárdalur (Kóngsstaðadalur) and from its head over Þverárjökull into Skíðadalur, and from there west into Kolbeinsdalur. It was also possible to travel 10 km further south along Skíðadalur to Sveinsstaðir and then through Vesturárdalur and over Ingjaldsskál corrie down into Kolbeinsdalur. There was also a route from the head of Svarfaðardalur southwest over Heljardalsheiði into Kolbeinsdalur, as well as one northwest through Skallárdalur, over Unadalsjökull, and then down through Unadalur to Hofsós on the eastern shore of Skagafjörður. There was also a trail from the head of Kolbeinsdalur northward along the Hákambar summits to Ólafsfjörður or Stífluvatn in the Fljót district.

7.2 Glaciers Between Eyjafjörður and Skjálfandi Bay

Between Eyjafjörður and Skjálfandi are dozens of small glaciers on three mountain ridges from 1000 and 1200 m high. The ridges ascend from two long valleys which are called Leirdalsheiði and Flateyjardalsheiði. The annual precipitation is well over 1000 mm on the highest of these mountains, and much of it falls as snow. The snowline is at a height of 800–1000 m, and the lowest glacier stretches down to an elevation of 500 m. Along the shoreline, corries can be seen with basins at a height of 200–300 m that have been gouged out by ice during the last glacial period, when the snowline was 500 m lower than it is today.

Above the Látraströnd shoreline, on the western side of Flateyjarskagi peninsula, there are glaciers on the eastern flanks of Kaldbakur, on the sides of Svínárhnjúkur (1083 m) and Þerna (1081 m), and below the eastern side of Skersgnípa (1081 m), about a dozen of them altogether, while there are also snow patches further north, though the mountain peaks there are all below a height of 800 m (Norðdahl 1983; Egilsson 2000). On the ridge between Leirdalsheiði and Flateyjardalsheiði valleys there are individual glaciers to the north of Blámannshattur (1241 m), in the highest corries of Lambárskálar (south of Lambárhnjúkur peak, 1027 m), and further north on Jökulbrekka slope below Mosahnjúkur (877 m), in all about thirty of them. The largest glaciers in this region, however, are the easternmost ones on the largest range of the Kinnarfjöll and Víknafjöll mountains at a height of 1000–1200 m. Three of them are the largest (about 2 km^2); the southernmost is Kotajökull, east of Vigga (1077 m), its meltwaters draining down through Vesturdalur and Kotadalur with the Purká to the sea at Náttfaravík. The cirque glaciers of Kambsjökull, north of Kambur (1211 m) and Grímslandsjökull, southeast of Kotahnjúkur (1074 m), face northwest to Flateyjardalsheiði, their meltwater draining into the Dalsá flowing north along Flateyjardalur. The Syðri-Jökulsá flows from Kambsjökull and the Ytri-Jökulsá emerges from Grímslandsjökull, 3 km further north. The small, northernmost ice carapaces are on the northern flanks of Skálavíkurhnjúkur peak (1128 m).

The land along the Látraströnd shoreline and the Leirdalsheiði and Flateyjardalsheiði valleys was poor and inhospitable and farmsteads were few and far between. Very wet northerly and northeasterly winds were normal, mist was common near the sea, winter snowfalls were heavy, avalanches frequent, and even sea ice encroached on the beaches, especially in the 19th century. There has been no farming in the region since the mid-20th century.

7.3 Drangajökull

In the northern part of the west fjords is the lowest snowline in Iceland, 600–700 m above sea level. There is a great deal of perennial snow there, with patches covering the lowland until well into the autumn in cold years. In this region are Snæfjall (793 m) and the Snæfjallaströnd cliffs, Jökulfirðir fjords, so-called because of their proximity to Drangajökull, and the Jökladalir valleys of Skjaldfönn and Kaldalón (Figs. 7.9, 7.10, 7.11, 7.12). There are about ten cirque glaciers in the west fjords at a height of 600–700 m, but the largest glacier of all is Drangajökull, an undulating sheet of ice 20 km long and 10 km broad, which covers a massif of lava layers 500–900 m high with a total surface area of about 145 km^2. Drangajökull has the sea on both sides of it, Ísafjarðardjúp fjord and Húnaflói Bay, and is the northernmost and lowest of Icelandic ice caps, descending from an elevation of 900 m down to 150 m, its average height being only just over 600 m. For comparison it is worth noting that, in the central highlands, the Sprengisandur route is at a height of 820 m at Nýidalur, and the Kjölur route is at its highest at 672 m above sea level.

Fig. 7.9 View of Drangajökull from the head of Reykjafjörður on Hornstrandir in the far north of the western fjords. Rising through the glacier are Hrolleifsborg (850 m) and Hljóðabunga (824 m). Friðjófur Helgason, July 2004

The elliptically shaped ice cap runs southeast to northwest, and rising from its central keel are four domes, the highest being the second most northerly and called Jökulbunga (925 m). There are three nunataks rising through the ice shield: Hrolleifsborg (851 m), the highest, from which there is the greatest view over the west fjords from Drangajökull, then Hljóðabunga, the northernmost (825 m), and between them Reyðarbunga, which was not visible before 1920, but now towers 100 m above the glacier. There is a meteorological divide on the glacier's high summit between Ísafjarðardjúp and the Strandir shoreline, so that while there might be mist on one side of the glacier, there could be bright sunshine on the other, and these conditions can be reversed almost immediately with a sudden change of wind direction.

The northwestern fjords of Iceland were fully inhabited during the first centuries of Iceland's settlement with many farmsteads at the foothills of Drangajökull, but once the Little Ice Age commenced they began to be abandoned over a wide area due to the advance of outlet glaciers down onto the lowlands and glacial rivers flooding and breaking up the land. These northern parts of the western fjords were on the boundary of the inhabitable world, especially considering the primitive farming methods of those times. There were few strips of lowland alongside the mountains, the valley heads were marshy and the topsoil shallow and stony. Winters were long and summers short. Sea ice could also besiege the land bringing fog and intense cold. Vegetation often grew there nonetheless, just ahead of the snow. Livestock usually consisted of some sheep and a very few cows. This was

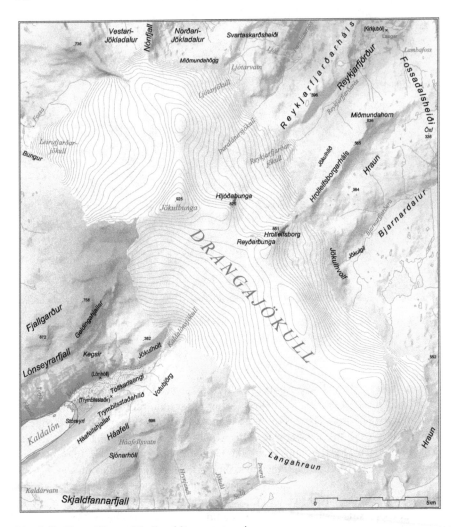

Fig. 7.10 Map of Drangajökull and its nearest environs

not agricultural land and farmers had to rely on grazing along the shore, fishing, seal hunting, gathering birds' eggs, and driftwood. Surviving was a hard, grinding task. There were difficult mountain tracks or long journeys by sea between farmsteads. Most of the farmers were tenants of the church or the Danish crown or other absent and wealthy landlords living far away, and any profits from the land did not benefit the local region, its inhabitants often having to move between one poor farm and hovel to another with their few belongings. The cold period of the Little Ice Age had a severe impact on the western fjords, and the consequent history of settlements along the Hornstrandir, in Árneshreppur, the Jökulfirðir fjords and on the Snæfjallaströnd shores bears witness to an exceptional battle for survival in an

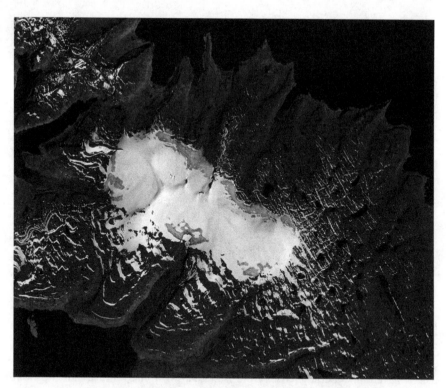

Fig. 7.11 Drangajökull (145 km²) is the northernmost and lowest ice cap in Iceland. There is only a short distance to the sea in Ísafjarðardjúp, the Jökulfirðir fjords, and the Strandir coastline. Annual precipitation on Drangajökull's summit is almost 3000 mm. Satellite image taken from a height of 820 km. SPOT

isolated, inhospitable and far-flung extremity of Iceland. Many farms were situated in very unstable environments. Folk first fled from the encroachment of Drangajökull as it advanced over what little vegetated land there was. Three farms next to the glacier were abandoned in the 17th century, and a further five in the beginning of the 18th century. All habitation near to Drangajökull finally ceased around the mid-19th century, though from social developments rather than changes in climate (Ólafsson and Pálsson 1981; Olavius 1964–1965; Erkes 1910; Keilhack 1934; Eyþórsson 1935; Iwan 1936; Grímsdóttir 1994).

Drangajökull is named after seven stacks ('drangar') at the edge of Drangafjall on the shores of Strandir, although the people living around the shores of Ísafjarðardjúp had previously called the glacier Lónsjökull. There used to be routes across the glacier from Ísafjarðardjúp north to the shores of Strandir, where driftwood was often to be found and collected. The most frequent way was up through Skjaldfannardalur and then over the southern end of Drangajökull and north to Drangar, but there was also a way from the Kaldalón valley over the middle of the glacier to Bjarnarfjörður. Journeys with horses over the glacier were difficult and

Fig. 7.12 Oblique aerial view south over Hornbjarg headland and cliffs to Drangajökull. Mats

dangerous and the journey by sea from Ísafjarðardjúp to the Hornstrandir was long
and in boats ill-suited for the open sea. Today tourists go on popular horse-riding
trips onto Drangajökull during the summer, and make quick tours onto the glacier
during winter in jeeps and skidoos, most often from the highway over
Steingrímsfjarðarheiði. The glacier is still not totally safe, however, because of
crevasses, and special care is needed in the declivity to the northwest of
Hljóðabunga dome to avoid three or four ice cauldrons, which are all in a row and
up to 50 m deep and about 50 m in diameter.

Drangajökull glacier is very susceptible to climate fluctuations and it advanced
considerably during the Little Ice Age. Árni Magnússon (1955) described one of its
advances while travelling in the vicinity of the glacier in 1710 when compiling his
Jarðabók ('Land Register'; Magnússon and Vídalín (1981)). Ólafur Olavius
reported in his *Ferðabók* ('Travelogue') of 1780 that rivers had broken up and
destroyed eight farmlands, five in the first half of the 18th century, when the
glaciers had reached their maximum extension. Þorvaldur Thoroddsen, in 1887,
relates how the outlet glaciers of Drangajökull had then remained static or retreated
for almost a whole century until they had begun to advance again in 1840; since
then they have retreated once more. It is worth mentioning here that the three main
outlet glaciers of Drangajökull are prone to surges and that their advances and
retreats are not in any direct correlation with climate changes (Fig. 7.13).
Drangajökull shrank considerably during the 20th century, indeed the snowline
during warm years was as high as 800 m, but the retreat of the main outlet glaciers

Fig. 7.13 Surges of
Drangajökull, 1700–1995
(Björnsson et al. 2003)

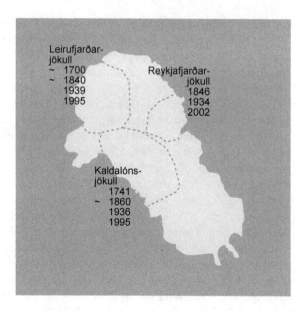

had been interrupted by short-lived surges that had nonetheless lasted for 3–4 years. Since regular monitoring and measuring began in 1931, Drangajökull's largest outlet glaciers have retreated by up to 1 km.

For a long time there was great uncertainty as to the actual size of Drangajökull due to large snowfalls in the western fjords, especially when the early scientists were visiting the area. The size of the glacier is unclear on Knoff's map of 1733, while Sveinn Pálsson (1945), in 1794, depicted it as being much too far south, and on Björn Gunnlaugsson's map of 1848 it reaches as far south as Steingrímsfjarðarheiði, which was twice the size it was in reality. Drangajökull currently has a surface area of 145 km^2 and has lost about 55 km^2 in a century. On average it is about 100 m thick, at its thickest about 200 m (Magnússon et al. 2004). Four main rivers flow from it: Selá, down through Skjaldfannardalur, Mórilla, through Kaldalón valley, Fjörðurinn, in Leirufjörður, and Reykjarfjarðarós estuary in Reykjarfjörður. There is some glacial tinting of the water in the Þaralátursós and Bjarnarfjarðarós estuaries during the summer.

7.3.1 Drangajökull's Outlet Glaciers

Kaldalónsjökull (or Lónjökull) creeps down into Kaldalón valley and a 5-km-long inlet (Lónsvík), extending northeast from Ísafjarðardjúp fjord. From its snout, at a height of 140 m, the river Mórilla í Kverk flows along the northern sides of the valley, having filled most of the inlet with glaciofluvial deposits and creating a lowland area stretching 4 km from the glacier to the head of the fjord, which is itself rather shallow. In the ground moraine in the glacier's forefield there are three curvilinear ridges between the mountainsides, the outermost one (called Jökulgarður) is about 3 km from the snout's present location. The farm of Trymbilsstaðir would have been on the southern side of Kaldalón, just within the Jökulgarður moraine, but was destroyed, probably by advancing ice, around 1600. The ruins had not emerged from beneath the glacier in 1710, but had become visible again by 1931. 'Transcribed oral histories report that inland from Jökulgarður there had been an excellent meadow, and so extensive that it took twelve men the whole summer to harvest the hay; it was thus called Tólfkarlaengi meadow' (Grímsdóttir 1994, p. 25). The farm of Lónhóll (Lón í Lónsvík) was on the northern side of Kaldalón, opposite Trymbilsstaðir, which Olavius (1780) relates as having been destroyed along with three other farms in a jökulhlaup in 1741; the last outburst flood from beneath the glacier was in the autumn of 1998, during which the Mórilla river trundled boulders weighing up to 10 tonnes as its flood wave surged forwards almost 1 km broad. The glacier also advanced in 1754 when Ólafsson and Pálsson (1981, Vol. 1, p. 389) travelled through Kaldalón, and noted that, where the snout was then, 'there had been grassy and vegetated land only twenty years previously.' They also mentioned how the glacier sometimes advanced and then retreated. In the middle of the 19th century, the glacier had withdrawn to the innermost moraine and by 1887 around 400–500 m further inland than that, and a further 340 m up until the year 1930, since when it has pulled back almost a further 1.5 km.

For centuries the main grazing lands for local farmers were at the head of Kaldalón valley. Today there are only two inhabited farms in the vicinity of Drangajökull, Laugaland and Skjaldfönn in Skjaldfannardalur, about 10 km from the glacier. The farms that survived the longest along Snæfjallaströnd shoreline were Bær, Tyrðilmýri and Unaðsdalur, Neðribær being the very last to be abandoned in 1995.

It is interesting to see the striated rocks that emerge from beneath the glacier at the snout of Kaldalónsjökull and the vegetation which takes root in the ground moraine: lichens, mosses, glacier buttercup, arctic poppy, and various willows. Pieces of colourful rhyolite rock can be found on the fluvial plain, too, making the area one of great natural beauty, which is why the physician and composer Sigvaldi Stefánsson associated himself with the valley and lagoon by taking as his artist's name Sigvaldi Kaldalóns. There is an easy hiking trail onto Drangajökull to the north of the outlet of the river Mórilla, just below Jökulholt, from whence there is a 7-km-long fell walk up 700 m to the summit of Jökulbunga (Figs. 7.14, 7.15).

7.3.2 Muddy Shores and Tepid Pools

Leirufjarðarjökull descends northwest toward Leirufjörður, the southernmost branch of the Jökulfirðir fjords, to a height of 140 m above sea level. Leirufjörður is similar to Kaldalón in that its glacial river, Fjörðurinn, has transported a great deal of glaciofluvial sediment, so that the glacier's forefield is one level outwash plain

Fig. 7.14 Approaching Kaldalónsjökull in the summer of 1958. Anon

stretching out to the sea with extensive shallows. The river flows in braided streams across the lowland plain, its channels frequently migrating. A few former terminal moraines lie across the lowland, the highest, Jökulbunga (925 m) towering at the head of Leirufjörður. The farm Öldugil was believed to have stood furthest inland along the fjord, but it had been abandoned in the 15th century because of floods. Árni Magnússon claimed the ruins of the farmstead were still visible near the glacier's snout in 1710, so that the glacier had then only recently advanced over the farm's home field.

Olavius (1964–1965, Vol. 1, pp. 159–160) describes the situation late in the 18th century as follows: 'There is nothing nice to say about Leirufjörður. It is a small recess about a mile in length and ¾ of a mile broad lying in a southeasterly direction. First of all, it is covered in snow all year, secondly, a dreadful glacier advances into it, with voluminous, milky-coloured rivers flowing from beneath it,

Fig. 7.15 Horses and men on their way up onto Drangajökull. Friðþjófur Helgason, July 2004

and sometimes there has been unrest in the glacier and it has surged, and it was said to have so totally destroyed the farmstead of Öldugil long ago that it would be impossible to rebuild it. Furthermore, the fjord is teeming with sunken rocks and it is said that there are no fish there except for trout, which is occasionally seen in the estuary.'

Þorvaldur Thoroddsen states that in the middle of the 18th century the glacier had reached the outermost moraine but had then retreated about 900 m over the next four decades. He also describes, however, sheer cliffs of ice and glacial collapses in the summer of 1883, which could indicate that the glacier had also recently surged. A hundred years after Thoroddsen had carved a mark in a rock in the glacier's snout in 1887, the terminus had retreated 2.3 km, and forty years later yet a further kilometre. The farm Leira was inhabited until the year 1926.

Reykjarfjarðarjökull, the largest outlet glacier from Drangajökull, descends into Reykjarfjörður on the northerly shores of Strandir, reaching an elevation of about 100 m above sea level. In earlier times it had been an important landmark for shark fishermen operating from Gjögur and Horn who called it Jökulsporður. The fjord gets its current name ('smoky fjord') from hot pools, one of which, Heimalaug, is now a concrete swimming pool. There are large buildings made from driftwood in the fjord and it has a runway for aircraft; it was one of the fjords that remained inhabited the longest, or until 1959, but is now only occupied during the summer by holiday-makers. Drangajökull is at the head of the fjord and the peaks of Hrolleifsborg and Hljóðabunga, rise up from the glacier, and between them the bald top of Reyðarbunga. There is a fair amount of lowland at the head of Reykjarfjörður and a sandy shoreline sporting lyme grass. For centuries this was a well-vegetated area for farming with grassy home fields, but with a worsening climate during the Little Ice Age, all agricultural activity was gradually abandoned. The farm of Knittilsstaðir was nearest the head of the valley in the 17th century, very near to where the glacier had reached when Thoroddsen was there in 1886, but by 1710 it had already been reduced to ruins, which the river later washed away. The farmstead of Fremrahorn had been engulfed by the glacier in 1706.

There had been a church at the farm of Kirkjuból on the eastern side of the estuary of the Reykjarfjörður glacial river, but the farm was abandoned just before the middle of the 18th century, about a decade before Eggert Ólafsson and Bjarni Pálsson visited the area in the summer of 1754. The river gradually broke up the churchyard and human bones were seen floating in the estuary for a long time afterwards, thus gaining the area a reputation for being haunted. Three decades later, in 1787, Olavius (1964–1965, Vol. 1, pp. 172–173) described the district thus: 'A voluminous, milk-coloured river flows along the valley; it emerges from beneath Drangajökull, which lies right across the valley head with three rocky peaks in a row in the bluey-white sheet of ice. The river has caused an enormous amount of damage, not only at Kirkjuból, where there used to be a church, but along the whole fjord inland, especially the outer meadows. Since the church-stead was destroyed, men have been forced to bury their dead wherever they thought best, and the home field of the farm Reykjafjörður has been chosen as a graveyard, just a short distance from the farmstead itself. Some of the local people have taken up the practice of keeping corpses in their outhouses over the winter and moving them to a church in the spring when the weather improves and bridle paths passable. All this clearly reveals the necessity of building a church in this area. It is said that the last incumbent priest at Kirkjuból was called Pastor Panti or Panteleon and that he had been very knowledgeable in the magic arts (*Arte magica*) or more likely the occult (*Arte physica occultiori*).' This is a reference to Pastor Pantaleon Ólafsson at Staður in Grunnavík, who existed sometime during the 16th century.

In 1846 Reykjarfjarðarjökull reached a curvilinear moraine that lay directly across the valley about 2.5 km from the ocean and at a height of only 20 m above sea level. A little later this glacial spur began to shrink, though very slowly at first, for after a decade it had only retreated about 20 m. Since then it has steadily retreated, a total of 2400 m up until 1930, and since then a further 2 km.

7.3.3   Poor Farmland, but Plentiful Driftwood

An outlet glacier from Drangajökull also descended to the lowland in Þaralátursfjörður, to a height of about 130 m above sea level. The glacial river Þaralátursós careens along a narrow valley basin bearing rocks and sediments. To quote from Ólafur Olavius once more (1964–1965, Vol. 1, pp. 171): 'Oral history has it that there was a farm in Þaralátursfjörður with the same name as the fjord, but it is said to have been destroyed long before anyone could remember by the encroachments of Drangajökull, which descends into the fjord. Moreover, the rivers that emerge from beneath the glacier spread gravel and mud everywhere, so that the area is now considered totally uninhabitable, indeed the land is completely rocky and barren. No one knows if fish have ever been caught in either of the Jökulfirðir fjords of Þaralátursfjörður or Leirufjörður, and this is blamed on the milky-coloured glacial waters flowing into and blending with the sea.' Þaralátsfjörður was in fact inhabited at a later date, though very intermittently, for indeed the farmland there was considered the poorest and most difficult on the Strandir coast and was called Snoðkot ('stubble cottage'). It was last continuously inhabited from 1922 to 1946 and its occupants were pioneers in buying large mechanical saws to cut up and exploit the driftwood that frequently accumulates along the Strandir shoreline.

There are fewer sources on the history of the other outlet glaciers from Drangajökull. A small glacial tongue extends onto the mountainsides above Bjarnadalur to the east, and yet another lies above Furufjörður, though a century ago it had descended to Skorarheiði moor. In its more prosperous years, so to speak, glacial water flowed from Drangajökull into Hrafnfjörður, which reaches furthest inland to the east of the Jökulfirðir fjords, though there is little sign of that happening today.

7.4   Gláma

The mountain track over the moors on the summit of Gláma is often mentioned in sources up until 1400, but very little is heard of it over the next three centuries (Fig. 7.16). The route crossed its northern part to the heads of Arnarfjörður,

Fig. 7.16 Oblique aerial view from above Ísafjarðardjúp south over the Gláma highlands. Foremost in the photograph, from *left to right*: Hattardalur, Mt. Vatnshlíð, and Seljalandsdalur. Mats

Dýrafjörður and Önundarfjörður. Sveinn Pálsson mentioned that, many centuries ago, there had also been a track called Fjallasýn ('mountain view') furthest inland from Ísafjörður round the western edges of Glámuheiði down to the Barðaströnd shoreline to the south. Glámuheiði reaches its highest point at 920 m, and with the slightest change in the snowline, there are fluctuations in the size of its circular snow patches. The summit of Glámuheiði reached above the snowline in cold years, so that all of the winter snowfall did not melt, but in warm years mountain ridges would appear though the snow as the summer progressed. During the Little Ice Age, an increasingly larger area of Gláma was covered in snow, indeed for a long time there was a difference of opinion as to whether there was a glacier or a permanent snow patch up on the moor. Gláma is named on Guðbrandur Þorláksson's map from 1590, but it is not clear from its depiction whether it is considered a glacier or not. Árni Magnússon, in a catalogue of his journey in 1710 (*Chorographica Islandica*) in connection with his *Jarðabók* ('Land Register'), states that Gláma is a glacier. Knoff drew a Glámujökull on his map of 1733. Eggert Ólafsson and Bjarni Pálsson (1772) say that Gláma is a glacier above Ísafjörður and the inner reaches of Arnarfjörður, which extends south to the boundaries of Barðaströnd County. Ólafur Olavius claimed that Gláma was mostly covered by a glacier. Björn Gunnlaugsson (1848) depicted an extraordinary large glacier from the head of Dýrafjörður to the southeast of Barðaströnd County, and indeed there was a great deal of snow cover in the last quarter of the 19th century when Gunnlaugsson was compiling his maps. It is also worth mentioning that Jónas Hallgrímsson crossed Gláma in August 1840 and later wrote in a letter (1989, Vol. 2, p. 52): 'Hard going in western Iceland, though Gláma not the worst of it as you're on a blessed glacier.'

The description of Kålund (1984–1986, Vol. 2, p. 176) of his journey across Gláma on 11 August 1874 is also often referred to: 'There are crevasses in some places, though no serious ones, a characteristic feature of them being the strange bluey-green colour that shines from their walls, especially in the deeper ones, for the crevasse walls are made of compacted snow or ice.' On the other hand, the botanist Stefán Stefánsson crossed Gláma in 1893 and pointed out that so much moss and so many plants grew on the moor that it was impossible to imagine it could ever have been a glacier. Þorvaldur Thoroddsen (1901) stated that the snowline on Glámujökull was at a height of about 600 m and its surface area four square miles, but added that the stretch of firn snow was variable depending on the season, either expanding and coalescing with snow patches nearby, or thawing so that rocky ridges and crests emerged. A few years later (1906 and 1914), he described Gláma as a firn zone, and on the DGS map of 1914, it is depicted as a firn zone of about 4 km^2 which did not melt during the summer. Furthermore, surveying measurements showed that a cairn, which Hans Frisak had stacked in 1805 on the highest point of Gláma, at 920 m above sea level, had not moved since. The German geologist Winkler crossed Gláma in 1914 and reported that there was no glacier there, no movement of the patch of firn snow, no crevasses, no glacial tongues, and that there were no glacial rivers on the moor. His compatriot Paul

Herrmann (1916) said much the same when he made a difficult crossing of Gláma in August of the same year. Snow now melts away during most summers on Gláma.

Björn Jónsson and Petrína Pétursdóttir Moving Household and Livestock across Gláma in the Spring of 1895

We set off up onto the glacier at four o'clock and then headed towards Dýrafjörður. We then passed over the high point of Gláma from where we could simultaneously see south to Breiðafjörður, west to Arnarfjörður and Suðurfjörður, and north to Ísafjörður, Snæfjallaströnd and Kaldalón. The weather was delightful on both of the days of our journey across Gláma, dead calm and bright sunshine without a cloud in the sky. But the heat of the sun was strong and tiring. The worst was having to put up with the glare from the snow. Travelling conditions in the snow were bad just after midday and later in the afternoon, especially when the land began to descend westward. The three of us rested in turn, either on the horse with a side-saddle, or sometimes on my father's horse with an ordinary saddle. But when the trek became more laboured there was little relief in sitting on either horse as they had to be sent ahead to trample a pathway for the cattle. Conditions got even worse as we approached the mountain Sjónfríð. We went along its northern side and down into the Botnsdalur depression. The saddled horses were then let lead in turn to break up and trample a way through the frozen snow. They had to jump to get over the frozen edges of the snow with each step. Thus a kind of corridor was formed along which the cows staggered as it was hard going for them. I thought I saw mother have tears in her eyes whenever her cows were having the most difficulty, but dad just gritted his teeth. He said the journey would gradually get shorter and cows had a hidden endurance. He changed the horses' saddles and their undercovers every now and then to rest the one which had been forging ahead for a while.

The cow that suffered the most on the journey was old Hryggja. She was the eldest cow, twenty winters, but had no calf. Mum was afraid we'd need to flay her on the glacier. The glacier undulated over a wide area and Hryggja lay down in the hollows, but once the train had disappeared over the next snow crest, she stood up and mooed loudly. Dad said that if she could moo so enthusiastically then she still had the strength to keep going. And he was proved right, too, for Hryggja survived the journey. She continued to live and milked fairly well as usual until the summer waned. And then she was put down.

Both men and dumb beasts were ever so glad when we had descended from the most snow to where there was a bit of level land above the steep sides. A little brook was flowing there and I could see a glint of joy in the animals' eyes when they saw it. The horses were unsaddled and most of the animals lay down and rested for a while, though they were up sooner than expected, as they began to nourish themselves on some half-green grass that had emerged from beneath the snow. The other folk had something to eat, but I had little appetite. My eyes and all my face were stinging from the glare of the snow in the sunlight.

Narrative of Guðrún Björnsdóttir (Hannesson and Eyþórsson 1953, Vol. 3, pp. 99–101).

7.5 Snæfellsjökull

The stratovolcano Snæfellsjökull (1446 m above sea level) is one of the most beautifully shaped mountains of Iceland and was for a long time the best known of all the country's glaciers (Figs. 7.17, 7.18, 7.19, 7.20). Snæfellsjökull can be found on nautical charts of the northern seas from the 16th century as it was an important landmark for sailors. It appears on Icelandic maps (also called Vesturjökull) from the days of Guðbrandur Þorláksson and was considered the highest mountain of Iceland until the early 19th century because, to seafarers, it seemed to rise so awesomely high from the shoreline. Originally the coned mountain was known as Snjófell, then later Snæfell, and when it became clear that the snow covered a glacier, the name was lengthened into Snæfellsjökull, often shortened to 'Jökull' in Icelandic, and the folk who live around its base are called 'Jöklarar'. Bárður Dumbsson, who claimed and settled land on Snjófellsnes peninsula, was the first to climb Snjófell, and after he disappeared from the human world by walking into the mountain, he became known as Bárður Snæfellsás, a heathen god, patron and protector of the local people around the mountain (Fig. 7.21). Bárður proved a saviour to many who called upon his aid (see *Bard's Saga*, Anderson 1997). From the early days of the settlement onwards, the locals around Snæfellsjökull have

Fig. 7.17 Oblique aerial view south over Snæfellsjökull from above the villages of Rif (on the *left*) and Hellissandur (on the *right*). In the summer of 1754, Eggert Ólafsson and Bjarni Pálsson ascended Snæfellsjökull from the church-stead Ingólfshóll (*lower centre*). The source of the river Hólmkelsá is in the southeast of the glacier, next to Jökulháls saddle. HB, March 1999

Fig. 7.18 Map of Snæfellsjökull

talked about supernatural beings and the 'hidden folk' protecting the glacier, all
tending to dissuade rather than encourage people from ascending it.

Eggert Ólafsson and Bjarni Pálsson hiked to the top of Snæfellsjökull in 8 hours
on 1 July 1754, beginning from Ingjaldshóll hill up a glacial tongue by Geldingafell
(824 m). They calculated the height of the glacier as 2154 m. Since then, no glacier
in Iceland has been climbed as often as Snæfellsjökull. At first this involved primarily foreign travellers. Lord Stanley's expedition in 1789 set off from Stapafell to
Jökulháls saddle, and using the triangulation method they calculated the height of
the glacier as being 1460–1470 m, though coastal surveyors later claimed it was
only 1436 m high in 1804. Other foreign travellers and explorers who ascended the
glacier included the Englishmen Bright and Holland, who in 1810 measured its
height as 1440 m, and the Scotsman Henderson, who climbed the glacier from
Stapi, arriving at the summit of Þríhyrningur on 25 May 1815. An unnamed
Icelander also ascended the glacier in 1821 and described the murmur of water in
deep crevasses. In 1910 the DGS calculated the glacier's height at 1446 m, and this
has remained its official height ever since.

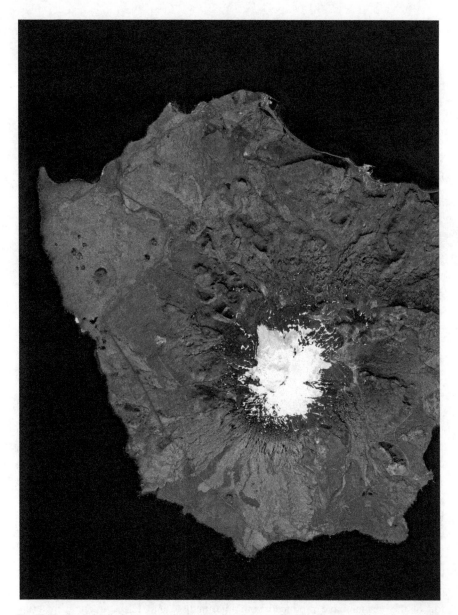

Fig. 7.19 Snæfellsjökull is a large stratovolcano with a white cap and broad stretches of lava. There is a road across Jökulháls saddle to the east of the glacier between Árnarstapi and Ólafsvík. Öndverðarnes peninsula is on the far *upper left*, and to the east of it are the villages of Hellissandur and Rif. SPOT

Snæfellsjökull became world-famous in 1864 with the publication of the science-fiction novel *Voyage au centre de la Terre* by the Frenchman Jules Verne (1828–1905), which relates the story of a German professor and his nephew Axel

Fig. 7.20 Snæfellsjökull. From the book *Danmark* by Em. Bærentzen 1856. NULI

who descend through the crater of Snæfellsjökull into the bowels of the Earth. In the first English version of *Journey to the Centre of the Earth* in 1871, the nephew becomes the English Harry Lawson (Verne 1864, 1871). The Icelandic Nobel prize-winner for literature, Halldór Laxness, translated the book into Icelandic in 1944, and used the mystique and attraction of the glacier in his own novel *Kristnihald undir Jökli* in 1968 (trans. *Under the Glacier*, 1972). Nowadays the glacier is said to be one of the seven global 'power centres' and even a landing place for aliens. The glacier and its environs are now within the boundaries of the Snæfellsjökull National Park established in 2001.

Snæfellsjökull is the main volcano at the westernmost point of a volcanic system 30 km long which has been very active over the last two million years. There have been no tectonic plate movements during any of its volcanic activity and so a high stratovolcano was not impeded from steadily being enlarged. The volcanoes at the top of the glacier have produced both intermediate and felsic volcanic material, while the volcanoes on the lowland have mostly had eruptions of basaltic lava. The southern and western flanks of the glacier cover the Holocene lava flows that have run uninterruptedly into the sea from Arnarstapi to Hellissandur. A very strange ropy lava has wound itself down from the cone-shaped volcano. To the north and east there have been flows of either palagonitic tuff or lava during interglacial periods. There are at least 20 lava fields in the region. Three pale layers of ash sediment in the topsoil on the northern side of the peninsula bear witness to

Fig. 7.21 Bárður Snæfellsás. A creation of the sculptor Ragnar Kjartansson. Björn Valdimarsson, 2006

tremendous explosive eruptions about 1600–1900 years ago, about 4000 years ago, and 8000–10,000 years ago. Snæfellsjökull still has all the signs of being an active volcano (Flores et al. 1981; Jóhannesson 1982a, b).

The top crater of the stratovolcano is covered by a glacier. It is shaped like a horse-shoe surrounded by high glacial walls to the south and east but open to the north and west. On the southeastern edge of the crater bowl, three peaks of acidic rock stand out, called Þúfur ('hummocks'). The central one is highest (1446 m), extremely steep and difficult to ascend, but the western peak (1442 m) is the largest in size covered in glacial ice and snow. The northern peak (1390 m) is the lowest, oblong in shape with a serrated crest. The glacial summit below the Þúfur peaks is at an elevation of 1430 m (Figs. 7.22, 7.23, 7.24). The largest outlet glaciers creep through a pass in the northwest of the crater. Blágilsjökull extends down to a height of 700 m and contains the source of the river Hólmkelsá (Hólmkela), which flows into the Rifsós estuary. It is the only glacial river on the northern side of Snæfellsjökull. What immediately caught the attention of Eggert Ólafsson and Bjarni Pálsson in 1754, was that only a small amount of the water which might be expected to drain from such a large glacier actually flowed in the glacial rivers. They thus stated that most of the volume of water must disappear through large holes at the base of the glacier and then flow underground into the sea. There are a few small streams emerging from beneath Snæfellsjökull's southern flanks.

Fig. 7.22 Drawing of Þúfur on summit of Snæfellsjökull by Eugène Robert; from the Gaimard (1840) expedition in 1835. NULI

Fig. 7.23 Travellers at Þúfur on Snæfellsjökull in 2005. In the forefront is the botanist Þóra Ellen Þórhallsdóttir. HB

The remains of an old volcanic crater, Þríhyrningur (1191 m), rises on the eastern edge of the glacier as a three-headed cinder-red peak. Hyrningsjökull descends to a height of 700 m in Hyrningsdalur, from which it draws its name. A meteorological and atmospheric research station was in operation for twelve months in 1932 on Jökulháls saddle, to the east of the glacier, to celebrate the 50th anniversary of the first International Polar Year of 1882–1883 (Zingg 1941; Flemming 1933; Laursen 1982; Kristjánsson and Jónsson 1998). Although primarily Danish and Swiss scientists were behind the project, there were associates from a total of 33 nations carrying out research on magnetic fields, northern lights, and climate in the northern oceans and arctic countries. Hólatindajökull descends from the western side of the crater bowl, from the palagonite peak of Sandfell (1217 m) in the direction of Hólatindar at an elevation of 1000 m.

Snæfellsjökull will be missed The location of Snæfellsjökull's outlet glaciers have been measured regularly from about 1930 and they have all become about 600–800 m shorter in response to climate change. Radio echo soundings indicate that only about a tenth of the stratovolcano itself reaches above a height of 1200 m, which is now the elevation of the firn line in an average year, though 40 % of it is above 1000 m. The surface area is around 10 km^2 and the total volume of the glacier has been estimated as near 0.5 km^3; the average thickness of its ice is thus only 50 m. It will probably not survive many more decades with a continually warming climate. At its thickest, the ice has been measured at 70 m within the crater and 100 m outside of it, though the outlet glacier descending north from the bowl might in some places prove thicker. The glacier has thinned out over the last

(a)

(b)

Fig. 7.24 Top a Snæfellsjökull as seen from Faxaflói Bay in 1860. Drawing by J.W. Bushby. Forbes 1860. NULI. **Bottom b** Snæfellsjökull as seen from the north. Drawing in Sir Richard Burton's *Ultima Thule*, 1875. NULI

few decades and become more uneven than it used to be, and the main outlines of the landscape beneath it are increasingly being exposed. Ridges can now be seen extending in all the main directions from the top crater. If there were to be an eruption within the crater, any consequent jökulhlaup would most likely flood toward the northwest.

Ascent of Snæfellsjökull by Eggert Ólafsson and Bjarni Pálsson on 1 July 1754

People thought our plan to walk up the glacier totally reckless. Furthermore it was considered absolutely impossible for various reasons. In the first place the route was so long and the mountain so steep that it was impossible to climb, secondly there were impassable crevasses in the glacier which no one could cross, and finally we were assured that we would go blind from the reflection of the sun off the glacier. Moreover, we were also told the story of two English seamen who had tried to climb the glacier many hundreds of years ago. They had certainly got all the way to the top, but then one of them became blind and so confused that he could not find the way back down, and he was never seen again. His obstinate insistence on continuing by himself had also led to his death. The other Englishman had been more careful and had slaughtered a sheep before he set off. He took the sheep's blood with him in a bag and let it drip during his journey up the glacier marking his tracks. This meant that, although he had lost his sight, he could still trace his way back down again as he could distinguish between red and white in spite of his blindness. He did not reach the high peak of the glacier, however. The most ignorant of local people also assured us of what we had so often heard before when journeying across Geitlandsjökull and entered Surtshellir cave... that there are certain beings of this earth, the 'hidden folk', dwarves, or especially the ghost of the heathen god Bárður Snæfellsás... who inhabit the mountain and would hinder our journey, for such visits were extremely unwelcome. But people believe this sort of thing about all such places that are awesome and difficult to reach. We took no notice of these persuasions for our longing to climb the mountain continued to grow apace, both in order to disprove these superstitions and also for other reasons. On the evening of 30 June the weather outlook improved. The wind dropped, there was a break in the clouds. The mercury rose in the barometer, and the thermometer rose also, showing that the air temperature was getting warmer...

We set off at one o'clock in the morning of 1 July. We had with us a compass, a mercury thermometer (Fahrenheit) and the previously mentioned barometer.... As is traditional in this country, we climbed in thin shoes, which are lighter than other shoes, and provide more trustworthy footing, especially on patches of ice. We also had with us a strong rope because of the crevasses in case one of us should fall into one, which sometimes happens on Jökulháls saddle.... We also took scarves with us to cover our eyes if the brightness became too powerful, and finally we had a sponge soaked in vinegar to sniff at if the air up on the glacier should prove too light and thin...

With some difficulty we managed to reach the edge of the glacier at Geldingafell with horses. The barometer had fallen by two inches since we had set off. We rode a little way up the glacier, which was without any crevasses and with not much of an incline. But it wasn't long before the glacier became uneven. Here we got off the horses and had them returned to Geldingafell, as it was safer and easier to find them again there than higher up on the glacier. The compass was now very uncertain and did not point in the right direction until after quite a while, though we could calculate our bearings from a watch that had been synchronised with the sun. The weather was most beautiful, calm and with hardly any mist, though thin trails of clouds covered the sun, but otherwise there was a clear sky. There was hardly any reflective glare from the glacier due to its unevenness. The coldness increased so much, however, that the heat from the sun did not match it, and the air grew rapidly thinner. We became weak although the slope was still gentle...

The higher up we reached, the greater the crevasses became. We got through it all safely up onto the glacial summit at the foot of the three peaks, or knolls, that arise above it. These peaks are all of equal height, about 50 fathoms. One of them is in the east, another in the west, and the third in the north. The two latter ones merge when seen from farmsteads below so that it appears there are only two peaks with a hollow between them, reminding one of a saddle. The peaks are called Jökulþúfur. They seem unclimbable, for indeed there was icing on them from the mist and all of the snow covering them was frozen hard. We finally managed to hack out footholds in the ice with ice axes and hunting knives and thus ascend the easternmost knoll, which is the highest and shaped like a sugar lump....

It was now 9 o'clock in the morning. The sun shone in a clear sky, but it was nonetheless so cold that we could hardly bear it, indeed the thermometer had dropped to 24 degrees [-4°C] which could be said to be the perfect winter cold in Iceland. There was no need to be amazed, therefore, that it would be so cold on glaciers during the winter when there was so much frost there in the height of summer. There are those who hold the opinion that the coldness of glaciers is caused by particles of saltpetre, but this seems a totally unnecessary explanation for this phenomenon.

The compass was now completely wrong. It pointed in one direction and then another. Sometimes it stopped at a totally erroneous direction, and if the compass were moved, it span around again until it stopped at yet another equally incorrect bearing.... The eastern knoll is only about 16-18 feet broad on top. The northern knoll is oblong in shape with a jagged crest, but the westernmost knoll is by far the largest. Its lower part is quite substantial and circular in shape, though very jagged further above. The glacier is steepest on its southern flanks and full of crevasses, the crevasses situated side by side all the way up and down it. We did not see the western side of the glacier, but on the northern side was an enormous transverse crevasse just a short distance from the summit knolls. This crevasse is so long it seemed to slice off a third of the glacier. It was also so deep that it appeared green in colour and its bottom out of sight. Below it the surface of the whole glacier was broken up by crevasses as far as the eye could see.

There was a strange snow cover on top of the glacier. It was neither slippery nor smooth, but it was as if some force had reorganised it, and although it was certainly uneven there was still some kind of order in the irregularity, so that they looked like a slated roof, or rather the arrangement of feathers on a bird. The ice slabs were about a foot in length, half as much in width and about $1-1^{1/2}$ inches thick. At the top they had groove marks lying north to south. They overlapped each other in that the southern part of each slab overlay the northern part of the next one below. The loose edges looked as if they had been chopped at leaving behind 3-4 blunt teeth. Many images and forms can be seen in snow in Iceland when the temperature is freezing and there is a gale-force wind, but they are usually rather small and always very irregular. But with such weather conditions on glaciers, and most often a northerly wind, the instability of the air so high up over the glacier must cause and shape these formations on the ice.

The view was most beautiful. A large part of Iceland could be seen, all of the southern region, all the glaciers and mountains in the central part of the country, Austurjökull, Hekla, the Geirfuglasker skerry off the coast of Reykjanes, Borgarfjörður and the mountains between northern, southern, and western Iceland, and finally all the land and mountains to the north of Breiðafjörður, along with the latter's innumerable islands and islets.'

(Ólafsson and Pálsson 1981, Vol. 1, pp. 160–64)

Ascent of Snæfellsjökull by James Wright in 1789

The physician and botanist James Wright (1770–1794), the great-grandfather of the philosopher Bertrand Russell, had the following entries in his diary relating to the ascent of Snæfellsjökull in 1789 during the expedition to Iceland of Sir John Thomas Stanley (1766–1850) [n.b. spelling as in original]:

Tuesday July 14th After refreshing ourselves with some provisions and Rum which we brought along with us, and leaving our Guide for whose assistance we had no longer any occasion, to take care of the horses, we began to ascend on foot each being provided with a pair of Coarse Woolen Stockings above his shoes to prevent him from Slipping. The snow at first was rayr [sic], soft, but soon became harder, so as to render walking upon it very agreeable, if it was not for the ascent, which soon put us all from complaining of the cold. We were soon encountered by very wide rents in the snow, in passing of which we found our poles of infinite service. Some of these rents were above 7 ft. in Breadth, and by a piece of Lava attached to a String I found one to be 42 and anoyr. 50 ft. in depth. In these we could see the different layers of every year to a considerable depth—assuming the most lively blue and Green tints—Mr. Colden and I led the Van, but we could hardly boast of our being good pilots—for we were often under the necessity of returning a great way back to find a narrower place in the many rents we met with—When we were within a quarter of a mile of the Top our sailor not having a Pole to assist him in ascending the Snow which was now become very steep and slippery, would proceed no further—At last after much labour, danger, and difficulty, we arrived wt. in about 500 ft. of the Summit. From this Mr. S. and myself began to ascend the remaining part of the hill on all Fours—to which we were obliged to have recourse, on account of the Steepness and Slipperiness, which last has become so considle. as to require our utmost caution and exertion to keep us from Falling, which if it had happened, would at once have put an effectual stop to all our measuring Schemes. We ascended this hill 'till we got within a few yards of the Summit

—when we were stopped by a Chasm, which from the Steepness of the place in which it was, appeared impossible to pass. This Chasm had on the Right, a Bridge of Snow over it of abt. a foot in thickness, whether to cross this at the Risk of our lives or content ourselves with what we had already done, was the subject of conversation betwixt Mr. Stanley and myself (for the rest remained in the same place where we last stopped). On the one hand it was confoundedly provoking to have got so near the Top of the famous Mountain—Where few, if any had been before, and after all to fail—on the other, the great danger wt. which it was attended, and giving way of the Snow over the Chasm, and slipping a foot and fetching way down the hill from such a fearful height and declivity made us Shudder with horror. In this dreadful suspense we remained for a few minutes wt. our feet in holes which we had dug out of the Snow, uncertain what to do. Those below called to us to desist, and on the Carpenter's being desired to fetch the Barometer up, his answer was 'that he had more regard for his Soul and Body than willingly to throw away both'. In the mean I was trying the depth of the Snow round abt. when in one part at the depth of 6 ft., I found or imagined I found the Solid rock or hard Ice resist my Pole—When I told Mr. Stanley of this we both resolved to proceed at all Hazards, and accordingly we set about digging holes in the Snow in which we placed our feet, raising ourselves gradually and bearing as lightly as possible over the Mouth of the Cavern, we got to the Top and congratulated ourselves on our resolution, patience and perseverance wc. had been attended wt. such success. We endeavoured to persuade those below to ascend but to no purpose, and as for the Carpenter neither threats nor promises had any effect on him—he swore he would not budge an Inch further for the whole world. It was 1 h.5m. a.m.

Wednesday July 15th
When we got to the Summit, From this great height we could observe the Sun Just about to rise his Beams beginning to appear from Behind a Hill in the Direction of NNE. Upon exposing the Thermr. the mercury fell to 27° [ca. −3 °C], a degree of cold which we found very inconvenient. There being no prospect of the Carpenter's ascending wt. the Barometer we began to descend, after writing our Mistresses names on the top of this high Mountain in the Whitest Snow (Emblematic of their Innocence and purity,) contenting ourselves with measuring the height to which those had got below, and guessing at the rest. We began to descend wt. as great Caution as possible, for the Carpenter assured us that we would find this much more dangerous than Ascending. This I had like to be very dearly convinced of, for when I was within about 30 yards of those below my pole slipt and away I slid. I was not very eager to stop myself forgetting that there was a great Chasm a little way before me until Mr. Benners called to me most earnestly to stop or I would be immediately in it. I attempted to Stop, but I was going with too much rapidity

for the snow to resist my pole. I must therefore have been either dashed to pieces against the Further edge of the Chasm or have tumbled to the Bottom of it, if I had not been prevented by a Lucky fall in which I had time to use the pole properly. Mr. Stanley who followed me was nearly in as much danger; for when he was about crossing the last Chasm which separated him from where we were all standing—Messrs. Colden and Crawford, desired him to come on showing him the best spot where to place his feet, and where they had crossed themselves. Fortunately however, before venturing he stuck his pole into it, when it immediately gave way, being a bridge of snow over a deep rent, less than a foot in thickness.

After we had got all collected together I proposed to Mr. Stanley to carry the Baromr. up to the Summit since the Carpenter refused, and measure it myself—this after some objections started by the rest was agreed to. I was accompanied by Messrs. Stanley and Benners. Mr. Colden and Crawford contenting themselves with admiring the Summit from where they stood. Indeed Mr. Colden had been up some way further before, and by a false step had very narrowly escaped getting into the chasm, by the assistance of Mr. Benners, this adventure took from him all relish for making any further attempts.

We found it much easier to get to the Top now than the first time, by reason of the footsteps which had then been made—We took the height of the Mercury in the Barometer and Thermr. exactly. The View now from our Elevated Situation was grand and striking in the extreme. To the West the other point of this high hill, presented itself at the distance of about 1000 yards. The Northern prospect was occupied by Old Ocean, whose Front was as quiet and unruffled as if he never intended to be in a Passion again—To the East high, peaked, misshapen Mountains (small however when compared with that on which we were standing) were seen overtopping each other as far as the Eye can reach, and to the South we saw the Sea, Stappen [Stapi] and its harbour, in which the John and anoyr. Vessel appear'd like two small specks. To the North east Hills beyond an Inlet of the sea Bound Horizon, above which the sun has now got up in all his glory—by which the shadow of Snæfields Jokul was thrown so finely on the surface of the deep to the South west, that it was some time before we could convince ourselves that it was not another hill which had before been concealed from us in a fog. After admiring this scene for some time we began to descend at 1 h.40 m. a.m.—at ½ past 3 we got to the place where the guide remained wt. our horses, on whom we arrived safe at the Merchants house in abt. 2 hours more—Campbell hailed the Boat, and while Coffee was preparing for us by Mr. Hialtin who had got up from bed for the purpose I took the Height of the Barometer in the same place where I had done it before ascending the Mountain, as likewise at High water Mark.

(Wright 1970, p. 79–82)

References

Anderson, S. (Trans.). (1997). Bard's Saga [Bárðarsaga Snæfellsás]. In V. Hreinsson (Ed.) *The complete sagas of Icelanders* (Vol. 2, pp. 237–266).

Bærentzen, E. (1856). *Danmark*. Copenhagen.

Björnsson, H. (1971). Bægisárjökull, North-Iceland. Results of glaciological investigations 1967– 1968. Part I. Mass balance and general meteorology. *Jökull, 21*, 1–23.

Björnsson, H. (1972). Bægisárjökull, North Iceland. Result of glaciological investigations 1967– 1968. Part II. The energy balance. *Jökull, 22*, 44–61.

Björnsson, H. (1979). Glaciers in Iceland. *Jökull, 29*, 74–80.

Björnsson, H. (1991). Jöklar á Tröllaskaga. In *Árbók Ferðafélags Íslands* (pp. 21–37).

Björnsson, H., Pálsson, F., Sigurðsson, O., & Flowers, G. (2003). Surges of glaciers in Iceland. *Annals of Glaciology, 36*, 82–90.

Burton, R. (1875). *Ultima thule; or, A summer in Iceland. With historical introduction, maps and illustrations*. London: William P. Nimmo.

Caseldine, C. (1983). Resurvey of the margins of Gljúfurárjökull and the chronology of recent deglaciation. *Jökull, 33*, 111–118.

Egilsson, V. (2000). Í strandbyggðum norðan lands og vestan. In *Árbók Ferðafélags Íslands*.

Einarsson, Þ. (1968). *Jarðfræði. Saga bergs og lands*. Reykjavík: Mál og Menning.

Erkes, H. (1910). Der Glámujökull. *Globus, 98*, 147.

Eyþórsson, J. (1935). On the variations of glaciers in Iceland. I (Drangajökull). Some studies made in 1931. *Geografiska Annaler, 17*, 121–137.

Eyþórsson, J. (1956). Frá Norðurlandsjöklum. Brot úr dagbók 1939. *Jökull, 6*, 23–29.

Flemming, J. (1933). Progress-report on the international polar year of 1932–1933. *Transactions, American Geophyics Union, 14*, 146–154.

Flores, R., Jónsson, J., & Jóhannesson, H. (1981). A short account of the Holocene tephrochronology of the Snæfellsjökull central volcano, western Iceland. *Jökull, 31*, 23–30.

Forbes, C. (1860). *Iceland: Its volcanoes, geysers and glaciers*. London: John Murray.

Gaimard, P. (1840). *Voyage au Islande et au Groenland exécuté pendant les années 1835 et 1836 sur la corvette «La Recherche»*. (Chapter on mineralogy and geology by Robert, E.) Paris: A. Bertrand.

Grímsdóttir, G. (1994). *Ystu strandir norðan Djúps. Un Kaldalón, Snæfjallaströnd og Strandir. Árbók Ferðafélag Íslands*. Reykjavík: Ferðafélag Íslands.

Häberle, T. (1991). *Spät- und postglaziale Gletschergeschichte des Hörgárdalurgebietes, Tröllaskagi, Nordisland* (Doctoral dissertation, Geographical Institute of the University of Zürich).

Hallgrímsson, J. (1989). *Ritverk Jónasar Hallgrímssonar* [Works of Jónas Hallgrímsson] 4 vols. In H. Hannesson, P. Valsson, & Y. Egilsson (Eds.). Reykjavík: Svart á hvítu.

Hannesson, P., & Eyþórsson, J. (Eds.). (1953). *Hrakningar og heiðavegir*. 3 vols. Akureyri: Norðri.

Herrmann, P. (1916). Die Gláma. *Mitteilungen der Islandfreunde, 4*(2), 31–38.

Iwan, W. (1936). Beobachtungen am Drangajökull, NW-Island. *Zeitschrift der Gesellschaft für Erdkunde zu Berlin., 3*(4), 102–114.

Jóhannesson, H. (1982a). Yfirlit yfir jarðfræði Snæfellsness. In *Árbók Ferðafélags Íslands* (pp. 151–174).

Jóhannesson, H. (1982b). Kvarter eldvirkni á Vesturlandi. In H. Þórarinsdóttir, Ó. Óskarsson, S. Steinþórsson, & Þ. Einarsson (Eds.), *Eldur er í norðri* (pp. 129–137). Reykjavík: Sögufélag.

Kaldal, I. (1978). The deglaciation of the area north and northeast of Hofsjökull, Central Iceland. *Jökull, 28*, 18–31.

Kålund, K. (1984–1986). *Íslenzkir sögustaðir*. 4 vols. (H. Matthíasson, Trans.). Reykjavík: Örn og Örlygur.

Keilhack, K. (1934). Riesiger Gletscherrückgang in Nordwest-Island von 1844-1915. *Zeitschrift für Gletscherkunde, 21*, 365–370.

Kristjánsson, L., & Jónsson, T. (1998). Alþjóða-heimskautaárin tvö og rannsóknastöðin við Snæfellsjökul. *Jökull, 46*, 35–47.

Kugelmann, O. (1989). *Gletschergeschichtliche Untersuchungen in Svarfaðardalur und Skíðadalur, Tröllaskagi, Nordisland* (Doctoral dissertation, Institute of Geography, University of Munich).

Laursen, V. (1982). The second international polar year (1932/1933). *WMO Bulletin, 31*, 214–222.

Laxness, H. (1968). *Kristnihald undir Jökli*. Reykjavík: Helgafell.

Laxness, H. (1972). *Under the Volcano* (M. Magnusson, Trans.). Reykjavík: Helgafell.

Magnússon, Á. (1955) [ca. 1702]. *Chorographica Islandica*. In Ó. Lárusson (Ed.), *Safn til sögu Íslands*. Second series. Vol. 1, (2). Reykjavík: Hið íslenska bókmenntafélag. pp. 269–388 (7–120).

Magnússon, Á., & Vídalín, P. (1981) [1702–1712]. *Jarðabók Árna Magnússonar og Páls Vídalíns*. 13 vols. (2nd edn.). In B. Melsteð & F. Jónsson (Eds.). Reykjavík: Hið íslenska sögufélag (Originally published in Copenhagen, 1913–21).

Magnússon, E., Pálsson, F., & Björnsson, H. (2004). Íssjármælingar á Drangajökli 8.-9. apríl 2004. *Jökull, 54*, 87–88.

Norðdahl, H. (1983). Late Quaternary stratigraphy of Fnjóskadalur central North Iceland, a study of sediments, ice-lake strandlines, glacial isostasy and ice-free areas. *Lundqua Thesis* 12 (p. 78) (Doctoral dissertation, University of Lund).

Ólafsson, E., & Pálsson, B. (1981). *Ferðabók Eggerts Ólafssonar og Bjarna Pálssonar um ferðir þeirra á Íslandi árin 1752–1757* (S. Steindórsson, Trans.). Reykjavík: Örn og Örlygur.

Olavius, Ó. (1964–1965) [1780] *Ferðabók: landshagir í norðvestur-, norður- og norðaustursýslum Íslands 1775–1777: ásamt ritgerðum Ole Henckels um brennisteinsnám og Christian Zieners um surtarbrand. Íslandskort frá 1780 eftir höfund og Jón Eiríksson.* 2 vols. (S. Steindórsson, Trans.). Reykjavík: Bókfellsútgáfan.

Pálsson, S. (1945). *Ferðabók Sveins Pálssonar. Dagbækur og ritgerðir 1791–1794.* J. Eyþórsson, P. Hannesson & S. Steindórsson (Trans. and Eds.). Reykjavík: Snælandsútgáfan.

Sigbjarnarson, G. (1983). The Quaternary alpine glaciation and marine erosion in Iceland. *Jökull, 33*, 87–98.

Steindórsson, S. (1949). *Lýsing Eyjafjarðar. Part I.* Akureyri: Norðri.

Stötter, J., & Wastl, M. (1999). Landschafts- und Klimageschichte Nordislands im Postglazial. *Geographischer Jahresbericht aus Österreich, 56*, 49–68.

Thoroddsen, Þ. (1892) Islands Jøkler i Fortid og Nutid. *Geografisk Tidsskrift. 11*, 111–146.

Thoroddsen, Þ. (1892–1904). *Landfræðissaga Íslands*. [Geological History of Iceland]. 4 vols. Copenhagen: Hið íslenska bókmenntafélag.

Thoroddsen, Þ. (1907–1911). *Lýsing Íslands*. 2 vols. Copenhagen: Hið íslenzka Bókmentafélag (Reprinted 1932).

Verne, J. (1864). *Voyage au centre de la Terre*. Paris.

Verne, J. (1871). *Journey to the Centre of the Earth*. English version. London: Griffith and Farran.

Víkingsson, S. (1978). The deglaciation of the southern part of the Skagafjörður district, Northern Iceland. *Jökull, 28*, 1–17.

Wright, J. (1970). Diary. In J. West (Ed.), Vol. 1 of *The Journals of the Stanley Expedition to the Faroes Islands and Iceland in 1789*. 3 vols. Tórshavn: Føroya Fróðskaparfelag.

Zingg, T. (1941). Die Polarstation Snaefellsjökull 1932/1933. *Année Polaire Internationale 1932–1933*. Participation Suisse (p. 22). Zürich: Station central Suisse de Météorologie.

Þórarinsson, S. (1937). The main geological and topographical features of Iceland. *Geografiska Annaler, 19*, 161–175.

Chapter 8
Vatnajökull and Glaciers of Eastern Iceland

Where the glacier meets the sky, the land ceases to be earthly,
and the earth becomes one with the heavens; no sorrows live
there anymore, and therefore joy is not necessary; beauty alone
reigns there, beyond all demands.

(Laxness, *World Light*, 2002 [1937], p. 453)

Abstract Vatnajökull (7800 km^2 and 2900 km^3) is the largest ice cap of both Iceland and Europe. This chapter contains a detailed description of the types, shapes, sizes, locations, and relevant scientific data of all of its main ice domes, ice flows, and outlet glaciers, as well as some of their recorded local histories, folklore and legends. Iceland's largest rivers originate from this glacier, and beneath it lie the country's most active volcanoes, geothermal areas and subglacial lakes, which drain in frequent jökulhlaups. Skeiðarársandur is the world's largest outwash plain in front of a presently active glacier. Grímsvötn is the most renowned subglacial lake, and Jökulsárlón on Breiðamerkusandur is the largest proglacial lagoon into which Breiðamerkurjökull calves. The around 30 outlet glaciers advanced several km after the settlement of Iceland until the end of the 19th century. Surges have been frequent in about ten of the main outlets, the largest in Brúarjökull. Skeiðarárjökull and Breiðamerkurjökull have eroded their beds up to 300 m below sea level. More than 80 volcanic sub- or proglacial volcanic eruptions have taken place over the last 800 years, the most catastrophic occurring in Öræfajökull, the fourth largest stratovolcano is Europe, in 1362, and with the Laki Fires of 1783–84. Nowhere in Iceland has the relationship between man, glaciers and rivers been as intimate as in the vicinity of Vatnajökull. Many expeditions onto this huge glacier are described along with some of the unique findings in the history of glaciology that they produced.

8.1   Vatnajökull

Vatnajökull is the largest glacier in Iceland and indeed in the whole of Europe (if the Severny Ice Cap in Novaya Zemlya in the Arctic Ocean is not included). It stretches from its western edge in the Tungnaáröræfi wilderness for 150 km to the

© Atlantis Press and the author(s) 2017
H. Björnsson, *The Glaciers of Iceland*, Atlantis Advances
in Quaternary Science 2, DOI 10.2991/978-94-6239-207-6_8

Fig. 8.1 Aerial photograph southward over the peaks of the Hrútfjöll mountains to Öræfajökull. The highest is Hvannadalshnúkur, while the Hnappar peaks mark the southern rim of the Öræfajökull caldera. The outlet glaciers Skaftafellsjökull (on the *left*) and Svínafellsjökull (on the *right*) begin their descents. HB

highest foothills of the valleys of eastern Iceland, and extends from farmsteads along the southern coast for 50–100 km northwards onto a highland plateau 900 m high (Figs. 8.1, 8.2 and 8.3). Vatnajökull covers a twelfth of Iceland, having a surface area of 7800 km^2. Most of the largest rivers of Iceland flow from this glacier, their catchment areas including about half of the country, stretching eastwards and anti-clockwise from the estuary of the Þjórsá in the south all the way to Skjálfandi Bay in the north. In southern Iceland, its rivers have piled up large outwash plains, while in the northern highlands they have gouged out ravines. The most active of Iceland's volcanoes lie hidden beneath Vatnajökull and they have caused sudden floods and dispersed volcanic ash all over the country from more than 80 volcanic eruptions recorded over the last 800 years. Jökulhlaups have also emerged from the volcanic centres every few years because of the geothermal heat beneath the glacier's bed, which is constantly melting ice, the resulting meltwaters accumulating in subglacial lakes until bursting forth in floods (Fig. 8.4).

Nowhere in Iceland has the proximity and relationship between man and glacier and its rivers been so intimate and difficult than along the coastline south of Vatnajökull. During the Little Ice Age, Vatnajökull's outlet glaciers advanced about 10–15 km and glacial rivers destroyed cultivated land and swamped farms from the settlement of Iceland, eleven centuries ago, with ice and mud. The glacier also came very close to reaching the sea and blocking routes east and west along the coast.

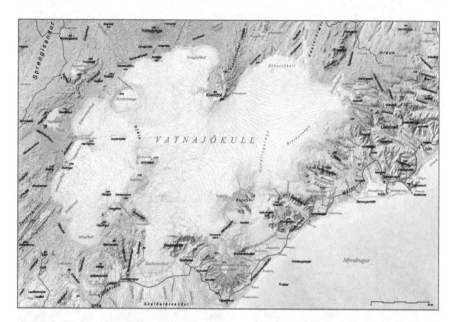

Fig. 8.2 Map of Vatnajökull and its environs

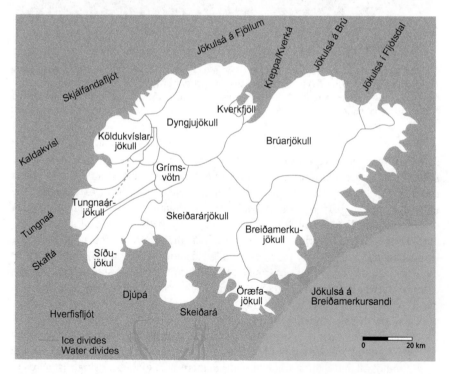

Fig. 8.3 The outlet glaciers of Vatnajökull, their main rivers and catchment areas. The main ice divide is from west to east, from one end right to the other (Björnsson 1982, 1986b, 1988; Flowers et al. 2003, 2005). Ten major rivers flow from the glacier, bearing on average 1000 m³/s of water

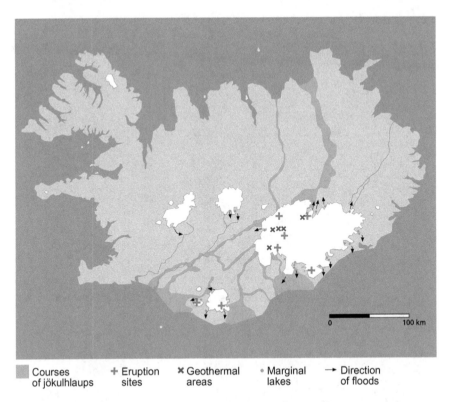

| Courses of jökulhlaups | + Eruption sites | ✗ Geothermal areas | • Marginal lakes | → Direction of floods |

Fig. 8.4 Origins and courses of jökulhlaups (Björnsson 1974, 1975, 1976, 1977, 2002). The *arrows* show every jökulhlaup that has burst out from marginal glacier lakes in the 20th century, but outburst floods have also taken other routes from geothermal sources and volcanic eruptions ever since the end of the last glacial period

When viewed from a satellite, Vatnajökull resembles a human brain (Fig. 8.5); its frontal lobe faces west, its occipital lobe east, and the convex parietal lobe looks north. On its outer edges, its ablation zones are like grey cortex, and its inner parts are the white accumulation zones. Its bulging pleats and grooves delineate the individual outlet glaciers. Vatnajökull is an undulating ice cap with four main domes that reach from 1400 to 2000 m high: Bárðarbunga (2000 m) is at the northwestern edge, Kverkfjöll (1920 m) are in the centre of the northern edge, Breiðabunga (1500 m) is furthest east, while Háabunga (1730 m) is in the centre of the glacier to the southwest of Grímsfjall (1720 m). Steep and short outlet glaciers descend outwards from these glacial domes to the margins, but from the centre of the ice cap, the ice flows into hollows and accumulates in broad, long, flat snouts with few crevasses, similar to the glaciers that lay over the northwestern part of Europe and North America during the last glacial period. The largest outlet glaciers are Dyngjujökull and Brúarjökull in the north. Five outlet glaciers descend to the west. Síðujökull and Skaftárjökull are the southernmost of these, rising eastwards to Pálsfjall, Háabunga and Grímsvötn; Tungnaárjökull is further north, stretching from

Fig. 8.5 A satellite image of Vatnajökull taken from a height of 700 km on 4 August 1999. Snow covers the accumulation zone, but volcanic ash and gravel have been dispersed over the ablation zones. The very dark traces of ash from the eruption of Jökulbrjótur (also known as Gjálp), are still clearly visible almost three years after it ended in October 1996. SPOT

the Tungnaárfjöll mountains and Langisjór to Kerlingar, while Sylgjujökull descends between Kerlingar and Hamarinn; finally, Köldukvíslarjökull extends to the southwest of Bárðarbunga to Hamarinn. All of these glaciers creep down from a 600- to 800-m-high plateau of palagonitic tuff formed during the last glacial period, while beneath the highest glacial domes rise the main volcanic centres (Figs. 8.6 and 8.7).

Steep, very crevassed, and long outlet glaciers descend southwards from the edge of the central highlands onto the outwash plains along the southern coastline at a height of 10–100 m. The largest of these glaciers is Skeiðarárjökull, which splays into a piedmont glacier as it reaches the lowlands. Breiðamerkurjökull accumulates ice from many valley glaciers, similar to those that can be seen in Alaska and Svalbard, and they have long medial moraines between them. The mountain massifs of Esjufjöll and Mávabyggðir rise up through the glacier. Both Breiðamerkurjökull and Skeiðarárjökull, two of Vatnajökull's most powerful and active outlet glaciers, have gouged out beds 200–300 m below sea level along a 25-km section on both sides of Iceland's highest mountain, Öræfajökull, creating the greatest height differential in the country: around 2300 m from the glacial beds to the Hvannadalshnúkur peak. Steep, much-crevassed, and almost impassable valley

Fig. 8.6 Computer-generated images of Vatnajökull and its subglacial terrain (Björnsson 1986a, 1988, 1996). Vatnajökull is 7800 km^2 in size and covers one twelfth of Iceland's surface area. Its main part lies on a 600–800-m-high plateau of palagonitic tuff formed during the last glacial period, its highest mountains reaching up to 2000 m. At its lowest point, the glacier has harrowed itself down into ancient oceanic sediment layers to 200–300 m below sea level. Beneath its western reaches is the most powerfully active volcanic region of Iceland, the reason why many of the country's highest mountains have been amassed there: Bárðarbunga, Hamarinn, Háabunga, Grímsfjall, and Kverkfjöll. The calderas in Grímsvötn, Bárðarbunga and Kverkfjöll are 500–700 m deep and 10 km in diameter. Mountain ridges extend from them that were formed from eruptions along fissures. Altogether, there are five volcanic centres discernible beneath Vatnajökull. The eastern part of the glacier covers the basalt massif of the eastern fjords formed during the Tertiary period and indented with deep valleys and fjords, for glacial erosion has long shaped the landscape here undisturbed by volcanic activity. Short and steep outlet glaciers descend toward the margins from the four main domed summits of Vatnajökull, but from the centre of the ice cap, ice flows into the hollows and accumulates in long, broad and flat snouts at a height of 600–800 m above sea level. From the edge of the southern highlands, long and steep outlet glaciers descend onto outwash plains at a height of 10–100 m

glaciers, similar to Alpine glaciers, descend from the southeast of Breiðabunga dome, many of them having also gouged deep beds in prehistoric outwash plains. The eastern part of Vatnajökull covers part of the basalt massif of the eastern fjords from the Tertiary period, and it is torn asunder by deep valleys, for glacial erosion has shaped the landscape there for centuries untroubled by volcanic eruptions.

Future Landscapes

Breiðamerkurjökull and Skeiðarárjökull, two of the most active of Vatnajökull's outlet glaciers, have excavated 20-km-long trenches beneath

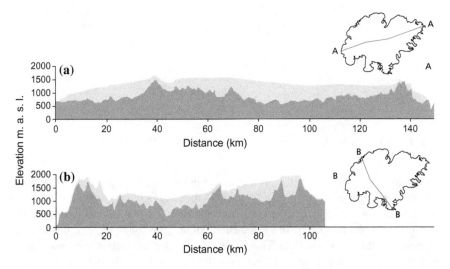

Fig. 8.7 Cross sections of Vatnajökull from Tungnaárjökull through Grímsfjall to the Goðahnjúkar peaks (*A-A*), and from Bárðarbunga through Grímsfjall to Öræfajökull (*B-B*)

them to a depth of 200–300 m below sea level on both sides of the highest mountain in Iceland, Öræfajökull, thus creating the greatest height differential in the country, there being 2300 m from the glacial beds up to Hvannadalshnúkur peak. Valley glaciers in the basalt area of southeastern Iceland have also ploughed down to depths below sea level. These troughs have mostly been formed by advancing glaciers during the Little Ice Age, although earlier ice-age glaciers have ploughed even deeper troughs into the offshore continental shelf, as can be seen in Skeiðarárdjúp, Breiðamerkurdjúp, and Hornarfjarðardjúp. The mountains of Vatnajökull formed indented bays in the southern coast of Iceland during the last glacial period, when the sea reached inland as far as Skaftafell, Mávabyggðir, Esjufjöll, and to the mountains of the southeastern districts of Suðursveit, Mýrar and Nes in Hornafjörður.

Only a tenth of Vatnajökull's bed rises above a height of 1100 m, which is the average height of the firn line along the southern flanks of the glacier. If Vatnajökull were to disappear now, it could not be reconstituted in the current climate. Vegetation would reach a height of 600–700 m, although there would always be snow for long periods of time on the highest mountains. Around three dozen lakes and tarns would come into being in the calderas and glacially eroded troughs. There is a valley up from Skeiðarárjökull, right across to Brúarjökull, which never reaches a height of 700 m above sea level.

The lowest pass between the Breiðamerkurjökull and Brúárjökull glaciers is at a height of about 800 m. Vatnajökull is currently sustained by its own altitude as over 70 % of its surface area is above an elevation of 1100 m.

8.1.1 Volcanic Activity and Jökulhlaups

Volcanic activity is most powerful and active in the western region of Vatnajökull, which is why so many of Iceland's highest mountains have been amassed there (Figs. 8.8, 8.9 and 3.12). There are two reasons for this. Firstly, beneath the northwestern part of Vatnajökull is the centre of an upwelling plume of magma that comes from the Earth's mantle. This is a kind of boiler installation with anomalous melting in the mantle so forceful that naturally light material flows upwards, melts, and penetrates the Earth's crust, creating shallow magma chambers; it also frequently breaks right through and erupts on the Earth's surface. Scientists call the upward flow of molten material a mantle plume and the crustal layer a hot spot. Secondly, in addition to the upward flow of molten magma, the material drawn up from the bowels of the Earth is channelled to the middle of Iceland because the divide between two of the largest tectonic plates of the Earth's lithosphere lies precisely there; and these plates are gradually moving apart from each other. Moreover, the rifting of the crust reduces the overburden pressure in the mantle, decreases the melting point of the rocks, and thereby increases melting in the mantle. The centre of the hot spot beneath the northwestern part of Vatnajökull is bounded by a cluster of volcanoes: Bárðarbunga, Kverkfjöll, Grímsvötn and Hamarinn, and the fissure swarms extending from them. The repeated eruptions form a localised 'bump,' on the Earth's crust. This is indeed how Iceland itself arose from beneath the ocean.

Altogether, a total of almost 80 volcanic eruptions in Vatnajökull have been noted over the last 800 years (Larsen et al. 1998; Björnsson and Einarsson 1990). Five volcanic systems can be discerned beneath the glacier, the location of the volcanoes having been ascertained from radio echo soundings, while seismic data has traced the areas of unrest associated with the movement of magma. The systems can be distinguished from one another by the composition of their volcanic material. A fissure chain from the main volcanic centre of Bárðarbunga extends for 100 km southwest along a ridge past Mt. Hamarinn and on to the Veiðivötn lakes area, and 50 km northeast to Dyngjuháls saddle. Grímsvötn is the main volcano on the fissure chain that seems to extend under Síðujökull (through Háabunga and Þórðarhyrna) to the Laki craters, and the composition of volcanic material indicates that the Kverkfjöll mountains are in the same volcanic system, though no mountain ridge is seen to reach them beneath the glacier. More than 30 eruptions have been traced to Grímsvötn over the last 400 years, and 60 in all since 1200. Grímsvötn,

Fig. 8.8 The volcanic centres and fissure swarms of Vatnajökull (see Fig. 3.12)

along with Bárðarbunga, is the most active volcano beneath the glacier. There have been 15–20 eruptions in Bárðarbunga and the Dyngjuháls saddle from about 1200, the last series of eruptions in the Bárðarbunga/Veiðivötn system being just after the middle of the 19th century (1862). There are very high levels of heat in Grímsvötn and Kverkfjöll causing the continual melting of ice. Resulting jökulhlaups cascade onto Skeiðarársandur plain and through the Jökulsá á Fjöllum river.

Between the Veiðivötn and Grímsvötn volcanic systems there is a swarm of fissures, named after the Fögrufjöll mountains, that extends beneath the glacier and could be connected to the so-called Lokahryggur ridge, which heads eastwards from Hamarinn to the northern part of Grímsvötn. There is geothermal heat in this ridge beneath the Skaftá cauldrons, which discharge meltwater into the river Skaftá, and it is

Fig. 8.9 A volcanic plume breaks through a 600-m-thick glacier along a 6-km-long volcanic fissure in Jökulbrjótur to the north of Grímsvötn in the beginning of October 1996. The meltwater from this eruption drained into Grímsvötn, and its surrounding layer of ice was uplifted 125 m (as can be seen in the *top right* of the photograph) until a jökulhlaup burst onto Skeiðarársandur a month after the eruption had begun. HB, November 1996

even believed there may be frequent small volcanic eruptions there that never reach the glacier's surface. The swarm of fissures from Askja disappears beneath Dyngjujökull, where no subglacial eruption has been known to have occurred. There was, however, an eruption in these fissures in the Holuhraun lava field, which is in the immediate forefield of Dyngjujökull, from September 2014 until February 2015 (see below). The Öræfajökull volcano is outside the plate boundaries of the rift belt, but is still considered active nonetheless, as indeed might be the volcano in the Esjufjöll mountains. A short distance to the northwest of Vatnajökull is the main volcanic centre of Tungnafellsjökull (Sæmundsson 1979, 1982; Jakobsson 1979a, b; Einarsson and Björnsson 1979; Einarsson and Sæmundsson 1987; Einarsson 1991; Brandsdóttir 1984).

Volcanic activity in Vatnajökull has been at regular intervals in bouts of 130–140 years. These bouts consist of a period of turbulence for 40–80 years followed by a rather longer and calmer period; the longest period between high points has been 160 years (1720–1880) and the shortest 100 years (1620–1720), the average being 130 years. The longest pause in volcanic activity in Vatnajökull was from 1938 to 1983, or 45 years. If only Grímsvötn is taken into account, however, there was a pause between 1540 and 1600, or for 60 years. The most recent eruptions in Grímsvötn since 1938 have been in 1996 (Jökulbrjótur, or Gjálp), 1998, 2004, and 2011. See Appendix L.

8.1.2 Cartographic History of Vatnajökull

A huge glacier in the southeast of Iceland appeared on Knoff's map of Iceland in 1734 (Fig. 4.10). It is not very clearly delineated, but its named constituent parts are Breiðamerkurjökull, Skeiðarárjökull and Knappafellsjökull, while Skaptárjökull is depicted as a separate glacier to the northwest of the main ice cap. As far as is known, this is the first time that Vatnajökull could be said to have appeared on a map. A Gríms Vatna Jökull is recorded on Peder H. Resen's *Atlas Danicus* from 1684–1687, but its outlines are not delineated (Resen 1991). Hans Hoffgaard's map of 1724 shows nothing other than Öræfajökull in southeast Iceland. Vatnajökull is not present on Þórður Þorláksson's map of Iceland from 1668, or on Guðbrandur Þorláksson's, from 1590, although the latter does mark its glacial rivers (Fig. 4.5): Jökulsá on Breiðamerkursandur outwash plain, Almannafljót (now Hverfisfljót), Skaftá, Skjálfandafljót, Jökulsá á Fjöllum, and Jökulsá í Fljótsdal, as well as several other unnamed rivers.

On the map of Iceland in Ólafsson's and Pálsson's (1981, Vol. 2, p. 87) travelogue of 1772 (Fig. 4.12), the ice sheet has been given the name Klofajökull, because 'Klofajökull draws its name from two enormous glacial sections that create a pass or bight between them, and from this glacial inlet flow three large rivers, Skjálfandafljót, Jökulsá í Axarfirði, and Jökulsá í Múlasýslu.' Ólafsson was here referring to the pass, or cleft, in the Kverkfjöll mountains, which is clearly visible from the Möðrudalsöræfi wilderness on the route between northern Iceland and the Fljótsdalur region in the east. We now know that only the river Jökulsá á Fjöllum emerges from that area.

Pálsson (1945, p. 446) agreed with this nomenclature: '… it is called Klofajökull because of the many spurs which extend from it in all directions, and a large number of mountain ranges that jut into and divide it up. It is also called Vatnajökull due to the almost innumerable rivers and streams that originate from it.' Pálsson compiled the first exact map of Vatnajökull, or Klofajökull, in 1794 (as well as Þrándarjökull and Snæfell), and Skaftárjökull was depicted as a part of the main ice sheet. The map of the northern verges of Vatnajökull was drawn up from the clear descriptions of Pétur Brynjólfsson, who had travelled along the so-called Vatnajökulsvegur highland trail and then later, with Pálsson himself, to Snæfell in the summer of 1794 (Fig. 8.10).

Many of Vatnajökull's outlet glaciers are named on Gunnlaugsson's map of 1848 (Fig. 4.21): Skeiðarárjökull, Öræfajökull, Breiðamerkurjökull and Heinabergsjökull, which overlooks the Mýrar district in Hornafjörður. The northern section of Vatnajökull was rather poorly delineated and Brúarjökull's snout is extraordinarily serrated, for indeed no one had ventured there to see it. Lake Grænalón was wrongly called Grímsvötn, just as Sveinn Pálsson had erroneously named it on his map.

The topography of the Vatnajökull maps did not improve until Þorvaldur Thoroddsen's expeditions around Iceland between 1881 and 1898. He later published his geological map in 1901, which marked the outlines of Vatnajökull on a

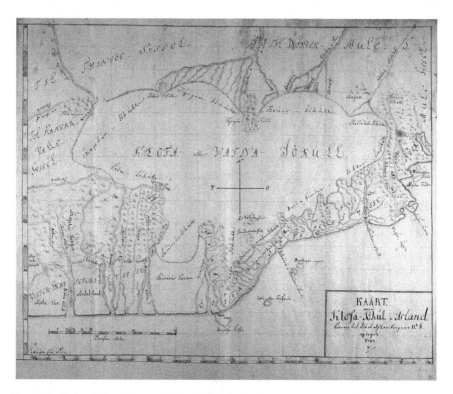

Fig. 8.10 Sveinn Pálsson's map of Vatnajökull from 1794. AMI

scale of 1:600,000 and gave exact descriptions of the outlet glaciers all around its margins. Until then, views on the locations of the sources of the glacial rivers had varied considerably. A detailed map of the glacier was first compiled when the Danish General Staff commenced cartographic work in 1901, and in 1905 a good map of the southern edge of Vatnajökull appeared on a scale of 1:50,000 (sheets 77, 87, 96 and 97) stretching from Skeiðarárjökull to Fláajökull.

Early Difficulties of Exploring and Surveying Vatnajökull

Scientific exploration of Vatnajökull began in the years 1902–04 with the surveying of the outlet glaciers and their southern outwash plains in Austur-Skaftafell County. During both summers, the surveyors made many trips on horseback, pulling sledges with equipment up onto Heinabergsjökull, Breiðamerkurjökull, and Öræfajökull, the furthest onto the glacier they reached being 14 km to the north of Miðfell. These surveying trips were feats of endurance and quite different from quickly crossing a glacier or popping light-heartedly up onto a glacial summit in good weather, e.g. on Öræfajökull. Local folk guided the Danish surveyors over roaring, fast-flowing streams and narrow pathways, and led caravans of packhorses through pathless wilderness and mountains, where sandstorms or mist often prevented them from seeing their way. They crossed wide, bridgeless rivers, sometimes on the backs of horses that had to swim wherever quicksands were expected, and traversed outwash plains where jökulhlaups could flood at any time. The surveyor Johan Peter Koch (1870–1928), later a captain and then colonel in the Danish air force, crossed Skeiðarársandur in the autumn of 1903, but then stayed on the plain with his men in tents from 20 April until the beginning of May 1904. Gales bent and broke the tent poles, sandstorms damaged the tents' canvas, tearing it open with the least pressure, and gravel got into everything, including the food chests, so that it had to be scraped off the butter. The camp site was then suddenly flooded by water, forcing them to pack up and move elsewhere as quickly as possible, for fresh in their memories was the eruption in the glacier of 1903, which had caused a jökulhlaup that had scattered icebergs all over Skeiðarársandur. They were obliged to set up many storage centres to which they needed to transport food and equipment, along with forage for the horses, and all the necessary apparel for long sojourns in all kinds of weather. The luggage of each surveyor was the equivalent of 12 cartloads. The horses calmly made their winding way between the network of crevasses, along slippery ice-ridges with gaping precipitous gorges on either side, clambering or jumping over the next crest of ice. If a horse slipped and fell, it calmly waited until it was pulled up by ropes, but if that failed, it simply died, surrounded by wet and sad explorers, who were probably also lost, as compasses were of little use when negotiating crevasses in a mist.

They had to find weight-bearing snow bridges to cross crevasses, and splash through areas of wet slush before they ascended onto the ice cap. Higher up on the glacier, came deep and loose snow, the horses wading through it up to their girths, so that it was best to travel at night or early in the morning before the hard crust of the night frost had been softened again. Koch, the leader of the surveyors on the glacier, said that the project had considerably strengthened both his muscles and his mind! The experience certainly came in good stead when he later went on difficult explorations of the Arctic region, including the northern part of the Greenland ice cap.

8.2 Skeiðarársandur Outwash Plain

Iceland is one stony, mountainous and glacier mickle land with uninhabitable moors, gravel plains and frequent wastelands, especially in its central highlands. (Stefánsson 1957, p. 1)

The inhabitants of Austur-Skaftafell County live on a narrow strip of harbourless land along the southern coastline in the foothills of Vatnajökull. More than twenty outlet glaciers stretch down from Vatnajökull onto the lowland and from them many diverging branches of glacial rivers which, when flooding, tear apart vegetated land and inundate outwash plains from one end of the glacier to the other. Jökulhlaups burst out from lakes and subglacial volcanic eruptions. The rivers were unbridged for a long time and proved difficult obstacles for travellers trying to reach farms situated on the vegetated land strips between them; the entire history of man and nature in this area has been moulded by the proximity of the glacier. The climate is mild with a lot of precipitation, though with no fewer clear, sunny days than anywhere else in Iceland. The districts along this coastline are called Skeiðarársandur, Öræfi, Breiðamerkursandur, Suðursveit, Mýrar and Nes. But to reach them from the west, one had first to cross the formidable natural phenomenon of Skeiðarársandur, the largest glacial outwash plain in the world.

Until 1974 the main national highway, Route 1, only reached as far east along the southern coast as Fljótshverfi, after which began the road-less and sometimes impassable terrain of Skeiðarársandur, which covers an expanse of 1000 km^2 and 1 % of Iceland's total surface area. Skeiðarárjökull suddenly comes into view once the massive promontory of Lómagnúpur is rounded from the west. The glacier splays out over a 40-km-broad, flat plain that gradually descends 4 m per kilometre on its 25-km journey to the sea (Figs. 8.11, 8.12 and 8.13).

Fig. 8.11 Skeiðarásandur and Skeiðarárjökull. While its snout is 30 km broad, at its narrowest the glacier is only 8 km wide. The sediment deposits bear witness to the glacier's advances over many centuries. Skeiðarársandur is probably the most active glaciofluvial outwash plain in the world today. Three main rivers flow through it: Núpsá (Súla) on the western side, Gígjukvísl (Gígja or Sandgígjukvísl) in the centre, and the Skeiðará outlet furthest to the east. HB, August 1998

Fig. 8.12 The many rivers of Skeiðarársandur were bridged during the years 1972–1974 and levees were raised to direct the flow of water. Outside of these levees, vegetation began to take root, though with great difficulty, especially when the rivers overflowed onto the plain. HB, August 1998

The only notably uneven spots are the shallow riverbeds and terminal moraines in Skeiðarárjökull's forefield, where it had reached its furthest extension at the beginning of the 20th century. The plain bears witness to the advances of Skeiðarárjökull and the sediment deposits of its glacial rivers over the past centuries; it is now quite probably the largest active outwash plain on Earth. It actually has a history about 9–10,000 years old, right from the beginning of this current interglacial period, from when the sea reached the Lómagnúpur promontory and large glacial rivers bore huge amounts of glaciofluvial deposits into the sea, piling up estuary islets that gradually became connected to each other as they extended southwards. Jökulhlaups due to volcanic eruptions were also always bearing mud and glacial till onto Skeiðarársandur. The overburden of the ice-age glacier was then lightened as it retreated and the land gradually began to uplift, though not as quickly as the sea level rose, so that the tideline reached further inland once more, the ocean covering most of the sedimentary deposits. In the middle of the plain, these deposits are 200–250 m thick, but less than half of this nearer to the glacier's terminus. Öræfajökull, the highest mountain in Iceland, towers above Skeiðarásandur on its eastern side. There are few landmarks on the plain itself,

Fig. 8.13 Aerial view south of Skaftafell banks along the Skeiðará to the sea. A protective levee on the eastern side contains the waters. The bridge over the Skeiðará is Iceland's longest at 964 m

though Ingólfshöfði rises furthest to the east, a rocky headland 76 m high. The plain's coastline is 46 km long.

There are three main rivers that flow from Skeiðarárjökull through Skeiðarársandur: Núpsvötn (Súla), at its western edge, Gígjukvísl (Gígja or Sandgígjukvísl) in the centre of the plain, and the Skeiðará furthest east. The usual summer runoff flows of these rivers are: Skeiðará 200–400 m^3/s, Gígja 20–70 m^3/s, and Súla 10–60 m^3/s. Skeiðará no longer runs beneath the bridge, however, but now flows along the front of the snout before joining Gígja. Their sediment deposits play a great part in the forming of Skeiðarársandur. The plain's rivers were bridged from 1972 to 1974 and levees were raised to the east of the Skeiðará's channel, thus enabling the completion of the national highway or ring road. The longest bridge was over Skeiðará itself, 964 m (880 m after repairs following a jökulhlaup). Outburst floods occur in all these rivers at fairly regular intervals, threatening travellers and destroying vegetation, farms and buildings and other manmade structures, one of the reasons why it took decades for these rivers to be bridged in the first place. Many of these jökulhlaups come from an accumulation of meltwater caused by geothermal heating beneath Grímsvötn in the centre of Vatnajökull, which eventually floods through a 50-km long waterway beneath the glacier. Floods also begin from lakes at glacial margins wherever Skeiðarárjökull dams valleys in which meltwater and rainfall have amassed. Grænalón is one of the largest of these lakes, and it discharges its water into the river Súla. Most of the outburst floods from subglacial eruptions in Vatnajökull have inundated Skeiðarársandur.

From 1920 until 1990, Skeiðarárjökull retreated 1–3 km, shrinking rapidly after 1930, the land at its terminus now being lower than at the moraines further out on the plain. Glaciofluvial deposits have not been able to fill in the glacier's former location and so lakes have formed at its snout and rivers have found their way through gaps in the moraines and flow through the plain in demarcated channels. When jökulhlaups occurred, water collected in these proglacial lakes and then flowed down the same channel. Vegetation took root on the plain over a wide area, and after the middle of the 20th century these jökulhlaups became less powerful than previously. Warm summers also increased the growing conditions for vegetation and so lyme grass hillocks and wetlands are widespread in the lower regions of the plain. Large areas are now covered by moss, and even birch wood copses are now growing on Skeiðarársandur. When Skeiðarárjökull still extended as far as the terminal moraines, the outburst floods swept across the plain in strong currents in many braided rivers and channels, giving vegetation very little chance of taking root (Fig. 8.14).

8.2.1 History of Skeiðarársandur

It is believed that, during the first centuries of the settlement of Iceland, the eastern part of Skeiðarársandur had been vegetated, but that the western part, called Lómagnúpssandur, had been relatively barren mudflats. There are no sources or

Fig. 8.14 Skeiðarárjökull stretches its long paw down onto Skeiðarásandur. Running south through the plain is the dry channel of the huge jökulhlaup of November 1996. This outburst flood rushed westward along the glacial snout into the river Gígja before slashing its way south through the centre of the half-vegetated plain. Old waterways can be seen outside of the new channel, as well as kettle holes from icebergs, which had been scattered all over the plain in previous jökulhlaups. HB, August 2006

evidence of settlements on the western side of the glacial outwash plain, but there were 15–18 farms on the eastern side, though they were situated further west than today's farmsteads in the Öræfi district. It is thought that no large river then ran from the eastern parts of Skeiðarárjökull, the main outlet for runoff water probably being through the centre of the plain. This is exactly where the riverbeds are highest now, indicating that this is where sediment layers have largely been deposited over a long period of time. Skeiðarársandur is not named on Þorláksson's map of 1590, and nor are any of its rivers. Núpsvötn rivulets are wrongly depicted as being to the west of Lómagnúpur. In the *Book of Settlements*, Rauðilækur, the chief residence of the district and later the location of a church, is said to be on lowland between Svínafell and Sandfell. Further north were the farms Jökulfell, Skaftafell and Freysnes. Birch trees and brushwood are very noticeable at Skaftafell, and it is possible that the name of the farm is derived from this ('skaft' = shaft or haft, usually made from wood). Freysnes was on low land to the east of the present channel of the Skaftafellsá, while the farm of Skaftafell was on the flat land below the heath. The farm Jökulfell was located at the foot of the mountain of that name on the northern side of Morsárdalur, and Bæjarstaðarskógur draws its name from a farm there with the most upright birch trees in Iceland. There was an annex-church at Jökulfell until 1343, which is seen as indicating there had been no outburst floods onto the eastern side of the plain during the first 4–5 centuries of the settlement of Iceland. All farms and communities on the plain were made uninhabitable by the eruption of Öræfajökull in 1362, and those of the Litla-Hérað area between Skeiðarársandur and Breiðamerkursandur were completely destroyed. Rauðilækur seems to have been inhabited again later, but it was finally and permanently abandoned in the early 15th century. Around 1540, however, it is recorded that the Skeiðará had begun to break up the farmer's lands in the Öræfi district, and this is the oldest source of the river's name. Since then, the Skeiðará has flowed up to the Skaftafellsbrekkur slopes and right alongside the district of Öræfi. There are many written sources for this and the river is depicted as the easternmost one on Skeiðarársandur on Þórður Þorláksson's map of 1668. During an outburst flood from an eruption at Grímsvötn in 1684–1685, a jökulhlaup careened with tremendous power through the eastern side of the plain, and it is reported that 200 sheep and 14 horses perished. Around 1700, the parish priest at Sandfell, Gísli Finnbogason, states that the Skeiðará was flowing in front of Skaftafell (Stefánsson 1982, p. 103; Þorsteinsson 1985; Björnsson 2003). On Knoff's map of 1733, the whole of Skeiðarársandur is riddled with large rivers with a great number of islets in between them. In the county reports of 1746, Stefánsson (1957, pp. 1, 10) states that the Skeiðará comes '… out of this large glacier at its eastern end…' Skeiðará is the main easternmost river on the plain according to Sæmundur Hólm's map of Skaftafell County from 1749, and Núpsvötn the westernmost, with a nameless river in the centre of the plain. According to Sveinn Pálsson's 1794 map of Vatnajökull, there are two rivers on Skeiðarársandur, Skeiðará and Núpsvötn.

River Skeiðará destroys cultivated land The farmer at Skafatafell, Einar Jónsson, complained bitterly about the damage to his farmland in a letter to the county administrator in 1756, saying that his hayfields had been torn apart, '… so that those ancient meadows are now nothing but mud and naked earth.' Jón, his son, wrote to the royal superintendent in 1787: '… with all that this large river, the so-called Skeiðará does at every floodwater, if it reaches closer to my farm of Skaftafell it will have to be abandoned' (Stefánsson 1982, p. 103; Þorsteinsson 1985; Björnsson 2003). The farmstead stood on the plain at the base of the Skaftafellsbrekkur slopes. The low fiscal value of the farmlands in the Öræfi district can be seen in a description by Páll M. Thorarensen (1801–1860) of the parish of Sandfell and Hof in the district's county and parish register from the 19th century (dated July 1839). Skaftafell is valued at 12 hundreds and Hof at 28, while other farms (Svínafell, Hofsnes, Fagurhólsmýri, Hnappavellir and Kvísker) are valued at less than 10 hundreds and are thus virtually barren farmlands by the standards of the time (Thorarensen 1997). The farmstead of Skaftafell was finally abandoned after the large jökulhlaup in 1861, and new houses were built much higher up on the slopes up to the heath, where there was once a shieling, and have been there ever since. The river then destroyed the home fields and farmlands on the lowland. Nonetheless, according to the land register compiled in 1769 by the Governor of

Iceland, Skúli Magnússon, the largest and best woodlands in southern Iceland were
to be found in this area, high up on the local mountainsides and summits (Tómasson
1980, p. 17).

Thus, by the first quarter of the 20th century, Skeiðarársandur had become a
barren desert. The destruction of vegetated land had begun with an eruption in the
14th century and a cooling climate, and had continued until the first quarter of the
20th century. But after a period of 550 years of devastation, circumstances changed
around the middle of the 20th century, as the glacier began to retreat from its
foremost terminal moraines, and jökulhlaups ceased flooding over all of the plain.
Outburst floods were also smaller than previously because the glacier was thinning
out and thus not damming up valley lakes, while there was also much less volcanic
activity in Grímsvötn during the period 1938–1996. There has thus been a great
increase in vegetation on the plain since the early more barren years of the 20th
century.

Hypotheses have emerged in recent times predicting that the Skeiðará river will
soon move its channels further west on Skeiðarársandur. This is because the eastern
side of the plain has gained additional layers of glaciofluvial sediments, and is
consequently about 20 m higher at the foot of the mountain to the west of
Skafatafell farm than it had been in the middle of the 19th century. A former farmer
at Skaftafell, Ragnar Stefánsson, noted in the mid-1980s (1987) that he thought the
land alongside the farm had risen by about 10–12 m since the mid-19th century,
including 1.5–2 m that he had personally witnessed. Sandgígjukvísl river has
grown in size during the 20th century, again possibly because the land on the
eastern side of the plain has gradually risen.

Fluvial deposits and quicksands

Rivers bear prodigious quantities of gravel and mud from beneath the glacier, night
and day, though mostly during the spring, dispersing it over the lowlands and
changing large areas of grassy land into steriles arenas (mud banks and gravel spits
in our language), both by flushing away the grass roots but more especially by
burying them with gravel and sand from swollen rivers and jökulhlaups. All this
means that the rivers cannot create fixed channels or courses wherever the land is
flat, but instead they either settle into deep, straight and narrow runnels, or they split
up into many broad and shallow streams, and all such changes could happen in one
and the same day, so that sandbanks are broken down in one place and piled up in
another. This depositing of sand and gravel causes many strange and extraordinary
creations in the waters which those unfamiliar with them could easily mistake for
some kind of water monster. Thus I felt very uneasy as I was crossing the Skeiðará
on 9 September 1793 and beheld an enormous wave–like the very worst of shoreline
breakers–suddenly rise from the deep a few dozen steps from the ford which I
proposed to use. This wave travelled slowly ahead of the current, becoming
extremely long one moment and then converging into a pyramid of water in another.
Sometimes it almost disappeared completely, but then returned, as if it were rolling
up into a dangerous rapid, and thus it continued towards the end of the sandbank,
where the river split into two branches and where I was waiting. It was then that I
saw at last that this was nothing more than a gravel reef which the force of the

current had torn up from the riverbed. Sometimes these flood waves charge straight along the river with the current, and I was once witness to this in the summer of 1794 in the Núpsvötn and Skeiðará rivers. In swift currents one can be unaware of these waves until they suddenly rear up quite high above the surface and seem to stand still for a few seconds, while all at once dozens of similar waves arise in a row directly behind the first one which has by now lowered considerably. But then at that very moment all these waves rush against the stream cascading into it like a large waterfall until they all completely disappear at once, only to begin all over again somewhere else. Many people who are familiar with these kind of waters will nonetheless not hear of anything other than that they are caused by water monsters which splash about for a moment against the current. But the real truth of the matter is as follows: the current first harrows out a deep hole in the riverbed, but because the gravel is so loose, the higher edge of this hole is swept immediately onwards. New holes are created further along, one after another, all along the riverbed. The gravel that has been swirled up from the holes causes the waves, which all seem to struggle against the current because each pile of gravel is continually collapsing into the new hole ahead of it, and receives even more gravel from the next hole behind it, although both the water and the gravel are borne downstream as usual. Such loose gravel can never be so cohesive or compressed by the water that men might cross it safely. Before you know it, the horses sink below the water and this is especially dangerous if occurring in the middle of a river where sandbanks cannot be seen in the water. It is then very expedient to cast oneself off the horse on the side ahead of the current and hold on for dear life to the mane or tail of the horse and let it struggle on as best it can to scramble onto the riverbank. In most instances, one shouldn't turn back if one gets into quicksand, no matter how bad it seems, but let the horse drag itself directly forwards, and those who are following should not hesitate to follow in the tracks of those who've gone before, for quicksands disappear as quickly as men or horses traverse them, even though they are just as bad both upstream and downstream. Those who are experienced in these rivers, can see beforehand from the shape of the current if there is quicksand in the riverbed or not, and seldom stop because of them where the current is very rapid or gravel has been stirred up in the manner I have described above. (Pálsson 1945, pp. 451–452)

Sveinn Pálsson's crossing of Skeiðarársandur in September 1793

The 1st [October]. Skeiðará was now in full flood and had become such a major obstacle, even though the storm had abated, that we had to give up after having tried for three days to cross the river on horseback and taking very great risks. We thus decided to take a diversion up over the glacier near the source of the river. Whenever such a journey is attempted, an experienced man must always be sent ahead to look at and decide whether the glacier is passable or not, for the route can change in such a short time that it can become impassable, even for a man on foot. We at last set off from the farm of Skaftafell at one o'clock in the afternoon and had to slide down with horse and baggage, almost half-dead with fear, a very steep mountainside to the north of the farm to get to Morsdal [Morsárdalur] valley because the Skeiðará flows by the mountainside so rapidly that it annually rips away pieces of its banks.

We arrived at the glacier at sunset right next to the source of the river. It was fearsome to see how it foams and bustles up, as if from the bowels of the earth, right by the edge of the glacier, instead of like other sources of rivers I have seen emerging from vaults beneath the ice. The edge of the glacier is probably much deeper here than the gravel plain ahead of it and so the river must gouge deep down in order to escape from under the ice to then burst directly upwards. The glacial snout, situated next to and along Mt. Jökulfell, has torn away parts of the lower mountainside as it advanced, pulverising bedrock cliffs, though it does not reach as high now as it once did. The part of the glacier which we had to cross was full of columns of ice with deep gullies between them. It was very dangerous to cross there, especially because of the slippery glacial ice, and the crevasses and holes were widely hidden from view under the muddy surface. We led the horses singly, one by one, through the deep, steep and mazy hollows, and it had become dark when we finally descended from the glacier. The journey across the glacier might have been very easy, though, if the horses had been roughshod. (Fig. 8.15). (Pálsson 1945, pp. 299, 301–302)

E.T. Holland's crossing of Skeiðarársandur after a huge jökulhlaup in 1861

When we came in sight of the Skeiðará, the roar of whose waters we had heard for a considerable distance, we saw a broad sheet of murky water, rushing across the sandy plain with indescribable velocity, the waters appearing to be raised above the level of the ground on which we stood. ...

The farmer, who had exchanged his second horse, which he had ridden from the Núpsvatn, for his favourite cream-coloured water-horse, led the way; zig-zagging from sand-back to sand-bank across two or three lesser streams, we soon reached the shore of the main channel. After a moment's halt the guide rode on into the stream.

CROSSING AN ICELANDIC RIVER.

Fig. 8.15 Crossing a glacial river. Copper engraving by Edward Whymper from a photograph by Frederick Howell. NULI

Scarcely had his horse advanced half a dozen steps into the water, when the force of the current all but swept him off his legs, and the guide had to turn him back to the shore. Finding that it was impossible to ford the river at this point, we left the island sand-bank upon which we were, and rode down stream through the water for some distance, picking our way through the shallows as well as we were able. We proceeded in this way for some distance without being able to find any place where the river appeared at all practicable.

At length we came to a shallow, where our guide pulled up to take a survey of our way. The water here was up to our horses' girths, and very swift; but it served as a sort of resting-place in the midst of the deeper waters round it.

Having satisfied himself as to our route, our guide again urged on his horse through the stream, and led the way towards the mid channel. We followed in his wake, and soon were all stemming the impetuous and swollen torrent. In the course of our journey we had before this crossed a good many rivers more or less deep; but all of them had been mere child's play compared to that which we were now fording. The angry waves rose high against our horses' sides, at times almost coming over the tops of their shoulders. The spray from their broken crests was dashed up into our faces. The stream was so swift, that it was impossible to follow the individual waves as they rushed past us, and it almost made one dizzy to look down at it. Now, if ever, is the time for firm hand on rein, sure seat, and steady eye: not only is the stream so strong, but the bottom is full of large stones, which your horse cannot see through the murky waters. ...

Not the least of the risks we ran crossing the Skeiðará, was from the masses of ice carried down by the stream from the Jökull, many of them being large enough to knock a horse over. ...

When we reached the middle of the stream, the roar of the waters was so great that we could scarcely make our voices audible to one another: they were overpowered by the crunching sound of the ice, and the bumping of large stones against the bottom. Up to this point, a diagonal line, rather down stream, had been cautiously followed; but when we came to the middle, we turned our horses' heads a little against the stream. As we thus altered our course, the long line of baggage-horses appeared to be swung round altogether, as if swept off their legs. None of them, however, broke away, and they continued their advance without accident; and at length we all reached the shore in safety. From the time that we first entered the river we were about an hour actually in the water, and it cannot have been less than a mile in breadth. (Holland 1862, pp. 36, 38–39, 40–41)

They leapt up onto the glacier

It must have been a terrible sight for the two men who were crossing Skeiðarársandur and got caught in the last jökulhlaup [1784]. Fortunately for them, the path they were following was right next to the glacier's edge, and as soon as they saw movement in the glacier, they leapt up onto it, just before the floodwaters burst through. They had to remain there for a few days, between life and death, on the shaking glacier amidst thunder and lightning, until they finally succeeded in crossing the glacier and reaching the nearest farmstead. (Pálsson 1945, p. 501). (Figs. 8.16, 8.17 and 8.18)

Fig. 8.16 A train of packhorses crossing Skeiðarárjökull. There were supposed to be only four horses in such a train

He decided to bind himself to the horse

It would have been in the spring of 1851 when a man was crossing the Skeiðará just as a jökulhlaup began. This was a matter of life and death, for the man who had come from the west saw that the river was virtually impassable. He then decided to

Fig. 8.17 Sheep herded over Skeiðarárjökull (**a**) and the braided rivers of Skeiðarársandur (**b**) before the latter were bridged in 1974. Jóhann Þorsteinsson

bind himself to the horse, so that the same fate should befall them both, and set off into the river. The horse was very reliable and succeeded in getting them both across, though those who were watching from Skaftafell thought it quite incredible. The man was warmly welcomed at Skaftafell by the farmer Þorsteinn Bjarnason, who examined the horse carefully and asked the man if he were willing to sell it. The

Fig. 8.18 Travellers descending the eastern side of Skeiðarárjökull in July 1937, Morsárdalur ahead of them. Ingólfur Ísólfsson

man answered that if the offer was high enough, the horse would be for sale. To which Þorsteinn replied: "It wasn't really my intention to make an offer for the horse, but if I were in your shoes, I wouldn't sell this horse but would try to ensure it was well taken care of for ever more." (Björnsson 1979, p. 47)

Kettle holes on Skeiðarársandur

The country, which lies between the present and the old course of the Skeiðará, is very remarkable. We found the greater part of the Sandr [sic] honeycombed, so to speak, with innumerable round quicksand holes. The largest of these were as much as thirty feet in diameter, and from ten to fifteen feet in depth. Many of them were half-filled with water, generally of the milky-white colour of glacial streams. In some of them the farmer pointed out what he said was Jökull ice underneath the water; but in most there was no water, but only wet quicksands, which bubbled up as we rode by them. These holes lay close to one another, and were separated only by narrow ridges of sand, often scarcely a foot in width at the top, but sloping outwards and widening towards the bottom. ... I think, however, that they were caused by the melting of the masses of ice, which were left deposited upon the plain on the subsidence of the floods caused by the eruption, and that the water coming from these masses of ice as they melted, soaked into the sand around them, and converted it into quicksand. (Holland 1862, pp. 41–42)

Report of Pastor Magnús Bjarnason of Prestbakki at Síða, 11 June 1903

The Skeiðará flooded last St. Urban's day, 25 May. The river had remained inside the glacier all winter, that's to say its bed was completely dry. But this is indeed how it often is before flooding. Men were thus always expecting a flood this spring and it was rather unpleasant to be crossing the almost 7-mile-broad Skeiðarársandur plain expecting a jökulhlaup to engulf you at any moment. The last time the river flooded, the winter of 1897, the man bringing the post from the east was almost caught in it, but narrowly escaped to the west over the Núpsvötn rivers just as they were becoming impassable. But now there was a man within half-an-hour's ride of Skaftafell—the farm is on a small ridge to the east of the plain—when the flood began, and it was an old man on a lean horse, which would not have been quick enough to flee, who, in God's mercy, had just crossed the plain. The jökulhlaup began on Monday before Whitsun, and by Whitsunday had begun to recede so that the post bearer crossed over the plain westward with the help of two excellent men, Þorsteinn Gunnarsson and Stefán Benediktsson, farmers at Skaftafell. But although the flood lasted no longer than that, it was nonetheless one of the largest and most frightful of jökulhlaups as it occurred at the same time as an eruption in Skeiðarárjökull (?) [nearer Þórðarhyrna] which often happens, to a greater or lesser extent, when there is an outburst in Skeiðará river. Flames arose north and west of the so-called Grænafjall mountain, which lies to the north of Eystrafjall, on Thursday 28 May, upon which the floodwaters increased as the glacier melted, searching for channels with such power that they burst open the front of the outlet glacier and surged forwards with blocks of ice that, according to the post bearer, were 60 feet high, and over a wide area there were others on top of them another 25–35 feet thick. This ice jam now covers many square miles of Skeiðarársandur, from the glacier down to the sea (a length of ca. 5 miles), and covers the plain in front of Öræfi, right up to the farm of Hof in the east, so thickly that it is hardly possible to get through except by going right up past the glacier. There were four main flood surges from the glacier as well as many smaller ones, the most easterly below Öræfajökull, and the most westerly at the so-called Sigurðarfit: and there was still 1/3 of the plain left west to Lómagnúpur, and there was a tremendous flood wave in the Núpsvötn rivers (the furthest west on the plain), although there was no consequent ice jam.

The most terrifying jökulhlaup was on Thursday and Friday night, when the eruption had begun and was at its peak. Water spurted up in high columns through cracks opening up in the glacier and all the ground and houses in Skaftafell and Svínafell in Öræfi shook so much in an earthquake that windows were broken in Skaftafell, and thuds and bangs were heard as far east as Hornafjörður, for there was a westerly wind. The evening and night sky were lit up by the volcanic flames that stretched high into the sky and from its smoke came flashes of lightning that brightened the land and rivers all the way out to Álftaver, and this was later than ten o'clock at night. A cloud of smoke lay over the Öræfi district, concealing the mountain tops above Sandfell and Hof in black billowing waves with flashes of fire and lightning streaking from them above the farmsteads. Even the earth itself shook when the glacier was cracking open and the explosions louder than the firing of any canon. All of this combined was so terrifying that the folk of Sandfell (Pastor Ólafur himself had recently just left to the west with two children) fled eastwards to Fagurhólsmýri. Fortunately, there was no consequent ash fall, and the cloud and eruption were much milder the next day, indeed the jökulhlaup itself was beginning to abate. ...

To illustrate the volume and power of the floodwaters, it can be stated that on the Saturday before Whitsunday, a lighter was unloading a cargo from a timber ship at Vík in Mýrdal in a dead calm. Suddenly the sea became turbulent, dark and muddy and with such violent currents westwards that the lighter had to immediately cast off from the ship being unloaded. (Þórarinsson 1974b, pp. 116–117).

Description of Skeiðará jökulhlaup of 1934 by Hannes Jónsson, post bearer and farmer at Núpsstaður

Jökulsá is a large and fast-flowing river that is seldom crossed on horseback during the summertime. The route then taken is over the glacier, which has uneven surfaces to traverse and sometimes takes a long time, up to 2–3 hours to cross it. Again, it is sometimes possible to go by the so-called glacial undersill and then right along by the 'spout', as we call it. This is along a ledge right up by the outlet of the river. It's just as if the water spouts straight out of the earth and is caused by the glacier's sinking into the earth that is being undermined by the water which then comes up to the surface in bursts of flume, breaking as much onto the glacier as away from it, sometimes forming piles of brash ice in a circle in front of the spout. Within the

Fig. 8.19 Hannes Jónsson, post bearer

circle of ice is an appalling whirlpool, which probably keeps it all together, but with a huge increase in the flow it can also sweep everything away, including even the glacial overhang itself.

It came to pass that I once crossed over there. On the way east, everything went well. There was only one crevasse or gap in the undersill, but the horses could step over it. But while I dwelled east of the river, the crevasse had widened, though not dangerously so as I thought, and indeed it wouldn't have been so bad if the horse I'd left behind, while I took the one with the postbags over, hadn't tried to follow immediately after, so that it was too late for me to get hold of it before it stepped over the gap, as it didn't know how to be afraid. The horse had stepped so unsurely onto the western edge that it slid back along the ice and fell into the gap lengthwise, the gap being so narrow that its torso was not that deep down so its back was only a foot below the gap's edge. The horse threw itself up so high that it could have pulled clear if the rim of the crevasse hadn't been so bevelled that its feet couldn't get a grip on the edge. I was not in a good enough position to be able to tug him far enough towards me, and it was impossible for me to hack at the edges because I had to hold onto its head and keep it up so it wouldn't get it stuck in the crevasse.

I saw nothing else for it but that I would have to bring this to an end with my pickaxe, though I felt really bad about finishing off the horse right there. But just as I was about to drive the pickaxe into its head, I got the idea of trying to let it flounder forwards and then jump from the crevasse down into the spout, which was then no more than 6–7 metres ahead. I managed to get him to the edge, after which we were both breathless from the effort. It was amazing that all its legs should still be all right. I then removed the bridle, but just as I let the horse free, I noticed one of those cursed masses of brash ice that I hadn't noticed before I let it go, and would not have done so had I seen it. The horse swam up to the icy circle and got its head up onto the piled ice, but was sucked down into the water, and I could see nothing for it but that it would be dragged under for good. But every now and again it got its head above water and gradually the waves seem to grow larger further down and so further on they billowed it over the ice wall into the awfully fast current of the Jökulsá, which swept it up onto the bank. The horse then ran a little bit from the river's edge and lay down and still until I got there. I saw nothing wrong with it except a few scratches on its legs and that it was shaking and shivering with cold. The horse was chestnut with a white under-lip called Laufi. (Valtýsson 1942–1953, Vol. 1, pp. 152–153).

On this event, Nielsen (1937a, pp. 66–67) wrote: 'It is well known that on the lowest part of the glacier little or no water flows on the surface ice, even when there is an outburst flood in full flow, and it was exactly at such a time on 31 March that Hannes of Núpsstaður crossed the glacier. He had been held up at Skaftafell ever since the jökulhlaup had begun, and by then he was becoming bored with being delayed and so determined to get home before Easter. It was by then impossible to cross in front of the glacier, for although he was always trying to do so, he could not make it across there with the horses. So now he had to rely on two horses as fast as he was, even though water was surging up from beneath the glacier and filling its cracks and crevasses.

Hannes got his postbags organised and departed. He was a man of small stature in appearance and of a slim build and was in his fifties at this time

(Fig. 8.19). Oddur of Skaftafell and another man from the farm accompanied him. They first had to take a large diversion up into Mosárdalur valley, for its mouth was completely under water, and then up onto the mountain until they came to the eastern margin of the glacier. A stream flowed alongside the glacier, but it was not such a hindrance that they could not cross it. They then proceeded out onto the glacier. The surface was more or less dry, but the ice was in constant motion, with widespread groaning and cracks appearing.

When they came to the middle of the glacier, Oddur and his companion turned back. There was nothing to relate about their return journey until they came to the eastern edge, where the stream was now a fast-flowing current between the glacier and mountain. They were then obliged to go much further north along it until they finally succeeded in getting off the glacier. As far Hannes, he continued his journey alone.

The Núpsá river flood plains were awash with water, due to a jökulhlaup along the river Súla, so Hannes had to take a long diversion northwards to get around it, but he reached his home at Núpsstaður, safe and sound, on the morning of Easter Sunday after a trek of 18 hours.

This trip was taken at a time when there were many calamitous jökulhlaups. There were huge reservoirs of swollen floodwaters beneath the glacier and the outwash plain before them had become an enormous water-course. The eruption in Grímsvötn had begun that very same day. Flashes of fire and thundering roars flickered and reverberated across the glacier, and a column of ash 13 km high, lined with streaks of lightning, rose up into the sky above the travellers,–but Hannes wasn't going to let such a trifle prevent him from setting off, because he needed to get home for Easter.'

A good river horse

A good river horse does not blunder straight ahead in a passionate attempt to be the first to cross over. It uses most of its energy to counter the current, taking great care to face it in such a way that it breaks upon its shoulders or chest with as little exertion as possible and making sure its torso is never turned sideways to the stream, like a wall. It has to move its feet very carefully, not putting its weight on any step until a secure footing has been found. Boulders are part of fast-flowing currents of rivers, especially near their outlet from beneath the glacier. Indeed, they're no small rocks that the floodwaters drag along with them beneath the glacier, as they pulverise mountainsides, break down cliffs and then add the debris to the current for further trundling. It is not pleasant to have a fast current high up on your flanks and then to stub your toe on a large stone and even less so to get a leg stuck between them or even have it broken ... To those unfamiliar with the sight, it may seem quite amazing to see a very strong horse seem to stand motionless against a fast-flowing current as if it were nothing. But it takes time for a horse with enough foresight to make sure of its footing as it tries to maintain its balance against the tremendous pressure of the flowing water, while also deciding where it is least dangerous to make its next step. If you should be riding such a horse, you must above all things avoid letting it sense that you're feeling nervous. (Benediktsson 1977, p. 125) (Fig. 8.20)

Fig. 8.20 Ragnar Stefánsson, park ranger at Skaftafell National Park, crossing Skaftafellsá in 1952. Lómagnúpur in the far background. Jack Ives

The Great Day of Rivers

The author Þórbergur Þórðarson's (1943, p. 78-80) description of crossing the Skeiðarársandur westwards.

After almost two minutes of riding I was presented with a vivid sight that I've never since been able to forget. Above grey, muddy waves a few steps in front of us, the sun glittered on brown billows, heaving waves and breakers, which were either rising and falling or rushing on at high speed in spuming waters. There was something extremely frenzied and mysterious in this vision, which formed such a striking contrast to the lifeless desert around it. When we got a little nearer, it was not unlike seeing an endless caravan of camels sprinting down the plain.

It all happened more or less in a flash. Without the slightest warning we had arrived at the bank of a sheet of water more reminiscent of an ocean than any river I'd ever seen. I have never been so startled. No one can possibly cross this heaving sea, I automatically thought out loud to myself again and again. We'll have to turn back. The whole surface of this troll-sized, loamy-brown cascade of waves rushed past us at breakneck speed, rising up into high billows here and there, before falling into deep valleys, weaving themselves in no time at all into huge spouts of water tumbling about in spuming breakers, spinning around in sucking whirlpools, and swirling up into plumes and heaving waters, and the roar of these frenzied waves was so loud that we could only hear each other with the greatest difficulty. And as I looked at the water flying all around me, it seemed as if all the surroundings were spinning in circles like a powerful carousel. It was immediately obvious, that no horse could possibly cope with the terrible conditions right here, where we had come upon them … Perhaps it might be possible to try one last time a little further upstream. …

The horses felt their way very cautiously in the dark brown current, step by step, their every movement revealing the tremendous experience they had so clearly learned to utilise from a knowledge born of previous crossings. The water did not suddenly deepen, yet slowly but surely it advanced higher and higher up their sides. The ford, if it can be called that, seemed to be situated in such a way that they had to direct the horses diagonally with the current. They thus advanced further and further downstream with the flowing water, and further and further away from the bank from where we were watching them. The water had risen to nearly half way up their sides, and from that point on it seemed there was no turning back to the same shore. I waited in suspense at every step to see how deep it would be when they next put down their front foot. A deep trench in this foam-laden flow would capsize the horses, and Iceland would be two hearty lads the poorer. But the water level remained the same. It was almost incredible that such a tempestuous sea was no deeper than it was.

They thus moved doggedly onwards at an angle ahead of the current quite a way downstream and across a good third of the river's width. They then turned their horses to the right and aimed straight ahead towards the western riverbank. With this change of direction, the horses were now moving completely athwart the fast-flowing current and right then the water began to ride up above the middle of their flanks. It was a lonely sight to see the horses with men on their backs, like little cones attached to them, swaying as they advanced into this broad, frantic sea, like overloaded boats low in the water, while you hold your breath on a distant shore, expecting to see them disappear into the deep at any moment. What amazing powers these blessed creatures must possess to be able to withstand such an avalanche of water pounding so relentlessly high up on their flanks!

When they had about a third of the river's width to go, they changed direction once more and now broke upstream through the powerful oncoming current. The rushing current spumed as it broke on the horses' chests and rushed along their flanks, while they seemed to advance stubbornly in short bursts. Slowly but surely, the ocean between them and the western bank grew less and less. And finally–thank God–we see them loom into view in bright sunshine on the other bank, almost directly opposite to where we are waiting. What seems the most incredible miracle in the world has been accomplished right before our eyes.

8.2.2   Skeiðarárjökull

Skeiðarárjökull (ca. 1300 km^2) is the largest outlet glacier to flow south from Vatnajökull as it descends from an elevation of 1650 m down to a height of 100 m (Figs. 8.21 and 8.22). Beginning from the west, its boundaries stretch from Grænafjall to the volcanoes of Þórðarhyrna, Háabunga, and Grímsfjall up to Grímsvötn; they then extend onto the Kverkfjallahryggur ridge before veering southeast to the Esjufjallahryggur ridge, and finally southwest to the Miðfell and

Fig. 8.21 Skeiðarárjökull is the largest outlet glacier on the southern side of Vatnajökull. It flows down to Skeiðarásandur between Færnes (on the *left*) and the peaks of Súla (on the *right*). Its eastern part has harrowed a trench 270 m below sea level. The medial moraines come from nunataks that rise higher above the glacier every year. The Lómagnúpur headland is visible in the far distance, beyond the Súla peaks. HB, August 1996

Fig. 8.22 Skeiðarárjökull veers south along the Skaftafellsfjöll mountains at Færnes, but the Skeiðará emerges at Jökulfell. The glacial snout descends from nunataks in the centre of the glacier and the irregular patterns of the tephra layers reflect the varying thickness and velocity of the ice flow. Furthest to the east is Kjós, above which rises Þumall peak. The Kverkfjöll mountains are visible on the other side of Vatnajökull. HB

Skaftafellsfjöll mountains marking the glacier's eastern side. From its 40-km-broad accumulation zone, the glacier continues through an 8-km broad panhandle between the peaks of Súla and Færnes before splaying out onto Skeiðarársandur like a half-circular tongue as an 18-km-broad piedmont glacier. It descends for 75 km on its way to Skeiðarársandur, with a distance of 40 km from its snout to its firn line, and 60 km to its ice divide on Dyngjujökull at a height of 1650 m.

The glacier advances southwards from its sources down three valleys that cut into the central highlands beneath the glacier. They are divided by two mountain ridges that lie in a southerly direction from the eastern side of Háabunga and Grímsfjall towards Þórðarhyrna. The westernmost valley is about 5 km broad between the sides of the volcanoes Þórðarhyrna and Háabunga. The centre valley extends south from the ice-dammed Grímsvötn along a 2-km-broad valley basin. The eastern valley, the largest, descends from Vatnajökull's main ice divide at Dyngjujökull and Brúarjökull southeastward to the Hermannaskarð pass. This is where the thickest and fastest-moving part of the glacier is, and its main ice flow descends to a snout. The three branches of the glacier converge where the three valleys meet, the height of the glacial bed being at about 300 m, while the glacier's surface is at an elevation of around 800 m. Further south, the bed descends steeply and rapidly and creates a 5-km-broad incline onto the eastern part, which on its 20-km-long way reaches as low as 270 m below sea level. About 13 km upstream from the terminus, the ice is 600 m thick. This valley, gouged out by an ice-age glacier, continues out to sea at Skeiðarárdjúp. Seismic reflection surveys reveal that it lies beneath Skeiðará's riverbed to the south of Skaftafell as a 100-m-deep valley in bedrock filled with sediment deposits. If the glacier were to disappear now, it would be replaced by a lake that would gradually become shallower and eventually vanish as the land begins to uplift after the tremendous overburden of the glacier had been removed.

This channel beneath the eastern part of the snout has been gouged out mostly during the glacier's advance during the Little Ice Age. Dispersed evenly over the whole of Skeiðarársandur, the sediment deposits would be the equal to a layer of gravel 20 m thick. The origin of this channel could explain why the Skeiðará succeeded in emerging from beneath the eastern side of the glacier in the 16th century, as it has done so ever since, after having previously always flowed into the centre or more westerly parts of the plain. Grímsvötn outburst floods flow first through the central valley and then through the eastern valley.

The larger section of the western part of the terminus is above sea level, although another section further downstream, 10 km long and 1–2 km broad, is below sea level. The eastern branch of the outlet glacier has been much more productive than the western one. The influx of ice to the western part of the snout is from a much smaller and thinner zone, and this difference in the amount of ice explains why the western part has retreated much more than the eastern part during the warm period of the 20th century.

Skeiðarárjökull is one of Iceland's most active glaciers Skeiðarárjökull has advanced since the settlement of Iceland, gouging out a deep bed for itself. The

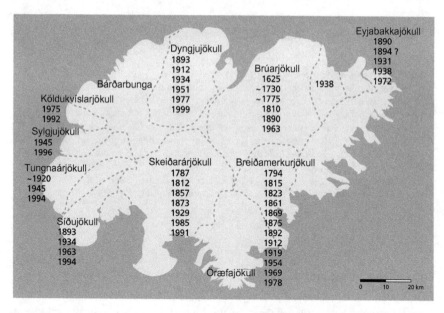

Fig. 8.23 Surges of outlet glaciers from Vatnajökull (Björnsson 1998b; Björnsson et al. 2003). The main outlet glaciers of Vatnajökull are flat with only slight inclines and move too slowly to maintain their mass balance, and so there are occasional surges. This has had a great effect on the hydrological conditions of about 3/4 of the glacier's surface area during the 20th century. The steep outlet glaciers descending to the southeast from Vatnajökull, and from the northwestern flank near Bárðabunga, are very active glaciers, although they do not surge

glacier's surface has often been seen cracking open in billowing waves of crevasses along its entire length, the terminus being shoved forwards dozens of metres, sometimes even surging for hundreds of metres, moving its river channels in the process. Its last surge, of about 1 km, was in March 1991, when, 10 km up on the glacier, a large area of crevasses was spotted, and the ice burst asunder all the way north to Grímsvötn. When this upheaval reached the terminus, it crept forwards, covering previously glacier-free terrain, whilst further upstream the glacier lost several dozen metres in height as the ice was transported forward (Fig. 8.23). For months there was little water in the Skeiðará, but conversely there was an increased volume of water in the Gígjukvísl streams in the centre of Skeiðarársandur. Water flowed across the plain through long, dry channels, and there was a danger it would cut through the national highway, but fortunately it amassed in old channels that had been bridged. Sediment deposits increased tenfold in the glacial rivers. The surge lasted for a few months, but when it came to a halt, water appeared once again from beneath the glacier in the same locations as previously. There had been a similar surge in 1929 when work was being done on laying telephone lines across Skeiðarársandur. The rivers on the plain then changed channels so frequently that the telephone engineers hardly knew what to expect. During the years 1784–1890, Skeiðarárjökull advanced four times about a kilometre, but retreated again in between times. These advances took 5–10 years and could probably be classified as surges.

The Skeiðarárjökull surge of 1727

Some trustworthy and observant men have told us that on the eve of Trinity Sunday in 1727, they had been travelling across Skeiðarársandur, when they saw to their horror that the glacier was rising and falling in waves like the sea breaking on the shore, even the very glacier's edges were ebbing and flowing. At the same time they also saw innumerable rivers spring from beneath it, both big and small, sometimes in one place, then in another, for many of them disappeared as quickly as they emerged. The men were expecting to die at any moment, but managed to save themselves on a high gravel hill which the glacier did not raze, though it only missed it by a few fathoms. Skeiðarársandur was impassable all that summer because of the innumerable rivers that continually burst forth from beneath the glacier, flooding the road without any warning. (Ólafsson and Pálsson 1981, vol. 2, p. 104).

8.3   The Subglacial Lake Grímsvötn

Lake Grímsvötn lies hidden beneath a thick sheet of ice right in the middle of Vatnajökull (Fig. 8.24). Its very existence is due to a powerful geothermal system, for the heat flux from its magma reaches the surface of the bedrock beneath the glacier. Water seeps down to the magma through faults, becomes heated, and then rises again to the base of the glacier, melting ice there before cooling and sinking once again. This cycle of water and heat has existed for centuries, the geothermal heat causing a constant melting of ice at the base of the glacier (Figs. 1.25 and 8. 25). Depressions have thus formed on the glacier's surface above the geothermal areas into which ice flows and melts, but because the overburden pressure surrounding the depressions exceeds the water pressure beneath it, the accumulating meltwater cannot drain away and so it is sealed in and trapped in a subglacial lake (Björnsson 1974, 1975, 1983, 1988, 1992, 2002; Björnsson and Kristmannsdóttir 1984). This is how Grímsvötn came into existence, a lake covered and enveloped by an ice sheet 250 m thick (Fig. 8.26).

Ice continually flows into Grímsvötn, and so its surface level constantly rises until the pressure forces the water to penetrate the glacial dam and to burst out as a jökulhlaup that rushes for 50 km beneath the ice before flooding out onto Skeiðarársandur. The ice-dam is at its narrowest, about 7 km broad, in the southeast of the lake, but because the glacier's surface slopes downwards all the way from Bárðarbunga, there is enough of an incline to channel water and ice into Grímsvötn for the ice-dam to be lifted in its southeastern corner, though jökulhlaups generally occur before this actually happens (Figs. 8.27, 8.28, 8.29 and 8.30). For further information see (Björnsson 1974, 1992, 1997, 2010; Jóhannesson 2002; Nye 1976).

Once the outburst flood has come to an end, the channels from the lake are closed and the volume of water begins to build up once more. Grímsvötn not only

Fig. 8.24 The ice sheet above Grímsvötn subsided in the autumn of 1996 in a similar manner as during the jökulhlaup onto Skeiðarársandur in 1938 (see Fig. 8.37). This picture was taken by Pálmi Hannesson as he flew over Vatnajökull with Agnar Kofoed-Hansen on the first ever monitoring flight over the glacier in 1938

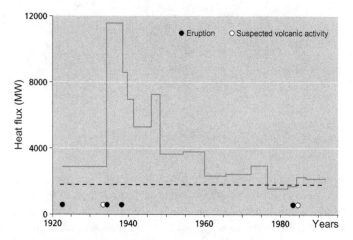

Fig. 8.25 Changes in geothermal activity in Grímsvötn (Björnsson and Guðmundsson 1993). Groundwater is continually borne upwards from hot bedrock (1500–2000 MW base current), but because of magma intrusion into the groundwater system, the level of the heat flux can vary as much as 10,000 MW between years. The overall heat flux energy during the period 1922–1991 was 8×10^{18} J, and is equal to the energy released during the coagulation and cooling of 2 km^3 of basalt magma. About 45 % of the total energy was extracted from the magma chamber, about 35 % from magma intrusions higher up in the geothermal system, and 20 % from the glacial bed during an eruption

Fig. 8.26 A skidoo speeds across the level ice sheet over Grímsvötn towards Grímsfjall, which marks the southern edge of the Grímsvötn caldera. HB

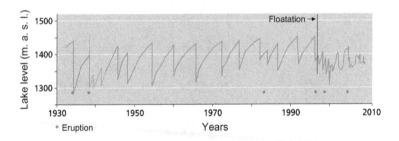

Fig. 8.27 Water levels in Lake Grímsvötn

rests on a powerful geothermal area, however, but also covers one of Iceland's most active volcanoes, which melts glacial ice during eruptions, and if meltwater from such an eruption, even one beyond the lake's catchment area, flows into the lake, its surface level can rise very rapidly, and this can also activate a jökulhlaup. An eruption at the base of Grímsvötn, on the other hand, melts its ice cover, but this does not cause a rise in water level or instigate a jökulhlaup if the volume of the volcanic material is relatively small in comparison to the volume of the lake and the eruption does not open a channel from it. Ice in the vicinity takes quite some time to flow into the lake in order to replace the ice that has melted there.

Altogether about 40 jökulhlaups onto Skeiðarársandur plain have been recorded, the first being in the 14th century (Þórarinsson 1974b). Most of them originate from

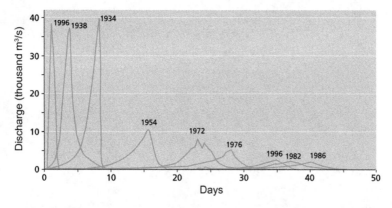

Fig. 8.28 Jökulhlaups onto Skeiðarársandur. Until 1934, outburst floods from Grímsvötn were at intervals of about ten years, in between which the water level rose around 150 m. The total volume of these jökulhlaups was 4–5 km^3 and the maximum flow rate was about 30,000 m^3/s. A typical jökulhlaup accelerates slowly, reaching its peak in about ten days, after which it rapidly subsides. Since 1938, the floods have come at a 4–6 year interval on average, have had a total volume of 0.5–3 km^3, and reached their peaks within one or two weeks. Until 1996 there were floods with a maximum flow rate of 10,000 m^3/s and bearing 30,000,000 tonnes of sediment deposits. The water level of Grímsvötn fell about 80–100 m during these jökulhlaups. Outburst floods from Grímsvötn flow 50 km along the base of Skeiðarárjökull on their way to Skeiðarársandur and into the Skeiðará, but the largest floods also emerge from other channels beneath Skeiðarárjökull into the Sandgígjukvísl and, finally, Súla rivers. Jökulhlaups run through the central valley to the south of the ice-dammed Grímsvötn before flowing beneath the eastern section of the snout and emerging at Jökulfell, gushing into the Skeiðará. Outburst floods run near to the water divides between Skeiðará and Sandgígjukvísl. The migration of the various waterway channels can cause floodwaters to flow due south and to spread out beneath the snout, as happens when jökulhlaups not only flow into Skeiðará but also into Sandgígja and Súla. These floods inundate the main regions of Skeiðarársandur. The glacial terminus begins to break up around the outlets, and icebergs, sometimes huge ones, are strewn across the plain, which is then filled with kettle holes once the icebergs have melted. These jökulhlaups also transport a great deal of sediment onto Skeiðarársandur, about 150,000,000 tonnes in 1934 for example. Some of the floods are connected to eruptions in Grímsvötn, as was the 1934 one. A minor eruption in 1983, however, did not initiate a jökulhlaup

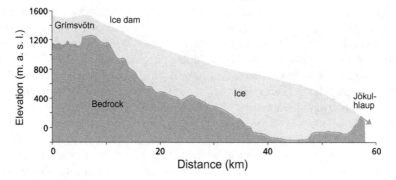

Fig. 8.29 Cross section of jökulhlaup channel from Grímsvötn to Skeiðarársandur

Fig. 8.30 The Grímsvötn jökulhlaup of 1996 (Flowers et al. 2004)

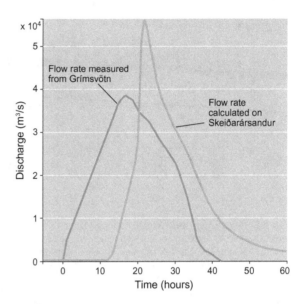

the Grímsvötn area, but outburst floods also accompany eruptions to the south and southwest of Grímsvötn, from Háabunga to Þórðarhyrna. Jökulhlaups from Grímsvötn delayed the building of a road across Skeiðarársandur for some time, and they still remain a constant threat to its bridges.

8.3.1 Grímsvötn: Volcanic Activity and Jökulhlaups

The Grímsvötn ice cauldron (depression) in the middle of Vatnajökull is about 15 km in diameter, and the lake lies hidden beneath it, covered by ice 240–300 m thick. The Grímsvötn volcano is a mountain range between 1400–1700 m high, and over an area measuring 18–20 km from west to east, and about 10–15 km from north to south, its elevation above 1100 m (Figs. 8.2 and 8.6). A ring of peaks encloses the depression in the mountains, which descends to a height below 1050 m, and where three connected calderas can be discerned. The total surface area within the outermost rims of the caldera is about 60 km². Its southern edge is marked by the precipitous, 300–400-m-high palagonite cliff of Grímsfjall, its two highest peaks, Svíahnúkur vestari (1700 m) and Svíahnúkur eystri (1722 m), rearing up through the glacier. Between the glacial tongues that descend northward from the mountain to the ice sheet of Grímsvötn, dark, palagonite cliffs can be seen with occasional narrow intrusive rock layers, or dykes (less than 1 m thick). Evidence of geothermal heat can mostly be seen at the northern foot of Grímsfjall, where there are large cauldrons and water has often been seen in holes in the ice. The western edge of the caldera is completely covered by the glacier, although near the end of the 1960s there have been glimpses of the Vatnshamar rock face as it rises, covered in snow, 100–200 m above the ice sheet. The northern and eastern edges of the caldera are beneath thick glacial ice, though there are numerous shallow ice cauldrons above the the caldera itself. The Grímsvötn pass in the glacier's surface is at the southeastern edge of the Grímsvötn cauldron, and beneath it is a mountain pass, which at its lowest reaches a height of 1150 m, and through which jökulhlaups from Grímsvötn flood onto Skeiðarársandur.

There is a shallow magma chamber beneath Grímsvötn, seismic data indicating it is at a depth of 4 km and is less than 1 km thick. The accumulation of magma in the chamber and the uplifting of terrain on Grímsfjall, both preceding and during the eruptions of 1998, 2004, and 2011, were closely monitored and recorded in detail by using precise and sensitive equipment. Immediately after the eruptions, the uplifted terrain sank again before beginning to rise once more. Many minor earth tremors are frequent in Grímsvötn, especially when magma forces its way into the chamber, but an intense series of earthquakes begins once the magma bursts out of the chamber and the mountain erupts.

Eruptions in Grímsvötn do not cause jökulhaups because the water level does not rise in the lake when the floating ice cover melts. The eruptions that have been closely monitored (1934, 1983, 1998, 2004 and 2011) have emerged through short fissures (0.5–1.5 km) at the southern edge of the caldera on the northern side of Grímsfjall and have only lasted for a few days, melted little ice, and the volume of volcanic material that has entered the lake has been insignificant. The volcanic plume forced its way through an ice sheet 50–200 m thick within an hour and created a circular opening about 0.5 km in diameter (Figs. 8.31, 8.32 and 8.33).

Fig. 8.31 Volcanoes in Grímsfjall in 1983. Sludge and blocks of ice were spewed out onto the ice cover as tephra is borne in ash clouds to the south and west. HB, May 1983

Fig. 8.32 Aerial view, from a distance, of eruption inside Grímsvötn in 1998. HB

Fig. 8.33 Aerial view, close-up, of eruption inside Grímsvötn in 1998. HB

After the eruptions, geothermal heat increased along Grímsfjall, and water holes remained open for years. When there is continuous volcanic activity within the caldera, the ice cover above the lake gradually becomes thinner, and it conversely thickens and the volume of stored water is diminished when there is a long pause between eruptions. The ice sheet thickened from 150 to 250 m during the years 1960–1990. There are examples of unrest in Grímsvötn lasting for as long as two years, such as during the Skaftá fires of 1783–1785.

Although eruptions within the caldera do not necessarily cause jökulhlaups, volcanic eruptions can still accompany outburst floods, for when the water level in the lake falls, the pressure on the shallow magma chamber decreases allowing magma to burst through. Such eruptions occurred at the end of jökulhlaups in the years 1922, 1934 and 2004, and there were many other examples of this in the 19th century. For further information see (Björnsson 1988; Guðmundsson 1989; Guðmundsson et al. 1995).

8.3.2   Eruptions to the North of Grímsvötn Cause Jökulhlaups

There are ridges a few hundred metres above their surroundings to the north of the Grímsvötn caldera. They reach a height of 1300–1500 m and bear witness to recent volcanic activity that has piled up land more rapidly than the glacier has managed to erode it. There was an eruption in a fissure in the centre of the caldera in 1938 that did not penetrate the glacier, but which nonetheless created a huge depression in its surface while also piling up a subglacial palagonite ridge with a total volume of 0.3–0.5 km^3. There was a further eruption in this ridge in the autumn of 1996, and this time it rose higher and extended further northwards. Another ridge heads north-northwest from the eastern edge of the main caldera in the direction of Bárðarbunga, though not quite reaching it. The most northerly volcanic centres of 1867 might have been from this ridge. From the southeastern corner of Grímsvötn, a long row of 10–15 mountain peaks heads northeast towards the Kverkfjöll mountains, some of which are over 1300 m high and are probably the volcanic centres of a fissure chain.

In 1938 and 1996 there was a fissure eruption to the north of Grímsvötn beneath the 500- to 700-m-thick glacier. In both instances the resulting meltwater flowed into the lake, where it accumulated and later burst out as jökulhlaups. In the first eruption, there was 2.7 km^3 of water in the outburst flood, and in the second, 4 km^3. The events leading up to the later jökulhlaup were closely monitored (Figs. 8.34 and 8.35).

An eruption of the so-called Jökulbrjótur (Gjálp) began on the evening of 30 September 1996. It erupted beneath glacial ice 550–600 m thick, according to seismic data. Two ice cauldrons and a shallow depression leading from them down to Grímsvötn were visible in the glacier from an aircraft the next morning. The volcanic plume finally broke through the glacier very early that same morning, 1 October, between 4 and 5 a.m., and explosions spewed volcanic ash, tephra, over the northern

Fig. 8.34 A 3.5-km-long and 50-m-deep ice gorge was formed above the fissure of the volcanic eruption to the north of Grímsvötn in the autumn of 1996. Meltwater flowed through this gorge until it disappeared into the glacier at its southern end. HB, November 1996

Fig. 8.35 A view southeast over Grímsvötn to the steep cliffs of Grímsfjall, 300–400 m high, and then down towards Skeiðarárjökull. The photograph was taken at the end of a jökulhlaup in November 1996, when a 250-m-thick ice cover above Grímsvötn subsided 175 m. HB, November 1996

regions of Iceland. A third ice cauldron then formed further north of the volcanic fissure, where the thickness of the glacial ice had been 750 m before the eruption had begun. A 3.5-km-long and 50-m-deep ice gorge appeared above the volcanic fissure south of the volcanic vent, meltwater flowing along its base until disappearing into the glacier at its southern end. The eruption lasted for two weeks, and no water seemed to have accumulated in the 6-km-long volcanic fissure. The meltwater that flowed from the volcanoes into Grímsvötn was at a temperature of 15–20 °C, calculated from the volume of the depression that had been created all along and above the ice tunnels, and from the measured flow of the water within them. The mass volume of the ice that melted in the eruption was calculated from the volume of the depression on the surface of the glacier above the volcanic fissure and the total volume of the meltwater that accumulated in Grímsvötn (Guðmundsson et al. 1997, 2002; Björnsson et al. 2001).

While the eruption was taking place, the mass volume of the depression on the glacial surface was continually monitored with a radar altimeter. Once it had ended, it was possible to measure through echo soundings the volume of volcanic material that had amassed at the volcanoes. The amount of volcanic material that reached Grímsvötn with melting water was derived from measuring the height of the lake's bed from seismic data. Another calculation, independent of the other, as to the volume of the water that flowed from the eruption sites, was made from the measurements of water levels in Grímsvötn. During the first four days of the eruption, 5000 m^3 of meltwater were discharged per second; the heat flux at the high point of the eruption was calculated as 10^{12} W. When the eruption ended on 13 October 1996, about 3 km^3 of ice had been melted, and after six weeks had passed from the beginning of the eruption, the total volume of melted ice was about 4 km^3, and is the equivalent of 1.1×10^{12} kg of magma having cooled from 1000 °C to 0 °C, assuming that all the heat was used for melting. A year later (in January 1998), the volume of melted ice had become 4.7 km^3.

The name Gjálp (a giantess) was taken from Nordic theology and was considered rather clever and appropriate. Professor Þórhallur Vilmundarson suggested to the current author, in the autumn of 1996, that the volcano might also be called Jökulbrjótur ('glacier-buster').

8.3.3   Eruption in Grímsvötn 2011, Iceland's Largest Since Hekla in 1947

A volcanic eruption began in Grímsvötn between 6 and 7 p.m. on 21 May 2011 and came to an end on 28 May. The eruption emerged from the southwestern corner of the lake, in the same place as the eruption of 2004. In the beginning an ash-laden volcanic plume rose up to a height of 17 km and within a few hours ash began to fall on populated areas south of Vatnajökull. The first signs that an eruption was likely had come about 5.30 p.m. with a strong sequence of small earthquakes in

Grímsvötn (all up to 2.5–3 on the Richter scale). About 7 p.m. the eruption had broken its way through the glacier's surface and the earth tremors decreased, but once the increasing volcanic activity reached a certain level it remained steady for the next 48 h, though with occasional sudden spurts. The eruption was at its most active the following night, during which it is estimated 10,000 tonnes of magma were being spewed out every second. The volcanic plume held at a height of 15–19 km well into the next day, but in the afternoon it only reached a height of about 10 km and the flow of magma had been reduced to around 2–5000 tonnes a second. The volcanic plume began to spread out at an altitude of 5 km and formed a 60-km-broad pyrocumulus cloud above the volcanic centres. The ash cloud was blown by a northerly wind, and flashes of lightning were thus most frequent (50–300 per hour during the first two days of the eruption), on the southern side of the volcanic centres, where the ash fall was also the greatest. During these first two days, most of the ash fell on a sector that reached from Vík in Mýrdalur to the east of Öræfajökull, the thickest layer of ash (up to 5 cm) at Fljótshverfi. There was virtually no visibility at all in the populated parts of this area for the first four days, which made life extremely difficult for both people and livestock, and was a severe threat to the safety of road travellers. The eruption then began to abate and ash only fell in the vicinity of the volcano. Finally, just a few white plumes of steam rose a few kilometres into the air when there were little explosions now and then as magma was injected into the water of vents that had formed above the subglacial craters. In between, thicker clouds rose up to a few hundred metres from small volcanic vents with rims of slag. There was no lava flow from the craters. Although the majority of the volcanic ash had been borne southward by northerly air currents, winds blew in other directions higher up in the atmosphere and dust from the ash was found over a wide area of Iceland except in the western fjords. Air traffic was disrupted slightly for 4 days domestically and in northwestern Europe, with around 900 flights being cancelled in all. There was no jökulhlaup, however.

Volcanic dust got into almost every building in the populated areas that suffered ash falls, and at the end of the eruption a huge cleaning up operation began. Common pastures and grazing land were badly affected, though in time the ash was flushed away by rain, sinking into layers of topsoil, and did not cause any permanent damage to vegetation except to moss in lava fields. The SCSI immediately began immediately to spread fertiliser to strengthen growth and to reduce dust storms. A few farms in Fljótshverfi and Landbrot could not produce hay that summer in 2011. The ash had settled so thickly in the grass that it swirled up during haymaking, making it impossible to complete the process.

The eruption of 2011 did not come as a surprise to geologists. Ever since the eruption of 2004, seismic data from GPS measuring points on Grímsfjall had revealed a continuous uplift and expansion of land of a few centimetres a year, and this indicated an injection of magma into the shallow chamber beneath Grímsvötn. There had also been an increase in seismic and geothermal activity in Grímsvötn for months previously. As soon as the eruption began, large movements were recorded on the equipment on Grímsfjall, 5 km to the east of the volcanoes. After two days the measuring point had subsided 25 cm and moved 50 cm northwest toward the centre of the Grímsvötn caldera.

8.3.4   History of Grímsvötn

Icelanders became aware very early on that Grímsvötn was a lake within Vatnajökull glacier. Such a conclusion might have been drawn from the jökulhlaups that flooded Skeiðarársandur, but it is also possible that someone had actually come across open waters in the glacier.

Grímsvötn is first mentioned in a letter from around 1600: 'With this must be added that men with a knowledge of nature are telling us quite bluntly of the immense and supernatural origins of fire from within the lake ... that is called Grímsvötn in our language, flinging out, higher than the highest mountains and with tremendous power and destruction, huge amounts of pumice and gravel that is catapulted and dispersed to the most far-flung parts of the country, damaging meadows and terrifying the inhabitants' (Þórarinsson 1974b, p. 8). Þorsteinn Magnússon later mentioned Grímsvötn in a description of the eruption of Katla in 1625 (*Relatio*, 15 September 1625), and his narrative is not least interesting in that it discusses how fire burns stone to ash: '... but it burns on the water like cod liver oil or the best kind of fat, and there is also a common rumour that to the north of Glómagnúpssandur [sic], next to Skeiðarárjökull, there is a lake called Grímsvötn, which flows with fire, ice and water like this glacier [i.e. Kötlujökull] and what so often happens is that the fiery eruption first bursts forth from the centre of the aforesaid Grímsvötn, from which flames burn like a pyre or bonfire' (Magnússon 1907–1915, p. 205). The idea of a lake erupting volcanically caused much amazement then, and was much discussed until into the 19th century. Grímsvötn is also referred to as a lake in the work of Bishop Gísli Oddsson's, *De mirabilibus Islandiæ* ('On the Wonders of Iceland') from 1637–1638, and there is an account of a volcanic eruption in Grímsvötn and Grímsfjall in Gríms Vatna Jökull glacier in the descriptions of Iceland in P. H. Resen's *Atlas Danicus* from 1684–1687. Árni Magnússon (*Chorographica Islandica*, 1702) related that the inhabitants of the Síða area in southern Iceland believed Grímsvötn to be northeast of Öræfajökull, 'but the glacier has now advanced over it and it is now beneath the glacier. ... It is said to have erupted from this very lake itself, its waters appearing to be burning' (Magnússon 1955, p. 21). In the lawman Þorlákur Markússon's *Íslandslýsing* ('Account of Iceland') from about 1730, there is the following statement about Grímsvötn: 'Men have seen 15 separate fires burning on the surface of the lake; what the cause of this can be, people find hard to say, except perhaps that it is caused by an overabundance of sulphur deep in the Earth, for whenever folk ride by these glaciers there is the most powerful odour of sulphur there could ever be' (Markússon 1932–35, p. 24).

The Story of Grímur of the West Fjords

This is the story of Grímur, who grew up in the western fjords, from whence he gained his soubriquet. He slew the killer of his father and fled to ask help of a widow who 'pointed out to him a lake and its rivers in the south where he could survive on fishing until a ship arrived at Ingólfshöfði headland. Grímur went south to the lake as advised and built for himself a hut from the plentiful woods around it and began to fish the lake and its rivers. It then so happened that whatever catch he made during the day, disappeared during the night. After this had occurred a few times, Grímur deliberately stayed awake one night to see who was responsible. During the night a giant came and picked up all the fish Grímur had caught, swung them over his shoulder, and walked away. Grímur followed him and stabbed him with a spear. The giant then quickened his pace and got home to the cave where he lived, the spear still in his wound; his daughter was there, and he told her that Vestfjarða-Grímur had given him the wound, before asking her to bury him in the cave, after which he died. … Later the giant arose and came to haunt Grímur in his cabin. … And so the next day Grímur went to the cave, dug up the remains of the giant, and burned them on a pyre. The giant's daughter offered no resistance, but merely stated that the waters by which Grímur now dwelled would at certain times burst into flame and burn down all the woods around it, and her spell has often betimes come true.' (Árnason 1961, vol. 1 pp. 161)

Eggert Ólafsson also had this to say: '*Grímsvötn* is the name of a lake in the highlands to the northwest of Skeiðarárjökull. It is connected to it in that when the latter belches forth in flame, Grímsvötn is also afire. It spews out ashes and flames and the plume of fire towers up out of the water, which does not put it out. Indeed, it seems the water helps rather than hinders its burning' (Ólafsson and Pálsson 1981, vol. 2, p. 104). The outlaw Fjalla-Eyvindur Jónsson knew Grímsvötn's location so well that it might even be supposed he had actually been there.

Grímsvötn is first depicted on Peter Raben's map of 1721, but its more or less correct location did not appear until Sæmundur Magnússon Hólm's map of 1777 (though he called the lake Súla). Sveinn Pálsson described outburst floods in the river Skeiðará very artistically, and he considered that they originated from the volcanic centres in Grímsvötn, where geothermal heat was constantly melting ice: 'Before such a flood descended, the whole of Skeiðarárjökull could be seen to shake. It either swells up or collapses upon itself, first slowly but then more speedily. … A deep moaning and whining is heard at first, and then comes a tremendous cracking sound and a rumbling of thunder. The glacier bursts into a thousand pieces, the floodwaters dragging the lesser icebergs along as they burst out everywhere from beneath the southern edge of the glacier like a tumultuous sea full of breakers rushing over the whole plain down to the ocean. This can continue for many days before the first tempestuousness abates. The flowing water now bears only small icebergs along with it, for the larger ones are already stranded, standing like small mountains here and there across the plain. The thunderous rumbling becomes less frequent, and the air begins to clear' (Pálsson 1945, pp. 500–501).

Sveinn Pálsson, however, believed that Grímsvötn was where the ice-dammed lake Grænalón is situated. He did not agree with the opinion of the inhabitants of Skaftafell County that Grímsvötn lay much further inwards and beneath the glacier. On Björn Gunnlaugsson's map, Grænalón is indeed designated as Grímsvötn, and the poet Jónas Hallgrímsson might have had a hand in this as he assisted O. N. Olsen in producing the map. Gunnlaugsson had calculated the bearings to what he called 'Grímsvötnmöckr' in a sketch in his diary and had indeed found the correct bearing of Grímsvötn in the middle of Vatnajökull. Þorvaldur Thoroddsen was not confident enough, either, to openly contradict Sveinn Pálsson's location of Grímsvötn, even though he could see no indication of volcanic activity at Grænalón from Mt. Björninn, inland from Fljótshverfi. The people of Skaftafell County, on the other hand, seemed to have always held the opinion that Grímsvötn was a volcanic centre within Vatnajökull itself. (Þórarinsson 1974b, pp. 20–21).

8.3.5 Grímsvötn Rediscovered

In the late summer of 1919, two Swedish students, Hakon Wadell (1895–1962) and Erik R. Ygberg (1896–1952), came across the ancient Grímsvötn in the centre of Vatnajökull. They ascended the glacier up Síðujökull on three draught horses with a sledge, a tent, sleeping bags, a compass, and a barometer. They said that the aim of their expedition was to explore the possible causes of the well-known jökulhlaups that emerged from beneath Skeiðarárjökull at sporadic intervals. It was very difficult to drag a sledge up to a height of 1500 m on Síðujökull, because the glacier was covered by a 10–20-cm-thick layer of tephra that had been blown there by westerly winds from Katla, 80–100 km away. Snow had smothered the ash deposits further up the slopes and from measuring the depths, the students could see how much it had snowed since the tephra had settled. This was how the first data on Vatnajökull's mass balance was attained. Wadell and Ygberg continued 2.5 km east of Pálsfjall and then onwards in an easterly direction through thick fog. Suddenly, the front horse stopped, refusing to budge or go any further forwards no matter how hard the students beat it. There was then a slight break in the fog and they discovered they were standing on the edge of a precipitous cliff. The horse had sensed the up-draught at Grímsfjall! 'When we came to the head of the horse, there appeared below us a crater comparable to the side of hell in size, though even that would have been expanding in these recent and most difficult times.' The crater welcomed them with '… the rumbling of ice rocks that continually crashed from the cliff walls, hundreds of metres high, down into the crater's basin where they melted in warm, emerald green water.' They realised that the volcanic source of the outburst floods had been found, though it is possible that they might, in fact, have been standing on the rim of a volcanically active ice cauldron, and that there was not just one but several volcanic vents ahead of them (Wadell 1920a, p. 3; b, pp. 300–323).

It is interesting that they could smell a powerful stench of sulphur in Grímsvötn and described the sulphur-smoking islands in the crater, which they toasted in cognac and christened Svíagígur ('Swedish crater'), and the two highest peaks of Grímsfjall, Svíahnúkur. They measured the size of Svíagígur and sketched an important map of the crater, believing it to be at least 7.5 km long and 5 km broad, with a surface area of 37.5 km^2. But there was so much silent disbelief surrounding their discovery that it never occurred to anyone that the volcanic centre of the eruptions in 1922 originated from Grímsvötn. Neither in newspaper reports, nor an article in a journal by Þorkell Þorkelsson, the director of the Meteorological Office, was the centre of the eruption named as Grímsvötn. At the instigation of Guðmundur G. Bárðarson, the geologist, who was then a teacher at the Akureyri grammar school, the inhabitants of the Mývatn area sent expeditions to explore the sources of this volcanic activity. The people in the north of Iceland still had vivid memories of the Askja eruption of 1875 and its consequences. Bárðarson was accompanied by Pálmi Hannesson, later rector of the Reykjavík Grammar School. They began by ascending Dyngjujökull, upon which tephra had fallen quite clearly from an eruption somewhere towards the centre of Vatnajökull. It was not until another eruption in 1934, however, when expeditions reached the eruption sites, that Grímsvötn was once again accepted as a volcano by Icelanders.

As for the Swedes, they continued from Grímsvötn eastwards across the glacier and began their descent from it at Heinabergsjökull in the southeast. They then got into an impasse among a number of crevasses and were forced to abandon their horses and sledge and walk to the nearest farm by themselves. When they later attempted, with the help of an Icelander, to retrieve their horses and belongings, they were caught in such a tremendous storm that they barely survived in their tent and had to give up, limping back down the glacier exhausted, having lost all their possessions. Two of the horses managed to make it down off the glacier by themselves, but the other two and the dog were either overwhelmed by the storm or swallowed by the glacier. Wadell and Ygberg spent most of their careers in the United States, Ygberg prospecting for metals, and Wadell as a cartographer. Ygberg never fully recovered his health after his misadventure on Heinabergsjökull, and they were both bitter that their discovery of the Grímsvötn volcanic centres was little appreciated and even doubted; it just seemed so unbelievable that there might be a volcano beneath the centre of Vatnajökull. Wadell published a map of Vatnajökull, which he claimed was mostly based on Thoroddsen's map and the descriptions of Daniel Bruun.

8.3.6   Research into Grímsvötn After the Eruption of 1934

Grímsvötn has been part of the history of glacial research in Iceland ever since, and expeditions were made to investigate its volcanic centres after the eruption there in 1934. The first group was well-prepared with food for two weeks under the leadership of Guðmundur Einarsson from Miðdalur. They ascended the glacier from

Fljótshverfi, by Hágöngur onto Síðujökull, and then proceeded on skis north of Þórðarhyrna. From then on there was a continuous layer of tephra that prevented the hauling of sledges and so the group walked the remaining 20 km to the eruption site, guided by the volcanic plume. 'Clumps of slag, pieces of lava and chunks of the crater bed lay in heaps and mounds, in some places as if raked together like haystacks, and the nearer we reached the edge of the crater, the more grandiose everything became. I felt I was standing on the slag heap of the Askja lava field' (Einarsson 1946, p. 53). Plumes of steam, yellow, green and steel-coloured, seethed and billowed up through a gaping wound in the ice sheet while black pillars of smoke were flung into the air from the pillar lava. Swirling clouds of steaming sulphur seared the lungs of expedition members as cinders rained down onto the glacier and hissing and sucking sounds were heard all around them as well as a loud rumbling from the deep. The expedition did confirm, however, that the volcanoes were indeed the same ones that the Swedes had found just over 14 years earlier. The group then succeeded in getting back to their tents before a storm struck, but the bad weather marooned them there for a further 72 h.

Another exploration was funded by the Carlsberg Foundation in April under the leadership of the Danish geographer Niels Nielsen (1937a, b), who had previously taken part in geological research in the highlands of Iceland. The Icelandic geologist Jóhannes Áskelsson also joined the group, for on the way home from the first expedition, he had met Nielsen at Vík in Mýrdalur and decided to return with him to Grímsvötn. Áskelsson was later to explore the Grímsvötn region more than any other scientist over the following twelve years. In this later Danish-Icelandic expedition, a train of 16 packhorses carried their equipment across Kálfafellsheiði and through Djúpárdalur inland to the glacial margin southwest of Hágöngur, where the horses were turned back. Once on the glacier, the members of the expedition harnessed themselves to the two Nansen sledges with all their equipment and hauled them literally on foot as skis and snowshoes were impractical on this glacial surface. When they reached Grímsfjall the eruption had ended, but a layer of tephra, as much as 25 m thick, covered the glacier and the ash was still at a temperature of 50 °C on the surface. Looking down from Grímsfjall, two circular, gaping-wide craters could be seen with perpendicular ice cliffs, in which huge clouds of steam arose from the edges of crater islets that reared up in the main crater, surrounded by water, while further away upturned icebergs floated, all tumbling over each other, as if they had been spewed there by the volcanic plume. Single, vertical towers of ice reached heights of 25–50 m. Further away, there were standard ice cauldrons and regularly patterned crevasses, although everywhere there were signs of instability and there was not a single moment of calm as flakes of ice cracked, blocks of ice collapsed, and avalanches and ash floods careened down the northern flanks of Grímsfjall. Gusts of wind whirled up the ash clouds and then, as if at the wave of a hand, uncontrollable bad weather set in, a blizzard and an ash storm with occasional intervals when it brightened up. Snow cascaded down onto the tents, and members of the expedition had to stand for hours in the centre of them to hold them up and prevent them from collapsing. Their stacked provisions were buried under snow and it proved difficult to find them. Just like members of previous expeditions, they learned at first hand not only the savagery of the outer forces of Vatnajökull, but also its unrestrained inner strength. Nielsen returned to research Grímsvötn in a fourth Danish-Icelandic expedition in April 1936. By then the volcanoes were hidden by snow and ice, the craters were frozen over, and there was no evidence of any geothermal heating to be seen, so that he concluded the volcanoes had cooled completely, that there was little geothermal heat between eruptions, and that the Grímsvötn depression was filled in by the glacier. The jökulhlaups from Grímsvötn were therefore caused by volcanic heat, but whereas no eruption was visible until the outburst flood was in full flow and was shortly about to end, it had to be assumed that the ice had begun to melt sometime before; energy had begun to melt the ice above the volcano and water had collected on the subglacial valley bed until the jökulhlaup burst through the glacier down onto Skeiðarársandur. Once the eruption breaks through the surface ice, the column of smoke rises into the air, so that the energy is no longer being used to melt ice. By measuring how much ice melted, the heat energy of the eruption might be evaluated, before it visibly reaches up into the skies.

8.3.7   Close Monitoring of Grímsvötn

Jóhannes Áskelsson made two trips to Grímsvötn in the summer of 1935 and discovered there was still heat in the crater. On the first trip, he ascended Skeiðarárjökull on foot along with the geologist and astronomer Trausti Einarsson, who compiled a map of Skeiðarárjökull and Grímsvötn from trigonometric measurements with contour lines marked at intervals of 40 m. Einarsson corrected the location of Grímsvötn, as depicted on Wadell's map, moving it further to the west, and also shifting Hágöngur mountain to the western side of Vatnajökull. The later trip was made in June, starting from the eastern side of Hoffell up onto Svínafellsjökull in Hornafjörður, then continuing along the eastern part of the Kverkfjöll mountains to Grímsvötn, before descending via Síðujökull. There were also two expeditions by foreigners in the summer of 1935. A group of Austrians ascended Dyngjujökull, and its leader, the geographer Franz Stefan, was the first person to descend onto the plain of the Grímsvötn depression and reach the volcanic craters of 1934. They then continued westwards on the glacier to Kerlingar, and from there southeastward to Pálsfjall, coming down from the glacier off Síðujökull, to the east of Hágöngur. They took photographs of the upheavals in Grímsvötn, and Rudolf Jonas (1909–1962), a physician, and Franz Nusser (1902–1987), a geographer, later published fine travelogues of their expedition. Nusser (1936, 1940) came to Iceland three times on research expeditions (1934, 1935 and 1939), described Dyngjujökull's glacial surge, and made a map of it as well as of Skaftárjökull and Síðujökull (on a scale of 1:25,000). Three explorers from Alpine countries also made an expedition that summer from Kálfafell in the Fljótshverfi district, calling their trip the International Vatnajökull Expedition. They were the Italian Andrea de Pollitzer-Pollenghi, the Austrian glaciologist Rudolf Leutelt

(1907–1940), and the German Karl Schmid (of the International Academic Ski Club of Stuttgart). In the beginning of their trip they were the first men to ascend Bárðarbunga peak, which has the best panoramic viewpoint over a great extent of Iceland: Geirvörtur, Þórðarhyrna (Vatnajökulsgnípa) and Grímsfjall were visible nearby, while further away were Öræfajökull, Snæfell, Kverkfjöll and Kistufell, and to the northeast was Ódáðahraun. Below them lay Tungnafellsjökull, Vonarskarð pass, a bare Sprengisandur plain, the river Þjórsá, Hofsjökull, and Þórisvatn, and, in the distance, the Laki craters, the Veiðivötn lakes, Mt. Hekla, the Ölfusá, and Þingvallavatn. They calculated the height of Bárðarbunga, using barometric data, as being 2080 m, thus being the second-highest peak of Iceland, and for a while they actually considered it the highest, at about 2123 m (Leutelt 1935, 1937; Þórarinsson 1980a, b). It was then sixty years since Watts had been so far north on the glacier. From Bárðarbunga, the Alpine colleagues went to Grímsvötn, and two of them descended to the volcanic crater of 1934. At its base, the water was warmish and there were constant plumes of steam reaching up into the sky while a continual rumbling could be heard as surging ice crashed down from the edges of the crater, disturbing the water in water holes. From Grímsvötn they continued their journey to the west of Vatnajökull.

There were many trips up onto Vatnajökull after the Grímsvötn volcanic eruption of 1934. During the next upheaval in 1938, there was a large jökulhlaup onto Skeiðarársandur that could be traced to Grímsvötn, and Jóhannes Áskelsson paid yet another visit to this destination (Áskelsson 1936a, b, 1959), while Steinþór Sigurðsson and Pálmi Hannesson flew with Agnar Kofoed-Hansen, the government's aeronautical adviser, over Vatnajökull and Skeiðarársandur taking photographs of the upheavals, both in Grímsvötn and to the north of it. This was the first time that an aircraft had flown over Vatnajökull for scientific research, and Gísli Gestsson, a museum curator, compiled a map of Grímsvötn from these photographs. This was thus also the first time that it was possible to calculate how large an area gradually inclined into the Grímsvötn depression. Around the same time, Danes took aerial photographs for cartographic purposes and Pálmi Hannesson often flew with observers over Vatnajökull during the following years when outburst floods from Grænalón were being monitored.

8.3.8 The Motor Age Begins at Grímsvötn

The number of ski trips onto Vatnajökull increased during the Second World War. The architect Skarphéðinn Jóhannsson and his companions hiked to Grímsvötn every summer from 1942 to 1945. The mathematician and astronomer Steinþór Sigurðsson, the engineer Einar B. Pálsson, the physicist Sveinn Þórðarson, and office manager Franz Pálsson, transported baggage on four packhorses to the glacier in August 1942 and travelled on skis to Pálsfjall and Grímsvötn in order to compile a map of its surrounding area. They found a store of provisions at Svíahnúkur, which Niels Nielsen had abandoned there in 1936, along with a letter to the

Ahlmann-Eyþórsson expedition that was doing research on the eastern part of the glacier. The food was rotten, for according to Jóhannes Áskelsson, there had been a considerable increase in temperature in the mountains since the stores had been left there six years earlier.

Áskelsson set off for Grímsvötn once again when another period of unrest began in 1945, and this was the last expedition to the area that was made only on skis, for at the end of the war the age of glacial research with motorised vehicles began. In July 1946, a jeep was driven 14 km up to the slush area on Dyngjujökull, and in August of that same year, Steinþór Sigurðsson, Sigurður Þórarinsson and companions travelled in two jeeps from the Mývatn district 18 km up Dyngjujökull, and then an American, 25-horsepower Eliason snowmobile towed two draught sledges and members of the expedition on skis to Grímsvötn; a total of 650 kg of material was transported over dry snow at speeds of up to 40 km/h. They were the first men ever to drive in motorised vehicles to Grímsvötn, and they then proceeded to the Kverkfjöll mountains and up onto Bárðarbunga, calculating the latter's height from triangulation measurements as being 1988 m. They returned the same way northwards into the Mývatn district (Sigurðsson 1947, 1984; Þórarinsson and Sigurðsson 1947). It had thus become clear that motorised vehicles could move fast on glaciers and expedite both the speed and efficiency of research. There is extant a film of this first snowmobile expedition onto the glacier. Sigurðsson and Þórarinsson also collaborated in beginning to calculate the mass balance of Grímsvötn, for it was now abundantly evident that it was continuous subglacial geothermal heating that was melting ice and causing water to amass perennially in Grímsvötn, and not simply whenever there was a volcanic eruption, as had been previously thought, though Áskelsson had actually suggested that the former was indeed the case after the eruption of 1934. Þórarinsson was the first Icelander to become an academically-trained glaciologist and had written his doctoral dissertation in 1937 on glacially-dammed lakes in Iceland, a model of which he applied to Grímsvötn: glaciers creep into depressions where geothermal heat melts them and the water continuously accumulates in marginal or subglacial lakes until it succeeds in lifting the dam and bursting forth in a jökulhlaup. Precipitation data showed that 0.7 km^3 of water was added every year to the Grímsvötn ice cauldron, the equivalent of the 7 km^3 volume of Skeiðarársandur jöklahlaups, which then, on average, occurred at intervals of every ten years.

The Skeiðarársandur Jökulhlaup of November 1996

The jökulhlaup onto Skeiðarársandur in November of 1996 was the most rapid outburst flood from Grímsvötn that has ever been recorded, with 3.2 km^3 of water flowing across the plain within a mere 40 h (Figs. 8.36, 8.37, 8.38 and 8.39). Moreover, it was seen to rise high enough to lift the ice barrier off its bed. The eruption of Jökulbrjótur (Gjálp) to the north of Grímsvötn caused this jökulhlaup, the meltwater from the eruption sites amassing in Grímsvötn for a whole month before bursting forth from the lake on 4–7 November 1996. The water level had risen so high by 4 November

Fig. 8.36 The ice tunnel through which the jökulhlaup from Grímsvötn flowed in November 1996 collapsed, forming a depression in the glacier's surface 6 km long, 1 km broad, and up to 100 m deep. HB, 7 November 1996

Fig. 8.37 A jökulhlaup swept away the bridge over the Sandgígjukvísl river on Skeiðarársandur outwash plain on 5 November 1996. The magnificent Lómagnúpur cliff face in full view straight ahead; electricity pylons on the left. HB, 5 November 1996

Fig. 8.38 The outburst flood from Grímsvötn in November 1996 dispersed chunks of ice over a wide area of Skeiðarársandur. Öræfajökull is in the distance. HB, 7 November 1996

Fig. 8.39 The jökulhlaup of November 1996 transported a large amount of sediment out to sea. The thicker elements sank to the ocean floor, but the more refined deposits floated on the surface and spread out from the shore in ripples forming this circular pool. HB, 7 November 1996

(1500 m above sea level, 60 m higher than had ever been measured previously), that the Grímsvötn ice dam gave way. Ice-quakes in the evening of that same day revealed that water was being discharged from Grímsvötn and, about 10.5 h later, flowing water was detected at Skeiðarárjökull's terminus and a flood wave a few metres high careened down the Skeiðará. By then the 0.6 km^3 of water that had accumulated beneath the glacier had uplifted it above the flood path, for melting ice from frictional heat could only have provided a very small space (0.001 km^3) for this water. A long crevasse above the flood path became visible, and a trench, about 6 km long, 1 km broad, and 100 m deep (0.3 km^3 in volume), appeared on the glacier's surface above the ice dam when the ice tunnel collapsed above the jökulhlaup's watercourse at the end of the outburst. If all the heat energy had been employed to melt the ice tunnel, the mean temperature of the glacial water had been 8 °C. This high water temperature in Grímsvötn is explained by the hot water flowing from the eruption sites in Jökulbrjótur.

When the jökulhlaup's flood wave reached the glacier's margin, it burst through ice a few hundred metres thick and rushed along the glacier's surface. Icebergs broke off from the terminus and were dispersed all over the plain. Supercooled water arose from a depth of 200–300 m to the outlet so that shards of ice were formed and suspended sediments were frozen. Altogether, the flood flushed away 15 km of the national highway over Skeiðarársandur, while its sediment deposits extended the plain's coastline by 900 m.

The discharge rate for this jökulhlaup was calculated by continually measuring how quickly the water level of Grímsvötn subsided; the surface area and water level of the ice-enveloped lake were already known. The flow rate increased linearly in time and reached a high point of 40,000 m^3/s within 16 h. The flow had completely abated 27 h later. While water was being discharged from the lake, its water level dropped about 175 m and the ice cover floating above it was reduced from 40 km^2 to 5 km^2 by the end of the jökulhlaup.

The rapid Grímsvötn outburst flood of November 1996 cannot simply be explained by the discharge of water along an ice tunnel from the lake, the tunnel walls being gradually enlarged as they were melted by the flowing waters. It is probable that a broad sheet of water had escaped from beneath the ice barrier and suddenly opened up a channel for the many waterways dispersed beneath the glacier. A wave of pressurised water had then rushed along the glacier's bed and uplifted the glacier itself. The water had amassed in all the subglacial waterways and not just at the opening of a tunnel higher up the glacier, as is typical of a Grímsvötn jökulhlaup. This would explain why the flood was so unusually swift.

There had also been an eruption to the north of Grímsvötn in 1938, the resulting meltwater (2–3 km^3) flowing into Grímsvötn and then through the Grímsvötn system of waterways as a jökulhlaup, reaching its peak in three days and abating within a week. A sudden rise of the discharge then indicated that floodwater had been above the melting point of ice, and the slow abating of the flood is explained by the water being dispersed along the glacier's bed before accumulating at the river's outlet. Similar jökulhlaups, reaching their peaks within 2–4 days following an eruption to the north of Grímsvötn, might also have occurred previously, or in 1861, 1867 and 1892.

8.4   The Stratovolcano Öræfajökull

The most magnificent glaciers of Iceland descend steeply to the lowlands from the 1800-m-high crater rims of the stratovolcano now called Öræfajökull. The crater rims enclose an ice-filled caldera 4–5 km in diameter and with a surface area of 14 km^2. Within the caldera, the ice is up to 550 m thick. Six of Öræfajökull's mountain peaks rise above 1700 m. Hvannadalshnúkur (2110 m), on the north-western edge of the caldera, is Iceland's highest mountain, its 200-m-high cliffs towering over the glacier (Figs. 8.2, 8.6, 8.40, 8.41 and 8.42). On the southern edge of the caldera are Rótarfjallshnúkur (1848 m) and the two Hnappar peaks (the western one at a height of 1851 m and the eastern one at 1758 m), indeed, the glacier was originally named Hnappafellsjökull after them. The Norwegian surveyor Hans Frisak climbed both the western peak of Hnappar and Rótarfjallshnúkur in 1813, both then considered the highest mountains in the country. In 1891, however, Frederick Howell saw that Hvannadalshnúkur was in fact higher, and so he and two Icelanders, Páll Jónsson and Þorlákur Þorláksson, climbed it, the first men ever to do so as far as is known. Finally, to the east are the summits of Sveinstindur and Sveinsgnípa (both around 1900 m high); the latter was climbed by Sveinn Pálsson in 1794. The stratovolcano is covered by a glacier above the height of 1000 m, but there are a few cliff faces that rise through the blanket of ice, and they have increased in number and height in recent years. The diameter of the mountain's base is 20 km, the surface area of its base 400 km^2, and its volume is estimated at 370 km^3, the ice sheet covering an area of 80–100 km^2. The summit of Öræfajökull has Iceland's greatest amount of precipitation, accumulating a 10–15-m-thick layer of snow every year.

Öræfajökull is the fourth largest stratovolcano in Europe after Etna in Sicily, Elbrus in the Caucasus mountain range, and Beerenberg on the Norwegian island of Jan Mayen. It is a singular volcano standing outside of Iceland's active volcanic zone. The mountain was amassed in the later part of the last glacial period from two ancient volcanoes, one in the Skaftafell mountains and the other in Breiðamerkurfjall. Its base is from palagonitic tuff and basalt more than 750,000 years old, although at its summit it has younger rock with the same magnetic direction as is current now. The ice-age glacier hemmed in the palagonite so that the mountain rose swiftly. Volcanic activity in the current interglacial period has been mostly in the form of explosive eruptions, but a few fissures have

Fig. 8.40 Aerial view from above the Esjufjöll mountains of Breiðamerkurjökull, southwest over the Mávabyggðir nunataks. In the far distance is Öræfajökull and its towering peak of Hvannadalshnjúkur. HB, September 1997

Fig. 8.41 Öræfajökull in the autumn, showing the clear outlines of the caldera at the top of the stratovolcano and the outlet glaciers extending from it. On the far side of Öræfajökull is Breiðamerkurjökull. HB, Autumn 2004

branched out from the volcano beneath the glacier, and a lava flow from one of them has reached the lowland next to Kvíárjökull. The lower flanks of the mountain have been severely eroded by the glacier. Ten steep outlet glaciers plunge 1700 m from the firn sheet on the volcano's summit down into valleys leading onto glacial outwash plains, the most notable of them being Skaftafellsjökull and Svínafellsjökull in the west, Falljökull and Kvíárjökull to the south, and Fjallsjökull in the east (Figs. 8.43, 8.44 and 8.45).

The fast-flowing rivers that swept down onto the lowlands from these glaciers made travel and communications between local communities very problematic until they were eventually bridged. It is extremely difficult to restrain these rivers in specific channels and they are never biddable for very long. For centuries the farmsteads in this area were isolated from other communities by the large rivers running through Skeiðarársandur, to the west, and Breiðamerkursandur, to the east (Figs. 8.46, 8.47, 8.48 and 8.49). But there have always been routes through the Öræfi district, indeed in olden times travellers sojourned there overnight because the two plains were never crossed in one and the same day. The people of the Skafatafell area would guide travellers westward over Skeiðarársandur, while the farmers of the Kvísker district would guide travellers eastward over Breiðamerkursandur, towards Reynivellir. The farmers of Núpsstaður would assist those travelling east to the Öræfi district. The river guides in these districts were

Fig. 8.42 The higher reaches of Öræfajökull. The Hvannadalshnjúkur peak is in the top left corner, and stretching up to it is the Hvannadalshryggur ridge. To its east are the outlet glaciers Virkisjökull and Falljökull, from beneath which the jökulhlaups of the 1362 eruption destroyed the farming community of Litla-Hérað where Skeiðarársandur is today. HB

Fig. 8.43 Oblique aerial photograph of Morsárjökull and Morsádalur. The glacier plunges in an icefall off vertical cliffs, though it is split asunder by the jagged massif. The glacier maintains itself from blocks of ice that fall off its edge. Skeiðarárjökull can be seen in the far distance. The *middle photograph* was taken from the Kristínartindar peaks in 1937. From the ogives in the glacier, it can be estimated that it advances at about 150 m per annum. In front of its snout there is a lake (*bottom photograph*). The glacier's bed is below sea level at the very end of the snout, and remains so for 3 km inland beneath the glacier. *Top photograph* **a** HB, October 1996. *Middle photograph* **b** Ingólfur Ingólfsson, July 1937. *Bottom photograph* **c** HB, July 2004

Fig. 8.43 (continued)

Fig. 8.44 Kvíárjökull. HB, September 1995

very experienced in finding fords and passable routes over the braided streams of the glacial rivers, or indeed over the glaciers themselves. There was a ferry boat over the Jökulsá on Breiðamerkursandur until it was bridged in 1967, opening up the way east to Hornafjörður. From 1948 to 1973, most cargo was transported by air to and from Fagurhólsmýri. Shortly after the Second World War, special off-road vehicles were used for journeys across the rivers, the most famous being the so-called 'river dragon.' All of the waterways of Skeiðarársandur were finally bridged in 1974, completing the last link in the ring road around Iceland.

Fig. 8.45 Fallsjökull. HB, July 1999. Both of these glaciers (Figs. 8.43 and 8.44) were within Hrollaugur Rögnvaldsson's original land claim. At the end of the 19th century, a glacier covered the spot from where these photographs were taken

Fig. 8.46 Clusters of farmsteads at the foot of Öræfajökull, Svínafell (on the *right*) and Freysnes (on the *left*). The lowest 6 km of Svínafellsjökull has ploughed its bed below sea level. Up until 1940, Svínafellsjökull and Skaftafellsjökull used to coalesce in front of Hafrafell (*far left*). Behind Svínafell is Hvannadalur. HB, August 2003

8.4.1 Öræfi's Beautiful Countryside and Unique Culture

Many believe the Öræfi district to be the most beautiful in Iceland with its black sands, mountains of varying shapes and colours, birch woods, vegetated and blooming mountainsides, cliffs and scree slopes, all bounded by the ocean to the south and the glacier itself with the heavens above it, the kingdom of the great skua

Fig. 8.47 In 1938, Svínafellsjökull could be seen right at the edge of the home field at Svínafell farmstead. Sigurður Þórarinsson

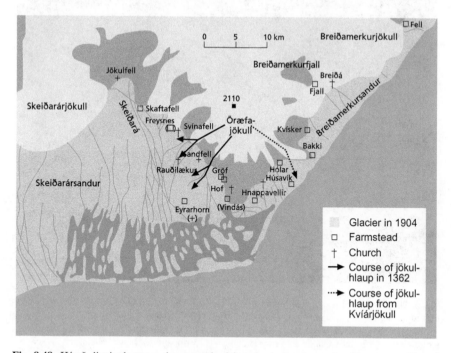

Fig. 8.48 Hérað district between the outwash plains. Ancient and new farming communities on Skeiðarársandur and Breiðamerkursandur and the courses of the jökulhlaups from the Öræfajökull eruption. Illustration based on map compiled by Þórarinsson (1958)

Fig. 8.49 Kvíárjökull, the largest outlet glacier to flow south from Öræfajökull, descends 500–600 m through a narrow canyon between Staðarfjall (1207 m) and Vatnafjall (995 m). Its lowest section has ploughed out a bed all of 100 m below sea level. In front of the glacier are the highest terminal moraines in Iceland, Kvíármýrakambur (173 m) and Kambsmýrakambur. When the glacier was at its fullest extent at the end of the 19th century, it rose above these moraines and chunks of ice would tumble down the outer flanks of Kvíármýrakambur. There are many moulins and ogives on the glacier. HB, August 1998

and the great black-backed gull. The climate here is mild and warm with a good deal of precipitation—though not too much—and occasional vigorous gales, which may occur in a few places when the air pressure suddenly drops and gusts of a northerly wind blow down the glacier's mountain passes. Öræfajökull is indeed so huge and awesome as it towers high above the shoreline that, when a National Park covering an area of 1600 km^2 was established at Skaftafell in 1967, the very first in accordance with new laws on the preservation of nature, most of it was formed by the glacier stretching north towards Grímsvötn.

A unique way of life had been preserved in the isolated community at the foot of Öræfajökull ever since the settlement of Iceland (Figs. 8.50, 8.51, 8.52 and 8.53). There had been more than twenty farms scattered over the vegetated lands between the plains, or in Litla-Hérað as it was called (to distinguish it from the Fljótsdalshérað region northeast of Vatnajökull), right up until well into the 14th century, and according to the *Book of Settlements*, the area of Ingólfshöfðahverfi was said to extend eastwards as far as the river Kvíá. The main church was at Rauðilækur. In 1362, however, the history of disasters began with a huge explosive eruption in the caldera of Knappafellsjökull after a very long pause. This spewed out 10 km^3 of tephra and 2.5 km^3 of hard rhyolite rock, the greatest pumice eruption in Iceland since 800 B.C. Settlements as far as 70 km to the east of the volcano suffered damage, although most of the tephra from the eruption was borne

Fig. 8.50 Svínafellsjökull in the Öræfi district; drawing by Auguste Mayer from 1836. NULI

Fig. 8.51 Stigárjökull in Öræfi; a drawing by August Mayer in 1836. NULI

out to sea, and traces of it have even been found in bogs in Scandinavia. Rhyolite
pumice from the 1362 eruption is also still visible over a wide area of the district.
Jökulhlaups from beneath the Falljökull, Virkisjökull, Kotárjökull, Rótarfjallsjökull
and Svínafellsjökull outlet glaciers, all flooded over Skeðarársandur, while
Kvíárjökull surged into the sea. These outburst floods and ash falls so badly
damaged the community, that it was 40 years before anyone settled in this district
again, which was then renamed Öræfi, or wilderness, from which the glacier has
drawn its name ever since. The revived community has been in four main clusters
of farmsteads, and there is still mostly desolate land between them, with gravel, clay
and bogs, where there was once the fully vegetated Litla-Hérað region. The church
was later moved to Sandfell, which remained a church-stead until 1914 and a
rectory until 1931. This farm again suffered badly in the eruption of Öræfajökull in
1727, when landslides fell on both sides of Sandfell and destroyed a significant part
of its land. The accompanying floodwaters indicate that the eruption had begun
within the caldera and then emerged through a fissure on its western side at the foot
of the glacier, above Sandfellsker. The fissure reached as low as 1100 m above sea
level, and six or seven columns of flame were seen within it, indeed fire and smoke
were visible until the following spring, though it is not believed there was any flow
of lava. The scree-sides of Sandfell still have little vegetation. The church is now at
Hof, as is the local school.

Fig. 8.52 Fjallsjökull descends from Öræfajökull between Ærfjall and Breiðamerkurfjall. Breiðamerkurjökull can be seen ending in Breiðárlón on the far right. Breiðá flows south from it into Fjallsárslón, and from there all waters flow out into the sea. To the northwest of Breiðamerkurjökull, rise the peaks Káratindur and Þuriðartindur and the Fjölsvinnsfjöll mountains, while the Mávabyggðir are high up on the glacier

On Svínafellsjökull in 1857

On the glacier is a large circular opening down which the surface water of the glacier cascades. It is impossible to see the bottom of this hole which, with its whitish, blue-greenish shiny walls, reminds one mostly of the entrance to the underworld. It is horrifying to lean over such a bottomless depth. We rolled a few large rocks down one such opening. It was a long time before they struck any obstacle, and then there was a sinister clunk, as one might imagine the sound of an earthquake to be like.

Sheep sent out to graze in the mountains in summer sometimes fall down these crevasses and this gives rise to many superstitions and all kinds of stories. This is because occasionally a sheep falls 50 to 60 or even 100 feet down into a crevasse and cannot be hauled back up, but its body later reappears on the glacier's surface quite undamaged and with its wool untouched; none of its bones are broken and its flesh is so fresh it might be eaten, though admittedly with a lot of salt. Sometimes only bones are found, and then usually because the carcass had not been found before the ravens had got at it. But even then the bones lie in a natural position, none of them broken or crushed, as long as the sheep had not suffered any bone fractures during its fall.

Fig. 8.53 The view west from Hali in the Suðursveit district. 'Breiðamörk plain was from ancient times, from the age of settlement, and for a long time after that, a beautiful region with grassy fields, widespread copses, and many farmsteads.' Eggert Ólafsson and Bjarni Pálsson (1981, Vol. 2 p. 106). HB, August 1998

Men in Iceland believe that sheep carcasses are returned to the surface because the glacier tosses and turns so much during an outburst flood that it turns itself upside down. But this is obviously no kind of real explanation; this fancy comes completely out of the blue, indeed the carcass must be completely crushed by such a tremendous upheaval. When a sheep reappears calmly and silently on the glacier's surface—"it lies there as if just freshly slaughtered"—the probable explanation is that the glacier is melting from without and that the melting has reached the depth to which the sheep originally fell. (Gadde 1983, pp. 82–83)

Otto Torell (1857) and his companions measured the velocity of the ice flow of Svínafellsjökull from 30 July until 5 August 1857. They positioned markers in the medial moraines in the centre of the glacier and painted a white cross on the mountainside to the north of it; then from the southern side, probably at a point about 1.5 km from where the terminus then was, between Skerholt and the foremost part of Hafrafell, they monitored how the markers moved downstream on the glacier, 25 cm per day for 6 whole days, 8–10 in. every 24 h. This was the first ever measurement of a glacier's velocity in Iceland.

The Height of Hvannadalshnjúkur Peak
In the spring of 1904, the Danish expedition led by Johan Peter Koch measured the height of Hvannadalshnjúkur with the triangulation method from the surrounding lowland; angles were measured from three exactly located

and known points on the lowland that were visible from the peak, which was then calculated to be 2119 m above sea level. In 1955 the peak's elevation was again measured with the triangulation method from the lowland, and proven to be slightly higher (2123 m). In the spring of 1993, members of the SIUI calculated the height as being 2111 m from GPS-measurements whereby the target location is pinpointed from at least four satellites in orbit around Earth. The margin of error was 5 m. Members of the GSI then climbed the peak in 2004, exactly a century after the Danes had measured its height, and once again it was calculated as being 2111 m high. The height of the glacier was last measured by employees of the NLSI in July 2005, and they came to the conclusion it was 2109.6 m high.

8.4.2 Devastating Eruption of Hnappafellsjökull in 1362

There is a reference to the eruption of 1362 in one of the annals from the Skálholt bishopric from the late 14th century: 'Fires appeared in three places to the south. They kept on from June to autumn. With a great deal of noise. All of Litla Hérað destroyed & much of Hornafjörður & the Lón district. Roads to the Alþing also destroyed. Knappafell glacier surged into sea. Up to thirty fathoms deep. With landslides, mud & dirt that levelled gravel plain. Kirk & parishes at Hof & Rauðilækur all razed. Sands piled up on plain. Icebergs stranded on them. Houses could hardly be seen. Ash fell in north of land. Could be traced all the way. Accompanying this, pumice seen in western fjords. Ships could hardly sail because of it [& even more widespread to the north]' (Sigurðsson 1857, vol. 1, p. 245).

There is one narrative stating that an outburst flood had destroyed all the buildings of the rectory at Rauðilækur except for the church. In the *Oddaverjaannáll* ('Annals of Oddi'), written about 1580, it is stated: 'Anno 1366: Fires arose in Litla Hjerad and destroyed the whole region: there once had been 70 farmsteads: no single living being left except one olde woman: and mare' (Þormóðsson and Grímsdóttir 2003, p. 177). Although this entry is marked 1366, the description must refer to the eruption of Öræfajökull in 1362.

Ísleifur Einarsson (1655–1720; 1918a, b), the sheriff at Fell in the Suðursveit district, wrote down the following oral history from people in Öræfi: 'According to the folk of Öræfi, from what they have heard [and from those who had heard it from their predecessors, and they from theirs, by word of mouth], the Öræfi district has been wiped out twice. [Some say three times]. On one occasion, the shepherd at Svínafell, by the name of Hallur, had been herding the sheep home and the women had begun milking them, when all at once a huge crack came in Öræfajökull so that they were truly amazed. Then after a little while, another crack came. And then the shepherd said: it would not be wise to wait for the third. He had then fled up into the Flosahellir cave – it is up on the mountainside to the east of Svínafell farm – and then the third crack came in the aforementioned glacier, and it all burst open and so much water and so many rocks gushed right out of it and through every gully that all the folk and livestock in Öræfi perished, except for this shepherd and one horse with a blaze on its forehead' (Einarsson 1918b, p. 51).

8.4.3   Eruption of Öræfajökull, 3 August 1727 to April or May 1728

Shortly after the church service at Sandfell on Sunday 3 August 1727, there were a few sharp earthquake tremors; everything became calm again for a while, but early on the morning of 4 August the tremors had become so powerful that everything loose was shaking and falling down in the local farmsteads (Hálfdanarson 1918–1929). Constant quakes and movement then began. At 9 o'clock there was a huge crack in the mountainside and almost immediately a large jökulhlaup burst out from beneath Falljökull, Virkisjökull, Kotárjökull and Rótarfjallsjökull. The flood that emerged from Berjagil gully to the east of Sandfell inundated the Kotá river plain and swept away Kotársel and all the sheep and horses before it as its waters spread out over the meadows of Sandfell and Hof. A short while later, there was a surge in the Virkisjökull outlet glacier that razed two of Sandfell's annex-farms. Much of the grassland was destroyed and two girls and a teenager were killed. The jökulhlaup rushed down between Svínafell and Sandfell so that the rectory was cut off, and the priest, Pastor Jón Þorláksson, fled up onto Dalskarðstorfa and sheltered in a tent. The floodwaters bore boulders, pumice stones, ash and icebergs. The ash fall made everything dark for three days, but it cleared up on the fourth day, after which there was little further ash fall. Volcanic ash covered the farmlands of Svínafell and Skaftafell, and there was no hay that summer in the districts nearest the eruption site. The ash fall was much less than in 1362, however, no more than 0.2 km^3. The scattered icebergs had still not melted when Eggert Ólafsson and Bjarni Pálsson passed through the area 29 years later in September 1756 (Ólafsson and Pálsson 1981, vol. 2, p. 105).

Niels Horrebow described the jökulhlaup from Öræfajökull in 1727 as follows: 'The water flowed from the mountain between two farmhouses called Hoff and Sandfeld [sic], which lie not above six miles from the foot of the mountain, and about the same distance from each other. It spread itself beyond these houses in the flat country, and washed through the lower house and dairy, and carried off all the milk, butter, &c. the people saved themselves by getting on the tops of the houses. The water did not rise so high, and only filled the inside. Numbers of the cattle from both farms were carried off, and some of them were afterwards found parboiled. It ran along the vallies [sic] and emptied itself into the sea; but the stones, sand and earth, carried away by the current were not any way equal in quantity, nor rushed on so violently as the former, though greater damages were sustained. No cattle were lost in the other eruption, nor any field of grass destroyed, the same being entirely demolished by former inundations: but the stream of the eruption we are now talking of, passed over fine fields, and destroyed the better part of the cattle that were grazing in them' (Horrebow 1758, p. 14).

And an entry in the annals of Mælifell reads: 'The glacier at Síða surged eastwards and razed three farmsteads, the people escaping except for three persons. Widespread damage to the home fields and meadows' (Þorsteinsson et al. 1922–1987, vol. 1, p. 637).

There is an interesting narrative on the upheavals in Öræfajökull in 1727 in the travelogue of Ólafur Ólafsson (Olavius), based on the account of Jón Þorláksson, the parish priest at Hof in Öræfi. 'When I was below the middle of Flögujökull looking up at the summit, it seemed to me the glacier grew higher and swelled out one moment and then collapsed and sank inwards the next. Nor had I been mistaken, for it soon came to pass what this sight boded. The following morning, Monday 8 August, we all not only often felt tremendous earth tremors, but also heard a horrifying cracking sound as loud as thunder. In all this commotion, all loose objects in the home were falling down, and everything seemed most likely to collapse, both the buildings and the mountain. ... At 9 in the morning, three tremendous cracking sounds were heard, much louder than the previous one, and accompanying them were jökulhlaups or eruptions, the final one the worst, carrying away all the horses and other livestock in its way in one fell swoop. After which the glacier itself slewed down onto the lowland, like molten metal poured out of a melting-pot. It was so high once it had surged down onto the level ground that I could see only a tiny part of Lómagnúpur promontory above it, about the size of a bird. Once this had happened, water began to pour out to the east of the glacier and destroyed what little grassland that was left there. ... The situation then changed yet again, the glacier itself surging forwards, bearing icebergs all the way to the sea, though the largest remained right at the foot of the mountain. The next thing, the air was filled with fire and ashes, with ceaseless cracking and crashing; the cloud of ash was so thick that it caused there to be no difference between night and day; the only visible light was coming from the edge of the fires that had arisen in five or six gashes in the mountain. For three continuous days, the parish of Öræfi suffered from these fires, floods and ash falls. It is not at all easy to describe all this exactly as it really was, for all the earth was black from pumice cinders and it was risky to go outside because of glowing stones that rained down from above, so that one had to hold buckets and tubs above one's head for protection' (Olavius 1964–1965, vol. 2, pp. 224–225). Jón later commented that the floodwaters had been so hot that the horses shied from them.

8.5 Breiðamerkurjökull and Breiðamerkursandur

The view of glaciers from the western parts of the Breiðamerkursandur is the most spectacular in Iceland (Fig. 8.54). Steep and awesomely crevassed outlet glaciers plunge down the eastern flanks of Öræfajökull, while from the main ice sheet,

Fig. 8.54 Nowhere in Iceland have glaciers and volcanoes moulded Iceland's terrain more over the last thousand years than on Breiðamerkursandur. View over Jökulsárlón toward Öræfajökull. Mávabyggðir cliffs rise up through Breiðamerkurjökull in the *top right* of the picture. HB, November 2008

Vatnajökull stretches forth its huge paw. This paw is Breiðamerkurjökull, Iceland's third largest outlet glacier, which descends southwards down to the coast where it calves into Jökulsárlón lagoon, now one of the country's most popular tourist attractions (Figs. 8.55, 8.56 and 8.57). An important road has traversed this area

Fig. 8.55 Breiðamerkurjökull, the third largest outlet glacier of Iceland, descends to the southern coast before calving into the Jökulsárlón lagoon. There has been a vital road route between the glacier and the ocean ever since the settlement of Iceland. Rising up through the glacier are the Esjufjöll and Mávabyggðir mountains, medial moraines extending from them all the way down to the terminus. Breiðamerkurjökull has ploughed itself a bed 200–300 m below sea level along a 25-km section. By the end of the 21st century, the glacier might have retreated so far that the Jökulsárlón would reach the Esjufjöll mountains. HB, August 1998

Fig. 8.56 Jökulsárlón lagoon came into being at the foot of Breiðamerkurjökull when the glacier began to retreat in the 1920s. Most of the glacial sediments are now deposited in the lagoon, but they do not compensate for the coastal erosion along the barrier beaches. The sea has slowly but surely encroached on the bridge and Iceland's ring road could be torn asunder within the next few decades. HB, August 1998

between the glacier and the ocean ever since the settlement of Iceland, and before the glacial rivers were bridged in 1967, they had been the most difficult obstacles on this route along the south coast (Fig. 8.58).

The cohabitation of Icelanders and glaciers for more than a thousand years, is nowhere better recorded than in the history of Breiðamerkurjökull. Some of the most important of man's early ideas about glaciers, indeed the very beginnings of Icelandic glaciology, can be traced to this one glacier. It might well be the source of the first ever recorded narratives in the *Gesta Danorum* of Saxo Grammaticus (early 13th century) concerning the movements of glaciers and their propensity to regurgitate human bodies and the carcasses of horses and sheep that had fallen into crevasses, as later reported by the Dane Horrebow (1758). Farmers along the southern verges of Vatnajökull knew that glaciers were created by snowfalls in the winter which did not melt in the summer, even though learned scholars had maintained, far into the 18th century, that glacial ice was formed from a mixture of saltpetre and water that flowed up to the Earth's surface from underground vaults. It was near the eastern side of Breiðamerkurjökull that Sveinn Pálsson first had the idea, near the end of the 18th century, that ice could creep down an incline as viscous matter, and he argued against the theory of Eggert Ólafsson that

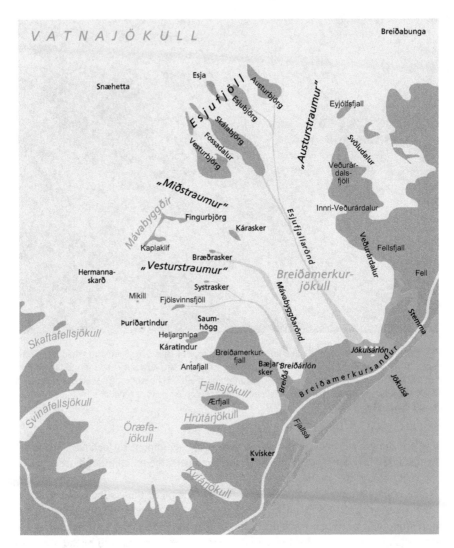

Fig. 8.57 Place names on Öræfajökull and Breiðamerkurjökull

Breiðamerkurjökull was a ground level glacier that sustained itself on seawater beneath its base (Pálsson 1945, p. 478).

Breiðamerkurjökull has also provided clear and important evidence of how a glacier advanced during the Little Ice Age (Figs. 8.59 and 8.60; Björnsson 1998a). Vegetated fields and woodlands were converted into gravel deserts and the glacier ploughed its bed to hundreds of metres below sea level. Glaciofluvial sediment deposits were transported to the mouth of the Jökulsá and formed a delta of estuary islets down to the sea. From the early part of the 18th until the end of the 19th century, the glacier advanced about 9 km. During the whole of this time, the snout

Fig. 8.58 The channel and banks of the Jökulsá on Breiðamerkursandur have been strengthened with boulders and two underwater levees that have been placed directly athwart the riverbed, 100 m to the north and south of the bridge. The levees reduce the effect of erosion on the riverbed and prevent icebergs from ploughing their way downriver. The icebergs are stranded above the northernmost levee and melt in the lagoon there, leaving only small bergs to float down to the estuary. If it ever seems likely that these defences will no longer suffice, a new road will be constructed across the filled-in lagoon about 400 m to the north of the present bridge. HB, 2008

lay right up to the glacial moraines where water gushed from it and spread out all over the plain in numerous streams that continually migrated between channels. By the end of the 19th century there were twelve named rivers running through Breiðamerkursandur. The largest was Jökulsá itself right through the centre of the plain and the most dangerous river in the country, a very trying experience for guides attempting to find a shallow, meandering ford across it while avoiding boulder-strewn beds, swift currents, floating icebergs, deep channels and quicksands. Whenever Jökulsá was completely impassable, its outlet was traversed over the glacier. This was always done when it was necessary to herd livestock between farms and areas. From about 1870 until the 1930s, there was a route marked by wooden posts across the terminus of Breiðamerkurjökull. The local farmers received a stipend from the government for maintaining this track and for providing guides across the glacier. The route was very circuitous due to crevasses, and steps and ledges had to be hacked out of the ice on steep climbs, ice ridges had to be reduced in size, and gaps bridged with planks of wood. Sometimes it was possible to cross the

Fig. 8.59 Jökulsá river on Breiðamerkursandur; drawing by August Mayer from 1836. NULI

river on the undersill, a low, smooth sheet of ice just above the river's outlet. This ice sill was formed during the winter at the front of the snout but began to break up once the flow of water increased with the spring and could thus be very treacherous.

Once Breiðamerkurjökull glacier began to retreat in the 1920s from the deep trench it had ploughed for itself while advancing in the Little Ice Age, rivers began to flow behind the foremost moraines that had been formed at the end of the 19th century, and a proglacial lake came into being right in front of the snout. Sediment deposits from beneath the glacier now settled mostly in this lake instead of being dispersed all over the plain as previously, and clear water now flowed through fixed and deep riverbeds all the way to the sea. Vegetation was thus left in peace to take root on the plain, but the rivers, on the other hand, became impassable for horses. Boats and rafts were thus resorted to, in order to ferry people and cargo across the main streams (Figs. 8.61 and 8.62) until a route across the plain for motorised vehicles was opened up in 1967 with a suspension bridge over the Jökulsá, thereby ending an important chapter in the history of transport and communications in the Öræfi district. But the powers of nature still had to be grappled with, only now in a different form. The fluvioglacial deposits from Breiðamerkurjökull now settled on the bed of the Jökulsárlón lagoon and no longer compensated for coastal erosion along the barrier beaches at the river's estuary. The sea has thus continually edged nearer and nearer to the bridge over the Jökulsá, threatening to cut through the ring road around Iceland within the next few decades.

Fig. 8.60 An illustration of the glacial outlet of the Jökulsá on Breiðamerkursandur. From Thienemann's (1824) travelogue of 1824. NULI

The Subglacial Terrain of Breiðamerkurjökull The western edge of Breiðamerkurjökull is marked by Breiðamerkurfjall, Öræfajökull and Hermannaskarð pass, while its northern boundary is defined by the Snæhetta summit and the Mávabyggðir and Esjufjöll mountains (Fig. 8.6). Veðurárdalsfjöll and Fellsfjall mark its eastern boundary (Fig. 8.57). Ice flows into the glacier through a pass from Breiðarbunga to the northeast and descends from a height of 1350–1800 m down to Breiðamerkursandur, which is believed to have been created by the constant dispersal of sediment loads since the end of the last glacial period. There is no greater snowfall anywhere on Vatnajökull than here, each year's precipitation equalling 3–7 m of water equivalent on the main ice cap. A large part of this precipitation settles as snow above an elevation of 1000 m in an accumulation zone which covers an area of 500 km². Ice creeps swiftly with much erosive power down a 410-km² ablation zone at an average speed of up to 1 m every 24 h. About

Fig. 8.61 Sheep ferried across Fallsárlón in a rowing boat on 14 September 1960. Hálfdan
Björnsson of Kvísker

Fig. 8.62 Hálfdán Björnsson, farmer at Kvísker, ferries a jeep across Jökulsá on
Breiðamerkursandur in 1965, two years before the river was bridged. Helgi Björnsson of Kvísker

10 m a year melts at the snout at a height of 50 m and a large volume of water is discharged from the glacier.

Breiðamerkurjökull (900 km^2) is divided into three main ice flows, which are divided by medial moraines from the Mávabyggðir and Esjufjöll mountains and the valley glaciers that descend from them. The western ice flow (160 km^2) descends for 18 km from Hermannaskarð and the steep northeastern flanks of Öræfajökull from its Þuríðartindur peak (a rhyolite pinnacle named after the farmer's wife at Skaftafell farm by the surveyor J.P. Koch in 1904). This flow passes Mikill (1472 m), Fjölsvinnsfjöll (1191 m) and Saumhögg before veering to the east along Breiðamerkurfjall and calving into Breiðárlón glacial lake (15 m above sea level). Its highland bed is incised by deep and narrow valleys. It has very little lowland and does not descend below a height of 100 m until nearly 4 km from the terminus. Three nunataks rear up from the ice: Kárasker furthest north, then Bræðrasker, and then 3 km to the south of it, Systrasker. Ice flows erode the nunataks on both sides, forming medial moraines on the glacier. The western ice flow is bounded in the north by Mávabyggðir and then Kárahryggur ridge (east of Fingurbjörg), which turns southeastwards and ends about 4 km to the north of Breiðárlón. Kárasker nunatak rears up from this ridge. The medial moraine from Mávabyggðir runs by Kárasker in the same direction down to the snout, joining another moraine from Bræðrasker further south.

Bræðrahryggur ridge extends from Kaflaklifur through the Bræðrasker nunatak and ends 5 km to the north of Breiðamerkurfjall. Between the Bræðrahryggur and Kárahryggur ridges is a narrow trench, and southeast from it the glacier has ploughed out a 7-km-long and 1-km-broad trench below sea level. Its bed begins to ascend again 2 km to the north of Breiðárlón.

Ice is borne from the western sides of Mávabyggðir and Hermannaskarð pass south to Mikill through a deep valley, between Systrasker and Bræðrahryggur ridge, which reaches down to a height of 100 m, where the ice is 650 m thick. Ice also flows from between Mikill and Þuríðartindur peak through a pass to the south of Systrasker to Fjölsvinnfjöll mountains. The glacier plunges from the cliff edge between Fjölsvinnsfjöll and Saumhögg and, with an icefall to the southeast of Saumhögg, it descends through a deep, narrow and craggy valley to the north of Breiðamerkurfjall, where the glacier is all of 550 m thick. With its tremendous abrasive force, the glacier has gouged out a channel below sea level which stretches down to Breiðárlón. A medial moraine, created from the rocky debris, which ice flows have eroded from the mountainsides of Systrasker and Fjölvinnsfjöll, is borne right into the lake.

The central ice flow (ca. 210 km^2) is about 26 km long. It reaches as far northwest as the ice divide with Skeiðarárjökull and as far north as Snæhetta summit, receiving ice from Fossdalur, while its eastern boundaries are marked by Skálabjörg and the Esjufjallarönd moraine and its western edge by the Mávabyggðarönd moraine. The glacier attains a thickness of 800 m midway between Mávabyggðir and Snæhetta above a trough 5 km long that reaches down to a height of 100 m. The ice flow then cascades from a 300-m-high edge to the northeast of Kárasker down to the lowlands, the ice being mostly 300–400 m thick at this point. At the foot of the slope, its bed extends into a depression that is 2 km long and below sea level. To the north of Kárasker, the ice covers mountains eroded by previous glaciers, but south from their foothills it covers a 10-km-long and 3-km-wide gravel plain that reaches to the end of the glacier's snout, though without reaching a height of 100 m.

The eastern ice flow (ca. 540 km^2) descends sharply from the Breiðabunga summit to the east of Esjufjöll mountains, all of 34 km onto the lowlands, where its snout breaks up in Jökulsárlón. Its boundaries to the west are marked by the medial moraines from the Austurbjörg and Skálabjörg cliffs, and to the east by the Veðurárdalur mountains. This ice flow is the largest of the three and its erosive force correspondingly greater than the others. On meeting Eyjólfsfjall, it descends steeply to below sea level before advancing along the remaining 20-km-long trough to the Jökulsárlón. The glacier is 800–900 m thick at the northernmost end of this trough and reaches an average depth of 175 m below sea level, though in certain places it is deepest at 300 m below sea level, just 2–3 km before it reaches the lagoon, where the ice is 300–400 m thick. The flow is at its narrowest just southwest of Innri-Veðurárdalur, or about 3.5 km broad, before it splays outwards again to a width of 7 km at its terminus. Although the ice flow heads southward into the lagoon, a part of it veers to the east toward the source of the river Stemma. Looking inland along the subglacial trough, a conical mountain 300 m above sea level and about 3 km to the south of Skálabjörg attracts attention, as it has resisted erosion while sediment deposits were being buried all around it. This nameless island in the trough is due northeast from a mountain spur that stretches to Kárahryggur ridge, and the glacier above it is 350 m thick.

Radio echo soundings cannot detect whether there is loose sediment layers or hard bedrock beneath the glacier, but it can be surmised that there are sediment layers, at least until the foot of the mountains. Refraction measurements on the Breiðamerkursandur plain near the glacier's snout, and reflection surveys in Jökulsárlón lagoon, reveal sediment strata on top of rock; these layers are about 40 m thick nearest to the sea, but 80 m thick nearer the glacier. To the south of the lagoon, however, the bedrock is at a depth of 130–140 m, and in the northern part of the lagoon it is even deeper, or 200 m, as it extends northwards beneath the glacier. It may well be that the deepest part of the trough reaches solid bedrock, but its sides are from unconsolidated sediments. The trough continues as a sediment filled glacial valley eroded into rock southeastward out into the Breiðamerkurdjúp area of the sea, almost all of 60 km to the edge of Iceland's basal shelf.

8.5.1 Glacial Advances and Land Destruction Since the Time of Settlement

During the settlement of Iceland (870–930 A.D.), Vatnajökull was much smaller than it is now. Valley glaciers might have then stretched from the Mávabyggðir and Esjufjöll mountains and Breiðabunga dome a short distance down onto the level, vegetated plains formed out of ground moraine from the end of the last glacial period 10,000 years ago. The lush prairie of Breiðamörk might have extended from the snout of today's glacier northward to the foot of the mountains. The gradient of the land would have been similar to what it is today around the troughs in the glacial bed. Breiðamörk would have then stretched directly east of Breiðamerkurfjall, about 5 km to the north of the present Breiðárlón, and extended to the east of Kárahryggur ridge for up 10 km further inland from today's terminus, and finally 15 km or even further north under the present eastern ice flow.

Breiðársandur (later Breiðamerkursandur) is mentioned in the *Book of Settlements*, and as Eggert Ólafsson and Bjarni Pálsson stated: 'Breiðamörk plain was from ancient times, from the age of settlement, and for a long time after that, a beautiful region with grassy fields, widespread copses, and many farmsteads' (Ólafsson and Pálsson 1981, vol. 2, p. 106). Þórður Illugi claimed land between the Kvíá and Jökulsá rivers and built and cultivated his farm at Fjall right by Breiðamerkurfjall around the year 900. The farmstead probably stood in front of the Bæjarsker cliffs. It is believed that Þórður's settlement claim included five more farms by the 14th century, one of them being Breiðá on the meadowlands to the southwest of Fjall and to the south of Breiðamerkurfjall. There had been a church there from fairly early on, Maríukirkja, and its wealth can be seen from the fact that its inventory of 1343 revealed other nearby farmlands as being in its possession, and that it had bequeathed the farm of Hólafjara to a Pastor Fjölsvinnur, after whom the Fjölsvinnsfjöll mountains on the glacier are now named.

There were flourishing settlements on the Breiðamörk prairie until it was covered with tephra ash from the eruption of Öræfajökull in 1362. By 1387 the church at Breiðá (later also called Breiðármörk) had neither livestock nor ecclesiastical ornaments (Þorkelsson 1921–23). It was still in use for the next 300 years, but always in a constant struggle with the inclement powers of nature: a worsening climate and the advance of Breiðamerkurjökull. From the late 14th until the early 20th century, there was an almost constant cold period, especially around 1600, now known as the Little Ice Age, which was probably the coldest period in Iceland since the end of the last glacial period. It is believed that around 1650 the firn lines on the southern flanks of Vatnajökull had been about 350 m lower than in the 11th century, or at a height of 750 m (Þórarinsson 1974a). Snow amassed on the glaciers and they began to creep forward, destroying both land and property. The ancient manorial home of Breiðá became a wretched hovel about which little is known, though one of the recorded farmers there was a Mikill Ísleifsson (1587), from whom the great pinnacle on the glacier draws its name. According to the records on abandoned farms in the district of Öræfi in 1712, Ísleifur Einarsson, the sheriff at Fell, stated that Breiðá had finally been abandoned in 1698: 'It has now been overcome by the glacier, its rocks and waters, though ruins of its foundations are still visible (Einarsson 1918b, p. 47).' Around that time, or in 1695, the farmstead at Fjall was also abandoned due to the glacier's advance, and just after 1700 it had been completely encircled by the glacier. The western ice flow of Breiðamerkurjökull and Fjallsjökull (which was then called Hrútárjökull) converged in front of Breiðamerkurfjall in the years 1700–1709. Nonetheless, sheep were still sent to graze on the mountain there in the summers, despite it being surrounded by glaciers (Jónsson 1914). The glaciers became separated once again in 1946, and an ice-free route opened up onto Breiðamerkurfjall. Jökulhlaups had previously flooded from a glacial lake by the mountain.

The large farmsteads of Fell, at the foot of Fellsfjall, and Brennhólar, near the river Brennhólakvísl, were at the easternmost edge of Breiðamerkursandur. Fell had 'a great deal of beautiful farmland, with two or three sublet farms within its boundaries ...' (Einarsson 1997, p. 137). The glacier and its rivers advanced on this small community. The Veðurá, Brennhólakvísl and Jökulsá rivers bore glaciofluvial sediments, which were dispersed and piled up in layers along the various migrating channels through the plain. There is nothing to indicate that the Jökulsá had flowed out of the lake in front of the glacier. According to Knoff's map of 1732, the river was about 9 km long (Nørlund 1944; Sigurðsson 1978), but when Eggert Ólafsson and Bjarni Pálsson crossed the plain in 1756 it was only around 7.5 km long and their descriptions indicate movement in the eastern part of the glacier. By then it had burst into long and narrow strips of ice running north to south, single seracs towering up in between them, and the easternmost margin of the glacier was a vertical wall 8–10 fathoms high. Water poured out from holes here and there all over the glacial wall, both high and low, each 1–2 feet in diameter. Less than 30 years later, in 1794, Sveinn Pálsson arrived to find the shattered eastern margins of the glacier advancing, and in the same year the farm at Brennhólar was abandoned. 'The glacier's terminus moved forwards about 400 m in one spell, from Whitsunday until into August, its edge swollen and bulging out like a turf wall just before it collapses because of water saturation. In some places the ice was so shredded it looked like filigree, in other places large icebergs were tumbling over each other. Small streams gushed out here and there from fissures and cracks' (Pálsson 1945, pp. 358, 478). After reading Sveinn Pálsson's descriptions, Sigurður Þórarinsson later came to the conclusion that Breiðamerkurjökull's snout had been within at least 7.5 km of the sea (Þórarinsson 1943).

The glacier's advance now began to accelerate. When Henderson crossed the Breiðamerkursandur in 1814, the eastern part of the glacier was again on the move, and he claimed that the source of the Jökulsá was no more than 2 km from the ocean. The route across the plain from the previous year was quickly disappearing under glacial ice, indeed the very tracks of their journey from just eight days previously had already been covered by ice. In front of the glacier was a mound of glacial till, gravel and grassy clods that it had piled up. Henderson's description of the power of the Jökulsá, which he had to cross, was precise and vivid, for the glacier was now so crevassed above it that the river was impassable along the undersill. The river was broad, ran in many channels, and had recently changed its watercourse. Henderson recollects riding across the river as follows: 'I had not gained one of the banks two minutes, when a huge piece of ice, at least thirty feet square, was carried past me with resistless force. The foaming of the flood, the crashing of the stones hurled against one another at the bottom, and the masses of ice which, arrested in their course by some large stones, caused the water to dash over them with fury, produced altogether an effect on the mind never to be obliterated' (Henderson 1818, pp. 246–247).

On Björn Gunnlaugsson's map of Iceland, the Jökulsá is about 2200 m long, this figure being based on the surveys of Scheel and Frisak in 1813 and Aschlund's in 1817 (Nørlund, 1944; Sigurðsson 1978). In 1820 the glacier surged forwards 1 km (on some days by up to 4–8 m), and much water poured out from it. Thienemann (1824) claimed that, according to the local population, the glacier had advanced in a similar fashion every fifth year. Over the next 15 years, the surge of the eastern ice flow increased even more, for in 1836 Gaimard estimated the distance between the source of the Jökulsá and its estuary to be no more than 400 m (Þórarinsson 1943). Þorsteinn Einarsson, the priest at Kálfafellsstaður, described the disturbances in the glacier and their effect on glacial rivers in the summer of 1855 as follows: 'If it then emerges [Jökulsá] further east or south, as e.g. this summer, it then gushes out of its usual channel and appears much more to the east, near Fell, where it is still flowing because of the current movements and advances of the glacier, but there are those who believe that it will soon flow back into its old channel' (Einarsson 1997, p. 132). The Swedish geologist Paijkull later stated (1866; 2014, p. 134) that four years previously, the glacier had advanced and ploughed deep into the plain, pushing up a high moraine ridge 30–40 feet high, before suddenly retreating again. He also mentioned reams of peat that the glacier had torn up and pushed before it. There appeared to be some abating in its advance, however, for Holland believed the glacier was as far from the sea as was presented on Gunnlaugsson's map, or about 2 km (Holland 1862). Nonetheless, it should be noted here that the shoreline always advanced at the same time as the glacier did, and this must always be taken into account when estimating the extent of the glacial surge by measuring the distance between the outlet of the Jökulsá and its estuary.

The farm of Fell was finally abandoned in 1869 when the glacier breached an ancient moraine that had protected the farmlands from time immemorial; it is possible that this moraine had been formed in the cold period about 2500 years ago (Thoroddsen 1892a, b, 1907–1911; Bárðarson 1934). The glacier bulldozed the moraine hills so vigorously that they were said to spill over each other like avalanches (Norðanfari, 1870; Thoroddsen 1907–1911). Indeed the glacier suddenly advanced that year a similar length as it had done in 1820 and almost reached the sea to the east of the Stemma river by the Fellsöldur hillocks, leaving a strip of land between them only 200 m broad. The rivers spilled over, covering with mud and stones the meadows of Fell, which in the mid-18th century had been considered the best hayfields in the Suðursveit district. The Brennhólakvísl and Veðurá rivers, which flowed from the glacial indent at Fellsfjall, had previously destroyed grassland when they had flooded the glacier's forefield (Einarsson 1997). The farm survived on an island for a while, but the river eventually razed that too. In 1873 Kålund believed that there were no more than 600 m between the sea and the outlet of the Jökulsá. Icebergs were borne along on swollen sheets of water, the brown-coloured crests of waves looking like rows of rocks from a distance. The river would also suddenly migrate between channels, bursting out of the glacier in new outlets, especially during jökulhlaups. There seemed to be no stopping the glacier's advance. Two years later, in 1875, it thrust forwards so quickly that the local farmers feared it would reach the sea and cut off the route over the Breiðamerkursandur, preventing farmers from the west from taking their wool produce to market at Djúpivogur on the east coast. The glacier ended in a high wall of piled up, shattered, icebergs of all shapes and sizes, ugly stacks of ice, or seracs, towering at its margins like the misshapen turrets and ramparts of giants' castles (Watts 1875, 1876). Nonetheless, the Norwegian geologist Amund Helland (1882) felt that the glacier was retreating by 1881, especially when the inhabitants of Reynivellir told him that the glacier had advanced even further six or seven years previously, had been higher and more uneven then, its surface studded with turrets and angular blocks of ice.

In 1891 Howell crossed the outwash plain before he climbed Öræfajökull, and maintained that there was a distance of 1600 m between the glacier and the sea (Howell 1892). A short time later, the glacier made one final attempt at reaching the ocean, and it is reckoned to have reached its furthest extent in 1894, stopping just 256 m short of the sea, its terminus 9 m above sea level (Thoroddsen 1907–1911). The last time there was any significant unrest in the area was in the years 1912 and 1919–20, according to the brothers at Kvísker farm, Flosi and Sigurður Björnsson, quoting their father Björn Pálsson as their source. In the summer of 1912, the Jökulsá changed its course, bursting out of the glacier in several different places and flowing east into the Stemma. Almost daily there were visible changes in the glacial snout and continuous rumbling noises could be heard from the glacier. As a typical example of the surge in 1912, Björn Pálsson of Kvísker recalls guiding men eastward across the plain and spending the night at Reynivellir. When he returned the following day, he found that the glacier to the east of the Jökulsá had already covered his previous tracks for at least 10 fathoms (ca. 20 m).

Breiðamerkurjökull glacier began to retreat in the beginning of the 20th century, slowly at first, but more rapidly after 1930, and it eventually became separated from Fjallsjökull in 1946, as mentioned above. Barren ground moraine, with no visible sign of the old farmsteads, first emerged from beneath the glacier. Later on, however, mounds of peat, the remains of centuries-old swards of grass, and branches of birch, began being borne into the light of day, and still are, all bearing witness to the vanished Breiðamörk and Breiðármörk, 'mörk' being an Old Icelandic word for woodlands. Of all the farms established within Þórður Illugi's original settlement, only one now remains: Kvísker.

Frantic competition with Breiðamerkurjökull

Some little difficulty was experienced in getting all into train, owing to the hurry all the famers of this locality were into get this year's wool to the store at Papós, which is situated four days' journey to the east; for tidings had been received that the ice of a portion of the Vatna Jökull, known as the Breiðamerkr [sic] had advanced to such an extent as to threaten the cutting off of all communication along the sea-shore, since the advance still continued. In consequence of this alarm every farmer was busy preparing the wool for market; steaming cauldrons were cleansing it from its grease, bands of sturdy Icelandic maidens were rinsing it in the clear water of the mountain streams–which are almost sure to be in close proximity to the farms in this part of the country–patches of white wool were drying upon the ground, while the male part of the community were measuring it in quaint wooden baskets, packing it into sacks, and forming bundles of equal weight to balance on each side of the pack-horses. It would be a very serious thing, indeed, if the road to Papós were to be intercepted, as it would compel the dwellers in this district to journey to Eyrarbakki before they could exchange their produce for the necessaries they require. Leaving Nupstað behind us, we set out for the advancing glacier, and turned our faces towards the snowy slopes of Öræfa. (Watts 1876, pp. 17–18)

Þorvaldur Thoroddsen's comments on Jökulsá on Breiðamerkursandur in 1894

Jökulsá is thought to be the most dreadful glacial river in Iceland. It has an evil reputation and a well-deserved one. The river was now completely impassable for both man and beast, indeed an ugly sight to behold; dark, muddy water spewing out from a gap beneath the glacier's edge, the water bubbling like one gigantic hot spring before bustling its way downriver to the sea in waves and waterfalls. Coal-black blocks of ice stand here and there in the shallows, while smaller chunks of ice swirl and skip along the crests of yellow, muddy waves. Where the river broadens it is possible to ford it on horseback, but it is always hard on the horses because of the cold and strong current. Everything has to go right if accidents are to be avoided. It must then often be crossed with innumerable zig-zags, sometimes with, sometimes against the current. An old man from Öræfi told me that once, when he was young, folk from Öræfi had come with their packhorses from Djúpivogur and had reached the river at such a bad time that they were struggling across it from early in the morning until late in the afternoon, and 17 loads had fallen into the water, though they were all eventually fished out of the river again except for three. A girl was swept off her mount and was rescued with great difficulty, and one of the horses was lost. Even the most experienced river guides have had more than enough of Jökulsá, and it is said that even the horses most familiar with it begin to shudder as they approach it. ... When the river is impassable, the glacier route is taken, though in fact glacier crossings have become more fashionable in recent years since the route over the glacier has been well maintained with planks bridging the larger crevasses. When conditions are at their best, it's possible to go the shortest route right over the front of the snout just above the river's outlet. It's called going on the undersill. Usually though, one has to go further up, scrambling over ridges and crevasses high up on the glacier; sometimes it happens that the crevasses are so many and so bad that men have to go far up onto the glacier and are scrambling over it for 5–6 hours or even longer. It has also happened that horses have fallen into

crevasses and have had to be put down right there and then. (Thoroddsen 1958–
1960, vol. 3, p.235)

Fatal accident on glacial route across Breiðamerkursandur in 1927
On 7 September 1927, Jón Pálsson, a teacher and the brother of Björn Pálsson
of Kvísker, died while crossing Breiðamerkurjökull. The river was impassable and so an attempt was made to cross it on the glacier above its outlet.
While waiting for a way to be cleared by pickaxe, Jón and the horses stood
about 20 fathoms from the glacier's edge. The men who were hacking at the
ice heard the glacier crack open very near to them. They were unhurt, but
when they looked around, Jón had disappeared along with all of the horses
except for two. They immediately hurried back to find the two horses
standing on an iceberg, crevasses on all sides; another horse was almost
underwater in a crevasse, crushed to death, while a fourth horse had fallen
into a crevasse and was also dead, with its hindquarters stuck between the ice
walls, its front half hanging in mid-air. Where Jón had been standing there
was a 20–30-m-deep chasm with water at the bottom (Valtýsson 1943–1949,
I, p.167). Three of the horses were rescued alive after a hard struggle, but the
man and the horse carrying the postbags had completely disappeared. Björn
Pálsson found the body of his brother in the glacier, along with the horse and
postbags, on 15 April the following spring.

8.5.2   The Creation of Proglacial Lakes and the Retreat
of the Glacier

Breiðárlón and Jökulsárlón were formed in front of the snout of Breiðamerkurjökull
when it retreated from a trench it had gouged out during the cold period after
Iceland was settled. Breiðárlón appeared early in the 1920s and the Breiðá river, 6–
7 km to the west of the Jökulsá river, flowed from it to the sea, until it changed
direction westward and flowed into the Fjallsárlón lake in 1954. Since then the
Fjallsá has borne all the water from the western ice flow of Breiðamerkurjökull to
the ocean through the Breiðá estuary. The Jökulsárlón lagoon, at the outlet of the
Jökulsá from the glacier, first became known in 1934, when there was about 1 km
from the glacier's terminus to the sea. The lagoon grew larger year by year because
more ice broke off from the snout and melted than had advanced with the glacier;
slowly but steadily the lagoon grew during the first half of the 20th century, but it
then suddenly increased in size just before 1960, and by 1975 it covered an area of
8 km^2 and had grown a further 2 km^2 by 1990. In 1992 Stemmulón (2 km^2)
became connected with Jökulsárlón, and since then the meltwater from all of
Breiðamerkurjökull's central and eastern ice flows has accumulated in Jökulsárlón.
The Vestari-Stemma, Eystri-Stemma, Brennhólakvísl and Veðurá rivers all flow

into Stemmulón, and water that floods annually into Stemmulón from the glacially dammed lake in Innri-Veðurárdalur valley also reaches the Jökulsárlón.

After 1990 the calving of icebergs into the Jökulsárlón lagoon speeded up and typically ice has broken off the terminus at the rate of about 600 m year^{-1}. At the same time the downglacier flow of ice compensated for almost half of this, so the net retreat of the glacier's margin has been just over 300 m year^{-1}, expanding the area of the lake about 0.5 km^2 a year on average (Björnsson et al. 2001); altogether 260,000,000 m^3 of ice broke off into the lagoon every year (260 × 10^6 m^3 year^{-1}) and all the icebergs melt in the lagoon. This requires energy equal to about 2500 MW; up to half of this, even two thirds, could be borne by the incoming tide when 10 °C warm sea water flows up the Jökulsá river into the lake at a typical rate of 500 m^3 s^{-1}; at low tides 600 m^3 s^{-1} of some 2 °C flow from the lake back to the sea. Moreover, ice melts due to the heat from the atmosphere and solar radiation. Since the mid-1990s, Jökulsárlón has grown rapidly, about 0.5 km^2 a year, and about 0.3 km^3 of ice has melted annually. In 2015 the lagoon covered an area of 25 km^2, and the glacier's calving front was about 8 km from the sea, where the lagoon's bed is at a depth of almost 300 m below sea level (Figs. 8.55, 8.56 and 8.63). Jökulsárlón will continue to grow in size over the next few decades as calving will increase in an unstable manner through the over-deepening at the base of Breiðamerkurjökull, its glacial bed being at an incline at the north of the lagoon. It has been calculated that Breiðamerkurjökull's terminus at Jökulsárlón will con-

Fig. 8.63 Breiðamerkurjökull calving into Jökulsarlón. HB, November 1998

tinue to retreat at the same rate as it did at the end of the 20th century if the glacier's mass balance remains the same as it has been over the last few decades. This means that Breiðamerkurjökull would have withdrawn completely from a 25-km-long Jökulsárlón by the end of the 21st century, and the whole of the glacier on the northern slopes of the Esjufjöll mountains would have disappeared by the end of the century after that (see also Nick et al. 2007).

The formation of Jökulsárlón was initiated by atmospheric climate warming in the 1930s. It then later grew larger due to thermal interaction with the ocean, and this was accelerated even more by the calving of Breiðamerkujökull into the lake, which makes it dynamically unstable because of the over-deepened basin of the glacier's bed. Jökulsárlón would thus be an interesting laboratory for studying glacier-ocean interaction as this is of vital importance in addressing the world-wide problem of understanding the contribution of glacial calving in Antarctica and Greenland to rising global sea levels.

8.5.3   Coastal Erosion Near Jökulsárlón Threatens the National Highway

After the emergence of the Jökulsárlón, sediment deposits from the glacier have mostly settled in the lagoon and do not compensate for the coastal erosion at the Jökulsá estuary. The river's channel has been shortened in the last decade by about 5–8 m a year, and thus within a few decades the national highway across the plain might be cut in two. In the autumn of 2004, the outer barrier beach was only 250 m from the bridge over the river, and about 1 km further east the highway was only 70 m from the shoreline, while at another point pylons carrying electricity lines were only 26 m from the sea. Land uplift due to the shrinking of Vatnajökull provides some compensation for coastal erosion. At the southern edge of the glacier, land is now rising at about 2 cm a year, and this rate will probably double by the middle of the 21st century, so that by 2100 the barrier beach might be 4 m higher than it is now and the speed of erosion cut by half. Engineers of the IRCA believe that the highway can be protected for much of this century. The channel and banks of the Jökulsá on Breiðamerkursandur have been strengthened with boulders and two underwater levees that have been placed directly athwart the riverbed, 100 m to the north and south of the bridge. The levees reduce the effect of erosion on the riverbed and prevent icebergs from ploughing their way downriver. The icebergs are stranded above the upper threshold and melt in the lagoon there, leaving only small bergs to float down to the estuary. If it ever seems likely that these defences will no longer suffice, a new road will be constructed across the filled-in lagoon about 400 m to the north of the present bridge.

8.6   Glaciers in Mýrar and Hornafjörður

A traveller driving speedily from Jökulsárlón lagoon eastwards into Hornajörður will probably experience the Mýrar district as a quiet and remote place which rarely makes the news. Austur-Skaftafell County is a sparsely populated region, its farms standing out singly or in small hamlets on the sea-level lowlands through which the national highway meanders gently, every now and then crossing a bridge over a glacial river flowing across the broad outwash plains. This country district's setting is very unusual and has perhaps no equivalent anywhere else in Iceland or perhaps even the world. Along this 30-km section there is only a short distance to the sea as one outlet glacier after another plunges down from Vatnajökull and sprawls out as it reaches the lowland (Figs. 8.64, 8.65, 8.66, 8.67, 8.68 and 8.69): Skálafellsjökull, Heinabergsjökull, Fláajökull, Viðborðsjökull, Hoffellsjökull and finally Lambatungnajökull. It is an incredibly short journey from this green countryside up onto Vatnajökull's main ice dome, a mere 15 km. In the Suðursveit district the lowland is just a narrow strip of land, though it broadens the further east one goes.

Nowhere else in Iceland do glaciers creep down into inhabited areas as they do in the Mýrar and Nes districts of Hornafjörður, and modern tourists find the views here enchanting as the powerful glacier infiltrates down mountain passes and valleys onto the green and lush meadows and countryside of the local farms and communities. Vatnajökull towers above and beyond these districts with its row of steep outlet glaciers between barren mountain ridges, a 30-km expanse between

Fig. 8.64 View from a cliff edge over the snout of Brókarjökull and along Kálfafellsdalur to the Steinasandur plain. HB, May 1995

Fig. 8.65 Brókarjökull plunges into the head of Kálfafellsdalur, though mountains otherwise cut off other glaciers from the lowlands of the Suðursveit district. HB

Fig. 8.66 The Flatey farmland in the Mýrar district. Late in the summer, innumerable bags of hay lie on the home field, while the glacier and its outlets are clearly visible in the distance. HB, 2007

Skálafell in the west to Hoffell in the east. Its highest points are covered with clear white snow, then come the grey valley glaciers as they spread down and outwards onto the lowlands, and at the very bottom are outwash plains, glacial rivers, and green bogs, all just a short distance from the sea.

It must be assumed that the Mýrar and Nes districts had been completely covered by vegetation, and had woodlands and wetlands when the settler Hrollaugur Rögnvaldsson first made his land claim in that area. It would have been dry up by the mountain range, but boggier nearer to the ocean. Outlet glaciers did not then encroach onto the lowland and harmless rivers flowed through well-defined channels.

At the beginning of the Little Ice Age, around 1400, snow began to accumulate on Breiðabunga year after year. The glacier grew in height and eventually one glacial tongue after another began to reach beyond the mountain passes and creep over vegetated land into the valleys, and then finally down onto the lowland. Rivers that had once meandered with clear water in fixed channels, became dark and muddy and began to spill over their banks, dispersing layers of gravel and stone beneath them. Cowherds and their cattle became cut off in meadows, outlying sheepcotes became isolated, sometimes for days, and some farms had to be

Fig. 8.67 The Mýrar district, Fláajökull in the background, the farm Lambleiksstaðir centre foreground. HB, August 1994

Fig. 8.68 The vegetated area of Mýrar in front of Skálafellsjökull, Heinabergsjökull and Fláajökull, which have ploughed out their beds below sea level. Heinabergsjökull and Skálafellsjökull formerly coalesced below Hafrafell around 1900. HB, September 2007

Fig. 8.69 Aerial view from above Stokksnes, westward over Höfn to Mýrar. Fláajökull, Heinabergsjökull and Skálafellsjökull glaciers are all visible, as is Öræfajökull in the far distance. In the foreground is Skarðsfjörður within the Austurfjörur barrier beaches. Further west is the Hornafjörður estuary, and beyond that the Suðurfjörur barrier beaches. HB, September 2007

abandoned as they could no longer harvest enough hay to survive, their fields cut off by impassable glacial rivers.

The glaciers grew in size, becoming thicker and longer, and sometimes dammed tributary valleys, preventing waterways from flowing onwards. This happened in Vatnsdalur and Heinabergsdalur and in many of the gullies in the Hoffellsfjöll mountains. This damming formed marginal lakes and when they overflowed they caused jökulhlaups, which flooded the local farms and their land.

In the 18th and 19th centuries, life in the Mýrar district was a constant battle with glacial rivers, with man the frequent loser in such a competition. A century ago, travellers would have found these rivers a very different phenomena than today, for then they became swollen, frequently migrating from channel to channel, either eroding away arable land or covering it with glacial till and rocks. Drifting sand and gravel from the outwash plains also damaged farmland. Jökulhlaups became an annual event, during which farmsteads with their sheepcotes, barns and stables were turned into little islands surrounded by water. Farms which had stood on the same spot for centuries were eventually abandoned or else moved elsewhere to avoid the encroaching waters. This was a district with much poverty; its farm buildings were poorly constructed, there was little driftwood, and the turf was barely usable as the peat was sandy. There was no upland grazing for sheep because glaciers now filled in the valleys above the farmsteads. The farms of the Nes district at least had some perquisites from trout and seal hunting, for in earlier centuries there were fishing outposts at Hálsaós estuary and on the isle of Skinney. The terrain gradually began to sink under the increasing overburden of the glaciers, wetlands became more widespread along the lowlands, and harbours on coastal skerries became silted up and impaired. As a result of this, itinerant fishermen from the north had ceased crossing the glacier for the fishing season in Suðursveit by the beginning of the 16th century. The expansion of the glaciers had blocked mountain trails, hampered journeys between different parts of the country, and prevented the upland grazing of sheep in the highlands. When glaciers had reached their fullest extent, near the end of the 19th century, they were far more visible than the mountain edges between them.

After an almost continuous cold period of five to six centuries, the climate began to become warmer just before the end of the 19th century. The glaciers retreated, marginal lakes formed at their snouts, and their river outlets harrowed out more durable channels that were hemmed in by defensive levees. Old and dry water-courses were visible and widespread. But where there had once been only gravel and sand in the middle of the 20th century, there are now fertile bog lands. Soil reclamation programmes have achieved a great deal in a very short time.

When Eggert Ólafsson and Bjarni Pálsson visited this area in the middle of the 18th century, and Sveinn Pálsson at the end it, the local grasslands had been badly damaged by the encroachment of the glacial rivers and the results of dust storms, especially on the Heinabergssandur and Hólasandur outwash plains (Pálsson 1945, p. 284).

From the *Book of Settlements*

'There was a man called Ketill to whom Hrollaugur sold land from the shores of Hornafjörður to Hamrar; he lived at Meðalfell and the people of Hornafjörður are descended from him. Auðunn the Red bought land from Hrollaugur from Hamrar to the further side of Viðborð; he lived at Hofsfell and raised a great temple there; all the Hofsfell folk are descended from him.

Þorsteinn the cross-eyed bought land from Hrollaugur from Viðborð along the south through Mýrar to the river Heinbergsá. His son was Vestmar, and the folk of Mýrar are descended from him.

Úlfur of Vors bought land from Hrollaugur south of Heinabergsá to the Hreggsgerðismúli promontory and lived at Skálafell.

Þórður Illugi, the son of Eyvindur Ship-Hook was shipwrecked on Breiðársandur; Hrollaugur gave him land between the Jökulsá and Kvíá rivers and he lived below Mt. Fell by the Breiðá river.' (Benediktsson 1986, 1(2), pp. 319–20)

By this time, Hrollaugur had no further land left except from Hreggsgerðismúli in the east to Jökulsá in the west.

Account from mid-18th century

Eggert Ólafsson and Bjarni Pálsson describe how Heinabergsjökull had destroyed all of the central part of Hornafjörður in its advance. 'A section of it stretched right down to the level lowland of the farm Heinaberg, which is still inhabited. The Kolgríma, Heinabergsá and Hólmsá rivers … have totally destroyed all the topsoil along their banks and are difficult to cross because of quicksands. Two small communities have survived here near the sea, nonetheless: Mýrar and Nes. There has been a lot of unrest in the glacier here, and jökulhlaups from it have flooded the local settlements …' (Ólafsson and Pálsson, 1981, vol. 2, p. 119). (Fig. 8.70).

8.6.1 Ancient Route Over Vatnajökull from Suðursveit

An ancient route north over the glacier began from Staðardalur in Suðursveit, to the east of the rectory at Kálfafellsstaður, inland to Skálafellsjökull to the east of Hálsatindur peak (905 m) in the area near to where the Jöklasel tourist centre is

Fig. 8.70 The farmstead of Dagbjartur Eyjólfsson at Heinaberg was built in 1890 and was abandoned in 1934. Daniel Bruun took the photograph in 1902. Reg. Arch. Skagf

today. Itinerant fishermen from northern Iceland descended from the glacier here on their way to the seasonal fishing station at Hálsasker near the estuary of the river Kolríma. They had crossed over from Brúarjökull, and in the Norðlingalægð depression they had then taken their bearings from the peaks to the north of the Heinabergsfjöll mountains (Fig. 8.71).

In the summer of 1932, six students from the University of Cambridge aimed at exploring the wilderness to the north of Vatnajökull and to measure the thickness of the glacier using seismic echo soundings, which were then a recent innovation in glaciological research. Two men from Austur-Skaftafell County, Þorbergur Þorleifsson from Hólar and Skarphéðinn Gíslason from Vagnstaðir, accompanied them

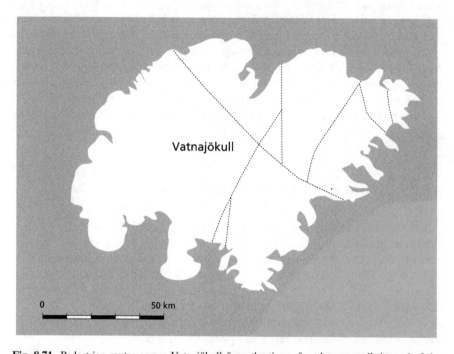

Fig. 8.71 Pedestrian routes across Vatnajökull from the time of settlement until the end of the 16th century. In previous centuries, routes across Vatnajökull were used for travelling between different parts of the country. There are sources for transglacial connections between Skaftafell in the Öræfi district in the south and Möðrudalur in the northeast and that migratory workers from northern Iceland crossed the glacier for the fishing seasons in the Suðursveit district in the 15th and 16th centuries. Men from Fljótsdalur in the northeast would approach Brúarjökull to the south of Snæfell and ascend it in a glacial indent to the west of Eyjarbakkajökull before traversing the Breiðabunga dome and then descending via Skálafellsjökull. Men coming from the Brúaröræfi wilderness would ascend Brúarjökull between the outlets of the Kverká and Kringilsá rivers. When they reached the centre of Brúarjökull, the routes diverged. Those heading for Skaftafell continued southwestward over the ice cap, while those going to Suðursveit crossed the Norðlingalægð depression and then descended southeastward down Skálafellsjökull. Men coming from Þingeyar County first crossed the Ódáðahraun lava field of central Iceland before ascending Dyngjujökull at Kistufell

to the glacier. They travelled the old fishermen's route into Kálfafellsdalur, north to the eastern Kverkfjöll mountains, and the same way back. They related that they had reached their highest point (1320 m) to the north of Heinabergsjökull before gradually descending northward. When it came to using the equipment for measuring the thickness of the ice, it malfunctioned, and their luggage, at 100 kg per capita, was far too heavy for a long journey across Vatnajökull. In a fine book on their expedition, they described the thick layers of slush and tephra and piles of sediment on Brúarjökull, and ice-dammed lakes, jökulhlaups and hot springs at Kverkfjöll mountains. They named a glacially-dammed lake just to the east of Kverkfjöll Þorbergsvatn, and they called the glacier directly to the west of it, by the eastern Kverkfjöll mountains, Skarphéðinsjökull. The lake Tómasarvatn, which Max Trautz had described in 1910, had disappeared, although terrace marks indicated that its water level had reached 70 m above its bed. The snout of Brúarjökull had become lower in height rather than having retreated. They returned from the glacier after a difficult sojourn ten days earlier than planned. Nonetheless, Skarphéðinn Gíslason was there waiting for them with the explanation that he had dreamt they would come on that day and at exactly the same place where they had started out from, and not descend from Öræfajökull, as they had planned, about 100 km further west. Such was the prescience of Gíslason of Vagnstaðir (Roberts 1933).

8.6.2 Glaciers to the East of Breiðabunga Dome

All the outlet glaciers of the Mýrar and other districts of Hornafjörður have their origins at Breiðabunga (1520 m), and descend for 20 km on the way east from its dome to the lowlands to a height of about 50 m above sea level. Breiðabunga is at its highest and broadest on its southern side at the beginning of Skálafellsjökull and Heinabergsjökull. It lies in a north-south direction and is supported by a mountain range north of the Kálfafellsfjöll and Heinabergsfjöll mountains, although separated from them by deep subglacial valleys. The range rises to a height of over 1200 m, but is incised by passes that reach as low as 850 m (Fig. 8.6).

To the northeast of the main range, the land descends onto a plateau 800–900 m high, though with a few peaks rising over 1000 m. The terrain begins to rise again as it continues into Sandmerkisheiði. Over this high plateau a narrow outcrop of Breiðabunga manages to bridge the gap between the highest mountains, and from it descends Fláajökull. On the other side of this ridge from Sandmerkisheiði there is another highland plateau until the terrain rises once more at Goðahryggur ridge, and from this bowl comes Hoffellsjökull.

On the high, undulating ice sheet of the glacier, low domes and broad hollows 300–400 m thick hide most of the rugged and wild terrain beneath them. Continuous ice flows creep forward and collect in the bowl-shaped hollows before plunging off the edges of the highland rim. Mountain ridges divide them up as they slink down passes between them, becoming individual ice flows as they descend through gullies and valleys. These flows are also broken up by nunataks, stretches of ice flowing on both sides of them before converging again below them. The medial moraines created by these nunataks mark the divisions between ice flows on their journey down to the glacial snouts. The glaciers tumble in thin sheets of ice from vertical cliffs, but when they meet hard rocky obstacles, they plough their beds deeper around them, reaching below sea level at their snouts, many of them ending in proglacial lakes and lagoons.

Skálafellsjökull descends eastwards through a valley that separates Kálfafellsfjöll mountains and the range beneath the southern part of Breiðabunga. When the ice flow meets Hafrafell it ploughs out a trench to below a height of 250 m and beneath its snout is a vale 3 km long below sea level. Heinabergsjökull divides into three tongues as it plunges from the sheer cliff edges and is split up by the nunataks of Snjófjall and Litlafell. As can be seen from the medial moraines extending from Snjófjall, most of the ice flows along the easternmost tongue, where the hemmed in glacier has gouged itself down to below sea level along a 6-km tract reaching a fair way back towards Snjófjall, about 2 km from where it meets Vatnsdalur. The valley bed beneath Fláajökull is below an altitude of 100 m along the whole of Fláfjall, and a vale 4-km long is actually below sea level, though not at the end of the snout. When the glaciers were at their furthest extent in 1880, the termini of Skálafellsjökull and Heinabergsjökull converged, but there was a glacier-free area about 1 km broad east towards Fláajökull. When Iceland was first settled, the glaciers might have stretched from Breiðabunga and Goðabunga down to the vegetated moors and mountain valleys at a height of 800 m. Old stories can still be heard about how it had once been possible to ride inland along the western foothills of Fláfjall where Fláajökull is today, for long ago there used to be vegetated lowland between Hafrafell and Geitakinn (Kristjánsdóttir 2000, p. 89).

The firn line (snowline on a glacier) was believed to have been at an elevation of about 1100 m on the southern flanks of Vatnajökull when Iceland was first settled, but by around the year 1500 it had fallen to below 800 m and subsequently remained at 700 m until the end of the 19th century. The glaciers on the highlands above Mýrar and Nes in Hornafjörður began to expand, for 70–80 % of this terrain was above 800 m in height. It is also clear, furthermore, that should these glaciers disappear, only a small glacier could survive on the mountain range beneath Breiðabunga in the current climate.

8.6.3 Mýrar Rivers Threatened Settlements

Mýrar was for a long time an isolated district because glacial rivers were great obstacles to any travelling. The river Kolgríma descended from Skálafellsjökull in the west to the sea at Hálsaós estuary, forming the western boundary line with Suðursveit district, while the Hornafjarðarfljót marked its eastern boundary with Nes. The Mýravötn, Heinabergsvötn, Landvatn, Hólmsá and Djúpá rivers, however, threatened the very settlements themselves, the jökulhlaups from Heinabergsjökull proving the most dangerous as they seemed likely to turn the district into a wasteland.

Heinabergsjökull dams Vatnsdalur further inland along the Heinabergsfjöll mountains, causing water to accumulate in the valley and there is an annual discharge that rushes for 8 km beneath the glacier and into the river Heinabergsvötn. Most of these outburst floods increase in size gradually over a week and then come to a sudden end. The Heinabergsvötn flowed south through Mýrar and into the Hólmsá until the mid-20th century, and Vatnsdalur jökulhlaups inundated a large part of the Mýrar lowlands, tearing up land so that farmsteads had to be relocated (Figs. 8.72, 8.73 and 8.74).

The first Vatnsdalur outburst flood of later times came very speedily, taking everyone by surprise, in November 1898 (Þórarinsson 1946). Its flow rate reached its high point (3000 m^3/s) within the first 24 h. Within two days, about 140,000,000 m^3 of water had been disgorged from the valley, its maximum flow rate similar to that of the jökulhlaup from Grímsvötn in 1965 (Kristjánsdóttir 1988, 1993, 2000, p. 84). Up until this flood in 1898, the ice dam of Heinabergsjökull had been so thick that Vatnsdalur was fully contained (with a surface area of about 2 km^2), and there was just a continuous, steady stream over the bedrock threshold (Vatnsdalsheiði) down into Heinabergsdalur.

As the 20th century advanced, the jökulhlaups decreased in size because Heinabergsjökull was thinning and the water level was not reaching as high as previously in Vatnsdalur. For a long time there was an annual flood from the valley, and as time passed, it came earlier and earlier in the summer. After 1940, it began to flood twice a year, a change again caused by the thinning of the glacial dam. After 1948 the Heinabergsvötn streams began to flow westwards into the river Kolgríma as the glacial snout retreated, and they have continued to do so ever since. A new bridge had then just been constructed over Heinabergsvötn, and it still spans a dry channel. The bridge over the Kolgríma from 1935, however, could not stand up to outburst floods, so that the national highway was disrupted for a few days every time there was a jökulhlaup. A new bridge spanned the Kolgríma in 1977, and since then floods from Vatnsdalur have not caused any damage.

By the mid-18th century, the snout of Heinabergsjökull advanced to block the mouth of Heinabergsdalur and thus dam its outflow of water, forming the so-called Dalvatn Lake. There were outburst floods from this lake into the Heinabergsvötn until 1920, after which they ceased because of the retreat of the glacial terminus. Jökulhlaups from Dalvatn came without warning because the water would suddenly burst its way through the glacial snout, although the floods never lasted for long. From the shoreline marks on the valley sides it can be clearly seen that the water level reached its highest at 140 m above sea level and that the lake was about 50 m deep with a volume of, at most, 19,000,000 m^3.

Although there are no records of any jökulhlaup from Vatnsdalur before 1898, it is almost certain that outburst floods have come from there in previous centuries once Heinabergsjökull had advanced in front of the valley entrance. As the glacier

Fig. 8.72 Heinabergsjökull descends between Heinabergsfjall and Hafrafell onto the Mýrar lowlands. Blocks of ice are scattered around the bottom of Lake Vatnsdalslón (on the *left*) which has recently been emptied by a jökulhlaup. On the *right* is Skálafellsjökull. HB, August 1998

Fig. 8.73 a Blocks of ice cover the bed of Vatnsdalslón after the jökulhlaup of 1938. Meingilstindur peak is at the end of Vatnsdalur. Sigurður Þórarinsson. **b** The bridge over the Kolgríma awash during the jökulhlaup from Vatnsdalur on 23 September 1957. The road was torn apart every time there was an outburst flood from the valley until a new bridge was built in 1977. Ingólfur Ísólfsson

Fig. 8.74 At the mouth of Heinabergsdalur, shoreline marks are still visible from the water levels of Dalvatn, which was formed when Heinabergsjökull dammed the valley. The lake disappeared after the glacial dam was breached in 1920. Þóra Ellen Þórhallsdóttir, 2007

thickened, the floods became larger until there was a constant flow of water from Vatnsdalur over into Heinabergsdalur. Vatnsdalur floods have played a major role in gouging the bed of Heinabergsjökull to below sea level.

The battle with the river Hólmsá Once Fláajökull (Fig. 8.75) began to retreat after 1930, the Hólmsá began to flow eastward along the glacier (instead of southward) and into Hleypilækur brook and thence into the Djúpá and east through the meadows alongside Hornafjörður, threatening the fields and pastures of 20 farms as well as the national highway. Early in 1937, while there was still little water in the river, an effort was begun to divert the current back into its old channel southward. This was done by the manual digging of a canal westward through the glacial moraines, while also aiming at damming its route eastward. The Icelandic Road Administration oversaw the project, hiring 50 men and 15 horses for two months' labour. Tents, a field kitchen, and a stable with drystone walls and mangers were raised, remnants of which are still visible today. Hay and other fodder was transported to the site, as indeed was freshwater at first until the stables were connected to flowing water through a hosepipe. Gravel for the dam was carried in horse-drawn wagons and unloaded into the channel, but as it became narrower the current became stronger and began to flush away the loads of landfill. Nor did bags of gravel suffice, and so finally, as a last resort, the gravel wagons themselves, loaded with rocks, were dumped in the channel. The current could not compete with

Fig. 8.75 A very ragged-looking Fláajökull descends onto the lowlands of Mýrar, the river Hólmsá (on the *right*) flowing directly from it to the sea, while the Djúpá (on the *left*) runs into the Hornafjarðarfljót. Defensive levees ahead of the glacier keep the rivers in check. HB, August 2003

them, so at last, after 48 h of constant struggle, the river channel was dammed. Altogether 4000 m³ of earth and gravel were used in the dam (Kristjánsdóttir 2000, p. 89).

Fláajökull continued to retreat, however, and its waters continued to find a way east along its margins. In 1946, and again in 1958–1960, levees were constructed at the edge of the glacier. At first a lorry was used to bring the levee materials, but they had to be shovelled onto the construction by hand. The arrival of bulldozers in the middle of the 20th century made building levees for controlling glacial rivers much easier. Finally, near the end of the last century, it became clear that there was no guarantee that any lengthening of these levees up at the glacial margins would prevent the river from flowing east along the margins, but by then a protective levee 1700 m long had already been raised all along the glacier's snout, though at some distance from it.

Turmoil in Hornafjörður

'At that moment, we were told to look at the glacier. An unusually loud noise of water, which we'd heard previously, now turned into the roar of an approaching storm. And what we now saw was one of the most unexpected and extraordinary sights I have ever seen. Up by the glacier there was one continuous, billowing, broad expanse of water that rushed down all the rivers and waterways of Mýrar, to the east of Lambableiksstaðir [Lambleiksstaðir] and as far west as the eye could see. A sheer wall of water was visible, surging onwards at frightening speed.' Gunnar Benediktsson's description of the jökulhlaup from Vatnsdalur valley in 1898. (Benediktsson 1944, pp. 184–85)

Abandoned farms in the parish of Einholt

All these farmsteads have been destroyed by encroaching water from Heinabergsjökull and Fláfallsjökull. ... The land was sodden from the glaciers, and a huge flood this summer bore icebergs a long way toward the sea, such large ones that you had to be on horseback to see over them. Most of these farms were destroyed almost a generation ago, not because they were nearer the glacier than other farms, but because the jökulhlaups (or rain and meltwater floods?) from the glaciers had deposited clay and gravel on the riverbeds so that they flooded the nearby fields ...

In those years, the glacial lakes were very full. The glaciers' snouts reached the furthest moraines, right up to the grass-covered level ground, and glacial tongues bulged out so high between the mountains that the latter seemed small in comparison. But in the year of Our Lord 1898, the glaciers had begun to shrink a little in height and to withdraw. The Heinabergsvötn streams were then very full of water, like all the glacial streams, and trundled sediments eastwards in front of Heinaberg and eastward into the old channel of the river Hólmsá and then then joined the Hólmsá itself east of Mýrar. ...

When I arrived above the home field, I saw that water was flowing on the surface of the ice. ... The folk then living at Einholt said that the water had come like a breaking wave and this had probably been caused by its being held up by the blocks of ice it had piled up ahead of it; the water then rose swiftly and so burst through the ice jam flushing the blocks further downstream, the process continually repeating itself. Folk said they heard a tremendous rumbling, crashing and breaking sound as the icebergs collided in the current and shattered each other. This is what the folk at home saw and heard, but I didn't have the time for such observations.

The glacial ice in the area of the outburst flood contained so many hidden gaps and chasms that it was cracked and unstable the whole winter. (Benediktsson 2000, pp. 67–70).

The farmer from Heinaberg

His name was Dagbjartur, and he asked us if we could give him a short lift. Poverty had stamped his features, they had hardened like a mask and in his eyes happiness and the joy of life had faded out long ago. He had once owned a farm of his own, but every year the river had taken away a bit more of his pastures, and finally had invaded even his *tun*. Now there were only a couple of stone walls left, and he himself lived at the farm from which he had just come and where he ate the bread of charity. (Ahlmann 1938, p.167; Fig. 8.70)

On 18 September 1919, Dagbjartur Eyjólfsson, a farmer at Heinaberg, accompanied the Swedes Hakon Wadell and Erik Ygberg back up onto Heinabergsjökull glacier to fetch their sledge and horses, which they had been forced to abandon when they got into an impasse in an area of crevasses on their way back from Grímsvötn. A sudden storm swept down on the three of them and raged for 48 h. The Swedes barely survived this in a tent, before stumbling down off the glacier, having lost all their baggage and animals. Two of Wadell and Ygberg's horses made it down from the glacier by themselves, but the other two and the expedition's dog perished. (Wadell 1920a, b)

8.6.4   Hoffellsjökull and the Hornafjarðarfljót River

Viðborðsjökull, Svínafellsjökull and Hoffellsjökull are the most northerly outlet glaciers that extend eastwards from the high dome of Breiðabunga on Vatnajökull (Figs. 8.76, 8.77, 8.78, 8.79 and 8.80). They are bounded by Sandmerkisheiði to

Fig. 8.76   The foremost snout of Hoffellsjökull ends to the east of Öldutangi. The glacial bed is up to 300 m below sea level, stretching back 10 km inland from the terminus. The Austurfljót used to flow through the furthermost terminal moraines, but since 2006 this channel has been dry and all the water has run through a pass at the northernmost point of Öldutangi, and flowed from there into the Suðurfljót. HB, August 2003

Fig. 8.77 Hoffellsjökull creeps down from an 800 to 900 m highland plain and divides at Nýjunúpar before converging once again at a bedrock threshold between Gæsaheiði and Múli, from whence it descends a sheer cliff and ploughs itself a bed below sea level. Snæfell is in the background, and further east are Goðahryggur ridge and Goðabunga dome. Mt. Herðubreið towers in the far distance (*top left*). HB, August 2003

the west and then the ice divide from Breiðabunga that stretches northeast to Goðahnúkur and then southeast along the Goðahryggur ridge. Viðborðsjökull descends sharply between Viðborðsháls saddle and Sandmerkisheiði, though only for a short distance. The Hofellsjökull and Svínafellsjökull glaciers extend from the high, undulating ice sheet down into a bowl-shaped area and then along a narrow valley before descending through a barely passable icefall between Múli, the westernmost spur of Hoffellsnúpur, and Gæsaheiði. Hoffellsjökull descends 330 m for 4 km through a gap 2 km broad. A little above the pass, the Nýjunúpar nunataks divide the ice flow, which then converges again below them, bearing a medial moraine from the nunataks which demarcates the ice flows all the way down to the snout. The moraine now ends at the glacial margin at Gjátangi, but it used to extend all the way down the centre of the tongue. The ice flow from the eastern part of the bowl extends between Múli and Nýjunúpar, seeking its ice from the Djöflaskarð pass and from as far north as Goðahnúkar, but the larger western ice flow transports ice from Breiðabunga and the ice divide in the centre of Eyjabakkajökull. The glacier breaks free from Gæsaheiði's restrictions to the south of the narrowing gaps, opposite Tungur, fanning out, and creeping down onto the lowland. It descends with Hoffellsfjöll mountains to its east, reaching as far as Efstafellsnes and Geitafell

Fig. 8.78 Viðborðsjökull descends between Sandmerkisheiði and Gæsaheiði. Sources of geothermal heat were to be found under the glacial dome at the head of Viðborðsdalur, but when the glacier retreated and the hot springs appeared from beneath it, they were buried by landslips. HB, August 2003

and ending in a proglacial lake at the snout's end, about 40 m above sea level. There is now water on both sides, as well as in front of, all the glacial tongues of Hoffellsjökull. The glacier previously extended itself so far as to split into two at Svínafellsfjall, the western part of the glacier given the name Svínafellsjökull. The moraine spur Öldutangi, which stretched northwards from Svínafellsfjall, then divided the glaciers, but it is now ice-free. Svínafellsjökull then ended in a lake, which is now separated from the glacier, to the west of Öldutangi, about 1.5 km from its snout. To the west of the snout are Viðborðsháls and Hálsaheiði.

The Austurfljót emerges from beneath the eastern snout of Hoffellsjökull, while the Suðurfljót, a much smaller river than Austurfljót, flows from Viðborðsjökull and the western snout of Hoffellsjökull, called Svínafellsjökull. The catchment area of the Hornafjarðarfljót and its tributaries therefore includes Viðborðsjökull, Svínafellsjökull and Hoffellsjökull. The river Hornafjarðarfljót had previously veered over onto Hoffellssandur every 30–50 years or so, but levees now restrain its currents.

Fig. 8.79 The farmstead of Stóralág in 2006, with Viðborðsjökull, Hoffellsjökull and Hoffellsfjall in the background. HB 2007

8.6.5   Expansion of Hoffellsjökull from the Time of Settlement Until the End of the 19th Century

Hoffellsjökull had originally flowed down the narrow gap between Múli and Gæsaheiði. Hoffellssandur was a level and vegetated lowland that reached Tungur at the front of Múli. The eastern part of Hoffellsjökull ploughed its bed as deep as 300 m below sea level, but its western snout, as Svínafellsjökull, splayed out much more thinly over harder bedrock. This helps explain, with the increasingly warmer climate of the 20th century, why Svínafellsjökull has retreated more rapidly than the eastern part from the glaciers' maximum extension from around the year 1890.

Stories of glacial changes in the Mýrar and Nes districts of Hornafjörður are best recorded at Hoffellsjökull. In the middle of the 18th century, the glacier extended along both sides of Svínafellsfjall (Stefánsson 1957; dated 21 July 1746). It probably did not reach right up to the sides of the Hoffellsfjöll, however, for in the latter half of the century it was possible to ride along the level, mud plain beneath the Hoffellsnúpar promontories (Jónsson 2004). According to the testimony of farmers in the beginning of the 19th century, there was even a slope from Hoffellssandur up to Múli. Thus it was not until the first half of the 19th century that

Fig. 8.80 Öræfajökull, Skálafellsjökull, and Heinabergsjökull mirrored in the peaceful morning calm of Hornafjörður. HB, August 1998

the glacier extended along Efstafellsnes as far as Geitafell, closing off the route to Tungur. The glacier attained its greatest expansion and thickness around 1890, though it still fell short by just less than 1 km of reaching Viðborðsjökull and merging with Svínafellsjökull, completely surrounding Hálsaheiði and Gæsaheiði.

The glacier's eastern edge was pressing so hard up against the Hoffellsfjöll mountainsides by the middle of the 19th century, that it dammed their gullies, preventing their streams from flowing onwards. Water thus collected in lakes which later burst out in floods onto Hoffellssandur, destroying a lot of local farmland. Northernmost of these lakes was Múlavatn, then came Gjávatn, and furthest south was Efstafellsvatn. The outburst floods were annual events, late in the summer, and grew in magnitude as the glacial dam became thicker and water levels had to rise to gain enough pressure to force a way beneath it. These lakes had become so large by 1840, that their jökulhlaups made rivers completely impassable as far downstream as the road from Bjarnanes in Nes to Holt in Mýrar. The western edge of Svínafellsjökull also dammed a gully on Gæsaheiði, and floods from there flowed into the Suðurfljót.

Lambatungnajökull had crept so far down Skyndidalur by 1700, that it cut off the route over the mountains between Hoffellsdalur and Fljótsdalur. Lambatungnajökull not only closed the way north to Skyndidalur, but parts of the glacier also crept out of it into Hoffellsdalur so that glacial water was discharged into the Hoffellsá, which flows down the eastern side of Hoffellssandur. Hoffellsá was a spring-fed river in 1746, but in 1894 Thoroddsen claimed it contained a considerable amount of glacial water (Thoroddsen 1895).

8.6.6   Retreat of Hoffellsjökull from the End of the 19th Century

When the DGS compiled its map in 1903, the centre of Svínafellsjökull had retreated about 420 m from its maximum extension around 1880. The eastern ice flow of Hoffellsjökull, on the other hand, had virtually remained static, although the glacier had thinned out. A low moraine ridge, Öldutangi, jutting northwards from Svínafellsfjall, which divided the glacial tongues, had just come into view. Around 1910 the shape of nunataks became visible at a height of 800 m and were given the name Nýjunúpar. By 1930 Svínafellsjökull had withdrawn a further 700 m and Öldutangi had become ice-free over a tract about 1 km long (Eiríksson 1932). The eastern snout of Hoffellsjökull had conversely advanced very slightly along the Geitafellsbjörg cliffs, while the western part of its snout, near to Öldutangi, had retracted about 300 m since 1890. In accord with the warming climate after 1920, the snout has retreated and the glacial ice thinned, especially during the years 1930–1960. Where the glacier had reached its maximum extension just over a century ago, with its ice 250 m thick just 3.5 km from its snout, there is only glacier-free terrain today. Other outlet glaciers have also withdrawn, and thus Lambatungnajökull stopped discharging glacial water into the Hoffellsá, which has now been a spring-fed river again since 1938.

8.6.7   Hoffellsjökull as Scene of Pioneering Glaciology in the Early 20th Century

When a Swedish-Icelandic expedition arrived in Hornafjörður in 1936, the impact of climate change on Vatnajökull's mass balance, glacial economy and meltwater discharge were all too visible. Svínafellsjökull had retreated about 1300 m since 1890, and even though Hoffellsjökull had actually only retreated a few dozen metres, it had thinned out considerably as a whole, and the expedition reckoned that it had lost a third of its volume over the 45-year period. The glacier had thinned by about 100 m at the mountain edge above Efstafellsvatn, 1.5 km from the snout, and by about 80 m further down at Múli, 7 km from the terminus. The Nýjunúpar nunataks that were first seen from the farmsteads in 1910, now rose 50 m above the glacier's surface. The ice dams of the glacial lakes had also become thinner, and the highest water level in Gjávatn was 45 m lower than the highest shoreline mark from 1890. Jökulhlaups had decreased in size during the 19th century because the proglacial lakes now filled up in May or early June, which was a month earlier than had been the case before 1900. There was no longer any danger of a major outburst flood like the one described in 1840. Sigurður Þórarinsson believed that the lakes at the margins of Hoffellsjökull had then contained a total of 15,000,000 m^3 of water, of which 10 million might have burst out in a jökulhlaup. More recent maps, however, indicate that the proglacial lakes might have contained up to 30,000,000 m^3 of water at their highest levels. The jökulhlaups in the months of May or June in the 1930s seldom reached a maximum flow rate of 150 m^3/s, but if summer meltwater was added in July, they might have reached a flow rate of up to 500 m^3/s. When the later Swedish-Icelandic expedition from Uppsala investigated the situation in 1951 and 1952, only the Gjávatn lake remained. Two jökulhlaups emerged from it during the first summer, the first one on 15–18 June, when its water level fell 25 m and its flow rate reached 240 m^3/s, and a second, much smaller flood, on 15 October. In the summer of 1952, there were a total of four small outburst floods which did very little damage.

One of the main tasks of the Swedish-Icelandic expedition of 1936–1938 was to investigate the very basis for the continuing existence of Vatnajökull: how it maintained itself as regards snow accumulation, and how much it shrank due to ablation. They discovered how much snow was added to the glacier's surface by digging channels into the winter snow cover. A layer of snow up to 8 m thick was measured on the ice sheet summit, the equivalent of 4000 mm of water. The summer ablation was calculated from measuring posts bored into the glacier. Ablation proved very little on the glacier's highest domes, but further down at the snouts the equivalent of a 10-m layer of water melted every summer, and a further 2 m during the winter, for the melting of ice continued all year round there. On top of this 12 m of meltwater, there was an additional 2000 mm of rainwater, so that in total the drainage from the snout was about 14,000 mm per annum, an amount equal to the runoff from areas with the greatest precipitation in the world. These were the conditions of the maritime climate on the island of Iceland in the middle of the North Atlantic in the late 1930s. (Ahlmann 1938; Ahlmann and Þórarinsson 1937–1943; Björnsson and Pálsson 2004)

Flashback: The Swedish-Icelandic expedition eighty years ago
Our story concerning glaciological research on Hoffellsjökull began on a sunny Sunday, 26 April 1936, when Guðmundur Jónsson, the farmer at Hoffell in Hornafjörður, welcomed a group of five Icelandic and Swedish scientists in his front yard. They had come a long way and were well stocked with apparel, provisions, and equipment for an eight-week research expedition onto Vatnajökull. The leading Icelander was Jón Eyþórsson (1895–1968), a meteorologist at the IMO, then just over forty years of age. He later became well-known for his glaciological and meteorological research, as well as being a writer and translator, and even a popular radio broadcaster. Six years previously, along with Helgi Hermann Eiríksson (1890–1974), an engineer and director of the Technical College in Reykjavík, he had begun measuring the displacement of Icelandic glacial snouts and these

measurements are still being carried out today. Leading the Swedish contingent was Hans Jakob Konrad Wilhelmsson Ahlmann (1889–1974), a professor from Stockholm, and with him were two of his young students. One of them was the Icelander Sigurður Þórarinsson (1912–1982), who had studied geography and geology at Stockholm University for three years. In 1934 he had already taken part in important research on Skeiðarársandur after the huge jökulhlaup from Grímsvötn, and had also investigated the consequences of a severe earthquake in Dalvík in northern Iceland that same year. Þórarinsson later became renowned all over the world as a glaciologist and volcanologist, and in Iceland as a rhymester as well. The other student was Carl Mannerfelt (1913–2009), then the champion hurdler of Sweden, who would later write a definitive doctoral thesis on the land formations caused by ice-age glaciers in Scandinavia. Later in life he became the CEO of Esselte and was on the boards of many other major Swedish companies. Also in the group was the Icelander Jón Jónsson from Laugar in the Biskupstunga area of southern Iceland and a policeman in Reykjavík, who had been in charge of transporting the horses from the shoreline up onto the glacier on Alfred L. Wegener's Greenland expedition in 1930–1931. The final member was the Swede Mac Lilliehöök, a ski-instructor and sledge-dog trainer, for there were four sledge-dogs on the expedition, provided by the Norwegian explorer Helge Ingestad; they were called Bonzo, Helgi, Úlfur and Sverrir. The dogs behaved badly on the back of the lorry on the way to Hoffell, and indeed they had been rather troublesome ever since they had come ashore, so much so that they had become rather famous in Iceland for their misbehaviour.

Hans Ahlmann did not understand spoken Icelandic, but felt that this isolated settlement and its people's modest reserve, had preserved an ancient way of life for much longer than any other part of the country. Guðmundur Jónsson was a proud and independent man, indeed his forefathers had lived right next to the glacier for centuries and knew it better than anyone. A large family lived at Hoffell and Ahlmann (1938, p. 5) saw in the eyes of the mistress of the house, Valgerður Sigurðardóttir, 'a sharpness and lustre which testified to a keen wit, an alert mind, and a conscious nobility of character.' Ahlmann remembered from the *Book of Settlements* that a temple, Goðaborg, had been raised on the mountain to the north of the farm, as he contemplated Hoffell towering over the magnificent farm site, with its home fields and barren mud plains dissected by streams. This had been a choice location for a manorial farm ever since the time of Iceland's settlement, but on this sunny day pioneers of the 20th century had arrived to do further research on the glacier, and although it was cool in the shade, they could feel the glacier's dome in the north was but a short distance away. The glacier and the barren outwash plain in front of them was a genuine ice-age view. Expectations were high.

Ahlmann had been a professor and director of the geographical institute of Stockholm University since 1929 and had set himself the target of establishing the mass balance of glaciers around the North Atlantic by measuring their accumulation and ablation levels. He also made weather observations in order to evaluate the role of individual meteorological features in the melting process. Ahlmann had investigated the Arctic glaciers of Svalbard in 1931 and 1934, and now considered himself ready to come to Iceland, where the climate was shaped by the battle between Arctic cold air and tropical masses of warm air, and where there could be snow in the summer and rain in the winter. And the expedition certainly had to endure some extremely bad weather, which only helped increase their understanding of the nature and existence of Vatnajökull. They knew that by garnering further data and knowledge of modern-day glaciers they might also increase the understanding of the history of the Earth. Ahlmann was also anxious to investigate the influence of climate change, which had become evident from 1920 onwards and had caused glaciers to shrink enormously, their general retreat clearly visible. The local folk in Hoffell were able to point out to them knolls and ridges that had recently appeared over a wide area from beneath the glacier. In the yard itself they saw a pile of grey trunks for firewood much larger than any tree that then grew in Iceland. The glacial river from beneath Hoffellsjökull had transported into daylight mangled and dismembered birch trees that the glacier had covered in the Iron Age, or from about 500 B.C. Now, at the beginning of the 20th century, they had appeared from beneath their icy armour, freed and rinsed out by melting waters. Indeed, men were beginning to speculate as to whether a warm, interglacial period was beginning, similar to that in the Stone Age, but their scientist visitors thought it more likely a short-lived respite and that it would be a long time before woodlands would grow again in Hoffellsdalur. The astute geographer Ahlmann also noted, conversely, that the sediment load borne by the glacial rivers out to sea might actually make it difficult to keep the shipping lanes into Höfn harbour open and free from silt; they themselves had arrived from Reykjavík by sea in the coastguard ship Ægir. The expedition members knew that three decades earlier, in 1906, the Icelandic geologist Helgi Pjeturss had illustrated how Hoffellsjökull had covered marine sediments containing shells from an interglacial period when the ocean was 50–60 m higher than it is today. A long time had passed since then, and the glacier was now a lot further from the sea.

The Swedes were also following in the footsteps of their compatriot Otto Torell (1828–1900), who had travelled around the country and used his observations from Iceland to gain further acceptance of the theory of the Ice Age and the idea that an ice-age glacier had once covered Sweden and northern Germany; the belief in the Great Flood had to be refuted. The Swedes had familiarised themselves with the journeys of their predecessors on Vatnajökull. They knew about the stories of people from northern Iceland who had crossed the glacier to gain employment during the fishing seasons south of Vatnajökull, probably on horseback, which would have been a day's journey in good weather (they later named the depression they believed these travellers had used Norðlingalægð after them). They also knew of the Englishman W.L. Watts' expedition on the western part of Vatnajökull, and his later crossing of the glacier in 1875 with a group from southern Iceland. The Swedes were also aware of the journeys of the very conscientious Danes, Daniel Bruun's trip up onto the glaciers from Hornafjörður in 1902, and the journeys of Johan P. Koch and his pioneering cartographic work for the DGS. Furthermore, they knew that Koch and the German Alfred L. Wegener had ascended Vatnajökull from the north and crossed to the Esjufjöll mountains in the south while proving the skill and endurance of Icelandic horses on glacial expeditions. Koch and Wegener easily succeeded in covering 150 km on the glacier in five days and so they took horses with them on the Greenland expeditions in the years 1912–1913, arriving at Queen Louise Land and crossing the Greenland ice cap. Ahlmann, on the other hand, preferred to have sledge-dogs with him as they had proved so reliable for Fritjof Nansen and Roald Amundsen. They were also aware of the misadventures of the Swedish students Hakon Wadell and Erik Ygberg, who had rediscovered Grímsvötn in 1919, but lost their horses at the end of their journey.

The expedition members had arrived in early spring, before the ablation period began, because they intended to investigate the glacier's mass balance systematically. They came well prepared for the sojourn on the glacier, where all kinds of weather could be expected. They had with them the low tents designed by Ahlmann and Mannerfelt and based on prototypes used by Admiral Richard Byrd on his trip to the Antarctica and which had withstood tremendous winds. They were only too well aware that bad weather had spoiled results for previous expeditions on Vatnajökull. The members could not maintain their physical strength and alertness while doing research unless they were able to enjoy sufficient rest in between their labours. They were certainly to become acquainted with how the air currents of the Atlantic and the cold of the glacier were to meet and react, there being few better places for experiencing such conditions (Fig. 8.81). Their tents withstood the wind, but proved to be less than waterproof as continual rainfall battered them and water seeped through the linings, causing the expedition members to suffer from dampness, coldness and frostbite. They duly bequeathed their camping site the name of Djöflaskarð, or Devil's Pass.

Their research took three years and the local farmers supported them wholeheartedly: Guðmundur Jónsson at Hoffell and his sons Helgi and Leifur, Skarphéðinn Gíslason at Vagnstaðir (who helped survey Heinabergsjökull), Sigurbergur Árnason at Svínafell, and Sigfinnur Pálsson at Stórlulág, who accompanied Sigurður Þórarinsson on his reseach trips in the area for three years after the main expedition had ended. Þórarinsson was also assisted by Ari Hálfdánarson, the farmer at Fagurhólsmýri, and his sons, along with the pastor Eiríkur Helgason at Bjarnanes.

Fig. 8.81 a People from Hornafjörður helped transport and set up the tents of the Swedish-Icelandic expedition in 1936. **b** They personally hauled baggage and equipment up the steep sides of Hoffellsjökull. **c** Once up on the glacier, the dogs took over the hauling of the sledges. **d** Carl Mannerfelt examining water as it spurts out almost vertically under high pressure from beneath Hoffellsjökull. This indicated that the glacial bed was lower than that of the outwash plain ahead of it. Sigurður Þórarinsson, 1936

(c)

(d)

Fig. 8.81 (continued)

Fifteen years after this Swedish-Icelandic expedition on Hoffellsjökull, there was another Swedish-Icelandic research expedition, this time to Hoffellssandur outwash plain. This group was from Uppsala University under the leadership of Filip Hjulström and included Åke Sundborg, who later became a professor at

Uppsala, and the Icelandic geologist Jón Jónsson (Hjulström et al. 1954–57; Arnborg 1955a, b). The sons of Guðmundur Jónsson, Leifur and Helgi, then lived at Hoffell and also gave their assistance to this expedition (Björnsson and Pálsson 2004).

The climate in 1936–38 similar to that at the beginning of the 21st century

The three years that the Swedish-Icelandic expedition was in Iceland had been demonstrably warmer than those in the first two decades of the 20th century. All the summers were warm, though especially the first one in 1936. Looking back over all three years, the expedition members calculated that Hoffellsjökull, together with Viðborðsjökull, had shrunk by the equivalent of 94,000,000 m^3 of water, or 0.30 m per year over their surfaces as a whole. Even so, they managed to expand a little over the next year, 1936–1937, due to an increase in snowfalls. The first year (1935–1936), they calculated the shrinkage to be about 364,000,000 m^3 and that the ablation zone's elevation reached up to 1200 m. The next winter, the snow accumulation had been average (643,000,000 m^3), but this was followed by a large amount of summer ablation (1,007,000 m^3). The ablation was less the following two summers, and thus the mass balance was decided by the winter snowfalls. During the winter of 1936–1937, there was a large total snowfall (reckoned to be 795,000 m^3), and ablation (604,000 m^3) failed to remove all of it, so that the glacier gained an additional 191,000,000 m^3 of water equivalent; the accumulation zone reached to below an elevation of 1000 m. In the final year, the glacier was reckoned to have shrunk about 111,000,000 m^3 and the equilibrium line had been around a height of 1090 m; the winter snow cover was 559,000 m^3 and ablation 670,000 m^3. Calculated as a depth of water covering the whole surface of the glacier, the shrinkage was measured as 1.17 m the first year, which then increased by a further 0.61 m the following year, but in the final year was reduced by 0.36 m. The average drainage rate from the glacier over these years was calculated as 32, 18.7, and 21.5 m^3/s. No measurement had hitherto been made as to the volume of glacial water borne by the Hornafjarðarfljót out to sea. Fifteen years later, the second Swedish-Icelandic expedition calculated glacial drainage more directly with hydrologic data from the rivers themselves. The average water drainage into Hornafjarðarfljót from 1951 to 1952 was recorded as 44 m^3/s, at least 2/3 of which came from the Austurfljót.

In addition to checking on the glacier's economy, transport, income, expenditure, and mass balance, the expedition also investigated glacial advances. The snow layers, which do not melt, creep forwards. By measuring the velocity of Hoffellsjökull, they reckoned that in 1937 about 200,000,000 m^3 of ice had advanced in a glacier 2.5 km broad between Múli and Gæsaheiði, where the elevation is about 800 m above sea level. The greatest speed was in the centre of the glacier, just over 2 m every 24 h, but the velocity was less at the margins, where it was on average 1.4 m every 24 h.

The glacier was eroding its bed while creeping forwards and from measuring the ground moraine from beneath the glacier, the expedition came to the conclusion that it scraped 5.5 mm off the glacial bed per annum, or the equivalent of 1 m every 180 years. Looking back over the 10,000 years since the last glacial period, such erosion would have equalled any land uplift due to shrinking glaciers, for the uplift in this area had been calculated as being 80–100 m over this period. (Björnsson and Pálsson, 2004)

8.7   Eastern Vatnajökull and Glaciers in the Lón District

Vatnajökull disappears from view when moving east from Hornafjörður, for its eastern outlet glaciers do not reach the lowlands. Although precipitation is high, the accumulation zones on the mountain ranges of Goðahryggur, Goðahnúkar and Grendill are small and thin and their valley glaciers short and steep and far from human habitation. They have played an insignificant role in the country's history, although there used to be an ancient route over the glacier from Hornafjörður to the Fljótsdalshérað region in the northeast. Meltwater from these glaciers flows into one river, Jökulsá í Lóni, and the glaciers themselves are frequently referred to as the Lón glaciers. The Jökulsá í Lóni meanders in braided streams through the lowlands similar to the river estuary in Hornafjörður, and it has often damaged vegetation and been difficult to cross.

The Stafafellsfjöll mountains in the Lónsöræfi wilderness are now a nature reserve; some of them reach a height of over 1000 m and are crowned with snow cover and glaciers. Lónsöræfi has stupendous scenery and is a hiker's paradise. The colourful mountains bear witness to volcanoes 5–7 million years old, savagely eroded by ice-age glaciers, and a ravine harrowed out by mountain rivers, a variety of waterfalls, birch copses in gullies, and green banks of grass in valleys, can all be found there. Lambatungnajökull is the southernmost and largest of the Lón outlet glaciers (Figs. 8.82 and 8.83). It heads east along Goðahryggur ridge and then down into Skyndidalur. The glacier is 15 km long and has gouged out the deepest bed of these glaciers, or as low as 250 m above sea level. Until the end of the 1930s, its snout extended so far that it discharged water into the Hoffellsá and from thence south into the Hornafjarðarfljót estuary. The author of a treatise on glaciers in 1695, Þórður Vídalín, explored this glacier when he was the local physician in Þórisdalur in Lón. Two short glacial tongues, Austurtungnajökull and Norðurtungnajökull, reach down from the highland ridge from Múlaheiði. Further north, the Axarfellsjökull, 10 km long, creeps eastward from Mt. Grendill, descending through a narrow opening. The northernmost Lón outlet glacier,

Fig. 8.82 View north over the easternmost part of Vatnajökull towards Mt. Snæfell (1833 m). Grendill (1570 m) is the highest point of the glacier here, and can be seen against the glacier's edge in the *top left* of the photograph. Outlet glaciers descend to the east, the one furthest away is Vesturdalsjökull and in front of it is the almost ice-free Geldingafell. HB, August 1998

Fig. 8.83 View south over the eastern part of Vatnajökull. In the foreground is Geldingafell and Vesturdalsjökull, the source of the Jökulsá í Lóni. The snow-capped peaks of Jökulgilstindar are in the distance (to the *left*) and to the west of them is a view down to the Lón district and the shore. The Sauðhamarstindur peak is in the *centre* of the photograph. HB, August 1998

Vesturdalsjökull, descends through a deep ravine, where the outlet of the Jökulsá í Lóni emerges.

There is also an ice flow from the Grendill and Goðahnúkar mountains northwestward, and at Hnjúkafell it converges with an ice flow that originates from the east of Breiðabunga and from the ice divide with Hoffellsjökull. Together these flows form the snout of Eyjabakkajökull which has ploughed its bed to 50 m below the level of land at Eyjabakkar, a wide expanse of vegetated land in the forefield of the glacial snout, providing rich feeding grounds for reindeer and pink-footed geese. Eyjabakkar are at an elevation of 650 m above sea level, higher than almost all other oases in the central highlands, but nonetheless with lush, marshy meadows and a luxuriant variety of flora. They are also the source of the Jökulsá í Fljótsdal, which flows 20 km from Eyjabakkar to Kleif through fifteen beautiful waterfalls and cataracts. There are sources claiming that the snout of Eyjabakkajökull has surged four times, in 1890 and 1972 (about 620 m, when ice from Grendill in the east pushed through), and in the years 1931 and 1938, when ice surged from its western branch.

In the glacial indent to the west of Eyjabakkajökull is the proglacial lake Háöldulón, from which jökulhlaups flood into the Eyjabakkar wetlands (Figs. 8.84, 8.85, 8.86 and 8.87). There are ice caves in the channels beneath the glacier that are a tourist attraction. Glacial water flows through the Blanda and Kelduá rivers to the south of Geldingafell, and from thence into Lake Lagarfljót. There is a much frequented hiking route over Eyjabakkajökull from Snæfell to the cabin at Geldingafell and then south to the Stafafellsfjöll mountains. There are a great number of snow patches and small glaciers on the mountains east of Vatnajökull. The glaciers on Jökulgilstindar peaks are the southernmost ones, while the Tungutindar peaks are further north, and the eastern arm of Hofsjökull extends between Hofsdalur and Víðidalur, the Hofsá flowing from the latter into Álftafjörður. Þrándarjökull lies between Hamarsdalur and Geithellnadalur and is rather larger and further north than Hofsjökull. There are small glaciers in the mountains to the west of the head of Fáskrúðsfjörður and in the mountains of Dyrfjöll and Fönn (Figs. 8.88, 8.89, 8.90 and 8.91).

Endless mowing near Eyjabakkajökull

In his *Lýsing Íslands* ('Description of Iceland'), Þorvaldur Thoroddsen states: 'A small mountain called Eyjafell lies ahead of the outlet glacier, slightly west of its centre, which divides the drainage channels so that streams emerge from the glacier in two main systems, the eastern streams bearing more water than the western ones. The grass swards on the islets between the streams are very thick and are so wet that in many parts only birds can cross them. In other large areas there are impassable quicksands and men have even seen reindeer disappear in a flurry of water and mud, and they are usually more than capable of crossing the wettest terrain. The streams sometimes flood over a large area so that they create for a short while a big lake (Thoroddsen 1958–1960, III, p.276). Elsewhere Thoroddsen writes: 'There could be endless

Eyjabakka-Jökull,   Snæfell.
hvorfra Eyjabakkaá kommer.   Nálhúshnjúkar.
Vatna-Jökull's Nordøstrand, set fra Fljótsdalsheiði í 40—50 km Afstand.
(Tegning efter Skitse af D. B.)

Ved Vatna-Jökull's Nordrand.
Udsigt fra Højderne Syd for Hrafnkelsdalur's Bund. Snæfell's sydlige Udløbere ses t. v.
Kverkfjöll ligger í Baggrunden ca. 40 km borte.
(Tegning efter Skitse af D. B.)

Fig. 8.84 *Top* Eyjabakkajökull, source of the Jökulsá í Fljótsdal, and Snæfell (on the *right*). *Below* Brúarjökull and the Kverkfjöll mountains. Drawings by Daniel Bruun. Skagafjörður Archives

mowing for hay on these islets, if it were only possible to utilise them.' (Thoroddsen 1895, p. 81)

Over wide areas of the central highlands of Iceland there are continuous vegetation lines at around a height of 600 m, with only smaller patches at higher elevations. Nowhere on Iceland's high plateau, except for Eyjabakkar and Vesturöræfi, can such an expanse of vegetated land be found above a height of 650 m. The Eyjabakkar wetlands are also considered special for being so lush at such an altitude above sea level. They are marshy meadows created by glacial rivers that are otherwise restricted to lowlands and only exist in two other places in the highlands of Iceland: at Hvítárnes by Langjökull, and in the Þjósárver wetlands to the south of Hofsjökull. Eyjabakkar are the feeding and moulting grounds of the pink-footed goose from June into September, and reindeer migrate there late in the summer.

Fig. 8.85 Eyjabakkajökull descends onto the Eyjabakkar wetlands, the source of the Jökulsá í Fljótsdal. The peak of Snæfell (on the *left*) is shrouded in clouds. HB, September 1996

Fig. 8.86 The now vegetated Hraukar mounds, which Brúarjökull bulldozed into existence during its surge in 1890. Þóra Ellen Þórhallsdóttir, 2007

Fig. 8.87 Foliated reels of earth on the Kringilsárrani tongue. HB, 2007

Fig. 8.88 Cirque glaciers and piles of moraine in the Dyrfjöll mountains. Guðmundsson (1996)

Fig. 8.89   Glaciers are still hidden in the shadowy corries of Reyðarfjörður. Ágúst Guðmundsson (1996)

Fig. 8.90   Hofsjökull in Lón (<5 km^2; high point 1069 m) is between Víðidalur in Lón and Álftafjörður. Its firn dome faces north, but ice tracts descend south between ridges. It is the source of the Hofsá in Álftafjörður. Snævarr Guðmundsson, August 2006

Fig. 8.91 Þrándarjökull (16 km²; 1248 m) is a single carapace glacier to the northeast of Vatnajökull. Snævarr Guðmundsson, August 2006

8.8   The Northern Margins of Vatnajökull

When looking west over the northern margins of Vatnajökull there is an ice-age view for as far as the eye can see (Figs. 8.92 and 8.93). A glacier more than 100 km broad stretches outwards from the ice divide 1600 m high on the ice sheet and advances 50 km northwards on a rugged, undulating high plain at about 700 m above sea level. The Kverkfjöll mountains divide this broad expanse into two ice shields. Dyngjujökull is the western one, and reaches as far as Bárðarbunga, covering young volcanic rock and lying over land strewn with sand and gravel. The eastern part is Brúarjökull, which lies above an eroded bed from the first part of the Ice Age, though in its forefield is a mixture of volcanic formations from the last glacial period and various glaciofluvial sediments. An active glacier and powerful glacial rivers are continually shaping the terrain. This area was part of the largest uninhabited expanse of Iceland, indeed of the whole of Europe (outside of Greenland and Svalbard). The Kárahnjúkar dam and hydropower plant have now effectively inserted a large wedge into this space, an intrusive area reckoned to cover about 3000 km².

There are many tributaries that flow from Vatnajökull's northern margins but they all converge into two main rivers, Jökulsá á Dal, otherwise known as Jökulsá á Brú (downstream in populated areas), and Jökulsá á Fjöllum, which descends

Fig. 8.92 Brúarjökull, and beyond it the Vesturöræfi wilderness and Snæfell (on the *right*). Englaciated layers of tephra, from volcanic eruptions centuries ago, emerge again towards the glacier's terminus. The ash layers furthest down are from the 12th century and the most distinctive ones, in the centre of the picture, are believed to be from the beginning of the 17th century. Most of the layers are from eruptions of Grímsvötn, though tephra strata on the glacier have also been traced back to eruptions of Öræfajökull in 1362, Veiðivötn in 1477, and Katla in 1625. HB, August 1999

200 km on its way to the ocean to the north of Iceland. Although these rivers' catchment area is the largest in the country, they are not the most voluminous rivers of Iceland, for the rain shadow area north of Vatnajökull has the driest weather of any part of the country. Nonetheless, nowhere else in Iceland does a glacier play a relatively greater part in the total discharge of water into glacial rivers. Outburst floods in the Jökulsá á Fjöllum and Jökulsá á Dal have often reminded the inhabitants of the Þingeyjar and Norður-Múli Counties of the existence of Vatnajökull. Eruptions in the main volcanic centres in the Kverkfjöll mountains and at Bárðarbunga, and fissure eruptions beneath Dyngjujökull, have caused huge jökulhlaups in the Jökulsá á Fjöllum river. The only outburst floods from Brúarjökull are from its proglacial lake.

The terrain to the north of Vatnajökull was little known or explored until well into the 20th century as it was so far away from the usual cross-country routes. It must be assumed that farmers had once reached as far as Brúarjökull while rounding up sheep, for the Brúaröræfi wilderness had been covered in vegetation from the glacier all the way down to the first early settlements. The Ódáðahraun lava field was also vegetated in the first centuries of Iceland's settlement, and so sheep could have grazed as far south as Dyngjujökull. Sources for any such

Fig. 8.93 Aerial view over Brúarjökull's snout towards Snæfell in August 1999. A jökulhlaup had recently burst from Lake Hnútulón (on the *left*), leaving large chunks of ice behind in its basin. HB, August 1999

journeys inland by inhabitants of Þingeyjar County do not go further back in time than early in the 19th century, however, when the so-called 'mountain brothers' (Jón, Þorlákur and Ólafur) had travelled further into the uninhabited interior than ever before and found unknown meadows; this was considered a daring and gallant achievement (Gunnarsson 1950). In 1830 five brave men from the Mývatn area went searching for outlaws in Ódáðahraun, armed with guns and swords. They did not find any trace of such men, but still steadfastly continued to believe in the existence of highland bandits (Jónsson 1945, vol.1, p. 218). Poor round-ups of sheep from their summer pastures in the Mývatnsöræfi wilderness also encouraged Þingeyjar County to send men south into the Ódáðahraun hinterland. In the autumn of 1855, four men from Mývatn went looking for pastures where sheep might spend the winters, but turned back after reaching south of the Dyngjufjöll mountains. Four farmers from Þingeyjar, later called 'the land-explorers', arrived at Dyngjujökull in 1880. They had travelled to the southeast of the river Skjálfandafljót and across the glacial snout to the north of Bárðarbunga and down to the Dyngjujökull fluvial plain to the east of Kistufell (Jónsson 1945, vol.1, pp. 274–278). At the Hvannalindir springs, at a height of 640 m, they found a meadow for 200 sheep and the ruins of a hut that might have been a former lodging place for the outlaw Fjalla-Eyvindur, when he lived at Hvannalindir in the 1760s, before he moved to the Eyvindarver wetlands on the banks of the Þjórsá. Angelica and willow shrubs flourish around the springs, and pink-footed geese and ptarmigan find a haven there. Hvannalindir was an oasis totally surrounded by barren land, which had remained, thanks to its isolation, completely undisturbed since Iceland was settled, for not even livestock had grazed there. One of the men on this expedition was the farmer and author Jón Stefánsson, who described the journey in the journal *Norðlingur* in Akureyri (1880). Another man on this trip was Jón Þorkelsson, a farmer at Víðiker and Halldórsstaðir in Bárðardalur, who made many journeys into the uninhabited hinterland and was the first man to describe in detail (1876) the volcanoes in Askja in 1875.

In previous centuries, the cross-country routes between distant parts of Iceland traversed Vatnajökull via Brúarjökull and Dyngjujökull. There are sources for connections between Skafatafell in the Öræfi district and Möðrudalur at Efri-Fjall in the northeast, and for itinerant fishermen from northern Iceland crossing Vatnajökull in the 15th and 16th centuries for the fishing season in the Suðursveit district in the south (Pálsson 1945, p. 300). Men from Fljótsdalur in the northeast arrived at Brúarjökull to the south of Snæfell, the most easterly part of the Vesturöræfi, and ascended the glacier in an indent to the west of Eyjabakkajökull. They then continued their way southwards along the eastern side of Breiðabunga before descending a trail to the terminus of Skálafellsjökull in the south. Those who came south to Brúarjökull on the western side of the Jökulsá á Brú, ascended Brúarjökull between the outlets of the Kverká and Kringilsá rivers, but once in the centre of Brúarjökull their paths diverged. Those going to Skaftafell continued southwest across the ice cap. Those going to Suðursveit, on the other hand, journeyed through the Norðlingalægð depression and then turned southeast to descend from Skálafellsjökull. Men from Þingeyjar County crossed the Ódáðahraun and ascended Dyngjujökull at Kistufell. Men from Hornafjörður came down from the glacier here in July 1926, west of the indent by Kistufell, after they had crossed Vatnajökull with a tent and a sledge from Hálsaheiði in Hornafjörður. They then continued onwards to Svartárkot in Bárðardalur before returning to Hornafjarður by the exact same route. The Englishman Watts, along with five Icelanders, also descended from Dyngjujökull to the east of Kistufell after a 12-day journey across Vatnajökull in the summer of 1875. They had covered a distance of 60 km on their way from Núpsstaður in the Fljótshverfi district, southwest of the ice cap. From Dyngjujökull they proceeded to cover a further 100 km in four days to reach Grímsstaðir in the northeast of the central highlands.

Johan Peter Koch and Alfred Lothar Wegener rode up onto Brúarjökull on the eastern side of the river Kreppa in July 1912. From there they continued for 65 km to the Esjufjöll mountains in Breiðamerkurjökull. This was a preparatory trip for their later expedition westward over the northern part of the Greenland ice cap (Koch 1912). Having traversed Vatnajökull himself, Koch maintained that there was no longer any reason to doubt the truth of the old stories of men crossing the glacier. Koch and Wegener had come from the settlements in the north across Ódáðahraun lava field east of the Hvannalindir springs. An expedition from Cambridge University in 1932 followed the ancient route of northern fishermen homewards from Suðursveit and descended from the glacier in the north at the eastern Kverkfjöll mountains. Two Scotsmen, J. H. Wigner and T. S. Muir, set off from the eastern side of Brúarjökull in the late summer of 1904. They traversed the glacier on skis, hauling a sledge behind them, and after passing by the Esjufjöll mountains they finally descended from the glacier to the south of Lake Grænalón. Wigner and Muir travelled 128 km on the glacier in 13 days, although the trip took 22 days altogether (Muir and Wigner 1953).

Watts' Tussle with the Glacier

At 6 P.M. [3 July 1875] we reached a steep ascent, where our compass twisted and turned about in the most eccentric fashion; the heavens became black as night to windward, the wind had risen, and was making the peculiar booming noise I have often remarked in these regions before a storm, and driving a blinding, pitiless, drifting snow before it, which eddied about the sleigh and wrapped itself around us, as if longing to enshroud and bury us in its frozen toils. But we had an idea of burying ourselves in our own fashion. "ósköp mikill stormur kemur bráðum" (A bad storm is coming on presently), said Eyólfur, sitting down for a moment on the sleigh, and clapping his feet together to knock off the snow which was clinging to his legs, and we were all of the same opinion. We were at the height of 6150 feet, so I ordered a hole to be dug, and the tent to be pitched. The snow was very hard and firm, even at the depth of four feet, and we cut out as clean a hole as if it had been in salt, but the wind drifted so much loose snow into it, that the men were obliged to hold up the tent to windward during its completion.

We had barely got ourselves snug and commenced breakfast, when the storm burst upon us, seeming to threaten the tearing up of the very snow in which we had taken refuge; and had not former experience taught us to fortify our tent well all round with banks of snow, I have no doubt it would have been the last we should have seen of that article of furniture. Being satisfied that all was snug, and that the worst which could happen to us was that we might be buried in a few feet in the snow, we went to sleep. When we awoke at mid-day the storm had subsided and the fog had lifted, showing three dark mountains to the north—doubtless Skjaldbreið, Herðubreið, and Dyngjufall.

[4-5 July]

We were speculating as to whether we should go on in spite of the still threatening aspect of the weather, when the fog returned, and the booming wind announced another storm to be close at hand. Presently it broke upon us; never before had I heard the wind make such an unearthly wail. It seemed as if every imaginable demon and all the storm spirits of that wild region had assembled to howl and make a united attack upon us. The light was fast becoming obscure, and we were getting fairly snowed up, but that made us all the warmer, all the more secure, and the shrieking of the storm was deadened by the friendly covering. We partook of some chocolate, smoked and sung, and finally slept again. At 8 P.M. the storm had somewhat subsided, and I sent out a man to clear away some of the snow from the roof of the tent to let a little light in. The snow had drifted nearly over the tent, and it took some hard work before we were dug sufficiently out to let in enough light to write by; outside there were 10° of frost, but we were comfortably warm in the tent. The air outside was so full of snow that we could not see a couple of yards in advance.

Another day showed us only a continuation of storm and snow which utterly prevented progress. We had now only about a week's provision left, so I again put everyone on half rations. The men were obliged to take turns in clearing away the snow, at intervals of every three hours, from the top of the tent, and before very long the tent had the appearance of lying at the bottom of a deep hole in the snow. We passed the time as best we could, by sleeping, eating, smoking, writing, singing, spinning yarns, and I occasionally amused the assembly by learning strings of Icelandic words by Mr. Stokes's method of mnemonics, and repeating them in order, either backwards or forwards, which puzzled the Icelanders not a little.

Before I started for the Vatna [sic] in 1871, I remember saying I should like to see one of its worst storms: I now had that gratification. Storms are interesting natural phenomena, but when prolonged indefinitely are, to say the least, tedious hindrances to progress; and now, lying upon the top of the Vatna Jökull, with the possibility of their lasting for a month, and provisions materially diminishing, their dreary monotony became intolerably oppressive, and after mature consultation we all came to the conclusion that if the weather did not clear in two days' time, we should leave all impedimenta behind, except provisions, instruments and my diary, and strike northwards, storm or no storm—"*sauve qui peut.*"

When we lay down and were fairly snowed over, the booming of the storm sounded as if it came from the interior of the mountain, and almost any familiar sound could be singled out from the hurly-burly in an exaggerated degree, without any great stretch of imagination. It stormed all night; the wind "Trolls" shrieked around us, the thunder of the storm roared through the, to us, dark midnight hours, surging upon the icy bosom of the Jökull, sweeping up its snowy slopes, bearing with it avalanches of snow-drift which had buried us several feet deep by morning. By 5 A.M. it lessened somewhat, the furies of the Vatna appeared to have given up the idea of overwhelming us, and the disheartened tempest sunk away in melancholy sobs, but a determined drift and south-west wind persevered in harassing us.

It was clear that we must now start forward, for not only was there a considerable amount of snow yet to be traversed, but a howling wilderness of volcanic sand, lava, and mountain torrents had to be crossed which lay between the north base of the Jökull and the nearest habitation. We could not remain in our present position, so deeply were we buried, and so difficult was it to get in and out of the tent; moreover the fury of the storm had beaten the snow hard, so there was no time to be lost. I served out a hearty meal, and as packing up under such circumstances seemed to demand some stimulant, I made some grog out of methylated spirit, for all our whisky was gone. This served to quicken our circulation, although it was far from being palatable, having, as my Icelanders said, "slæmr dropi," or a bad after-taste, and no wonder, as the first taste was not suggestive of an agreeable sequel. (Watts 1876, pp. 46–51)

The Vatnajökull Expedition of 1904

The first expedition onto Vatnajökull in the 20th century was undertaken in the late summer of 1904 when two Scotsmen, J.H. Wigner and T.S. Muir, climbed up onto the eastern part of Brúarjökull from the Maríutungur grasslands. They headed in the direction of the Esjufjöll mountains, the highest point of Breiðamerkurjökull, from whence they continued south-westward to Grænalón and descended from the glacier near the place where Watts had done so thirty years previously. Ögmundur Sigurðsson, a teacher at Flensborg school in Hafnarfjörður, had misgivings about this trip, and later commented that 'it must have been the most unpleasant journey in the world.' Ögmundur had been a travelling companion on Thoroddsen's highland expeditions for 14 summers, though he had not gone on many glacier trips. The arrival of these unknown travellers at Núpsstaður caused quite a stir and some amazement, but with the help of one of the Danish surveyors who spoke English, they were able to explain their journey. The Scots' equipment was much lighter and better-suited than those of Watts and his companions when they had walked across the glacier. Wigner and Muir each pulled their own baggage-sled and used skis, except for when crossing sheer ice or where they had to wade through slush or layers of muddy ash. They had endured snowstorms and rain, had to rely on the compass for long stretches, but felt they could have held their bearings better if there had been three of them, as had been their original plan. They had not been aware of any magnetic interference to their compass from bedrock, which had confused Watts on the western parts of the glacier, but they had miscalculated distances, being deceived by the clear mountain air and the undulating sheets of firn without any discernible horizons. This had other compensations, nonetheless: 'Monotonous though it was the view was most extraordinarily impressive, and had an indefinable beauty of its own, and has probably left a more permanent impression on our memories than any purely pretty landscape could' (Wigner 1905, pp. 440–441). No other member of a glacial expedition in Iceland had written such praise before. Watts let it suffice to praise a glorious day only when it provided a chance to get moving again. But the Scots had also had to hunker down, snowed in for days, relying on their calico tent, and spending their time reading Don Quixote, Omar Khayyam, and Charles Dickens' Great Expectations. They graphically described the hellish weather conditions on Vatnajökull and took many photographs, for Wigner was indeed a qualified photographer and surveyor; Muir was a geologist. Probably the most important result of their expedition was that they could not find any volcanoes around Grænalón or its environs, thus indicating that the lake could not be the source of jökulhlaups during an eruption beneath the glacier.

8.8.1 The Ancient Vatnajökulsvegur Highland Trail

The Vatnajökulsvegur highland trail linking southern to eastern Iceland across the highlands centuries ago, was either through Vonarskarð pass, or by going to the north of Tungnafellsjökull and then by the western edge of Dyngjujökull across the Gæsavötn area. From there the route continued over Dyngjuháls between Trölladyngja shield volcano and Vatnajökull to the north of Kistufell, and then continuing eastwards across Urðarháls, or by a shorter route over the glacial snout southeast of Gæsahnjúkur (1215 m), descending on the western side of Kistufell. The Vatnajökulsvegur trail then lay east for 30 km along the terminus of Dyngjujökull, across the Síðdegisflæður streams and Holuhraun lava field, the Jökulsá á Fjöllum river and spurs of the Kverkfjöll mountains to Hvannalindir springs to the northeast of Kverkfjöll, and from there north over Brúaröræfi wilderness on the shortest route to inhabited areas. The Kreppa river was forded just north of its convergence with the Kverká and the journey then continued onwards through the Grágæsadalur, Fagradalur and Vesturdalur valleys; the trail never came nearer to Brúarjökull than at a distance of 10 km. The Vatnajökulsvegur was rarely used all the way. It is said that the lawyer Árni Oddsson (d. 1665) had travelled this route alone on horseback when hurrying the shortest way between Vopnafjörður to the Alþingi at Þingvellir in the summer of 1618 with documents, having defended the case of his father, Bishop Oddur Einarsson, against the governor of Iceland, Herulf Daa, before the king in Copenhagen. In 1794 Pétur Brynjólfsson (1770–1798) from Brú travelled west across the Sprengisandur highland route, and was the source for Sveinn Pálsson's descriptions of the northern margins of Vatnajökull. Forty-five years later, or in 1839, Björn Gunnlaugsson and Pastor Sigurður Gunnarsson travelled along Vatnajökulsvegur in fine summer weather when working as surveyors looking for mountain routes for the Mountain Route Society which had been founded in 1831 by Bjarni Thorarensen. In 1840 Pastor Gunnarsson (1949, 1950) again used this route on a calamitous journey with the Danish geologist Jørgen Christian Schythe. In August 1912, two months after the above mentioned Vatnajökull expedition of Koch and Wegener, the forestry commissioner Agner F. Kofoed-Hansen crossed Vatnajökull alone, or as he himself reported it, with three good horses, enough provisions, 40 lb of oats for the horses, a sleeping-bag made of water-proofed canvas, and two woollen blankets (Kofoed-Hansen 1912, pp. 41–42; Kristinsson 1985). Kofoed-Hansen left from Brú in Jökuldalur and crossed the highlands to habitations at Skriðufell in Þjósárdalur in four whole days and nine hours, having slept for only twelve hours en route. Most of his travelling days had been sunny, warm, and calm, the only bother having been a sandstorm in the marshes near Holuhraun lava field and fog at the Sóleyjarhöfðavað ford over the Þjórsá. The summer of 1912 was clearly a good time for travelling alongside the northern edge of Vatnajökull.

8.8.2   Further Expeditions to the Northern Margins

Þorvaldur Thoroddsen, along with Ögmundur Sigurðsson, explored the northern
edges of Dyngjujökull in 1884, and for the first time a complete picture of this
region was drawn up and the superstitions and folklore concerning outlaw camps
north of Vatnajökull were finally refuted. But sheep pastures were not the only
object of these explorations. In August 1923, northern Icelanders, under the leadership of the geologist Guðmundur G. Bárðarson, travelled from the Gæsavötn
rivers up onto the northern flanks of Bárðarbunga in search of volcanoes. They
discovered pumice on the glacier, which they believed to be from a volcano further
into the centre of Vatnajökull, but could make no further progress south due to bad
weather, which forced them to take shelter below the Dyngjuháls saddle (Jónsson
1945, vol. 1, pp. 370–371; Hannesson 1958). They thus found no volcanoes and it
seems they were completely unaware of the Swedes' discovery of Grímsvötn in
1919.

The first trips onto Vatnajökull in which motorised vehicles were used was in the
ascent of Dyngjujökull, to the west of Holuhraun lava field, in the summer of 1946.
In the beginning of July a jeep was driven 14 km up the snout until it could go no
further because of deep slush. A month later two jeeps managed to drive 18 km on
ice up to the snowline, and from there motor sledges were used to cross the snow to
explore Grímsvötn, the Kverkfjöll mountains and Bárðarbunga dome (Sigurðsson
1947, 1984).

Sveinn Pálsson was the first natural scientist ever to reach Brúarjökull. Together
with Pétur Brynjólfsson, he climbed Snæfell in 1794 and compiled a map of the
whole northern margin of Vatnajökull that was not improved upon until the 20th
century (Pálsson 1945). Björn Gunnlaugsson and Þorvaldur Thoroddsen drafted an
outline of the glacial margin on the verbal evidence of men familiar with it
(Thoroddsen 1892a, b, 1905/6, 1907–1911, Thoroddsen 1913–15). Travellers
headed for Brúarjökull early in the 20th century. The Dane Daniel Bruun and Elías
Jónsson, a farmer at Aðalból in Hrafnkelsdalur, rode on horseback along the eastern
side of Brúarjökull to the south of Snæfell in order to investigate the credibility of
stories of journeys across the glacier to Hornafjörður in previous centuries (Bruun
1902). German geologists also increased the knowledge of the Brúarjökull snout
and the Kverkfjöll mountains during the years 1910–1912. The German Max Trautz
made a map of the glacial tongue between the rivers Kreppa and Jökulsá á Fjöllum
which was a great improvement on previous ones. He climbed the eastern
Kverkfjöll in 1910, the first man to do so as far as is known, and thus had a superb
view over Brúarjökull's broad terminus; he named the proglacial lake at its snout
Tómasarvatn in honour of his travelling companion Tómas Snorrason. Trautz
returned to the northern margins of Vatnajökull two years later and came to the
conclusion that Brúarjökull was 11 km further north than on Thoroddsen's geological map. The glacier had surged by up to 10 km in 1890, six years after
Thoroddsen had first drawn up his outline of the glacial margins (Trautz 1914).

In the interwar years, and right up until the Second World War, Germans explored the terminus of Brúarjökull, for they believed that it had a similar glacial shape, environment and conditions as those in northern Germany at the end of the last glacial period 15 thousand years ago (Woldstedt 1939; Todtmann 1952, 1953, 1955a, b, 1957, 1960). The pioneer of Icelandic research in this region was Pálmi Hannesson, who visited the area in the summer of 1933. The Danes also saw similarities between the region and the Jutland Heath. It is also worth mentioning that the Swedish Icelandophile and expert on the theory of the Ice Age, Gunnar Hoppe, also described land formations there, which he believed might explain the shaping of landscapes in many other countries that had been covered by ice-age glaciers (Hoppe 1953, 1995). Geologists still return to Brúarjökull to examine and study subglacial sediments that might explain the causes of surges (Schomacker et al. 2006). Since the 1970s onwards, there has also been much extensive geological and glaciological research carried out in this area because of plans to harness the glacial rivers flowing north from Vatnajökull.

8.8.3 Brúarjökull

Brúarjökull (ca. 1600 km^2) makes up one fifth of the surface area of Vatnajökull and is the broadest and flattest outlet glacier in Iceland. It has an average thickness of 445 m and its total volume of ice is estimated to be 728 km^3. This is also just over one fifth of all the ice of Vatnajökull. Brúarjökull is 40 km long from the Kverkfjöll mountains eastwards to the Maríutungur slopes south of Snæfell. Southwest from its snout is a 45-km-long and 20-km-broad level bed, a continuation of the Brúaröræfi wilderness. At its lowest, the land is at a height of 520 m near the outlet of the Jökulsá á Brú. The bed of the southeastern level plain rises over a 10-km stretch up to the 1000-m-high dome beneath Breiðabunga, where there is a 1400–1500-m-high ice divide from which outlet glaciers descend into the Mýrar district in Austur-Skaftafell County. The ice divides with Breiðamerkurjökull run to the west of Breiðabunga along the centre saddle of Norðlingalægð hollow (1350 m), where the glacial bed only reaches a height of 800 m, until it begins to slope southwards to Breiðamerkurjökull. The ice divides then rise to the highest glacial dome (1600 m) above a range of mountains (1300 m), which stretch 15 km north of the Esjufjöll. Further west are the ice divides with Skeiðarárjökull at a height of 1550 m above a 5- to 10-km-broad pass into a basin, which does not even reach an elevation of 700 m. The ice is 900 m thick above the pass marking the divide between Brúarjökull and Skeiðarárjökull, and Brúarjökull then creeps 55 km from an elevation of 1500 m down to a height of 600 m at the Brúaröræfi wilderness (Figs. 8.93, 8.94, 8.95, 8.96 and 8.97). A 1625-m-high ridge heading south from the Kverkfjöll mountains marks Brúarjökull's ice divide with Dyngjujökull to the west. About 2 km before the convergence of the Brúarjökull, Dyngjujökull and Skeiðarárjökull glaciers, the ice is 900 m thick. One half of

Fig. 8.94 Author with Þórhallur Helgason on the grassy meadows on the eastern flanks of Sauðafell in the Brúaröræfi wilderness that now form the bed of the Hálslón reservoir. In the distance is Fremri-Kárahnjúkur peak (on the *right*) after which the dam is named. Hvannstóðsfjöll mountains are on the left. Þóra Ellen Þórhallsdóttir, 2006

Fig. 8.95 View along the vegetated banks of the Jökulsá á Dal opposite Lindir at Háls. This whole area is now at the bottom of the Hálslón reservoir. HB, 2007

Fig. 8.96   The Jökulsá á Dal flowing through Hafrahvammagljúfur. The river bed is now dry due to the dam at Kárahnjúkar. HB 2007

Fig. 8.97   Kirkjufoss waterfall in Jökulsá í Fljótsdal. It has now dried up with the coming of the Kárahnjúkar hydropower station

Brúarjökull's surface is above an elevation of 1200 m, as is just under 5 % of its bed, indeed only half of it reaches a height of 750 m.

The Kverká and Kreppa rivers flow from the western flanks of Brúarjökull and eventually merge with the Jökulsá á Fjöllum just southeast of Herðubreiðarlindir springs; the Kverká runs into the Kreppa a short distance to the west of Grágæsadalur. Jökulhlaups flood into the tributary sources of the Kreppa from Þorbergsvatn, a lake named after Þorbergur Þorleifsson from Hólar in Hornafjörður by British students. Whenever the glacier's snout lies alongside Kverkárnes and Kverkárhnúta, it forms two other lakes, which discharge water into the Kverká. One is called Hnútulón, and the other, about 4 km southwest of it, is unnamed.

The Kringilsá flows from the centre of the Brúarjökull snout, while further east, the Jökulkvísl runs through the Maríutungur. Both rivers used to flow into Jökulsá á Dal, but now run into the Hálslón reservoir. There is an isolated stretch of vegetation on the Kringilsárrani tongue between the Kringilsá and Jökulkvísl rivers, while to the west of the reservoir is the Brúaröræfi wilderness; the Vesturöræfi wilderness is to the east of the river and west of Snæfell. These wildernesses are the main homeland for reindeer in Iceland, providing them with their breeding and grazing areas in the spring and summer. They are also important breeding grounds for the pink-footed goose.

The Jökulsá á Dal harrowed its bed down into a stack of sediment layers for 4000 years on its way from the glacier north of the Kárahnjúkar peaks (Fig. 8.98). This stack had been formed when the sediment load filled up a lake in front of a retreating snout at the end of the last glacial period. After that particular glacier had disappeared, a small stream ran through the ground moraine and helped vegetation take root. Once Vatnajökull began to advance again, however, the glacial stream began to harrow its bed down into the sediment layers forming terraces of strata; these terraces were formed in stages as each hindrance to the creation of the Hafrahvammagljúfur canyon was eroded away.

Hafrahvammagljúfur, below the Kárahnjúkar peaks, is one of the deepest canyons in Iceland, beginning a short distance below the convergence of the Jökulsá á

Fig. 8.98 Sediment layers in the channel of the Jökulsá á Brú, on the boundaries of the Kringilsárrani and Vesturöræfi. They are now all submerged beneath the Hálslón reservoir

Dal and Sauðá rivers. The canyons along Jökuldalur become deeper and broader the further north it reaches. Due to the tremendous amount of erosion and huge sediment loads, there are no waterfalls in the Jökulsá á Dal. It transports 10,000,000 tonnes of mud, sand and gravel every year, more than any other river in Iceland, and in Héraðsflói Bay it has formed great gravel banks called Eyjar, which boast a large and colourful variety of flora and fauna. The river is 150 km long from its source in Vatnajökull to where it enters the sea, its catchment area being 2610 km^2, and its average flow rate 152 m^3/s.

8.8.4 Enormous Surges of Brúarjökull

Over a period of several centuries, Brúarjökull has surged every 70–90 years, its movements being the greatest glacial upheavals of modern times. The glacier surged forwards in 1963, and previous to that in 1890 and 1810. Entries in Ólafsson and Pálsson's travelogue indicate that Brúarjökull had also surged in 1720 and 1625 (when there was an outburst flood in the Jökulsá á Brú and its water level was 20 ells higher than usual and destroyed the bridge) (Ólafsson and Pálsson 1981, p. 111). In Brúarjökull's forefield there are three very distinct terminal moraine belts piled up by the surges. The most well-known of these, the so-called Töðuhraukar mounds, stretch right across the Kringilsárrani spur and were rapidly bulldozed into existence when the glacier surged 10 km across vegetated land in 1890 (Þórarinsson 1964, 1969; Björnsson et al. 2003).

In the years 1963–1964, Brúarjökull surged forwards 8 km at the outlet of the Jökulsá á Dal, though not as far on both sides of it (Fig. 8.99). The surge lasted for about six months, its greatest speed being measured at 4 m/h. The front edge jerked forwards every second and the earth shook in step with the glacial movements. A glacial tongue up to 200 m thick stretched over land that had previously been glacier-free, covering a total area of 160 km^2. A volume of just over 60 km^3 of ice was transported from the accumulation zone down to the glacier's ablation zone. The highest part of the glacier had subsided by up to 100 m. The sediment loads increased considerably in all the glacial rivers, the total sediment transport in Jökulsá á Brú in 1964 being 25,000,000 tonnes (at an average of 6.5 kg m^3). This was the equivalent of an erosion speed of 12 mm per annum over a surface area of 1500 km^2 during the unrest. The sediment load slowly but surely decreased until 1980.

As in previous surges (1890), the glacier advanced right up to Kverkárnes and Kverkárhnúta, thus forming a glacially dammed lake, Hnútulón, at a height of 750 m above sea level (with a surface area up to 2 km^2). At first the ice dam was so thick that there was a continuous flow of water over a pass into the Kverká, but shortly after the mid-1960 s the dam became thinner as the glacier retreated and water began to burst out from beneath the ice dam in jökulhlaups. There were ten such floods from Hnútulón from then until the turn of the 20th century. Sveinn Pálsson's description of this lake in his travelogue suggests that the glacier's

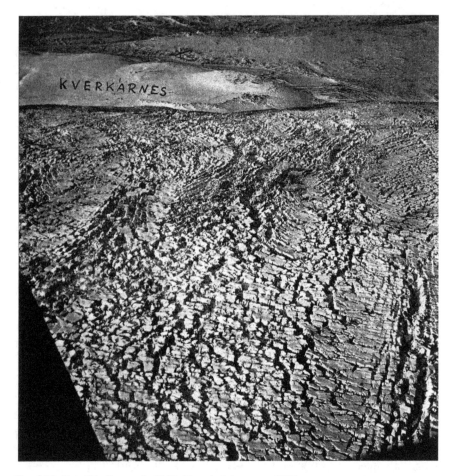

Fig. 8.99 Brúarjökull surge in 1964. Magnús Johannsson 1964

terminus, at the end of the 18th century, had been in a similar location as it was at the end of the 20th century.

A Vatnajökull surge in 1890
Description of Þorvarður Kjerulf, physician at Ormarsstaðir, Fell

This last winter, men became aware that there was some unusual disturbance in Vatnajökull at the end of the Jökuldalur valley and Fljótsdalsöræfi wilderness, especially the area where water flows into the Jökulsá í Fljótsdal, Jökulsá á Dal [Brú] and Kreppa rivers, for right at the beginning of winter, when there is usually little or no glacial colouring in these rivers, there was suddenly a lot of glacial clay in them, and this increased so much that, after the festivities [Christmas] when a bucket full of water had settled, almost half of it was glacial sludge ... groans and crashing noises were also heard as winter passed into spring, and at the weekend of the 14th week of the year there was a great flood of icebergs and a continuous flow of voluminous waters in the Jökulsá á Dal for several days. Soon thereafter, two men, Elías, the farmer at Vaðbrekka, and Jón Þorsteinsson from Aðalból in Hrafnkelsdalur, went hunting reindeer in the so-called Vesturöræfi—that is to say between Jökulsá á Dal and Snæfell—and found that the glacier had surged into the Vesturöræfi as far as the Sauðá river at approximately $1^{1/2}$ miles from the where the main glacier had previously been ... On the western side of the Jökulsá á Dal, the glacier had advanced onto the so-called Hraungarðir on the Kringilsárrani tongue, an old pushed-up moraine with glacier beneath it, which was now covered in grass fertilized by sheep dung and considered the best grazing land in the Brúaröræfi. The foremost edge of the glacier had slewed beneath these moraines so that they were now sprawled on top of the glacier's snout, which was 30 fathoms high at a guess. These moraines were at least 80 years old and the late Einar, who was born at the turn of last century and had lived so long at Brú, had remembered this surge. The glacier has melted in the years since then, and grazing ground formed above the layers of stone and gravel that had been left on top of the glacier which lay beneath it. When I heard all this, I went inland to the so-called Hvannstóðsfjöll mountains in Brúaröræfi because I had been told that this was the best viewpoint over the glacier and the wilderness; the inmost summits of these mountains are in line with the Snæfell and Herðubreið mountains, west of the inmost slopes of Laugarvalladalur ... This summit is at a guess (from barometric measurements) about 2700 ft. high above sea level. There is a very good view from there over all the wilderness between the Kverkfjöll and Snæfell mountains; I was also lucky enough to have good weather there that day (24 August), bright and clear and with no mist on the mountains. A more hellish but beautiful sight I have never seen; the glacier was cracked open and broken up down a 6-mile section, from the Kverkfjöll mountains in the west eastwards toward the centre of Vesturöræfi, and as far inland into the main glacier as I could see, probably no less than 3 to 4 miles; this was the glacier's condition over an area of 25 to 30 square miles. The glacier looks mostly like a series of cliff faces over which lies a half-melted autumn rime and over a wide area black bedrock and chasms are visible—or rather it's as if you could imagine seeing breaking waves frozen in the ocean. On top of every ice block lay stones and gravel, and in between them glimpses of blue-green glacial walls and pitch dark glacial crevasses. The glacier had disintegrated into conical blocks of ice, not very thick, but incredibly high, some of them probably 100 fathoms tall or more, especially further into the glacier. The moraine at the edge of the glacier will have been about 20–40 fathoms high, and in one place, east of the Kverkárrani tongue, it must have been near 100 fathoms high. The glacier was still in motion, for large blocks of ice were falling off the rock here and there, and a rumbling in the glacier could still be heard. (Kjerulf 1962, pp. 47–48)

Rolls and pancakes

On 27 July [1890], a jökulhlaup with large floating blocks of ice began in the Jökulsá á Brú. When men later went to check the situation further inland at the glacier, it appeared that Brúarjökull was so broken up in crevasses that in some places the bedrock beneath the glacier was visible between the huge cliffs of ice; the glacier swept away all the moraines on Kringilsárráni. ... There was also movement in Eyjabakkajökull in the autumn of 1890, which is nearer, to the west of

Þjófahnjúkar summits, and much smaller, though it pushed forward considerably into the grasslands. It so capsized everything in its vicinity that in 1894 there were still signs of its upheaval. In front of the glacier everything was jumbled together, soil and swathes of grass had been mixed in with the ice, mud, clay and gravel of the moraines, all mashed into one. In front of these moraines, now 20–70 feet high, are low ridges of soil and gravel all riddled with ice; apart from these, there were other clear and widespread signs of how enormous the pressure had been; the force of the glacier had rolled up these layers of earth into folds that accompanied the glacier like pleats on a dress, the waves of earth gradually becoming smaller the further away from the terminus they were; turf had also been wrapped up like giant pancakes containing sand and gravel; these rolled up earthen reels are all sprouting with grass, and there are swards of luxurious grass all the way up to the glacier. High on top of these moraines, in among the ice, gravel and clay, there were many earthen knolls in 1894 with grass growing out of them. This shows how a surging glacier has ploughed deep into the earth and soil, capsizing everything in its way. (Thoroddsen 1907–1911, vol. 2, pp. 13–14; Fig. 8.87).

8.8.5 Kárahnjúkar Hydropower Plant

The Jökulsá á Dal river has now been dammed at Kárahnjúkar with the resulting creation of the Hálslón reservoir which is 25 km long and 2 km broad (57 km^2), its highest water level rising to 625 m above sea level (Figs. 8.100 and 8.101). The three dams at Kárahnjúkar (193 m high), the Desjará (60 m) and Sauðárdalur (25 m) stand out like sore thumbs in their surroundings. The effects on the environment of the hydropower plant will be enormous, and many of them have yet to become apparent. The power plant splits asunder a vast uninhabited expanse of unified landscape to the north and east of Vatnajökull that was truly unique in the world. Hálslón reservoir divides the region into two and prevents reindeer from migrating between the Vesturöræfi wilderness, the Kringilsárrani tongue, and the terrain round Sauðafell. Moreover, the reservoir has submerged one fourth of the Kringilsárrani nature reserve, part of the Töðuhraukar mounds, the sediment strata to the south of Kárahnjúkar, and some especially well-vegetated and fertile moorland, the home of some rare plant species, as well as important geological and historical remnants. In a report on the environmental impact of the power plant, it was assumed that fluctuating water levels of the reservoir could reach a maximum height of about 70 m, but would on average be about 40–45 m high. Along the shores of the reservoir there would thus be a broad belt of land that would either be dry or submerged and where vegetation would soon cease to exist. What remains is a bare topsoil unprotected from erosion by wind and water. Right now, the moorland to the east of the reservoir has abundant vegetation, rich in plant species, and wetlands with thick layers of topsoil. It is estimated that within the fluctuating water levels of Hálslón there are approximately 30,000,000 tonnes of topsoil that could be eroded and blown away within the next few years. In addition to this

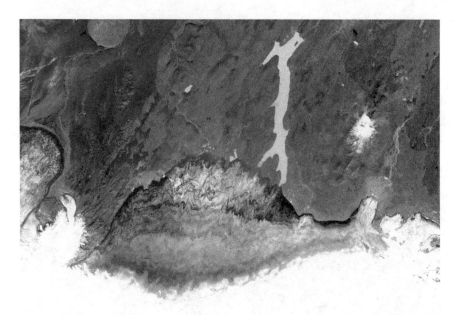

Fig. 8.100   Satellite image of the northern margins of Vatnajökull. Furthest west is Dyngjujökull and the tributary sources of the Jökulsá á Fjöllum, which merge to the south of Vaðalda (*top* of the picture). Then come the Kverkfjöll mountains and the very distinct Kverkjökull, and to the northeast of them the Kverkhnjúkar peaks. In the *centre* of the image is Brúarjökull, from which the Kreppa and Kverká flow and later merge to form the Jökulsá á Fjöllum. The Kringilsá runs northeast from the centre of the glacial snout into the Hálslón reservoir, as does the Jökulkvísl further east. Then comes the Eyjabakkajökull which creeps down into the Eyjabakkar wetlands and from which flows the Jökulsá í Fljótsdal to the east of Snæfell. The vegetated Vesturöræfi wilderness extends from Snæfell to the Hálslón reservoir, the Maríutungur grasslands southernmost at the glacier's snout. What is left of the nature reserve of the Kringilsárrani can be seen west of the reservoir nearest to the glacier. Hálslón divides into two one of the previously greatest expanses of wilderness left in Europe and a unique landscape. The reservoir flooded the Jökulsá á Dal canyon and overflowed onto lushly vegetated heaths and wetlands above a thick topsoil. Its water levels fluctuate much more than any other reservoir in Iceland (40 m or more) and in time nearby vegetation will die and the topsoil (all of 30,000,000 m^3, just on its eastern banks, according to the figures of Landsvirkjun, the Icelandic Power Company) will be left unprotected from erosion by wind and water. When the level in the reservoir is low in the spring and early summer, there is the danger of dust and sandstorms damaging or smothering the vegetation along the Kringilsárani tongue and in the Vesturöræfi wilderness. SPOT

potential dust mine are the sediment loads from Brúarjökull that settle on the banks of the reservoir. After the Icelandic National Planning Agency had rejected the plans for the building of a dam and power plant due to an unacceptable environmental impact, the decision was challenged and referred to the then Minister of the Environment, Siv Friðleifsdóttir, who agreed to the proposed dam with a few extra conditions, including the building of a protective wall or dyke to try and prevent soil erosion and dust storms.

Fig. 8.101 Near the Hálslón reservoir 22 June 2009. The water level had then fallen by 45 m, revealing a 30 km^2 area covered with glacial till. This was once vegetated land. Note the dust storm. Ómar Ragnarsson

Now that the power plant has been built, there is very little water in Hafrahvammagljúfur canyon, through which a clear river flows into Héraðsflói Bay. Its water flow in late summer is only half of what it used to be. The flow and currents of the rivers in the Úthérað region have completely changed from previous levels, meaning that wetlands along the river banks could dry out and islets in the estuary disappear because there is no longer any glaciofluvial sediment load to compensate for erosion. The flora and birdlife of the area might also be radically altered as the transport of nutrition and suspended sediments along the river to the sea are reduced, especially those containing certain minerals that are the staple nutrition for diatoms, which form the basis of the marine food chain.

Water flows out of the Hálslón reservoir at 107 m^3/s along a tunnel 40 km long from Fremri-Kárahnjúkur to the penstock of the power plant's turbines in Teigsbjarg, from where it is discharged into the Jökulsá í Fljótsdal and later the lake and river Lagarfljót. Changes in vegetation and shore erosion are expected along the Lagarfljót banks due to the increased flow of water. The Kárahnjúkar power plant also receives water from the Jökulsá í Fljótsdal, which was dammed at the Ufsarlón reservoir. Around 20 waterfalls in the Jökulsá í Fljótsdal will now have very little water, although it is expected that for a few days each year, especially during the summer, water will be allowed to flow out of the Ufsarlón overspill and thus briefly cascade down the river channel. Some of the better known waterfalls that will disappear from the Jökulsá í Fljótsdal are Eyjabakkafoss (10 m high), Tungufoss (10 m), Kirkjufoss (40 m), Faxi (20 m) and the Ytri-Gjögurfoss (20 m).

The volume of water harnessed from the Hálslón reservoir for the power plant is now 2100 gigalitres. If Brúarjökull glacier were to surge, which might well take place within the next 20–30 years, the glacial snout could reach as far as 6–8 km into the southern end of the reservoir, although the architects of the Kárahnjúkar power station claim that it will be possible to respond to and cope with this. It has been estimated that sediment deposits from the glacial rivers will silt up the reservoir in four centuries, though it might be silted up much earlier than that, should there be an increase in sediment load from rapid discharges of glacial meltwater over the next 100 years. There is certainly enough glacial till beneath Brúarjökull, and the sediment load will be determined by the meltwater flow rates. Once the Hálslón reservoir is full of sludge and the glacier has disappeared, rainwater will have to be accumulated in the highlands and new sites for reservoirs sought to provide water for hydropower plants.

As far as the participation of the Icelandic people in this issue is concerned, the nation was split into two ranks, one for and one against the Kárahnjúkar hydropower plant. The Icelandic National Planning Agency came out against the development because of its negative environmental impact and lack of clear and reliable data. The Minister of the Environment declared this decision null and void at the end of the year 2001 and a bill allowing the construction of the dam and power station was passed by the Alþing (parliament) in the spring of 2002. Moreover in 2005 the laws concerning the environmental impact of industrial or energy developments were changed so that the Icelandic National Planning Agency is now only an advisory body and can make no final executive decision as to whether the impact of such developments have an acceptable or unacceptable impact on the environment.

8.8.6 Dyngjujökull

The boundary between Dyngjujökull (ca. 1000 km^2; Fig. 8.102) and Brúarjökull in the east of the Kverkfjöll mountains heads southwestward toward Grímsvötn at a height of 1620 m. About 20 km southwest of Kverkfjöll, the boundary then veers west across the main keel of Vatnajökull (at 1675 m) and continues along a 12-km-long stretch of the ice divide with Skeiðarárjökull, this part of Vatnajökull being 700–900 m thick with its bed at 800–900 m above sea level. About 8 km to the northeast of the Eystri-Svíahnúkur peak of Grímsfjall, the ice divide then extends northwestward along a 1300–1400-m-high ridge to Bárðarbunga (2009 m). It continues directly over the summit of the dome and then along its northern edge before finally descending to the glacier's margin at Dyngjuháls, about 6 km to the west of Kistufell. The southwestern flank of Bárðarbunga descends to Köldukvíslarjökull, though on the dome's slope there is an ice divide for 7 km next

Fig. 8.102 The muddy and sandy Dyngjujökull last surged in 1999. A little way from the steep margins of its snout are the terminal moraines indicating the extent if its surge in 1977. HB, October 1999

to the catchment area of the Skaftárkatlar cauldrons. The ice is up to 800 m thick here, as the glacier bed reaches below an elevation of 1000 m. The greater part of Bárðarbunga is within the catchment area of Dyngjujökull. Over 70 % of the glacier's surface, but only a fifth of the glacier's bed, is above a height of 1200 m. Dyngjujökull descends along a 35-km-broad and 40-km-long stretch from the ice divide with Skeiðarárjökull to mud flats and gravel plains at a height of 730 m. The glacier surged in 1893, 1912, 1934, 1951 and 1974, and it advanced again about 1250 m during the winter of 1999–2000.

Jökulsá á Fjöllum Jökulsá á Fjöllum is the second-longest river in Iceland (206 km) and its catchment area is as large as the whole of Vatnajökull (about 8000 km^2), making it the largest in the country, with almost a fifth of it (1600 km^2) beneath the glacier (Fig. 8.103). Nonetheless, the river is only the fourth most voluminous waterway in the country (average flow rate about 250 m^3/s), which means that the terrain it runs through is rather dry. During volcanic eruptions, however, it can be transformed into by far the largest river in Iceland, for its catchment area beneath the glacier includes both of the craters in the Kverkfjöll mountains, the fissure chain leading southwest from them to Grímsvötn, the northern mountainside of the Grímsvötn volcano, and the eastern part of Bárðarbunga. In addition to all this, there are active fissure swarms beneath the centre of Dyngjujökull itself.

The most voluminous tributaries to the Jökulsá á Fjöllum emerge from beneath the glacier in an indent just to the west of Kverkfjöll mountains at a height of about

Fig. 8.103 In the foreground is a flood basin, with additional deposits from meltwater floods, left behind by an ice-age glacier in the central Icelandic highlands. The table mountain, or tuya, Herðubreið (1682 m) and the palagonite ridge Herðubreiðartögl (to its *left*) were formed during a volcanic eruption beneath a glacier. The river Jökulsá á Fjöllum is visible in the *centre*

800 m. Meltwater overspills into the braided channels of Síðdegisflæður to the east of Urðarháls saddle and as far as Kverkfjöll, and seeps down into the sand-eroded but porous lava before flowing north as groundwater, appearing in springs far away from the glacier. Glacial till can be seen in the marshes and this can be whipped up by the wind and carried long distances. There is a confluence of tributaries into one channel to the east of Mt. Upptyppingar, between the Krepputunga lava field and Mt. Vaðalda, which then flows north to Herðubreiðarlindir springs through flat, barren gravel plains and lava fields along the eastern side of the Ódáðahraun lava desert. The famous outcast Fjalla-Eyvindur had the worst time of his life at these springs, for his fire went out in the middle of winter.

The Jökulsá á Fjöllum then merges with the Kreppa across from Mt. Herðubreið, just southeast of the Herðubreiðarlindir. The Kreppa has its origins in the western parts of Brúarjökull and flows to the east of Krepputunga before being joined by the Kringilsá at the head of Grágæsadalur. Kreppa and Jökulsá have thus flowed almost parallel for 15 km before they reach and merge to the north of Upptyppingar. The Jökulsá á Fjöllum then runs northward for 60 km along the flat eastern edges of Mývatnsöræfi and to the west of Möðrudalsöræfi until it reaches the Hólssandur plain. The river draws its name ('á fjöllum') from the Hólsfjöll mountains. At the edge of the highlands, the river plunges down huge waterfalls through a 25-km-long canyon, Jökulsárgljúfur, which is up to 120 m deep and 500 m broad. The river, the canyon, and its immediate environs are a nature reserve and national park.

The canyon is believed to have been formed in catastrophic jökulhlaups from Vatnajökull about 7100, 4600, 3000 and 2000 years ago. A further huge outburst flood is believed to have occurred at the end of the last glacial period about 9000 to 8000 years ago (Sæmundsson 1973; Tómasson 1973; Elíasson 1977; Waitt 2002). The largest of these jökulhlaups could have attained a maximum flow rate of 1,000,000 m^3/s, which would have been larger than any outburst flood from Katla, and might have been triggered by volcanic eruptions beneath Dyngjujökull or the

Fig. 8.104 Jökulsárgljúfur canyon, gouged out by catastrophic floods from Dyngjujökull. HB, July 2004

dams of proglacial lakes being breached. The volcanoes could have been in Bárðarbunga or the Kverkfjöll mountains, or in fissures beneath Dyngjujökull, or furthest north of the Grímsvötn volcano. Large floods are least likely from the Kverkfjöll, however, as the glacier is at its thinnest there.

The highest and broadest ledge of the Jökulsárgljúfur canyon has been gouged out in the largest catastrophic flood (Fig. 8.104). The river itself now flows through the narrowest and deepest part of the canyon. Raging floodwaters had burst through a row of cinder craters in Vesturdalur, sweeping away all loose material and leaving behind only the Hljóðaklettar cliffs, the internal structure of the volcanic fissures that were opened up about 9000 years ago. Ásbyrgi, the famous horse-shoe shaped canyon 1 km broad, 3.5 km long, and with cliffs 100 m high, was also gouged out by catastrophic floods further north, in the Kelduhverfi district, about 2 km to the west of the present channel of the Jökulsá.

The waterfalls in the Jökulsárgljúfur canyon are, in order of descent from the highlands: Selfoss (10 m high), at an elevation of 320 m above sea level, then Dettifoss, the most powerful waterfall in Europe (45 m high, about 100 m broad, and with an average flow rate of 182 m^3/s), then Hafragilsfoss (27 m), and finally Réttarfoss, the last before the lowlands. These waterfalls will all disappear in their current shape and form if plans go ahead to divert the Jökulsá á Fjöllum from its present channel, just south of Herðubreiðarlindir, into a reservoir in Arnardalur. The water would then be redirected from there through a tunnel into another reservoir in

Jökuldalur and finally through a headrace tunnel to a power plant before being discharged into the Lagarfljót.

Once the river has passed through the Jökulsárgljúfur canyon, it flows through the lowlands of Kelduhverfi, transporting about 5,000,000 tonnes of sediment to the ocean at Axarfjörður, where the river is thus also called Jökulsá í Axarfirði. In its journey through Kelduhverfi, most of the volume of water diverges into other offshoots, including the Sandá and Bakkahlaup rivers. There was once a fjord quite far inland here, but it has gradually been filled in with glaciofluvial deposits and its vegetated flatlands are important breeding grounds for birds. Floods in this section of the glacial river have often damaged the meadows and home fields in Kelduhverfi and Axarfjörður. Jökulhlaups during winter might have been caused by iceberg-dams being breached. Seven fairly large glacial floods in the Jökulsá á Fjöllum have been cited in annals from the years 1477 to 1729, all believed to have been triggered by volcanic eruptions, though the exact location of the volcanoes is unknown. In 1477 there was an eruption in the fissure swarm from the Veiðivötn lakes to Bárðarbunga, in 1684–1685 an eruption in Grímsvötn, which also caused a jökulhlaup in the Skeiðará, and in 1726 there appears to have been an eruption in the northern part of Dyngjujökull, near to Grímsvötn. The greatest outburst flood was in August to September of 1717, and this caused tremendous damage in Kelduhverfi, ripping up land and reputedly covering the fields with pumice. It is quite feasible that such a jökulhlaup would have reached a flow rate of several thousand cubic metres per second (similar to a fairly big outburst flood from Grímsvötn). Accompanying the eruption was an ash fall that spread from Eyjafjörður to the inland valley area of Hérað in the northeast that delayed haying and prevented grazing. There were five jökulhlaups in the Jökulsá á Fjöllum during the winter of 1729, but it is not certain that these were triggered by volcanic activity (Þorsteinsson et al. 1922–1987; Thoroddsen 1897; Þórarinsson 1950). In the years 1902–1903 there were outburst floods in both the Jökulsá á Fjöllum and Skjálfandafljót, and these might have been due to an eruption on the ridge on the northern side of Bárðarbunga, for both of them had jökulhlaups again during the 1934 eruption of Grímsvötn. In the eruption of Jökulbrjótur in 1996, the volcanic fissure stretched northward, almost reaching all the way to the water catchment area of the Jökulsá á Fjöllum.

Þorvaldur Thoroddsen's arrival at the crevassed and black Dyngjujökull in 1894

Between the Kistufell and Kverkfjöll mountains there is a broad hollow in Vatnajökull from which a continuous outlet glacier, undoubtedly one of the largest in the country, has descended onto the plain. It is hard to imagine a more horrible sight when looking over this glacier. Indeed, no one would think it was a glacier to begin with, because it has borne so much gravel, sand, scree and rocks that all its lower reaches look as black as the darkest-grey aa lava. It was steepest just to the

east of Kistufell, where the glacier's edges are vertical, pitch black, shattered and
sundered with innumerable gullies, ravines, and horrifyingly deep gaps and chasms.
There is one ice tower after another in the glacier with crevasses in between and they
are 30 m high or more and all of them black with a coat of gravel and with enormous
boulders strewn around them. There is total upheaval here, like the ruins of buildings
after an earthquake, everything pitch black and muddy brown like the dregs of
grounded coffee. The glacier is flatter further east and the ice towers not quite as
perilous, though the glacier is still bursting apart, black crests of ice and small seracs
all on top of each other, covered by remnants of moraine, glacial till and huge
boulders; in every hollow there are brown muddy pools, into which sink clayish
sludge which flows downstream; in some places there are small fountains of water,
blue crevasses and little waterfalls that disappear into the depths of a pitch black
darkness. (Thoroddsen 1958–1960, vol. 1, p. 367)

When the Jökulsá dried up, a jökulhlaup was to be expected
'In 1717 there was an eruption in the Kverkfjöll mountains and there was a
large jökulhlaup in the Jökulsá, which was said to have created the river
Stórá. Oral tradition has it that there was a small creek with duckweed that ran
westward from Byrgi to Keldunes peninsula, and the jökulhlaup flooded this
creek all the way to Lake Víkingavatn, inundating all the meadows, its waters
reaching as high as midway up Skemmuhóll by the Víkingavatn farmstead. In
those days a boat was needed to get from Víkingavatn east to Ás. The water
level remained high for a long time, but once it receded the Stórá had been
formed. The Jökulsá dried up the day before the jökulhlaup, and it is reported
that the farmer at Skógar had said an outburst flood was thus to be expected.
He had his six-oared boat moored to the farmstead's front door, moved
everything that could be damaged up into the loft, and asked everyone in the
home to stay awake all night long, which they did. It was a bright night, and
after midnight folk saw a high wall moving across the plain, and it was indeed
the jökulhlaup. It did not reach Skógar, however, as it divided into two a little
higher up, though it did wash an iceberg up onto the front yard. Before this
outburst flood, there had been fine meadows below the upland slopes and
many farms on the moor above the lowlands had their hayfields below. All
these fields were destroyed and most of the farms were abandoned. The ruins
of many farmsteads, their drystone walls, and their bridle paths descending
onto the plain, can still be seen. Before the flood, the eastern part of
Kelduhverfi had been considered the better and more populated area than the
western part, but several of the farms to the east of the river were destroyed
and remnants of them are still visible. Old home field walls and very long
round-up and boundary walls can be seen over a wide area. It is unlikely,
however, that all these farms were wiped out by the flood of 1717, many of
them probably being destroyed much earlier. There are entries in the annals
concerning a large jökulhlaup in the Jökulsá in 1655. ... There have
undoubtedly been many outburst floods previously even though there are no
records of them. The community of farmsteads has probably been dissolved

slowly but surely ever since the olden days as the river destroyed the grasslands below them.' (Thoroddsen 1958–1960, vol. 1, pp. 322–323).

8.8.7 Volcanoes Beneath Dyngjujökull

Dyngjujökull is surrounded by volcanoes on three sides, in the Kverkfjöll mountains, on the northern flanks of Grímsvötn, and in the eastern part of Bárðarbunga, while beneath it is a fissure swarm that might be connected to Askja (Fig. 8.8).

The Kverkfjöll mountains The Kverkfjöll mountains (Figs. 8.105, 8.106, 8.107, 8.108) are Iceland's second-highest volcano after Öræfajökull glacier as they rise, partially covered by ice, more than 1000 m above their nearest surroundings, the highest point being Jörfi on their eastern side (1933 m). Parallel palagonite ridges and rows of peaks that have been piled up, mostly on fissure chains during the last glacial period, extend in both directions from the main volcanic centre. The Kverkfjallarani spur heads northeastward and is 30 km long, while to the southwest there are three parallel ridges, which stretch altogether over an almost 10-km-broad area beneath a 350–400-m-thick glacier, only just failing to reach the Grímsvötn mountain range by 2–5 km. There is a clearly water-eroded channel down the

Fig. 8.105 Kverkfjöll mountains. Sketch by Walther von Knebel and Hans Reck. NULI

Fig. 8.106 The Kverkfjöll mountains in winter. Fresh snowfalls have covered traces of geothermal heating in Hveradalur (on the *left*) and beneath the so-called Gengissig cauldron. Kverkin is in the *centre* of the photograph and Snæfell in the distance (see also Fig. 8.119). HB, February 1995

whole of Dyngjujökull through this gap between the ridges and mountain range, raising the question as to whether this might be the source of catastrophic jökulhlaups in the Jökulsá á Fjöllum. Could a tremendous volcanic eruption have split asunder the mountain ridge that connected the Kverkfjöll fissure swarm and Grímsvötn? The chemical composition of the volcanic material seems to indicate that the volcanic systems of Kverkfjöll and Grímsvötn are of the same origins. Another ridge, 20 km long and more than 1200 m high, heads due south from the Kverkfjöll, and volcanic activity in its eastern flanks might be the cause of jökulhlaups in the rivers Kreppa and Jökulsá á Brú.

The main volcanic centre of the Kverkfjöll has two oval-shaped calderas, both about 10 km long and 7 km broad. The more southerly one is 400–500 m deep, covered by a glacier, and its axis lies southeast to northwest. At its lowest, its basin reaches a height of 1200 m, though its narrow southern rim is at an elevation of almost 1700 m, and covering it is a dome of firn snow that travellers on the glacier call Brúðarbunga (after a very respectable pregnant woman who was on the glacier during an early expedition of the IGS). A canyon runs southwestward for 6 km from the lowest part of the caldera's western rim that might have been formed by erosive floods triggered by subglacial eruptions. These floods veered southeast from the end of the canyon beneath the centre of Dyngjujökull and then continued into

Fig. 8.107 One of the country's most powerful geothermal areas is in the Kverkfjöll mountains, the second highest volcanic centre in Iceland. Geothermal heat melts glacial ice and the meltwater accumulates in lakes in Hveradalur (on the *left*) and Gengissig (on the *right*). The IGS has a cabin on the ridge between them. Kverkjökull descends northwestward through Kverkin and ahead of it is the cabin of the Touring Association of Vopnafjörður, Fljótsdalshérað and Húsavík, built and named after Sigurður Egilsson (1892–1969) from Laxamýri in the Reykjahverfi district of Þingeyjar County

the Jökulsá á Fjöllum. The three other ridges head towards Grímsvötn from the southerly caldera south of the canyon.

The more northerly caldera stretches from southwest to northeast, and its bed is lowest at a height of 1600 m. Its western caldera rim is only 100 m high, but on its eastern edge an ice-free peak, Jörfi, rises above the glacier to a height of over 1900 m. Skarphéðinsjökull descends eastwards through a pass in the caldera's rim, and is named after Skarphéðinn Gíslason from Vagnstaðir in the Suðursveit district. A steep and crevassed Kverkjökull (ca. 6 km²) descends to the northwest through a ravine in the caldera's northern rim and then through a deep valley with high and sheer cliff walls down to a height of 900 m. The ravine is called Kverk, from which the mountains draw their name, and it divides the range into the Austurfjöll (1933 m) and the slightly lower Vesturfjöll. Evidence of geothermal heat is visible over a wide area to the east and west of the northern caldera. There is also geothermal heat at the base of Kverkjökull and the largest hot spring stream in Iceland, which melts an ice tunnel to create its own channel, flows from beneath this glacier. Further geothermal activity is present on the higher flanks of Jörfi, and to the east of the Kverkfjöll mountains is a 2-km-long ravine, Hveragil, in which there

Fig. 8.108 Skarphéðinsjökull descends eastward through a pass in the most northerly caldera of the Kverfjöll mountains. The peak Jörfi (1933 m) is to the left of the glacier. Dyngjujökull is in the distance. HB, August 2006

are hot springs and pools with temperatures from 40 to 60 °C as well as patches of rich vegetation.

The most geothermal activity is in the Vestur-Kverkfjöll mountains in a fault that stretches northeast to the west of the northern caldera. This is one of the most powerful high-temperature areas in Iceland (2000 MW, 10 km^2) and is in the southernmost part of Hveradalur, which is at an elevation of 1600–1700 m, is about 3 km long and almost 1 km wide, and is full of large fumaroles and solfataras. The bedrock here is saturated with geothermal heat. Glámur, one of Iceland's largest fumaroles, is in Þrengsli in the northernmost part of Hveradalur, while in the southern part of the valley is a fairly large, frequently ice-covered lake, which sometimes empties, probably westward beneath Dyngjujökull, into the tributary sources of the Jökulsá á Fjöllum. Neðri-Hveradalur, also called Hveraskál ('hot spring basin'), begins to the northwest of Þrengsli and extends in the direction of Dyngjujökull. There is another lake in an ice cauldron (humorously called Gengissig, 'devaluation'), which is about 600 m broad and 100 m deep. It is just within the western rim of the northern caldera, to the east of Hveradalur, and just to the east of the cabin of the IGS, which was built in 1997 at a height of 1700 m. There was an enormous explosion of steam in this ice cauldron in 1959, although the cauldron itself is much older, for it can be seen in aerial photographs from 1944 (Þórarinsson 1950, p. 127). There have often been outburst floods from this cauldron lake, and they have flowed beneath Kverkjökull into the river Volga and thence onwards into Jökulsá á Fjöllum. Recorded jökulhlaups have occurred in 1974, 1977, 1985 (smallest), 1987 (largest), 1993, 1997 and 2002. Sediment layers in the glacier's forefield also indicate that really huge floods have previously emerged from beneath Kverkjökull.

8.8.8   The Askja Swarm

A swarm of fissures that might be connected to the Askja volcano disappears beneath the centre of Dyngjujökull (Fig. 8.109). A row of several individual, probably palagonite peaks can be discerned beneath Dyngjujökull extending 30–35 km, heading in a northeast to southwest alignment towards the northern part of the Grímsfjall volcano. Eruptions along this row of craters trigger jökulhlaups in the Jökulsá á Fjöllum. The highest peak is 1050 m high, and about 20 km from the glacier's terminus it rises 300 m above its environs beneath a glacier 300 m thick. Its eastern side is precipitous and stands out from all the other summits, which are more rounded, indicating that there has been no glacial erosion there for a long time.

Fig. 8.109 Lake Askja in the Dyngjufjöll mountains attracted many scientists in the beginning of the 20th century, while they were also exploring the northern margins of Vatnajökull. HB, August 1996

Fig. 8.110 View south over Bárðarbunga, one of the largest calderas in Iceland, full to the brim with ice. The upheavals in Vatnajökull in 1996 began with an earthquake in the northern rim of this caldera. In the distance can be seen the volcanic centres in the glacier and Grímsvötn, in which the meltwater from the eruption accumulated (on the *right*). In the forefront is Rjúpnabrekkujökull. HB, November 1996

8.8.9   The Bárðarbunga Caldera

Bárðarbunga is one of Iceland's largest volcanoes (Fig. 8.110). It rises 900 m above its surroundings from a height of 1000–1100 m at its northern and southern flanks. In the centre of the mountain is a 700-m-deep caldera and there are 11 km in between its rims in the southwest and northeast, and 7–9 km between its north-western and southeastern rims, all enclosing an area of 80 km^2. The bottom of the caldera is at an elevation of 1100 m, and its highest rims are at 1850 m in the north, at 1650 m in the west, at 1750 m in the south, and are at their lowest at about 1450 m in the east. There are three passes in the caldera's rim, one in the southwest at about 1450 m, providing an outlet for Köldukvíslarjökull, another in the northeast at around 1400 m, and the lowest is in the middle of the eastern rim at about 1350 m, from which a long, narrow valley extends and which might be a conduit for jökulhlaups from the centre of Dyngjujökull.

The glacier is up to 800 m thick within the caldera (where there is a total of 43 km^3 of ice), but it is at its thinnest, about 150 m, above the northern rim; it is

250 m thick above the western rim, and 200 to 400 m above the southern rim, but the ice is at its thickest as it flows 300-400-m-thick over its eastern rims. In the southern part of Bárðarbunga, the glacier reaches a thickness of 700 m, but thins out to 400 m nearer the northern edge of Grímsvötn. The ice in the Grímsvötn hollow is from 250 to 350 m thick.

Bárðarbunga is believed to be at the centre of a volcanic system that extends southwestward for 150 km, as a fissure swarm, from its southwestern flanks through Köldukvíslarjökull in the direction of Mt. Hamarinn and on to the Veiðivötn lakes and Torfajökull, and northeastward along a ridge towards Dyngjuháls saddle and Trölladyngja shield volcano (Sæmundsson 1979; Larsen 1984). The volcanic system thus stretches for almost 200 km and is the largest in Iceland. Evidence of 27 eruptions along the Bárðarbunga to Veiðivötn volcanic system have been found (Larsen et al. 1998). There have been no recorded eruptions of Bárðarbunga in historical times, but there was probably a subglacial eruption in its northern part, south of the Dyngjuháls saddle, in 1902, when a jökulhlaup burst into both Jökulsá á Fjöllum and Skjálfandafljót. There might also have been an eruption in this region in 1797, as well as early in the 18th century, when outburst floods destroyed farmlands in Kelduhverfi (Þórarinsson 1974a, b; Jónsson 1945, vol. 2, p. 299).

No magma chamber has been found beneath Bárðarbunga and the geothermal activity there is insignificant, although there is some activity only a short distance away in Vonarskarð pass and beneath the lake Hágöngulón. Two ice cauldrons were formed in the most southeasterly part of the Bárðarbunga caldera during the Jökulbrjótur eruption in 1996, indicating there had been a small eruption there. Earthquakes were frequent in Bárðarbunga from 1974 right up until the volcanic eruption in the autumn of 1996. On 29 September that year, there was a major earthquake (5.4 on the Richter scale) at the northern rim of the Bárðarbunga caldera and this marked the beginning of a series of earthquakes, the original epicentre of which moved south from Bárðarbunga when magma broke through the Earth's crust half way towards Grímsvötn and erupted, causing continuing unrest and tremors.

Eighteen years later the Bárðarbunga volcano injected magma laterally from its centre again, and this time it emerged in a fissure at 750 m above sea level about 10 km north of Dyngjujökull after travelling about 40 km in ten days along a dyke at a depth of around 5 km beneath the glacier. The eruption began on 31 August 2014 and lasted for 180 days until late February 2015, producing a lava field of 85 km^2 named Holuhraun or Flæðahraun. The magma propagation started with intense seismic activity of a magnitude up to 4.5 in the northeast of the Bárðarbunga caldera, and when the eruption started the earthquakes increased in frequency at the caldera rims, with more than 70 quakes of a magnitude above 5. During the eruption, the caldera floor subsided by 40–50 m. Hardly any melting took place at the glacier bed and no jökulhlaups occurred. Nor did the eruption produce any significant amount of volcanic ash, but it did release large volumes of sulphur dioxide, which temporarily affected the air quality in many parts of Iceland, depending on weather conditions. Air traffic was only disrupted in the immediate vicinity of the eruption (IES and IMO websites 2016).

8.9   The Western Margins of Vatnajökull

The western side of Vatnajökull glacier stretches 60 km southward from the Bárðarbunga dome to the sheep pastures at Skaftártunga, creeping downward for 50 km from an elevation of 1600–2000 m to a wilderness at a height of 600–800 m (Figs. 8.111, 8.112 and 8.113). The landscape features parallel palagonite ridges from the last glacial period extending in a southwest-northeast direction. There are also signs, however, that the terrain has been shaped by more recent volcanic activity as one lava field follows another with individual craters and long volcanic fissures. The land nearest to Vatnajökull has been moulded by the glacier itself during the Little Ice Age, by its consequent retreat during the 20th century, and by

Fig. 8.111 Aerial view south along the western margins of Vatnajökull. Nearest is Köldukvíslarjökull, Hamarinn and Hamarslón, then comes Sylgjujökull, and ahead of it the Tröllahraun lava field (from 1862) and the Kerlingar mountains, then Tungnaárjökull, and finally, in the far distance, Mýrdalsjökull. HB, August 1996

Fig. 8.112 View south over Síðujökull at Hágöngur (1120 m), the source of the Djúpá. HB, August, 1996

Fig. 8.113 The sources of the Skaftá river (on the *left*) on the boundary between Skaftárjökull and Tungnaárjökull. Lake Langisjór (on the *right*) lies between the Fögrufjöll and Tungnaárfjöll mountains. Mýrdalsjökull is visible in the distance. HB, August 1996

the glacial till that has been dispersed in streams on the gravel plains there. Meltwater from the glacier and thawing snow seep into the porous layers of lava and merge with flowing groundwater. Glacial till has managed to solidify the glacial river beds in the lava so that they flow in channels high above the groundwater level. There is no wilderness more barren in Iceland, with hardly a blade of grass visible, than the one close to the western reaches of Vatnajökull. It has many sand and dust storms in the later part of summer, even though there is a great deal of precipitation. A few lakes do exist there, however, the water held by the thick palagonitic tuff, and moss grows along their banks. This volcanically-formed topography is thus one of contrasts, with its extraordinary mixture of black, barren, gravel plains, scattered lakes, and moss-green lava fields and palogonite ridges.

The mountains along the western margin of Vatnajökull dissect its broad mass into outlet glaciers named after the rivers that originate from them (Köldukvísl, Sylgja, Tungnaá, and Skaftá). Northernmost is Köldukvíslarjökull between Bárðarbunga and Hamarinn, to its south is Sylgjujökull as far as the Kerlingar mountains, then comes Tungnaárjökull down to the Tungnaárfjöll mountains, then Skaftárjökull, and finally and southernmost, Síðujökull, from which flow Hverfisfljót, Brunná and Djúpá. Fljótsoddi point and the inhabited district of Fljótshverfi, much further southwest, are named after the river Hverfisfljót, and it is from Fljótsoddi that the airplane Jökull took off after having been dragged down from the top of Bárðarbunga following the Geysir air disaster in the autumn of 1950. To the east of Síðujökull is a nameless glacial tongue, along which runs the water divide between the Djúpá and Núpsvötn rivers. The glacial snout to the west of Grænafjall (1063 m) is called Grænalónsjökull, and it once extended into Grænalón, but the water now flows into the lake through mud flats. There are two isolated drainage basins within the glacier that accumulate water in lakes until it is released in outburst floods into the Skaftá.

Beneath the northwestern reaches of Vatnajökull is the centre of the hottest spot in Iceland, and this zone is bounded by the main volcanoes of Bárðarbunga and Hamarinn at the glacier's edge and Grímsvötn in the middle of the glacier. These three main volcanoes have piled up mountain ranges in the passing of time from which stem long, very straight, palagonite ridges along fissure swarms. In the northern area, all the terrain is higher than 800 m, but further south, beneath Tungnaárjökull and Skaftárjökull, the land is much lower and its topography in many ways similar to that in the Tungnaáröræfi wilderness. The forms of the glacier's surface reflect its subglacial ridges and terrain.

Bárðarbunga dome towers above the northwestern corner of Vatnajökull and ice flows from it in all directions. The main ice flow descends east and southeast from the caldera down Dyngjujökull, while ice from the southern sides of the caldera flows towards Grímsvötn and the Skaftárkatlar cauldrons. The ice on the western part of Vatnajökull is densest here, or 700–800 m thick. Outlet glaciers descend steeply to Vonarskarð pass and its flat gravel plains and islets, to the northwest of Bárðarbunga. This pass marks the water divide between the catchment areas of northern and southern Iceland: the southernmost source of the Skjálfandafljót in the north and the northernmost origins of the Köldukvísl, which flows south. To the west of Bárðarbunga, the ice flows downwards into Köldukvíslarjökull (ca. 300 km^2).

8.9.1 Jökulhlaups and Magma Flows from Bárðarbunga Volcano

A powerful jökulhlaup, similar in nature to a Katla outburst flood, might rush down Dyngjujökull during an eruption within the Bárðarbunga caldera. A magma flow below the caldera might also cause an eruption outside of it (as in 2014), but it would depend on the location of the volcanoes as to whether a jökulhlaup would enter the Jökulsá á Fjöllum, Skjálfandafljót, Köldukvísl, Tungnaá, or Skaftá rivers, or head for Grímsvötn and from there down to Skeiðarársandur, as indeed happened in the Jökulbrjótur eruption in the autumn of 1996. A fissure eruption on Dyngjuháls saddle might be triggered by a magma flow below Bárðarbunga, and such a flow might have caused the eruptions in the years 1902–1903 when a jökulhlaup also flooded the Skjálfandafljót and Jökulsá á Fjöllum rivers. Indeed, the only real possible source for outburst floods in the Skjálfandafljót are eruptions in the northern part of the Bárðarbunga volcano.

A bedrock ridge extends southwest from Bárðarbunga beneath Köldukvíslarjökull and heads for the northwestern side of Hamarinn, where it reaches its highest point (1573 m). To the east of the ridge, the terrain is mostly above an elevation of 1000 m, descending to a height of 950 m due south of Bárðarbunga, where the ice is all of 700 m thick, though only 300 m thick above the ridge itself. Bárðarbunga rises to such a height that it directs ice flows and water over the ridge down onto a tongue of the Köldukvíslarjökull, and the glacier divides along the ridge into two almost equal-sized parts. Kaldakvísl flows from the Köldukvíslarbotn basin in the southernmost part of Vonarskarð pass through ground moraine, gravel plains and the Hágönguhraun lava field into Hágöngulón, into which the river Sveðja, emerging from the Hamarskriki indent, also flows. Kaldakvísl was diverted into the Þórisvatn reservoir in 1972, though up until then clear water flowed from Þórisvatn through the Þórisós into the Kaldakvísl.

The chain of fissures from Bárðarbunga to Hamarinn continues beneath the lowest part of Sylgjujökull and in a southwesterly direction beneath Tungnaárjökull, about 2 km east of the Kerlingar mountains. Syðri-Kerling (1339 m) has a pointed ridge, but Nyrðri-Kerling (1207 m) has a rounded summit. A few individual peaks, rising higher than 800 m, can be discerned along the palagonite ridge, and beneath Sylgjujökull there is a cone-shaped mountain which rises above 1000 m. If there were an eruption at this site, a jökulhlaup would rush down the Sylgja and out into the lava field. The fissure swarm connecting Torfajökull, Veiðivötn, Vatnaöldur and Heljargjá ravine, disappears beneath Sylgjujökull and Köldukvíslarjökull, even though no signs of it are visible in the subglacial topography. The fissure reaches Bárðarbunga and becomes visible again near Kistufell, from whence it continues to Trölladyngja volcano. The last eruption along this fissure swarm was in the Tröllahraun lava field in 1862.

The main volcano, Hamarinn (1570 m), gets its name from its precipitous western cliffs, where the mountain reaches its highest point. Hamarinn stretches 8 km eastward beneath the glacier and its summit is circular in shape, covering a surface area of 60 km^2 at a height of 1200 m, although the centre of its crater drops to a height of 1150 m.

A ridge extends southwest from the eastern side of Hamarinn towards the Fögrufjöll mountains at Lake Langisjór, where it is called the Fögrufjallahryggur, as it is on a fissure swarm named after the Fögrufjöll (Sæmundsson 1979). There has been no volcanic activity in this section of the fissure swarm outside of the glacier since the last glacial period. This ridge, which often reaches a height of almost 800 m, also marks the water divide between the catchment areas of the Tungnaá and Skaftá rivers beneath the glacier. The Skaftá flows in two main branches from below Vatnajökull, the western one flowing from the southernmost part of Tungnaárjökull and from Skaftárjökull just to the east of Fögrufjöll, while the eastern one flows from the northernmost part of Síðujökull. An outlet glacier, glacial tongue, and valley are all named after the Skaftá, as indeed are the greatest volcanic eruptions and lava flows in the history of Iceland since its forefathers first claimed land and settled there.

The Tungnaá has a catchment area of 120 km^2 within the glacier, which reaches all the way to Hamarinn. If Tungnaárjökull were to retreat 3–4 km at the river's basin, Tungnaá could be a groundwater river again, as indeed it is believed to have been until about the year 1600. The basin descends from the east of Fögrufjallahryggur into the main part of Vatnajökull, where the ice is more than 600 m thick. To the west of the ridge, the land is 100 m higher than to the east, for indeed Tungnaárjökull has had much less time to carry out glacial erosion than the old, main glacier. Moreover, the dispersal of subglacial sediments has uplifted the terrain a little at the glacier's snout.

8.9.2 Skaftárkatlar Ice Cauldrons on Lokahryggur Ridge

Hamarinn is more complex than the other volcanoes beneath the glacier in the sense that, due east from it, right across the normal path of the fissures in the Earth's surface, there is a 1100–1250-m-high spur that reaches almost all the way to the north of the Grímsvötn volcano. There is a hypothesis that Hamarinn and the ridge running east from it are indeed the main volcanoes, which, together with the fissure swarm on Fögrufjallahryggur, form a separate volcanic system (Björnsson and Einarsson 1990). All along this spur is one of the most active earthquake areas of Iceland with frequent, though not very large, tremors. The quakes are caused by the expanding or shrinking of a magma chamber at a depth of 2–3 km below it (Einarsson1991; Einarsson and Brandsdóttir 1984; Alfaro et al. 2007). An eruption on this ridge could have triggered a jökulhlaup in the Skaftá, Hverfisfljót, and even

Köldukvísl rivers, which would then have continued into the Þjórsá. Above the magma chambers are three geothermal areas where ice is constantly melting, so that ice cauldrons have formed in the glacial surface above them. Beneath two of them, 10 and 15 km northwest of Grímsvötn, meltwater accumulates until it bursts out into the Skaftá; the cauldrons are thus named Skaftárkatlar. The hollows in the ice are almost circular, are 1–3 km in diameter, and at their centres the glacier is 300–400 m thick. There have been at least 40 outburst floods from below these cauldrons since 1955 that have forced their way beneath the glacier for 30 km to the east of Fögrufjallahryggur all the way into the Skaftá. Some of the floodwater runs west through a pass in the ridge about 8 km to the east of the Tungnaá basin. This pass descends to a height of 500 m, but the pressure of the main glacier's overburden nonetheless forces this water up to an elevation of 650 m so that it emerges from beneath Tungnaárjökull to the north of Lake Langisjór and flows eastward from there along the glacier's margins into the Skaftá. There are even examples of these floods spurting a little water into the Tungnaá.

There is usually a two- to three-year interval between jökulhlaups from one of these cauldron (Figs. 8.114 and 8.115). The centre of the eastern cauldron sinks about 100–150 m with every outburst flood and it discharges about 200–350,000,000 m^3 of water. The cauldron to the west of it sinks about 50–100 m, and discharges 50–150 × 10^6 cubic metres of water. The catchment area of the eastern cauldron has been estimated as 29 km^2 and the western one as 20 km^2. The

Fig. 8.114 Eystri-Skaftárketill at the end of a jökulhlaup in 1982. The ice cauldron was 150 m deep and had a diameter of 3 km between the outermost rings of crevasses. To the southwest of the cauldron there is a fissure in the glacier's surface above the flood's subglacial watercourse. The outline of the Vestari-Skaftárketill cauldron can be seen in the distance. HB, May 1982

Fig. 8.115 A jökulhlaup in the Skaftá rushes beneath Ásabrú in Skaftártunga. The bridge is built on the Skaftáreldhraun lava field, but the river is constantly eroding the banks supporting it. HB, September 2003

combined geothermal power is about 1500 MW. Jökulhlaups from beneath the Skaftárkatlar can be both slow and swift. The water flow from the eastern cauldron usually begins quickly but then abates slowly; its maximum flow rate of 200–1500 m³/s is reached within 1–3 days, and the flood is over within 1–2 weeks. The fast early rate of acceleration indicates that the temperature in the water in the subglacial lake is above melting point. A constant flow of water into the Köldukvísl might also come from the third geothermal area on the ridge. There are signs that there was an eruption in the vicinity of the Eystri-Skaftárketill cauldron in 1910 (Jónsson 1986, p. 13).

The nature of geothermal areas beneath glaciers
In geothermal areas beneath a glacier, meltwater seeps down into leaky layers of basalt rock towards magma at a depth of 2–3 km in the Earth's crust. The magma is at a temperature of 1200 °C. As the water is reheated it rises once more to the glacial bed, melts ice there, cools, and sinks once again. Thus a cycle of geothermal fluid is maintained that is probably mostly water, 300–340 °C hot, which absorbs dissolved material from the bedrock, so that the

chemical composition of the water determines its exact heat. The fluid is probably at boiling point when it squirts up as water and steam through the many small vents spread all over the glacier's bed. Geothermal fluid boils when it reaches a height where the pressure of the overburden of the water above it is lower than the steam pressure in the rising fluid. At 300 °C the steam pressure is 86 bars, and so the fluid thus boils when it has risen high enough that the pressure of the water tower above it is lower than 86 bars. Once it has boiled, the fluid then cools itself, so that the steam pressure is always in balance with the overburden of the water above it. At 40 bars, the boiling point is 250 °C. The heat flux released constantly melts ice and the meltwater either drains away or accumulates in a subglacial lake within which water currents rise, the movement of the water bearing heat from the lake bed up to the ice (Björnsson et al. 1982; Björnsson and Stefánsson 1987).

When the water level in a glacial lake subsides during a jökulhlaup, there can be an explosion of steam due to reduced pressure in a similar way that a sudden lowering of the water level in a hot spring's bowl can trigger an eruption of the geysers Geysir and Strokkur in Haukadalur. If the pressure in a geothermal area falls from 40 bars to 25 bars, the boiling point of the fluid is lowered from 250 to 225 °C, but as the temperature of the bedrock surrounding the fluid remains steady at 250 °C, the fluid is suddenly 25 °C above the boiling point so that it expands in an explosion that extends into the porous groundwater system. Such disturbances can often cause increased geothermal activity for a few months after a jökulhlaup, though this does speed up the cooling of the shallow magma chamber. For geothermal activity to be prolonged, there needs to be a new and constant injections of magma into the chamber.

There is always plenty of meltwater beneath glaciers, and if it should be located above a shallow magma chamber, a powerful geothermal area will be the result. If a subglacial lake is also formed, which is emptied by outburst floods, there is increased fluctuations in the water pressure that pumps heat out of the magma chamber. Fluctuations in the overburden also pressurise fractures in the bedrock making waterways more porous. Furthermore, the drop in pressure after the sudden lowering of water levels after a jökulhlaup can also cause a volcanic eruption if there is a consequent rupture in the roof of the shallow magma chamber (Þórarinsson 1974a, b; Sigmundsson et al. 2004). There might have been as many as a dozen such small eruptions beneath the ice cauldrons of Vatnajökull since 1986 without any trace of them being visible on the glacier's surface.

From the mid-1970s, a continuing unrest, similar to that known to occur during a volcanic eruption, has been repeatedly detected on seismographs after a Skaftá jökulhlaup has reached its climax. Earthquakes increase when a flood bursts out of a subglacial lake, which is why the ridge running east from Hamarinn, Lokahryggur, is named after the Norse god Loki: 'Then Skadi got

the spring-fed vegetation along its banks might be destroyed. The bed of the outlet channel from Langisjór would be 20 m lower than the height of the lake's current water level (642 m), which in turn means that the lake's surface level might be lowered accordingly. Every summer the reservoir would be filled with meltwater from the western tributaries of the Skaftá, and in most years the surface levels would fluctuate between 10–15 m, with a maximum of 17 m. Vegetation along the banks would be covered by glaciofluvial sediments and destroyed. The lake-reservoir's water level would be carefully monitored to ensure it does not reach its present one, causing water to exit naturally out of Útfall again. It is assumed, nonetheless, that a Skaftá jökulhlaup could indeed fill the lake-reservoir and its water thus rise 6 m above the current level, inevitably resulting in an outflow of water through Útfall for a limited time. The overflow point would therefore need strengthening with a cement covering to prevent the present waterfall's thin palagonite threshold from giving way and bursting under the pressure. Langisjór would thus become a muddy glacial lake which could no longer be used for fishing. On the other hand, diverting Skaftá would prevent the Skaftáreldahraun lava field from being flooded with sediment deposits and preserve the highlands and lowland settlements from sand and dust storms.

The diversionary reservoir (Norðursjór) could be silted by glaciofluvial deposits within thirty years, and the whole of Langisjór, as far south as Útfall, within 80 years. It would then be necessary to build a tunnel through the Fögrufjöll mountains south of Fagralón in order to divert water from jökulhlaups into the Skaftá whenever necessary. It has been estimated that the whole of Langisjór will be half-full of glaciofluvial sediments within 150–250 years. Where there are now hiking trails, roads would have to be built to facilitate work on the dams, and a new route would be needed to Útfall. There are three possible routes to the dams at the northern end of Langisjór: alongside Mt. Breiðbakur, along the western banks of Langisjór, or along the Hverfisfljót and through the extraordinary natural wonders of the Lakagígar craters. Dams, reservoirs, canals, roads and other constructions would permanently change the scenic landscape and have a drastic effect on the value of the area as a major recreational and tourist destination. The effects of the proposed Skaftá diversionary reservoir can in no way be compared to the annual fluctuations of Langisjór's water levels when glacial water managed to flow into the lake for a short period early in the 20th century.

8.9.4   Síðujökull and the Grímsvötn Volcano

The western boundary of Síðujökull are marked by the Fögrufjallahryggur ridge, and its northeastern one by the volcanoes of Grímsvötn and Háabunga (1742 m). Its ice divide with Skeiðarárjökull is also there, and to the east the glacier flows southward past Þórðarhyrna (1668 m), Geirvörtur (1436 m) and Hágöngur (1118 m). It advances for 35 km while descending to a height of 600 m towards

Kálfafellsheiði. Hverfisfljót flows westward from the glacier in many braided streams along Fljótsoddi point, which gives the river its name. Further east, the Djúpá has its outlet near Mt. Hágöngur. Level gravel plains extend northward for 5 km beneath the snout of the western margin of the glacier at a height of 600 m, but to the east they stretch all the way to Hágöngur. Rather intermittent ridges, 700–800 m high, rise to the north of the plains and head northeastward into the highlands to the southwest of Grímsvötn.

There is a large volcanic centre beneath Síðujökull, which is 10–20 km wide and reckoned to stretch for 100 km from the main volcano in Grímsvötn southwest of the Lakagígar craters (Fig. 8.117). It includes Háabunga, Þórðarhyrna, Pálsfjall, Geirvörtur, Hágöngur, Rauðhólar and Hálsgígar craters. Þórðarhyrna is probably a separate volcano apart from this system as it contains acidic bedrock in Eystri-Geirvörtur and Pálsfjall. There only seems to be geothermal activity in one place, beneath an ice cauldron 1 km to the east of Pálsfjall, the water from which probably flows into the Hverfisfljót (Figs. 8.118 and 8.119).

The Eldborg fissure swarm, better known as the Lakagígar craters, extends beneath Síðujökull just to the southeast of an indent which separates it from

Fig. 8.117 The southern end of the Laki craters. The photograph was taken from the summit of Laki (812 m). HB

Fig. 8.118 An ice cauldron to the east of Pálsfjall (1335 m) indicates there is geothermal heat beneath the glacier. HB, October 1996

Skaftárjökull. There is a theory which suggests that, during the Skaftá fires eruption of 1783–1785, magma had rushed 85 km southwest from Grímsvötn before surfacing in the Lakagígar fissure craters, thus claiming Grímsvötn as the source of the famous eruption. During this upheaval there were simultaneous eruptions both in the Lakagígar craters and Grímsvötn and jökulhlaups flooded into the Skaftá. Indeed there was also a jökulhlaup along the Þjórsá, the floodwaters borne there along either the Tungnaá or Köldukvísl.

A further fissure swarm can be seen from the lava at the glacier's margin at Rauðhólar. This lava flow had come into existence when the glacier was smaller than it is now, but the ice has since advanced over it, although the continuation of this ridge beneath the glacier cannot be detected in any subglacial topography (Þórarinsson 1974a, b; Þórðarson 1990, 1991).

A deep valley stretches to the north of Pálsfjall (1332 m) to the foot of the highlands at Háabunga, for the land is lower there than at the glacier's snout and so the ice is up to 850 m thick. During an eruption to the southwest of Grímsvötn, a jökulhlaup might have flooded through this valley and emerged in the Djúpá, Hverfisfljót and Skaftá rivers. There were certainly floods in all of them during the eruption of 1753, which damaged meadows and dispersed boulders that can still be seen along the banks of the Djúpá (Ólafsson and Pálsson 1981; Þórarinsson 1974a, b). In an eruption near Þórðarhyrna in 1903, and probably in 1887 too, there was a huge outburst flood across the Skeiðarársandur. An eruption on the eastern side of

Fig. 8.119   The Hverfisflót in full spate in summer at Seljaland in the Fljótshverfi district. The river has flowed through the Brunahraun lava field since 1784. HB, 2004

Þórðarhyrna and on the eastern edge of Háabunga would discharge its meltwater into the Súla river, and from an eruption southeast of Þórðarhyrna, a jökulhlaup would flood the Núpsvötn rivers.

Grímsvötn and the Skaftá Fires

Grímsvötn is the most active volcano in Iceland and the Skaftá fires of 1783–1784, the greatest natural disaster in Iceland's history, can be traced back to this source. No volcano has erupted as often since Iceland was settled in the 9th century, and only Katla has produced more volcanic material. It has been estimated that about 20 km^3 of magma has emanated from the Grímsvötn volcanic system, and 2/3 of this came from the so-called Laki or Skaftá fires (Þórðarson 2003; Þórðarson and Self 2001; Þórðarson et al. 2003; Þórðarson and Larsen 2007). Moreover, Grímsvötn is also known for an even greater upheaval, for about 10,200–10,400 years ago, at the end of the last glacial period, and over a period of 100–200 years, there were three or four gigantic explosive eruptions in Grímsvötn which dispersed more than 15 km^3 of tephra over an area of 1,500,000 km^2 (fifteen times bigger than Iceland). The resulting layers of ash fall are named after Lake Saksunarvatn in the Faroe Isles and have proved significant in dating sediment deposits both on land and in the ocean bed between Greenland and Norway. It is the role of Grímsvötn in the Skaftá fires, however, which is the central point of this section.

From May 1783 until May 1785, or for two years, there were upheavals in the Grímsvötn volcanic system, including the Skaftá fires, which lasted for eight months. Lava flowed for 65 km from a row of craters and covered 0.5 % of Iceland, burying 18 farmsteads and much vegetated land. Ash falls and poisonous gases were borne all over the country, crops failed, livestock were killed, and starving people died from hunger and disease. The eruption spewed 120,000,000 tonnes of sulphur dioxide into the atmosphere, and after reacting with water vapour this created 200,000,000 tonnes of a sulphuric, acidic haze (Þórðarson et al. 1996). This haze spread out from the eruption over almost the whole of Earth's northern hemisphere, causing a reduction in sunlight, a long period of cold weather, and acid rainfall. The widespread effects of the greatest eruption of lava in human history on the climate and ecological systems of a large part of the world were to be felt for three years. Trees in Holland and Great Britain were defoliated and there was widespread crop failure throughout northern and western Europe. Mortality rates increased by 25 % in France and Great Britain (and probably further afield too). There was a very cold winter in North America and a great number of people died in western Alaska.

The Skaftá fires began with earthquakes and explosions and a fountain of ash rising 15 km into the skies. A black cloud was borne down from the mountains and there was a huge ash fall on the farms and communities. Such was the force of the eruption, that volcanic dross and bits of pumice were falling in Rangárvellir County, 110 km from the volcano. Loud rumblings and a stink of sulphur accompanied the ash fall, while heavy downpours of bluish rain tasted like a mixture of saltpetre and sulphur and burned bare flesh and vegetation. The surge of magma at the site of the eruptions was up to 6000 m^3/s, and the volcanic plumes rose from the vents to a height of between 800 and 1400 m and were visible in inhabited areas through the thick clouds of smoke as the lava flowed through ravines and over heaths. This eruption of lava lasted for eight months, from 8 June 1783 until 7 February 1784. There were about a dozen rounds of eruptions and the lava flowed from ten different fissures, which altogether had a length of 25 km, and a total of about 135 craters, 40–70 m high, all piled up along the fissure from either coagulated clumps or strips of loose volcanic dross.

The eruption first started in the Hnútagígar craters on a fissure to the southwest of Mt. Laki and lava flooded south along the Skaftá channel, the river drying up on the third day. Lava filled up the Skaftárgljúfur canyon and burst out of it on 12 June, sprawling over wetlands and green valleys. By the fifth day the lava flow had reached the lowlands having travelled 40 km. On 18 June there was an increased flow of lava from the canyon and the first farmsteads were buried beneath it. On the lowland, a lava flow spread eastwards in the Skaftá river channel; another flow went westwards along Skaftártunga and a third moved south towards Meðalland. These three flows are called the Eldhraun lava field. By the end of July, activity was reduced in the fissures to the southwest of Laki, but it had increased to its northeast, and a week later lava flowed down the Hverfisfljót channel, filling up its canyon within 24 h and covering sparsely vegetated fluvial plains; the Hverfisfljót now runs in a different channel than it did then. The most easterly lava flow was called Brunahraun.

On Sunday 20 July 1783, one and a half months after the eruption had begun, it was still in full spate with loud thunderous sounds and earth tremors. The flow of lava was approaching the village of Kirkjubæjarklaustur, where the inhabitants could already feel its heat. But the stream of lava came to a stop 2 km from the village near the small, steep, rocky hill of Systrastapi at a place now called Eldmessutangi, or 'fire sermon point'. This was on the very day that Pastor Jón Steingrímsson (1728–1791) preached his famous 'fire sermon,' vowing his congregation would repent of their sins, for right from the start he had considered the eruptions as a punishment from God for the excesses, laziness and sinfulness of his parishioners. After his sermon, the eruptions did indeed begin to abate in the craters to the south of Laki, although they increased in the northern part of the fissures, from which emerged the Brunahraun lava field.

There had been good weather in the years preceding the eruption, but the haze soaked up all solar radiation. A cold and wet winter thus followed the Skaftá fires of 1783–1784, a few fjords were frozen over and sea ice also reached Iceland's shores. A direct influence of the eruption could be seen on the climate and ecological systems for two to three years. Poisonous ash fell over almost all the country and there were poisonous gases in the air. Grass turned yellow and crops failed or were polluted. All over Iceland fluorine poisoning in the grass caused dental and skeletal fluorosis in livestock (deforming their teeth, jaws, bones and joints). Carbon dioxide and sulphuric compounds blended with the moist atmosphere to induce acid rainfalls. Harvest failures meant livestock died from a lack of hay and fodder all winter. People also died from starvation and pestilence. Aid from Copenhagen arrived late and its distribution was difficult due to the lack of horses. Two years after the eruption, the number of cattle in Iceland had been reduced by 50 %, the number of horses by 60 %, and the number of sheep by up to 80 %. A quarter of the Icelandic population died, almost 10,000 people, and in some districts this meant up to 40 % of the inhabitants. A total of 18 farms were enveloped by the lava, and about 30 other farms were abandoned for some time (Einarsson et al. 1984; Witze and Kanipe 2014).

In the summer and autumn of 1783, winds in the higher atmosphere bore the haze from the Skaftá fires over all the world to the north of 30° latitude, the smallest particles reaching as high as the stratosphere. The sun was described as blood-red and trees were defoliated. Within two days the haze had reached the Faroe Isles, Scotland and Norway, and by mid-June it was above all of Western Europe, reaching as far as the Balkans by 24 June. In July it reached Russia and from there it moved on to Siberia and the Altai mountains of China. The effects on ecological systems were most evident in Scandinavia, Great Britain and the western mainland of Europe. Recent research has indicated that an unusual number of people died from diseases in England in the autumn of 1783 until February of 1784. The rice crop failed in Japan resulting in a dreadful famine, and new research has even linked dearth and famine in Egypt to the haze. Benjamin Franklin described the haze and cold in Paris and Thomas Jefferson wrote about the long and cold winter in North America, temperatures falling to almost 5 °C below the average, and New York harbour having to be closed for ten days because of ice. It has been estimated that the average temperature in the northern hemisphere fell by 1 °C for three years. There have even been theories suggesting the Skaftá fires helped speed up the coming of the French Revolution in 1789! (Witze and Kanipe 2014).

This was the effect of the Skaftá fires more than 230 years ago, and it is becoming increasingly clearer that if there were to be anything like a similar eruption in Iceland today it would still cause chaos in our technologically dependent world, in spite of all our material progress. There are many things yet to be learned from previous natural disasters when having to address possible aftermaths of earthquakes, volcanic eruptions, tsunamis, and air pollution in the future, especially when modern technology proves ineffective, as for example in the case of the disruption of air traffic across the North Atlantic and Europe during the 2010 eruption of Eyjafjallajökull.

8.9.5 Surges Along the Western Margins of Vatnajökull

In the years 1992–1996, outlet glaciers on the western margins of Vatnajökull, from Síðujökull to Köldukvíslarjökull, surged forwards (Figs. 8.120, 8.121, 8.122, 8.123, 8.124, 8.125, 8.126, 8.127 and 8.128). For many years the glaciers had only moved half the amount necessary in order to transport its accumulated ice and maintain an equilibrium in its mass balance. They retreated and became increasingly steeper with every passing year. In the summer of 1992, however, Síðujökull

Fig. 8.120 Aerial view north over Tungnaárjökull towards the Kerlingar mountains. Sylgjujökull reaches as far as Hamarinn (on the *right*) and beyond it is Köldukvíslarjökull. The patterns made by layers of tephra are reminiscent of annual rings in tree trunks, and the traces of ash fall furthest down the snout are from the 12th century. Fairly low down on the glacier are tephra layers from the eruptions of Öræfajökull in 1362 and Veiðivötn in 1477. In the *centre* of the glacier are ash fall layers from the Katla eruptions of 1625 and 1755 (at the *top*). HB, August 1994

suddenly increased its velocity to more than the necessary speed for maintaining equilibrium in the lower part of the glacier, before slowing down again in the winter. The following summer it began to advance rapidly once more, however, and this time it retained a constant, higher speed. The first signs of a surge were crevasses around Pálsfjall in the autumn of 1993, but by the beginning of January 1994 a 70-m-high ice wave was visibly travelling down the snout at a speed of 20 m every 24 h. It formed an arc between the glacier's edges so that its speed seemed even greater in the centre of the wave than at its margins, where it encountered greater resistance than on the glacial bed. The wave was uplifted, but did not fracture except when advancing over unevenness on the glacier's bed, such as the nunataks a short distance from the sources of the Djúpá. Once the wave reached the terminus, the snout itself moved forwards and muddy water began to emerge from the glacier. Djúpá, which had previously had very clear waters, had become a dull grey colour in the Fljótshverfi district by 4 January 1994. In the beginning of February, the colour of the Hverfisfljót darkened and glacial tinting became evident in the Brunná a month later, and by then the whole margin of the

Fig. 8.121 Surge beginning in Tungnaárjökull. Ice rises in waves above the palagonite ridges beneath the glacier. Grooves can be seen in the ground moraine of the glacier's forefield, which were created centuries ago when the glacier had previously crept forward. HB, October 1994

Fig. 8.122 During the surge of 1994, it was as if Síðujökull was floating forwards as water issued from beneath it over a wide area. HB

Fig. 8.123   I am a glacier, and I will engulf all traces of your existence. HB

Fig. 8.124   Aerial view over a surging Sylgjujökull in 1996, and south along the Jökulgrindur hills ahead of Tungnaárjökull. HB, October 1996

Fig. 8.125 This Tungnaárjökull surge has advanced quite a distance. HB, 1996

Fig. 8.126 Síðujökull did not create any terminal moraines when it surged in 1994. HB, March 1994

Fig. 8.127 Lines of tephra clearly show the varying layers of Síðujökull in the huge gashes caused by its being ripped apart during the surge of 1994. HB, 1994

Fig. 8.128 Þórðarhyrna rises above the serried seracs of Vatnajökull. HB, October 1994

glacier was on the move, water pouring everywhere from beneath its terminus. There were tension crevasses on the upper parts of the glacier, and further down, as the glacier spread out, splayed or radiating crevasses were formed, many kilometres long. There was a widespread pattern of tension and radiating crevasses creating regular rows of ice towers, or seracs. Higher up on the glacier, these seracs were of the same height, but as they drew nearer to the snout some of them rose higher than the others. Where the glacier advanced over individual mountain peaks, it broke up completely, huge blocks of ice all tumbling together.

The glacial surge had mostly come to an end by May, but well into autumn the glacier began visibly to subside around Pálsfjall. This glacial upheaval covered, in all, an area of 500 km^2, reaching its highest point just below the western flanks of Grímsfjall. The surge advanced about 1200 m and only needed a further 100–200 m for the glacier to have reclaimed the territory it had retreated from at the end of its surge in 1964, all of 30 years earlier. At its greatest speed, the snout had been moving 100 m every 24 h (Björnsson et al. 2003).

Water was discharged into the Hverfisfljót during the Skaftá jökulhlaups of 1994 and 1995, and into Djúpá as well in 1995. It is worth remembering, in the light of this, that Thoroddsen had noted a smell of sulphur emanating from the Hverfisfljót when he came to the river after a glacial surge in 1893. Thoroddsen described the sheer vertical walls of the glacier's terminus as 100–120-m-high glacial cliffs.

8.9.6 History of the Exploration, Cartography and Toponymy of Western Vatnajökull

The western edges of Vatnajökull were so little known for such a long a time because few people ever travelled there. Farmers rounded up sheep from pastures north of the Fljótshverfi, Síða and Skaftártunga areas, but the lack of vegetation did not entice sheep right up to the glacier. Folk from Skaftafell County went as far as the Veiðivötn lakes, but were reluctant to journey further as there was no grazing for their horses. There were long-persistent rumours, nonetheless, that outlaws lived up by Síðujökull and Skaftárjökull and north of Tungnafellsjökull. Sheep round-up men from the Skaftafell area gave the moss-clad Fögrufjöll mountains their name, and in 1878 they first saw the lake, which Thoroddsen later called Langisjór, when he was the first man to explore the region scientifically in 1889 and 1893. There was a legend, though, that during the time of settlement a certain Gnúpa-Bárður had travelled from Bárðardalur in northern Iceland all the way along the glacier's margins to the Fljótshverfi district in the south. There were no known routes from the Vonarskarð pass in the west up onto Köldukvíslarjökull and then eastward, lengthwise, across the whole of Vatnajökull, though Daniel Bruun always believed that such a route had once existed (Bruun 1914).

Maps of the western side of Vatnajökull were for a long time inaccurate. On Knoff's map from 1734, Skaftárjökull is depicted as an individual glacier, and it was not until Sveinn Pálsson's map appeared that it was depicted as part of Vatnajökull or Klofajökull. Pálsson called the whole western edge of Vatnajökull Skaftár- and Síðujökull, though Gunnlaugsson called it Skaftárjökull or Síðujökull. Gunnlaugsson never visited the wilderness to the east of Þórisvatn, but he did go through Vonarskarð pass in 1839. On his map from 1848, Vatnajökull was made to reach too far west, and the glacial rivers Tungnaá, Skaftá and Hverfisfljót are shown to emerge from almost one and the same place in the glacier. Thoroddsen corrected these errors on Gunnlaugsson's diagram of the highlands west of Vatnajökull after he had visited the area in 1889 and 1893. No one else ventured into this region after Thoroddsen's expeditions until the summer of 1926 when Frank le Sage de Fontenay (1880–1959), historian and Danish ambassador to Iceland, described the barren lands between the Tungnaá and Kaldakvísl rivers. Le Sage de Fontenay (1926, pp. 119–129) had studied the history and culture of the Middle East in his younger years and was very attracted to this part of the country's uninhabited interior during the years he lived in Iceland. He was the originator of the names Heljargjá gorge and Galdrahraun lava field (now called Tröllahraun), but other names with which he endowed the landscape have not survived. The larger of the Kerlingar mountains then rose up through the glacier as a nunatak and reminded the scholar of the Middle East of the Sphinx encircled by glacial clouds, while the lower mountain seemed to him like a camel crossing the desert. Fontenay thus called a tributary stream of the Tungnaá Úlfaldakvísl ('camel stream'); he named the two lakes in front of the glacier Sigríðarvatn and Gunnlaugsvatn (in honour of his travelling companion, Gunnlaugur Briem, later a high-ranking member of the civil service). The Hraungígur crater in Veiðivatnahraun is called Fontur in memory of this ambassador. There was no further increase in knowledge of the area west of Vatnajökull until surveying began there in the 1930s.

Hiking routes on Síðujökull and vehicle access to Tungnaárjökull Until the middle of the 20th century, most expeditions onto Vatnajökull were made from the lowlands of southern Iceland, but then a ford passable by vehicles was found across the Tungnaá river, the so-called Hófsvað, and thus a route for vehicles was opened up through the Tungnaáröræfi wilderness into the Tungnaárbotn basin and up over Tungnaárjökull onto Vatnajökull. The Englishman W.L. Watts began his Vatnajökull expeditions from the Fljótshverfi district. He first came to Iceland in 1871 and hiked from Núpsstaður in Fljótshverfi onto Síðujökull and as far as just north of Grænafjall. His associate on this journey was the English geologist John Milne. In 1874 Watts, with Páll Pálsson from Seljaland and two other men from Páll's local area, Bjarni and Jón, ascended Síðujökull and travelled to the north of the present Skaftársig ice cauldron, where they turned back. On this expedition they discovered the nunatak which Watts named Pálsfjall (1329 m) in honour of his travelling companion, who subsequently gained the soubriquet Glacier Páll. A mountain that Watts called Vatnajökull's House was later named Þórðarhyrna

(1668 m), after the geologist Þórður Vídalín, at the suggestion of the geologist Jóhannes Áskelsson. Watts finally succeeded in crossing Vatnajökull with Glacier Páll and four other Icelanders in the summer of 1875. A further four companions accompanied them as far as the Pálsfjall nunatak. Just a short distance from where Watts and his team ascended the glacier, two Scots, J. H. Wigner and T.S. Muir, unexpectedly descended from Vatnajökull in 1904, having crossed it from Brúarjökull in the east, much to the surprise and astonishment of the local farmers, as has been related earlier in this chapter.

In 1890 the English ophthalmic surgeon and volcanologist Tempest Anderson ascended Skaftárjökull (Anderson 1897, pp. 216–221). Anderson travelled widely in Iceland in 1890 and 1893, and there is a valuable collection of photographs from these Icelandic sojourns and travels.

The Swedes Hakon Wadell and Erik R. Ygberg also set off from Fljótshverfi when they rediscovered Grímsvötn in the centre of Vatnajökull in 1919. After the Grímsvötn eruption of 1934, there were many expeditions north onto the glacier from Fljótshverfi via Síðujökull in order to investigate the volcanoes and the sources of the jökulhlaups onto Skeiðarársandur. These journeys often began from Kálfafell and then proceeded north through Djúpárdalur onto Síðujökull, but sometimes they began on Skeiðarárjökull too. The aircraft Jökull was also transported down Síðujökull after it had been dug out of the snow on Bárðarbunga in 1951, as has been mentioned previously.

Once the way to Tungnaárjökull had been opened for motorised vehicles, the IGS built a cabin at Jökulheimar in the Fremri-Tungnaárbotnar basin in 1951, and a further larger house was added on in 1966. It is easy to cross the Tungnaá in vehicles from there and then to ascend the glacier at the southern end of the Jökulgrindur ridge before continuing directly across to Grímsvötn, a journey of about 45 km. Just to the north of the cabins at Jökulheimar is Heimabunga knoll. When Thoroddsen was in the area in 1898, the Tungnaárjökull reached both this knoll and the palagonite ridge of Jökulgrindur, and when the first cabin was built, snowmobiles needed to drive 1.7 km east from the summit of Nýjafell to reach the glacier's terminus; the journey from the cabin to the glacier is twice as long today. During the summer months, it is a three-hour drive from Reykjavík to Jökulheimar.

8.10   The Origins of Vatnajökull

Glaciers began to expand on five or six of the largest mountain massifs in the southern part of Iceland while the climate was cooling about 3000–4000 years ago (Fig. 8.129). At first there were ice caps on Öræfajökull and in the highlands to the north of Mávabyggðir and Snæhetta as well as on Háabunga dome and Grímsfjall. Ice caps also formed on the summits of the Kverkfjöll mountains, on the Bárðarbunga and Breiðabunga domes and on the Goðahryggur ridge. By about

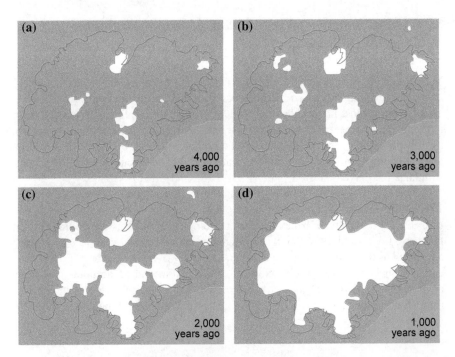

Fig. 8.129 Formation of Vatnajökull glacier. Results of numerical models illustrating the probable formation of Vatnajökull. About 3000–4000 years ago there were ice caps on Öræfajökull and in the highlands that lay to the north of Mávabyggðir and Snæhetta, and also on Háabunga dome, Grímsfjall, the summits of the Kverkfjöll mountains, Bárðarbunga, Breiðabunga, and the Goðahryggur ridge. These glaciers expanded and merged to form one Vatnajökull ice sheet. Vatnajökull was at its largest at the end of the 19th century

2000 years ago, the glaciers might have extended far enough to have merged into one continuously linked ice field stretching from Öræfajökull in the south to Háabunga, Grímsfjall and Bárðarbunga in the northwest, as well as northward beyond the Esjufjöll mountains and to Breiðabunga in the east. There were still individual glaciers, however, in the Kverkfjöll mountains and on Goðahryggur ridge, but by the time of the settlement of Iceland, all the glaciers had coalesced into one enormous ice sheet. The margins of Vatnajökull were nonetheless 10–20 km further back from where they are today. No outlet glaciers crept down onto the southeastern lowlands from Öræfajökull and all the way east to Hornafjörður until the later part of the five-centuries-long Little Ice Age, i.e. in the 17th and 18th centuries. By the end of the 19th century, on the other hand, Vatnajökull had become larger than it had ever been since the end of the last glacial period.

8.11   Mass Balance of Vatnajökull Since the Beginning of the 20th Century

Vatnajökull has shrunk considerably during the 20th century, its snouts retreating about 2–5 km, and its total surface area being reduced by about 300 km³ (∼ 10 %). This has been most noticeable on the outlet glaciers descending southwards, their termini having retreated 100 m every year and having sunk in elevation by as much as 8 m per annum, mostly because of the complete failure of the ice flows from above to compensate for the melting ice at the end of their snouts. Since 1995 this shrinkage has increased even more (Figs. 8.130 and 8.131).

The mass balance of Vatnajökull has been calculated exactly every year since 1992 while detailed weather observations began there in 1995. This means that the connection between mass balance and meteorological conditions can be traced over this period. Up until 1994 the glacier increased its thickness by up to 1 m a year, calculated as the water equivalent distributed equally over the glacier's surface. The equilibrium line was at an elevation of 1000–1100 m and its accumulation zone covered 70–80 % of the glacier's surface area. The mass balance was in equilibrium over the glacial year of 1994–1995, income equalling expenditure, and the

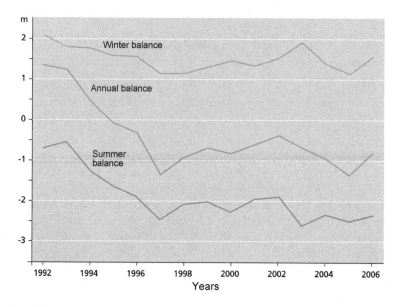

Fig. 8.130 Mass balance of Vatnajökull. Average mass balance of Vatnajökull during the years 1991–2006 (Björnsson et al. 1998, 2002, 2013; Björnsson and Pálsson 2008). The figures show the water equivalent in metres distributed equally over the glacier's surface. The winter mass balance was often greatest early in the 1990s, for then the summers were also colder than they have been more recently. The mass balance has been negative ever since 1996, a loss of almost 1 m of water per annum. The large volume of summer meltwater in 1997 was partly due to an ash fall of tephra that had been dispersed across the glacier in the autumn of 1996 and which reduced the reflection of solar radiation from the glacier's surface

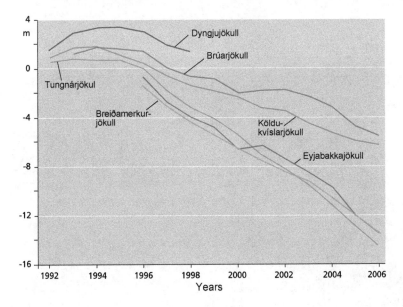

Fig. 8.131   Mass balance of some outlet glaciers of Vatnajökull. The average annual mass balance of some outlet glaciers of Vatnajökull during the years 1991–2006. The figures show cumulative values. Since the glaciers' mass balances began being generally negative in 1996, Tungnaárjökull has lost 15 m and Brúarjökull 5 m of water equivalent. Eyjabakkajökull in the northeast has shrunk at a similar rate to Vatnajökull's southern outlet glaciers

equilibrium line at a height of 1100 m in the southern part of the glacier and at 1200 m in the northern part; the accumulation zone constituted about 60 % of the glacier. Since then the mass balance has been negative every single year, the equilibrium line has risen up to a height of 1200 m on the southern snouts and up to 1400 m in the north. The accumulation zone has become less than half of the glacier, and during the warmest years it only constitutes 20–30 % of its surface area. From around the mid-1990 s Vatnajökull has lost, on average, 0.8 m of water every year. For every 100 m the equilibrium line rises, the glacier's mass balance is reduced by 0.7 m per year. From 1994 until 2006, Vatnajökull lost a total of 9.2 m (water equivalent) or the same as 84 km^3 of snow and ice (which is six times the winter mass balance). Vatnajökull has lost 2.7 % of its total volume in just one decade.

8.12   The Future of Vatnajökull

It has been estimated that by the end of the 21st century Iceland will have become 3 ° C warmer than at the end of the 20th century, and that during the following century, the 22nd, the temperature will rise even faster. Computerised models of the response

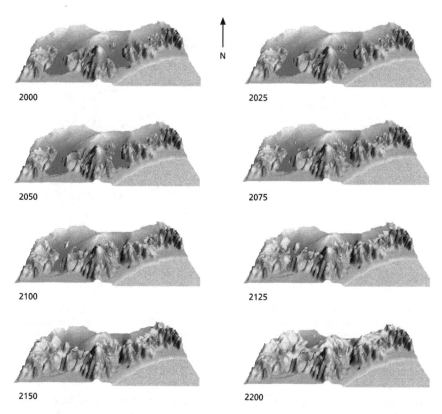

N

2000   2025

2050   2075

2100   2125

2150   2200

Fig. 8.132 Future possible shrinking of Vatnajökull (Aðalgeirsdóttir 2003; Aðalgeirsdóttir et al. 2003, 2005, 2006a, b; Flowers et al. 2003, 2005; Marshall et al. 2005). Predicted reduction in size of Vatnajökull if current estimations of climate change prevail. In about 200 years the situation might arise whereby there will only be glaciers on the highest knoll of Öræfi and on the mountain ranges between Grímsvötn, Bárðarbunga and the Kverkfjöll mountains. The southern side of Vatnajökull will shrink the fastest. Glacial lakes will form in front of most of the glacial snouts in the Skaftafell counties, and within 100 years the Jökulsárlón lagoon on Breiðamerkursandur might reach as far inland as Esjufjöll

of glaciers to changes in mass balance indicate that after 200 years there will only be glacial caps on the highest knoll of Öræfajökull and on the highlands between Grímsvötn, Bárðarbunga and the Kverkfjöll mountains. Vatnajökull could lose a quarter of its current volume within the next fifty years, though its northern part will survive a bit longer. The total drainage from the glacier will increase over the next fifty years and be maintained at a higher level than today well into the 22nd century before rapidly abating due to the glacier having shrunk so much (Fig. 8.132).

As the glacier thins out, its water divides will be moved, sometimes several kilometres, and the sources of rivers at its margins and snouts will also migrate. The pressure of the glacier's overburden will no longer be sufficient to force water over mountain ridges so that streams and rivers will begin to flow along the base of ridges to the glacier's margins instead. The terrain beneath the glacier will then dictate the location and flow of these waterways much more than it does today. Some rivers will grow in size while others will be diminished. As the ice dams of glacial lakes grow thinner, jökulhlaups will become more frequent but less powerful. New dammed, glacial lakes might form at the margins and outburst floods begin to flow from them. The shrinking mass balance of glaciers could also mean fewer glacial surges as ice no longer builds up in the accumulation zone. Ice will calve into glacial lakes that form at the glacier's snouts and these lakes will rapidly grow in size. Where the ice will be at its thinnest, ridges and mountain summits will rise up through the glacier as nunataks; ice will survive longest in the valley basins.

There will be an uplifting of land once the overburden pressure of a glacier is reduced. Should Vatnajökull disappear in its entirety, the terrain beneath it would uplift by just over 100 m, half the height at its margins, and about 20 m at Höfn in Hornafjörður and about 5 m at Kirkjubæjarklaustur. Even if the ice were to disappear suddenly, the land would still need an entire century to rise to its fullest extent (Sigmundsson 1990, 1991; Einarsson et al. 1994; Imsland 1994; Pagli et al. 2007; Pagli and Sigmundsson 2008; Árnason et al. 2009).

An increase in volcanic activity can be expected in the fissure swarms of Grímsvötn, Bárðarbunga and Kverkfjöll, perhaps even simultaneously in many volcanic systems, while Vatnajökull continues to shrink. There will then also be an increasing danger of outburst floods onto Skeiðarársandur, to the southwest of Vatnajökull, and all the way clockwise to the north and into the Jökulsá á Fjöllum. These jökulhlaups would threaten farmstead communities and villages, especially as they might suddenly burst forth and flood areas that have not been inundated for a long time, perhaps even centuries. Any tephra ash fall would also be dispersed all over the outlet glaciers and increase their rate of shrinking.

The southern part of Vatnajökull is the one which would begin to change first and foremost. The Jökulsárlón lagoon on Breiðamerkursandur would expand the most rapidly. Up until 1930, Breiðamerkurjökull extended as far as a terminal moraine that it had bulldozed into existence during the Little Ice age, but since then it has retreated about 5 km, bringing the Jökulsárlón lagoon into existence, with a present surface area of 25 km². Within a short period of time the glacial snout will retreat beyond the supporting bedrock threshold of a basin upon which it has rested. This important support will then no longer be functional, for the valley basin deepens increasingly the further north it stretches; the snout may thus still creep forwards, but as the bedrock threshold no longer supports it, huge blocks of ice will continually calve at its terminus and float onwards, the glacier thus breaking up and retreating, perhaps for several kilometres, over the next few years. The Jökulsárlón lagoon would then reach as far as the Esjufjöll mountains long before the terrain up to Mávabyggðir became ice free.

Proglacial lakes will form in front of most of the outlet glaciers descending onto the lowlands of Austur-Skaftafell County. The glaciers in Mýrar and Nes in Hornafjörður will disappear and glacial runoff water to the southeastern coast will then only flow through Breiðamerkurjökull.

Once Skeiðarárjökull has retreated 2 km, a lake will form at its snout similar to the Jökulsárlón on Breiðamerkursandur. All the rivers from the glacier will probably flow into this lake, and from there the elevation of land on Skeiðarársandur will determine the further course of its channels. Lake Grænalón will also become larger at first, but the jökulhlaups emerging from it, and from Grímsvötn too, will later become less powerful and less frequent than today, and eventually they will cease altogether. Skeiðarárjökull will survive longer than Breiðamerkurjökull because it receives its ice from an ice divide from far away at Grímsvötn, Dyngjujökull and Brúarjökull, while Öræfajökull will soon be isolated from the main glacier of Vatnajökull.

Ridges beneath Tungnaárjökull will direct water into the Skaftá and eventually the Tungnaá will no longer be a glacial river, although it will still receive groundwater that seeps through Vatnajökull's bed.

Glacial runoff water will soon disappear from Skjálfandafljót and all of the meltwater from Bárðarbunga will flow southward. The rims of the Bárðarbunga caldera will rise up through the glacier, though ice will remain within its centre and on its flanks. Indeed, Dyngjujökull will be the outlet glacier to survive the longest on the northern edges of Vatnajökull precisely because of its proximity to Bárðarbunga. The Kverká and Kreppa will become groundwater rivers to the east of the Kverkfjöll mountains, and glacial runoff will flow into the Jökulsá á Brú. Jökulsá í Fljótsdal will disappear along with Eyjabakkajökull, because a ridge will eventually cut the river off from the main glacier. After 100 years, there will be a large but shallow lake at the terminus of Brúarjökull, in which all the water from the glacier will accumulate and then flow into the Jökulsá á Dal.

The glacial rivers of the northern margins of Vatnajökull will survive the longest because the land elevation is highest there. Once glacial water ceases to run into the Jökulsá á Dal, runoff water will still flow into Jökulsá á Fjöllum. In about 200 years' time, all glacial tinting might have vanished from the rivers of Iceland.

References

Aðalgeirsdóttir, G. (2003). *Flow dynamics of the Vatnajökull ice cap, Iceland.* VAW/ETH Zürich, Mitteilungen. 181. Doctoral dissertation.

Aðalgeirsdóttir, G., Guðmundsson, G., & Björnsson, H. (2003). A regression model for the mass-balance distribution of the Vatnajökull Ice Cap Iceland. *Annals of Glaciology, 37*, 189–193.

Aðalgeirsdóttir, G., Guðmundsson, H., & Björnsson, H. (2005). The volume sensitivity of Vatnajökull ice cap, Iceland, to perturbations in equilibrium line altitude. *Journal of Geophysical Research, 110*, F04001. doi:10.1029/2005JF000289.

Aðalgeirsdóttir, G., Björnsson, H., Pálsson, F., & Magnússon, E. (2006a). Analyses of a surging outlet glacier of Vatnajökull ice cap, Iceland. *Annals of Glaciology, 42*, 23–28.

Aðalgeirsdóttir, G., Jóhannesson, T., Björnsson, H., Pálsson, F., & Sigurðsson, O. (2006b). The response of Hofsjökull and southern Vatnajökull, Iceland, to climate change. *Journal of Geophysical Research, 111*(F03001), 2006. doi:10.1029/2005JF000388.

Ahlmann, H. (1938). *Land of Ice and Fire: A Journey to the Great Iceland Glacier* (K. Lewes & H. Lewes, Trans.). London: Kegan Paul, Trench & Trubner. (Swedish original: *På skidor och till häst i Vatnajökulls rike.* Stockholm: Norstedt. 1936).

Ahlmann, H., & Þórarinsson, S. (1937–1943) Vatnajökull. Scientific results of the Swedish-Icelandic investigations 1936-37-38. *Geografisker Annaler, 19*, 146–231; *20*, 171–233; *21*, 39–66, 171–242; *22*, 188–205; *25*, 1–54.

Alfaro, R., Brandsdóttir, B., et al. (2007). Structure of the Grímsvötn volcano under the Vatnajökull icecap. *Geophysical Journal International, 168*, 863–876.

Anderson, T. (1897). The Skaptár Jökull. *The Alpine Journal., 18*, 216–221.

Arnborg, L. (1955a). Hydrology of the glacial river Austurfljót. *Geografiska Annaler, 37*, 185–201.

Arnborg, L. (1955b). Ice-marginal lakes at Hoffellsjökull. *Geografiska Annaler, 37*, 202–228.

Árnadóttir, Þ., Lund, B., Jiang, W., Geirsson, H., Björnsson, H., Einarsson, P., et al. (2009). Glacial rebound and plate spreading: Results from the first countrywide GPS observations in Iceland. *Geophysical Journal International, 177*(2), 691–716.

Árnason, J. (1961). *Íslenzkar þjóðsögur og ævintýri,* 6 vols. Böðvarsson, Á. and Vilhjálmsson, B. (Eds.) Reykjavík: Þjóðsaga, Hólar.

Áskelsson, J. (1936a). On the last eruptions in Vatnajökull. *Rit Vísindafélags Íslendinga, 18.*

Áskelsson, J. (1936b). Investigations at Grímsvötn, Iceland, 1934–1935. *The Polar Record, 11*, 45–47.

Áskelsson, J. (1959). Skeiðarárhlaup og umbrotin í Grímsvötnum 1945. *Jökull, 9*, 22–29.

Bárðarson, G. (1934). Islands Gletscher. Beitrage zur Kenntniss der Gletscherbewegungen und Schwankungen auf Grund alter Quellenschriften und neuesten Forschung. *Rit Vísindafélags Íslendinga, 16.*

Benediktsson, G. (1944). *Hinn gamli Adam í oss: ritgerðir.* Reykjavík: Víkurútgáfan.

Benediktsson, G. (1977). *Í flaumi lífsins fljóta. Bernsku- og æskuminningar.* Reykjavík: Örn og Örlygur.

Benediktsson, J. (Ed.). (1986). *Íslendingabók, Landnámabók. Íslenzk fornrit* (Vol. 1). Reykjavík: Hið íslenska fornritafélag.

Benediktsson, K. (2000). Þegar ég var 17 ára. *Skaftfellingur, 13*, 67–74.

Björnsson, H. (1974). Explanation of jökulhlaups from Grímsvötn, Vatnajökull, Iceland. *Jökull, 24*, 1–26.

Björnsson, H. (1975). Subglacial water reservoirs, jökulhlaups and volcanic eruptions. *Jökull, 25*, 1–14.

Björnsson, H. (1976). Marginal and supraglacial lakes in Iceland. *Jökull, 26*, 40–51.

Björnsson, H. (1977). The cause of jökulhlaups in the Skaftá river, Vatnajökull. *Jökull, 27*, 71–78.

Björnsson, H. (1979). Glaciers in Iceland. *Jökull, 29*, 74–80.

Björnsson, H. (1982). Drainage basins on Vatnajökull mapped by radio echo soundings. *Nordic Hydrology, 1982*, 213–232.

Björnsson, H. (1983). A natural calorimeter at Grímsvötn: Indications of volcanic and geothermal activity. *Jökull, 33*, 13–18.

Björnsson, H. (1986a). Surface and bedrock topography of ice caps in Iceland mapped by radio echo soundings. *Annals of Glaciology, 8*, 11–18.

Björnsson, H. (1986b). Delineation of glacier drainage basins on western Vatnajökull. *Annals of Glaciology, 8*, 19–21.

Björnsson, H. (1988). Hydrology of ice caps in volcanic regions. *Rit Vísindafélag Íslands, 45.*

Björnsson, H. (1992). Jökulhlaups in Iceland: Characteristics, prediction and simulation. *Annals of Glaciology, 22*, 141–146.

Björnsson, H. (1996). Scales and rates of glacial sediment removal: A 20 km long and 300 m deep trench created beneath Breiðamerkurjökull during the Little Ice Age. *Annals of Glaciology, 22*, 141–146.

Björnsson, H. (1997). Grímsvatnahlaup fyrr og nú. In H. Haraldsson (Ed.), *Vatnajökull. Gos og hlaup 1996* (pp. 61–77). Reykjavík: Vegagerðin.

Björnsson, H. (1998a). Frá Breiðumörk til jökulsands: mótun lands í þúsund ár. In G. Árnason (Ed.), *Kvískerjabók* (pp. 164–176). Höfn í Hornafirði: Sýslusafn Austur-Skaftafellssýslu.

Björnsson, H. (1998b). Hydrological characteristic of the drainage system beneath a surging glacier. *Nature, 395*, 771–774.

Björnsson, H. (2002). Subglacial lakes and jökulhlaups in Iceland. *Global and Planetary Change*, 35, 255–271.

Björnsson, H. (2010). Understanding jökulhlaups: from tale to theory. *Journal of Glaciology, 56* (200), 1002–1010.

Björnsson, H., Björnsson, S., & Sigurgeirsson, Þ. (1982). Penetration of water into hot rock boundaries of magma at Grímsvötn. *Nature, 295*, 580–581.

Björnsson, H., & Einarsson, P. (1990). Volcanoes beneath Vatnajökull, Iceland: Evidence from radio echo-sounding, earthquakes and jökulhlaups. *Jökull, 40*, 147–168.

Björnsson, H., & Guðmundsson, M. (1993). Variations in the thermal output of the subglacial Grímsvötn caldera, Iceland. *Geophysical Research Letter, 20*, 2127–2130.

Björnsson, H., & Kristmannsdóttir, H. (1984). The Grímsvötn geothermal area, Vatnajökull, Iceland. *Jökull, 34*, 25–50.

Björnsson, H., & Pálsson, F. (2004). Jöklar í Hornafirði. In H. Björnsson, E. Jónsson, & S. Runólfsson (Eds.), *Jöklaveröld* (pp. 125–164). Reykjavík: Skrudda.

Björnsson, H., & Pálsson, F. (2008). Icelandic glaciers. *Jökull, 58*, 365–386.

Björnsson, H., Pálsson, F., Guðmundsson, M., & Haraldsson, H. (1998). Mass balance of western and northern Vatnajökull, Iceland, 1991–1995. *Jökull, 45*, 35–58.

Björnsson, H., Pálsson, F., & Guðmundsson, S. (2001). Jökulsárlón at Breiðamerkursandur, Vatnajökull, Iceland: 20th century changes and future outlook. *Jökull, 50*, 1–18.

Björnsson, H., Pálsson, F., & Haraldsson, H. (2002). Mass balance of Vatnajökull (1991–2001) and Langjökull (1996–2001), Iceland. *Jökull, 51*, 75–78.

Björnsson, H., Pálsson, F., Sigurðsson, O., & Flowers, G. (2003). Surges of glaciers in Iceland. *Annals of Glaciology, 36*, 82–90.

Björnsson, H., Rott, H., Guðmundsson, S., Fischer, A., Siegel, A., & Guðmundsson, M. (2001). Glacier-volcano interactions deduced by SAR interferometry. *Journal of Glaciology, 47*(156), 58–70.

Björnsson, H., Pálsson, F., Guðmundsson, S., Magnússon, E., Aðalgeirsdóttir, G., Jóhannesson, T., Berthier, E., Sigurðsson, O., & Þorsteinsson, Þ. (2013). Contribution of Icelandic ice caps to sea level rise: Trends and variability since the Little Ice Age. *Geophysical Research Letters, 40*, 1–5. doi:10.1002/grl.50278.

Björnsson, S. (1978). Hlaupið í Jökulsá á Breiðamerkursandi árið 1927. - Athugasemd. *Jökull, 28*, 90.

Björnsson, S. (1979). Öræfasveit. *Árbók Ferðafélags Íslands*.

Björnsson, S. (2003). Skeiðarársandur og Skeiðará. *Náttúrufræðingurinn, 71*, 120–128.

Björnsson, S., & Stefánsson, V. (1987). Heat and mass transport in geothermal reservoirs. In J. Bear & M. Corapcioglu (Eds.), *Advances in transport phenomena in porous media* (pp. 145–183). Dordrecht: Martinus Nijhoff.

Brandsdóttir, B. (1984). Seismic activity in Vatnajökull in 1900–1982 with special reference to Skeiðarárhlaups, Skaftárhlaups and Vatnajökull eruptions. *Jökull, 34*, 141–150.

Bruun, D. (1902). Ved Vatna Jökulls Nordrand. Studier af Nordboernes Kulturliv. IV (Vol. 1, Special edition of *Geografisk Tidsskrift*). Copenhagen: Nordiske Forlag.

Bruun, D. (1914). Islænderferder til Hest over Vatna-Jökull i ældre Tider. *Geografisk Tidsskrift, 22*, 4–13.

Einarsson, G. (1946). Eldgos við Grímsvötn. *Fjallamenn* (pp. 47–62). Reykjavík: Bókaútgáfa Guðjóns Ó. Guðjónssonar.

Einarsson, Í. (1918a). Skrá frá 1712, eptir Ísleif sýslumann Einarsson, um eyddar jarðir í Öræfum, ásamt skrá eptir Jón sýslumann Helgason um eyðijarðir 1783 í Lóni, Nesjum og Fellshverfi. J. Þorkelsson (Ed.), *Blanda 1918–1920* (pp. 39–58).

Einarsson, Í. (1918b). Jarðabók Ísleifs sýslumanns Einarssonar um Austur-Skaftafellsþing, er hann gerði 1708 og 1709 í umboði Árna Magnússonar. J. Þorkelsson (Ed.) *Blanda 1918–1920* (pp. 1–38).

Einarsson, P. (1991). Earthquakes and present-day tectonism in Iceland. *Tectonophysics, 189*, 261–279.

Einarsson, P., & Björnsson, S. (1979). Earthquakes in Iceland. *Jökull, 29*, 37–43.

Einarsson, P., & Brandsdóttir, B. (1984). Seismic activity preceding and during the 1983 volcanic eruption in Grímsvötn, Iceland. *Jökull, 34*, 13–23.

Einarsson, P., & Sæmundsson, K. (1987). Upptök jarðskjálfta 1982–1985 og eldstöðvakerfi á Íslandi. Map in Þ. Sigfússon (Ed.), *Í hlutarins eðli* (p. 270). Reykjavík: Menningarsjóður.

Einarsson, P., Sigmundsson, F., Hofton, M., Foulger, G., & Jacoby, W. (1994). An experiment in glacio-isostasy near Vatnajökull, Iceland, 1991. *Jökull, 44*, 29–39.

Einarsson, Þ. (1997) [1855]. Kálfafellsstaðarsókn ár 1855. In J. Jónsson & S. Sigmundsson (Eds.), *Skaftafellssýsla. Sýslu- og sóknalýsingar Hins íslenska bókmenntafélags 1839–1873* (pp. 125–143). Reykjavík: Sögufélag.

Einarsson, Þ., Guðbergsson, M., et al. (Eds.). (1984). *Skaftáreldar 1783–84: ritgerðir og heimildir*. Reykjavík: Mál og Menning.

Eiríksson, H. (1932). Observations and measurements of some glaciers in Austur-Skaftafellssýsla in the summer 1930. *Rit Vísindafélags Íslendinga, 12*.

Elíasson, S. (1977). Molar um Jökulsárhlaup og Ásbyrgi. *Náttúrufræðingurinn, 47*(3–4), 160–179.

Flowers, G., Björnsson, H., & Pálsson, F. (2003). New insights into the subglacial and periglacial hydrology and Vatnajökull, Iceland, from a distributed physical model. *Journal of Glaciology, 49*(165), 257–270.

Flowers, G., Björnsson, H., Pálsson, F., & Clarke, G. (2004). A coupled sheet-conduit mechanism for jökulhlaup propagation. *Geophysical Research Letters, 31*(L05401), 2004. doi:10.1029/2003GL019088.

Flowers, G., Marshall, S., Björnsson, H., & Clarke, G. (2005). Sensitivity of Vatnajökull ice cap hydrology and dynamics to climate warming over the next two centuries. *Journal of Geophysical Research, 110*, F02011. doi:10.1029/22004JF000200.

Fontenay, F. (1926). Ferð til Vatnajökuls og Hofsjökuls sumarið 1925. *Andvari, 51*, 99–144.

Gadde, O. (1983). *Íslandsferð sumarið 1857* (G. Erlingsson, Trans.), Akranes: Hörpuútgáfan. (Swedish original: *En Färd till Island sommaren 1857 efter anteckningar och brev av Nils O:son Gadde*. Stockholm: LTs Förlag, 1976).

Grönvold, K., & Jóhannesson, H. (1984). Eruption in Grímsvötn 1983: Course of events and chemical studies of the tephra. *Jökull, 34*, 1–11.

Gunnarsson, S. (1949) [1877]. Miðlandsöræfi Íslands. In J. Eyþórsson & P. Hannesson (Eds.), *Hrakningar og heiðavegir* (Vol. 1, pp. 214–244). Akureyri: Norðri.

Guðmundsson, M. (1989). The Grímsvötn Caldera, Iceland: Subglacial topography and structure of caldera infill. *Jökull, 39*, 1–19.

Guðmundsson, M. (1996). Ice-volcano interaction at the subglacial Grímsvötn Volcano, Iceland. In S. Colbeck (Ed.), *Glaciers, ice sheets and volcanoes*. CRREL Special Report 96-27 (pp. 34–40).

Guðmundsson, M., Björnsson, H., & Pálsson, F. (1995). Changes in jökulhlaup sizes in Grímsvötn, Vatnajökull, Iceland, 1934–1991, deduced from in situ measurements of subglacial lake volume. *Journal of Glaciology, 41*(138), 263–272.

Guðmundsson, M., Sigmundsson, F., & Björnsson, H. (1997). Ice-volcano interaction of the 1996 Gjálp subglacial eruption, Vatnajökull, Iceland. *Nature, 389*, 954–957.

Guðmundsson, S., Guðmundsson, M., Björnsson, H., Sigmundsson, F., Carstensen, J., & Rott, H. (2002). Three-dimensional glacier surface motion maps at the Gjálp eruption site, Iceland, inferred from combining InSAR and other ice displacement data. *Annals of Glaciology, 34*, 315–322.

Gunnarsson, S. (1950) [1876]. Um öræfi Íslands. In J. Eyþórsson & P. Hannesson (Eds.), *Hrakningar og heiðavegir* (Vol. 2, pp. 196–209). Akureyri: Norðri.

Hálfdanarson, E. (1918–1929) [1727]. Frásögn síra Einars Hálfdanarsonar um hlaupið úr Öræfajökli 1727; "Af jöklinum, er öræfin hljópu." *Blanda, 1*, 54–59.

Hannesson, P. (1958). *Frá óbyggðum - ferðasögur og landlýsingar*. Reykjavík: Bókaútgáfa Menningarsjóðs.

Helland, A. (1882). Om Islands Jøkler og om Jøkulelvenes Vandmængde og Slamgehalt. *Archiv for Mathematik og Naturvidenskab, 7*, 200–232.

Henderson, E. (1818). *Iceland; or the journal of a residence in that island during the years 1814 and 1815*. Edinburgh: Oliphant, Waugh & Innes.

Hjulström, F., Sundborg, A., Arnborg L., & Jónsson, J. (1954–57). The Hoffellssandur, a glacial outwash plain. *Geografiska Annaler*. 36(1–2), 135–189; 37(3–4), 170–245; 39(2–3), 143–212.

Holland, E. (1862). A tour in Iceland in the summer of 1861. In E. Kennedy (Ed.), *Peaks, passes, and glaciers: Being excursions by members of the Alpine Club* (pp. 3–128). London: Longman, Green, Longman & Roberts.

Hoppe, G. (1953). Några iakttagelser vid islandska jöklar sommaren 1952. *Ymer*, 73, 241–265.

Hoppe, G. (1995). Brúarjökull. *Glettingur*, 5(2), 38–41.

Horrebow, N. (1758). *The natural history of Iceland* (J. Anderson, Trans.). London: A. Linde.

Howell, F. (1892). The Öræfa Jökull, and its first ascent. *Proceedings of the Royal Geographical Society*, 14, 841–850.

IES & IMO websites. (2016). http://earthice.hi.is; http://en.vedur.is/earthquakes-and-volcanism/articles/nr/2947; http://earthice.hi.is/bardabunga_holuhraun; http://futurevolc.hi.is. Accessed 1 May.

Imsland, P. (1994). Twentieth-century isostatic behaviour of the coastal region in southeast Iceland—Extended Abstract. In *Proceedings of the Hornafjörður International Coastal Symposium, Iceland, June 20–24, 1994* (pp. 513–518). Reykjavík: Hafnamálastofnun ríkisins.

Jakobsson, S. (1979a). Petrology of recent basalts of the Eastern Volcanic Zone, Iceland. *Acta Naturalia Islandica*. 26. Reykjavík: Náttúrufræðistofnun Íslands.

Jakobsson, S. (1979b). Outline of the petrology of Iceland. *Jökull*, 29, 57–73.

Jóhannesson, Hk. (1983). Gossaga Grímsvatna 1900–1983 í stuttu máli. *Jökull*, 33, 146–147.

Jóhannesson, Hk. (1984). Grímsvatnagos 1933 og fleira frá því ári. *Jökull*, 34, 151–158.

Jóhannesson, H. (1994). Coastal erosion near the bridges across Jökulsá á Breiðamerkursandi in southeastern Iceland. In Viggósson, G. (Ed.), *Proceedings of the Hornafjörður International Coastal Symposium in Iceland. 20–24 June 1994*. Höfn, Hornafjörður: Organizing Committee.

Jóhannesson, T. (2002). Propagation of a subglacial flood wave during the initiation of jökulhlaup. *Hydrological Sciences Journal*, 47, 417–434.

Jonas, R. (1948). *Fahrten in Island. Mit Beitragen von Franz Nusser*. Vienna: L.W. Seidel.

Jónsson, E. (2004). Í veröld jökla, sanda og vatns. In H. Björnsson, E. Jónsson & S. Runólfsson (Eds.), *Jöklaveröld. Náttúra og mannlíf* (pp. 11–86). Reykjavík: Skrudda.

Jónsson, F. (1914). Tvö heimildarrit um byggð í Öræfum með athugasemdum. *Afmælisrit Dr. Phil. Kr. Kålund bókavarðar við safn Árna Magnússonar 19. Ágúst* (pp. 34–47). Copenhagen: Hið íslenska fræðafélag.

Jónsson, J. (1986, December 9). Eldgos í Vatnajökli 1910. *Suðurland*. 34(17), 13.

Jónsson, Ó. (1945). *Ódáðahraun* (3 Vols.). Akureyri: Norðri.

Kjerúlf, Þ. (1962). Vatnajökull hlaupinn (Brúarjökull 1890). *Jökull*, 12, 47–48. [Originally published in *Ísafold*. 17(81), 321 (1890)].

Koch, J. (1912). Den danske Ekspedition til Dronning Louises land og tværsover Nordgrønlands Indlandsis 1912–1913. Rejsen tværsover Island i Juni 1912. *Geografisk Tidsskrift*, 21, 257–264.

Kofoed-Hansen, A. (1912). Fágætt ferðalag. *Suðurland*, 3(11), 41–42.

Kristinsson, S. (1985). Lítil samantekt um Vatnajökulsleið. *Múlaþing*, 14. Sögufélag Austurlands.

Kristjánsdóttir, U. (1988). Byggðin varin - ánni veitt. In G. Þorkelsson & E. Jónsson (Eds.), *Jódynur - hestar og mannlíf í Austur-Skaftafellssýslu* (pp. 167–174). Akureyri: Bókaforlag Odds Björnssonar.

Kristjánsdóttir, U. (1993). Hólmsá. *Skaftfellingur*, 9, 69–82.

Kristjánsdóttir, U. (2000). Vatnsdalur og Vatnsdalshlaup. *Skaftfellingur*, 13, 81–90.

Larsen, G. (1982). Gjóskutímabil Jökuldals og nágrennis. In H. Þórarinsdóttir, Ó. Óskarsson, S. Steinþórsson, & Þ. Einarsson (Eds.), *Eldur er í norðri* (pp. 331–335). Reykjavík: Sögufélag.

Larsen, G. (1984). Recent volcanic history of the Veidivötn fissure swarm, southern Iceland—An approach to volcanic risk assessment. *Journal of Volcanology and Geothermal Research*, 22, 33–58.

Larsen, G., Guðmundsson, M., & Björnsson, H. (1998). Eight centuries of periodic volcanism at the center of the Iceland hot spot revealed by glacier tephrastratigraphy. *Geology, 26*(10), 943–946.

Laxness, H. (2002) [1937], *World Light* (M. Magnusson, Trans.). New York: Vintage International.

Leutelt, R. (1935, June 23). Gengið á Bárðargnýpu á fyrsta sinn. *Lesbók Morgunblaðsins, 10*(25), 198–199.

Leutelt, R. (1937). Bergfahrten auf Island. *Die Alpen, Monatsschrift des Schweizer Alpenclub., 13*(4), 136–145.

Magnússon, Á. (1955) [ca. 1702]. *Chorographica Islandica.* In Ó. Lárusson (Ed.), *Safn til sögu Íslands.* Second series (Vol. 2). Reykjavík: Hið íslenska bókmenntafélag.

Magnússon, E., Björnsson, H., Pálsson, F., & Dall, J. (2004). Glaciological application of InSAR topography data of W-Vatnajökull acquired in 1998. *Jökull, 54*, 17–36.

Magnússon, E., Björnsson, H., Dall, J., & Pálsson, F. (2005). Volume changes of Vatnajökull ice cap, Iceland, due to surface mass balance, ice flow, and sub-glacial melting at geothermal areas. *Geophysical Research Letters, 32*(5), L05504.

Magnússon, Þ. (1907–1915) [1625]. Relatio Þorsteins Magnússonar um jöklabrunann fyrir austan 1625. In Thoroddsen, Þ. (ed.), *Safn til sögu Íslands* IV (pp. 200–215). Copenhagen and Reykjavík: Hið íslenska bókmenntafélag.

Markússon, Þ. (1932–1935). Kaflar úr Íslandslýsingu Þorláks Markússonar. Þorsteinsson, H. (ed.) *Blanda* (vol. 5, pp. 22–36).

Marshall, S., Björnsson, H., Flowers, G., & Clarke, G. (2005). Simulation of Vatnajökull Ice Cap Dynamics. *Journal of Geophysical Research, 110*(F3), 2005. doi:10.1029/2004JF000262.

Muir, T., & Wigner, J. (1953). Gönguför yfir þveran Vatnajökul fyrir hálfri öld. *Vísir. Christmas issue*, 7–8, 25–29.

Nick, F., van der Kwast, J., & Oerlemans, J. (2007). Simulation of the evolution of Breiðamerkurjökull in the late Holocene. *Journal of Geophysical Research, 112*(B01103), 2007. doi:10.1029/2006JB004358.

Nielsen, N. (1937a) *Vatnajökull. Barátta elds og ísa* (P. Hannesson, Trans.). Reykjavík: Mál og menning. *Vatnajökull.* (Danish version: *Kampen mellem Ild og Is.* Copenhagen: Hagerup).

Nielsen, N. (1937b). A volcano under an ice-cap. Vatnajökull, Iceland, 1934–36. *The Geographical Journal., 90*, 6–23.

Nørlund, N. (1944) *Islands Kortlægning. En historisk Fremstilling.* Geodætisk Instituts Publikationer. 7. Copenhagen: Ejnar Munksgaard.

Nusser, F. (1936). Bericht über die österreichische Island-Vatna-Jökull-Expedition. *Polarforschung, 6*(1), 2–3.

Nusser, F. (1940). Der Dyngjujökull auf Island in den Jahren 1935 und 1939. *Jahresber. des Archivs für Polarforschung im Naturhistorischen Museum in Wien, 3*, 4–20.

Nye, J. (1976). Water flow in glaciers: jökulhlaups, tunnels and veins. *Journal of Glaciology, 76*, 181–207.

Oerlemans, J., Björnsson, H., Kuhn, M., Obleitner, F., Pálsson, F., Smeets, P., et al. (1999). A glacio-meteorological experiment on Vatnajökull, Iceland. *Boundary-Layer Meteorology, 92*, 3–26.

Ólafsson, E., Pálsson, B. (1981). *Ferðabók Eggerts Ólafssonar og Bjarna Pálssonar um ferðir þeirra á Íslandi árin 1752–1757.* (S. Steindórsson, Trans.). Reykjavík: Örn og Örlygur.

Olavius, Ó. (1964–1965). *Ferðabók: landshagir í norðvestur-, norður- og norðaustursýslum Íslands 1775-1777: ásamt ritgerðum Ole Henckels um brennisteinsnám og Christian Zieners um surtarbrand. Íslandskort frá 1780 eftir höfund og Jón Eiríksson* (2 vols.) (S. Steindórsson, Trans.). Reykjavík: Bókfellsútgáfan.

Pagli, C., & Sigmundsson, F. (2008). Will present day glacier retreat increase volcanic activity? Stress induced by recent glacier retreat and its effect on magmatism at the Vatnajökull ice cap, Iceland. *Geophysical Research Letters, 35*, 1–5.

Pagli, C., Sigmundsson, F., Lund, B., Sturkell, E., Geirsson, H., Einarsson, P., et al. (2007). Glacio-isostatic deformation around the Vatnajökull ice cap, Iceland, induced by recent climate warming: GPS observations and finite element modeling. *Journal Geophysical Research, 112,* B08405.

Pálsson, S. (1945). *Ferðabók Sveins Pálssonar. Dagbækur og ritgerðir 1791–1794.* (J. Eyþórsson, P. Hannesson, S. Steindórsson, Trans. and Eds.). Reykjavík: Snælandsútgáfan.

Resen, P. (1991) [1667]. *Íslandslýsing.* (J. Benediktsson, Trans.). *Safn Sögufélags.* 3. Reykjavík: Sögufélagið.

Rist, S. (1983). Floods and flood dangers in Iceland. *Jökull, 33,* 119–132.

Roberts, B. (1933). The Cambridge expedition to Vatnajökull 1932. *Geographical Journal, 81,* 289–313.

Sæmundsson, K. (1973). Straumrákaðar klappir í kringum Ásbyrgi. *Náttúrufræðingurinn, 43,* 52–60.

Sæmundsson, K. (1979). Outline of the geology of Iceland. *Jökull, 29,* 7–28.

Sæmundsson, K. (1982). Öskjur á virkum eldfjallasvæðum á Íslandi. In H. Þórarinsdóttir, Ó. Óskarsson, S. Steinþórsson, & Þ. Einarsson (Eds.), *Eldur er í norðri* (pp. 221–239). Reykjavík: Sögufélag.

Schomacker, A., Krüger, J., & Kjær, K. (2006). Ice-cored drumlins at the surge-type glacier Brúarjökull, Iceland: A transitional-state landform. *Journal of Quaternary Science, 21,* 85–93.

Sigmundsson, F. (1990). *Seigja jarðar undir Íslandi, samanburður líkanreikninga við jarðfræðileg gögn.* M.Sc. thesis. Reykjavík: Háskólaútgáfan.

Sigmundsson, F. (1991). Post-glacial rebound and asthenosphere viscosity in Iceland. *Geophysical Research Letters, 18,* 1131–1134.

Sigmundsson, F., Sturkell, E., Pinel, V., Einarsson, P., Pedersen, R., et al. (2004). Deformation and eruption forecasting at volcanoes under retreating ice caps: Discriminating signs of magma inflow and ice unloading at Grímsvötn and Katla volcanoes, Iceland. *Eos Transactions, 85*(47), Suppl. 608 (Synopsis of paper delivered at fall meeting of American Geophysical Union).

Sigurðsson, F. (1990). Groundwater from glacial areas in Iceland. *Jökull, 40,* 119–146.

Sigurðsson, H. (1978). *Kortasaga Íslands frá lokum 16. aldar til 1848.* Reykjavík, Bókaútgáfa Menningarsjóðs og Þjóðvinafélagsins.

Sigurðsson, J. (1857). *Íslenskt fornbréfasafn, Diplomatarium Islandicum* (Vol. 1). Copenhagen: Hið íslenska bókmenntafélag.

Sigurðsson, St. (1947). Á Vatnajökli með vélsleða og jeppa. *Fálkinn, 20*(5), 4–5.

Sigurðsson, St. (1984). Grímsvatnaför. Útvarpserindi frá 1942. *Jökull, 34,* 180–185.

Stefánsson, R. (1982). Skeiðarárhlaupin. Margvíslegar afleiðingar þeirra. *Skaftfellingur, 3,* 99–118.

Stefánsson, R. (1983). Skeiðarárhlaupið 1939. *Jökull, 33,* 148.

Stefánsson, S. (1957) [1746]. Austur-Skaftafellssýsla. In B. Guðnason (Ed.), *Sýslulýsingar 1744– 1749. Sögurit* (Vol. 28, pp. 1–23).

Steinþórsson, S. (1977). Tephra layers in a drill core from the Vatnajökull ice cap. *Jökull, 27,* 2–17.

Sturluson, S. (1987). *[12th century] Edda* (A. Faulkes, Trans. and Ed.). London: J.M. Dent.

Thienemann, F. (1824). *Reise im Norden Europas, vorzuglich in Island in den Jahren 1820 bis 1821.* Leipzig: Carl Heinrich Reclam.

Thorarensen, P. (1997) [1839]. Sandfells- og Hofssóknir í Öræfum. In Jónsson, J. and Sigmundsson, S. (Eds.) *Skaftafellssýsla. Sýslu- og sóknalýsingar Hins íslenska bókmenntafélags 1839–1873.* Reykjavík: Sögufélag. pp. 145–153.

Thoroddsen, Þ. (1892a). Islands Jøkler i Fortid og Nutid. *Geografisk Tidsskrift, 11,* 111–146.

Thoroddsen, Þ. (1892–1904). *Landfræðissaga Íslands* (Geological history of Iceland) (4 Vols.). Copenhagen: Hið íslenska bókmenntafélag.

Thoroddsen, Þ. (1895). Ferð um Austur-Skaptafellssýslu og Múlasýslur sumarið 1894. *Andvari, 20,* 1–84.

Thoroddsen, Þ. (1897). Ferðir um Norður-Þingeyjarsýslu sumarið 1895. *Andvari, 22,* 17–71.

Thoroddsen, Þ. (1905/06). Die Gletscher Islands. Island, Grundriss der Geographie und Geologie. *Petermanns Mitteilungen, 152,* 1–161; *153,* 162–358.

Thoroddsen, Þ. (1907–1911). *Lýsing Íslands* (2 Vols.). Copenhagen: Hið íslenzka Bókmentafélag.

Thoroddsen, Þ. (1913–1915). *Ferðabók: skýrslur um rannsóknir á Íslandi 1882–1898* (4 Vols.). Copenhagen: Hið íslenska fræðafélag.

Thoroddsen, Þ. (1915). Vulkaniske Udbrud i Vatnajökull paa Island. *Geografisk Tidsskrift, 23,* 118–132.

Thoroddsen, Þ. (1958–1960). *Ferðabók* (J. Eyþórsson, Ed., 4 Vols., 2nd ed.). Reykjavík: Snæbjörn Jónsson.

Todtmann, E. (1952). Im Gletscherrückzugsgebiet des Vatna Jökull auf Island, 1951. *Neues Jahrbuch für Geologie und Paläontologie.* Stuttgart (pp. 401–411).

Todtmann, E. (1953). Am Rand des Eyjabakkagletschers 1953. *Jökull, 3,* 34–37.

Todtmann, E. (1955a). Kringilsárrani, das Vorfeld des Brúarjökuls, am Nordland des Vatnajökull. *Jökull, 5,* 9–10.

Todtmann, E. (1955b). Übersicht über die Eisrandlagen in Kringilsárrani 1890-1955. *Jökull, 5,* 8–10.

Todtmann, E. (1957). Kringilsárrani, das Vorfeld des Brúarjökull, am Nordrand des Vatnajökull. *Neues Jahrbuch für Geologie und Paläontologie, 104,* 255–278.

Todtmann, E. (1960). Gletcherforschungen auf Island (Vatnajökull). *Universität Hamburg Abhandlungen aus dem Gebiet der Auslandkunde* (Naturwissenschaften). Bd. 65, Reihe C, Bd. 19.

Tómasson, H. (1973). Hamfarahlaup í Jökulsá á Fjöllum. *Náttúrufræðingurinn, 43,* 12–34.

Tómasson, H., & Vilmundardóttir, E. (1967). The lakes Stórisjór and Langisjór. *Jökull, 17,* 280–299.

Tómasson, Þ. (1980). *Skaftafell. Þættir úr sögu ættarseturs og atvinnuhátta.* Reykjavík: Þjóðsaga.

Torell, O. (1857). Bref om Island. *Öfversigt af Kgl. Vetenskaps-Akademiens förhandlingar, 18,* 325–332.

Trautz, M. (1919). Am Nordrand des Vatnajökull im Hochland von Island. *Petermanns Mitteilungen, 65,* 122–126, 223–229.

Valtýsson, H. (1942–1953). *Söguþættir landpóstanna* (3 Vols.) Akureyri: Norðri. (Section on Hannes of Núpsstaður (1918–1942) Vol. 1, pp. 152–153, 158).

Wadell, H. (1920a, June 20). För yfir Vatnajökul. *Morgunblaðið.* 7(188), p. 3.

Wadell, H. (1920b). Vatnajökull. Some studies and observations from the greatest glacial area in Iceland. *Geografiska Annaler, 4,* 300–323.

Waitt, R. (2002). Great Holocene floods along Jökulsá á Fjöllum, North Iceland. In I. Martini, V. Baker & G. Garzón (Eds.), *Flood and megaflood processes, recent and ancient examples* (Vol. 31, pp. 37–51). (International Association of Sedimentologists' special publication).

Watts, W. (1875). *Snioland or Iceland, its Jökulls and Fjalls.* London: Longman.

Watts, W. (1876). *Across the Vatna jökull or, scenes in Iceland.* London: Longman.

Wigner, J. (1905). The Vatna Jökull traversed from North-East to South-West. *Alpine Journal, 22* (168), 436–448.

Witze, A., & Kanipe, J. (2014). *Island on Fire: The extraordinary story of Laki, which turned eighteenth-century Europe dark.* London: Profile Books.

Woldstedt, P. (1939). Vergleichende Untersuchungen an islandischen Gletschern. *Jahrbuch der Königlich Preussischen Geologischen Landesanstalt zu Berlin, 59,* 249–271.

Þórarinsson, S. (1943). Oscillations of the Iceland glaciers in the last 250 years. *Geografiska Annaler, 25*(1–2), 1–54.

Þórarinsson, S. (1946). Í veldi Vatnajökuls. *Lesbók Morgunblaðsins* 21(series continued in issues 16, 17, 18, 19, 20, 33, 34, 35).

Þórarinsson, S. (1950). Jökulhlaup og eldgos á Jökulvatnasvæði Jökulsár á Fjöllum. *Náttúrufræðingurinn, 20,* 113–133.

Þórarinsson, S. (1953a). Some new aspects of the Grímsvötn problem. *Journal of Glaciology, 14,* 267–274.

Þórarinsson, S. (1953b). The Grímsvötn-expedition June–July 1953. *Jökull, 3,* 6–22.

Þórarinsson, S. (1954). Athuganir á Skeiðarárhlaupi og Grímsvötnum 1954. *Jökull, 4,* 34–37.

Þórarinsson, S. (1955). Mælingaleiðangurinn á Vatnajökli vorið 1955. *Jökull, 5,* 27–29.

Þórarinsson, S. (1956). On the variations of Svínafellsjökull, Skaftafellsjökull and Kvíárjökull in Öræfi. *Jökull, 6,* 1–15.

Þórarinsson, S. (1958). The Öræfajökull eruption of 1362. *Acta Naturalia Islandica, 2*(2), 99 pp. Reykjavík: Náttúrugripasafn Íslands.

Þórarinsson, S. (1964). Sudden advance of Vatnajökull outlet glaciers 1930–1964. *Jökull, 14,* 76–89.

Þórarinsson, S. (1965). Changes of the water-firn level in the Grímsvötn caldera 1954–1965. (Breytingar á yfirborði Grímsvatna 1954–1965). *Jökull, 15,* 109–119.

Þórarinsson, S. (1969). Glacier surges in Iceland, with special reference to the surges of Brúarjökull. *Canadian Journal of Earth Sciences, 6*(4), 875–882.

Þórarinsson, S. (1974a). Sambúð lands og lýðs í ellefu aldir. In S. Líndal (Ed.), *Saga Íslands* (Vol. 1, pp. 29–97). Reykjavík: Hið íslenska bókmenntafélag.

Þórarinsson, S. (1974b). *Vötnin stríð. Saga Skeiðarárhlaupa og Grímsvatnagosa.* Reykjavík: Bókaútgáfa Menningarsjóðs.

Þórarinsson, S. (1980a). Enn um Skeiðarárhlaup. *Jökull, 30,* 74.

Þórarinsson, S. (1980b). Þú stóðst á tindi. *Jökull, 30,* 81–87.

Þórarinsson, S., & Sigurðsson, S. (1947). Volcano-glaciological investigations in Iceland during the last decade. *The Polar Record, 33–34,* 60–64.

Þórarinsson, S., Sæmundsson, K., & Williams, R. (1973). ERTS-1 image of Vatnajökull. Analysis of glaciological andvolcanic features. *Jökull, 23,* 7–17.

Þórðarson, Þ. (1990). *The eruption sequence and eruption behavior of the Skaftár Fires, 1783–85, Iceland: Characteristics and distribution of eruption products.* M.Sc. dissertation, The University of Texas, Arlington.

Þórðarson, Þ. (1991). *Skaftáreldar 1783–1785. Gjóskan og framvinda gossins.* Essay in 4th year of geology at University of Iceland. Reykjavík: Háskóli Íslands.

Þórðarson, Þ. (2003). The 1783–1785 A.D. Laki-Grímsvötn eruptions I: A critical look at the contemporary chronicles. *Jökull, 53,* 1–10.

Þórðarson, Þ., & Self, S. (2001). Real-time observations of the Laki sulfuric aerosol cloud in Europe during 1783 as documented by Professor P. van Swinden at Franeker, Holland. *Jökull, 50,* 65–72.

Þórðarson, Þ., Larsen, G., Steinþórsson, S., & Self, S. (2003). The 1783–1785 A.D. Laki-Grímsvötn eruptions II: Appraisal based on contemporary accounts. *Jökull, 53,* 11–48.

Þórðarson, Þ., & Larsen, G. (2007). Volcanism in Iceland in historical time: Volcano types, eruption styles and eruptive history. *Journal of Geodynamics, 43*(1), 118–152.

Þórðarson, Þ., Self, S., Óskarsson, N., & Hulsebosch, T. (1996). Sulfur, chlorine, and fluorine degassing and atmospheric loading by the 1783-1784 AD Laki (Skaftár Fires) eruption in Iceland. *Bulletin of Volcanology, 58,* 205–225.

Þórðarson, Þb. (1943). Vatnadagurinn mikli. *Tímarit Máls og menningar, 4,* 69–84.

Þorkelsson, J. (1921–23). Kirkjustaðir á Austur-Skaptafellsþingi. *Blanda, 2,* 246–268.

Þormóðsson, E., & Grímsdóttir, G. (Eds.). (2003). *Oddaannálar og Oddverjaannáll.* Reykjavík: Stofnun Árna Magnusson á Íslandi. 59.

Þorsteinsson, P. (1985). *Samgöngur í Skaftafellssýslum.* Höfn in Hornafjörður.

Þorsteinsson, H., Jóhannesson, J., Vilmundarson, Þ., & Grímsdóttir, G. (Eds.). (1922–1987). *Annálar 1400-1800.* Reykjavík: Hið íslenzka bókmenntafélag.

Appendix A
Overview of the History of Glaciers and Glaciology in Iceland

Settlement	Place names in the *Book of Settlements* bear witness to glaciers and glacial rivers in Iceland
10th C.	The household labourers of Egill Skallagrímsson are amazed by the white colour of glacial rivers
934	Volcanic eruption in Eldgjá gorge linked to Katla's volcanic centre causes outburst flood
End 11th C.	First portrayals of the movement of glaciers are recorded in *Gesta Danorum* from around 1200 by Saxo Grammaticus; the descriptions could well be from Iceland
Mid-13th C.	*The King's Mirror* correctly relates the connection between glaciers and climate
End 13th C.	The Saga of *Bárður Snæfellsás* relates how Snæfellsjökull is created from fresh snowfalls on firn snow, bearing witness to an understanding of the nature of glaciers
1362	Volcanic eruption in Öræfajökull (Hnappafellsjökull). Litla-Hérað between Skeiðarársandur and Breiðamerkursandur and all the farms on the lowlands to the south of Skeiðarárjökull are destroyed by tephra ash falls and flooding. The jökulhlaup's maximum flow rate reaches about 100,000 m^3/s. The tephra ash fall caused much more damage than the floods. This was Europe's largest volcanic eruption since Vesuvius in 79 A.D
1477	Huge volcanic eruption in the Veiðivötn lakes area. Lava flow believed to have travelled along fissure from Bárðarbunga volcano. Similar eruption occurred there in 871
1570	The names Snæfellsjökull (Vesturjökull) and Eyjafjallajökull (Austurjökull) appear on foreign nautical maps of the northernmost parts of Europe
1573	Journeys of itinerant fishermen from northern Iceland over Vatnajökull to Suðursveit district cease. The harbour on offshore skerries destroyed by overburden of advancing Vatnajökull causing land to subside

© Atlantis Press and the author(s) 2017
H. Björnsson, *The Glaciers of Iceland*, Atlantis Advances in Quaternary Science 2, DOI 10.2991/978-94-6239-207-6

1590   Bishop Oddur Einarsson's *Íslandslýsing (Account of Iceland)* published, in which glaciers are reported to be expanding because of a deteriorating climate. Also includes first contemporary description of a volcanic eruption of Katla

1590   Glaciers outlined for the first time on Bishop Guðbrandur Þorláksson's map of Iceland in 1570, published in the map collection of Abraham Ortelius in 1590, the first map known to delineate glaciers

1598–1608   Grímsvötn appears for the first time in written records in a letter from the bishopric of Skálholt

1612   Volcanic eruptions in Eyjafjallajökull and Katla followed by outburst floods

1625   Sheriff Þorsteinn Magnússon names Grímsvötn in his publication *Relatio*. His summaries of Katla volcanic eruptions are the most important descriptions of volcanic outbreaks since Pliny the Younger recorded the eruption of Vesuvius in 79 A.D

1625   Brúarjökull glacier surges

1637–1638   Grímsvötn considered a lake in Gísli Oddsson's work, *De mirabilibus Islandiæ (On the Wonders of Iceland)*

1660   A Katla jökulhlaup described by Pastor Jón Salómonsson. Fishing boats are no longer launched from Víkurklettir cliffs and Skiphellir cave in Mýrdalur because of huge glaciofluvial deposits

1684–1687   Gríms Vatna Jökull appears on Peder Hansen Resen's *Atlas Danicus* though it is not delineated. Report of volcanic eruption at Grímsfjall

1695   Þórður Þorkelsson Vídalín, the rektor of Skálholt cathedral, composes his *Dissertationcula de montibus Islandiae chrystallinis*, the most knowledgeable and important thesis on glaciers anywhere in the world at the time. Vídalín reports that more glaciers advance than retreat in Iceland. Also sets forth his 'frost expansion' theory as to how glaciers move. His work remains unpublished until a German translation appeared in 1754. Vídalín reports the common people's belief that 'ice mountains' are created from snow and not saltpetre

Ca. 1700   Settlement farmlands on Breiðamerkursandur plain destroyed by the advance of Breiðamerkurjökull

1702–1712   Collector of manuscripts Árni Magnússon describes the expansion of glaciers in *Chorographica Islandica*; it also contains one of the first scholarly descriptions of a jökulhlaup. Cites oral memories of journeys across Vatnajökull

1712   Ísleifur Einarsson describes the encroachment of glaciers and their rivers in Austur-Skaftafell County

1721   Lake Grímsvötn named on Admiral Peder Raben's map of Iceland

1721   A glacial outburst flood from Katla sweeps away farm on Mýrdalssandur below Hjörleifshöfði. A flood wave from the jökulhlaup crashes ashore in the Westman Isles

1727–1728	A volcanic eruption in Öræfajökull. An account of this by Jón Þorláksson is published in a work by Olavius in 1780
1734	The Norwegian surveyor Thomas Hans Henrik Knoff charts a map of Iceland showing all the country's main glaciers. This is the greatest improvement in the mapping of Iceland since the days of Bishop Guðbrandur Þorláksson in 1570. At first a state secret, it was not made public until two years later
1746	Sheriff Sigurður Stefánsson describes daily life in the vicinity of glaciers in Austur-Skaftafell County
1750–1757	Natural scientist Eggert Ólafsson and physician Bjarni Pálsson explore Iceland and relate their findings in a travelogue which was published in Danish in 1772, and in an Icelandic translation in 1943 and 1981. They claim glaciers are larger than they were in the time of settlement and are still expanding. Ólafsson and Pálsson climb Geitlandsjökull, Snæfellsjökull and Mýrdalsjökull. They believe glacial rivers flow underground from Snæfellsjökull. They describe crevasses and ogives, and are the first to give accounts of seracs and 'elf pools.' They also collect flora fossils at Brjánslækur in the western fjords, Ólafsson believing they are the remains of formerly large forests
1755	Volcanic eruption of Katla produces greatest amount of tephra since the Settlement. Höfðabrekkujökull formed from mud and blocks of ice on the lowlands
1772	Eggert Ólafsson and Bjarni Pálsson's *Ferðabók* published in Danish, edited by Jón Eiríksson
1780	Ólafur Ólafsson (Olavius) publishes travelogue including accounts and information of farmsteads that have been abandoned due to jökulhlaups and advancing glaciers
1783–1784	Skaftá Fires. Lava flow believed to have originated from the Grímsvötn volcano
1789	Lord Stanley's expedition to Iceland. Snæfellsjökull is climbed and its height calculated
1791–1794	Physician Sveinn Pálsson, supported by a grant from the Danish Academy of Natural Sciences, carries out research on the natural environment of Iceland during the summers. He composes his *Ferðabók* from his diaries, and this contains his theses on volcanoes and glaciers (1795). Pálsson's writings are first published in their entirety in 1945, and his 'Treatise on Glaciers' appears in an English translation in 2004. Sveinn categorises glaciers, explains the formation of kettle holes and describes glacial surges. He climbs onto Eyjafjallajökull in 1793 and draws maps of all the country's main glaciers. Pálsson hypothesises on the viscosity of glaciers to explain their ability to move. He and two of his companions ascend Sveinsgnípa summit on Öræfajökull from the farm Kvísker. Pálsson

	also climbs both Hekla and Katla. He also sees and describes the surge of Breiðamerkurjökull
1800	Langjökull advances into Lake Hvítárvatn
1801–1818	Danes carry out surveys along shores of Iceland and draw up first detailed map of coastline
1814–1815	Ebenezer Henderson travels around Iceland and writes about his journey. No foreigner travelled as extensively around the country as he did. He notes that the climate had worsened since the time of the settlement
1823	Volcanic eruptions in Katla and Eyjafjallajökull
1831–1844	Björn Gunnlaugsson surveys and maps Iceland. Completes *Uppdrætti Íslands* in 1848 on a scale of 1:480,000
1837–1840	Notions of an Ice Age born. De Charpentier presents his 'glacial hypothesis' and Louis Agassiz and Karl Friedrich Schimper follow up with the theory of an Ice Age
1846	The German geologist Wolfgang Sartorius von Waltershausen and chemist Robert Wilhelm Bunsen and the Danish geologist and General, Haagen V. Mathiesen travel around Iceland. Mathiesen is the first to put forward the idea that most of Iceland had been covered by a glacier. Waltershausen, however, was not a supporter of the Ice Age theory. Waltershausen is the first scientist to analyse Icelandic 'móberg,' calling it palagonite
1849	Björn Gunnlaugsson's Map of Iceland published, based on his surveying during the years 1831–1844
1850	The Norwegian geologist, Theodor Kjerulf, a supporter of the Ice Age theory, travels around Iceland. He is the first to measure the height of snowlines in Iceland
1857	The Swedish scientist Otto Martin Torell travels around Iceland and is the first to measure the movement of glaciers (Svínafellsjökull in Öræfi district). Torell supports the Ice Age theory and proved that an ice-age glacier had extended from Scandinavia into northern Germany
1859	The botanist Oswald Heer analyses the plant fossils collected by Jónas Hallgrímsson and Japetus Steenstrup and reveals they show that deciduous woods had existed in Iceland 6 million years ago
1860	Volcanic eruption of Katla
1861	One of the greatest outburst floods known to history disgorges from Grímsvötn.
1865	The Swedish geologist Carl Wilhelm Paijkull produces the first geological map of Iceland, based on Gunnlaugsson's map. Paijkull believed the land had been covered by a glacier in ancient times. He also draws the attention of foreign scientists to the phenomenon of Icelandic jökulhlaups
1867	A huge jökulhlaup and a volcanic eruption at Grímsvötn

1875	William Lord Watts crosses Vatnajökull with Icelanders from Skaftafell County. Watts discovers an unknown mountain and calls it Pálsfjall. Watts had previously ascended Vatnajökull in 1871 and 1874
1881	The Norwegian geologist Amund Helland comes to Iceland. He writes about glaciers and their rivers, and was the first to measure sediments loads in glacial rivers (from Vatnajökull); he also evaluates the erosive power of glaciers and describes jökulhlaups. Helland believes the smell from the river Jökulsá on Sólheimasandur to be caused by geothermal heat beneath the glacier. He arranged for part of Sveinn Pálsson's treatise on glaciers to be published in Norway (1883)
1882	Geologist Þorvaldur Thoroddsen begins his research expeditions in Iceland, which continued annually until 1898
1890	Brúarjökull makes one of the greatest surges known in recent world history
1850–1900	Main glaciers of Iceland reach their furthest extensions during the Little Ice Age
1899	Geologist Helgi Pjeturss reveals that the Ice Age in Iceland was not continuous, but had been interrupted by a warm phase. He demonstrates that palagonite is from the last glacial period and that there had been at least one warm phase between glacial periods when glaciers disappeared from most of Iceland
1901	Thoroddsen's geological map is published on a scale of 1:600,000
1907	German geologist Walther von Knebel (1880–1907) presumed drowned, along with his colleague Max Rudloff, in Lake Askja
1910	Geologist Helgi Pjeturss points out that the ice cap from the last glacial period had not retreated without interruptions
1915–16	Helgi Pjeturss presents strong evidence that glaciers in Iceland are now larger than during the time of settlement
1918	Eruption of Katla
1919	Swedish students Hakon Wadell and Erik R. Ygberg discover location of Grímsvötn in Vatnajökull and make a sketch-map of it. The location is forgotten until Grímsvötn erupts again in 1934
1930	Meteorologist Jón Eyþórsson and engineer Helgi Hermann Eiríksson begin regular measurements of the positions of glacial termini. These measurements were later under the auspices of the IGS and the supervision of hydrologist Sigurjón Rist, and later geologist Oddur Sigurðsson. Glaciers begin to shrink rapidly
1934	Jökulsárlón glacial lagoon begins to form on Breiðamerkursandur
1934	Huge jökulhlaup in the river Skeiðará. In March an eruption in Grímsvötn begins and its location is rediscovered

1936–1938   Swedish-Icelandic expeditions, led by geologist Hans Ahlmann and meteorologist Jón Eyþórsson, check the relationship between glacial mass balance and weather conditions on Vatnajökull. Other members of the expeditions include Sigurður Þórarinsson, Carl Mannerfelt, Mac Lilliehöök and Jón Jónsson from Laugar. Ahlmann compiles a map of the altitude of firn lines in Iceland

1937   Danish Air Force takes oblique photographs of Iceland, including its glaciers

1938   Scientists flew over Vatnajökull for the first time to study jökulhlaup from Grímsvötn

1940   Sigurðar Þórarinsson publishes an overview of the size and retreat of glaciers all over the world and the consequential changes in oceanic volumes

1942   Expedition led by Steinþór Sigurðsson surveys Grímsvötn. Members included Einar B. Pálsson, Franz Pálsson and Sveinn Þórðarson

1943   Sigurður Þórarinsson publishes a comprehensive catalogue of all glacial changes in Iceland since the 16th century

1943   Geologist Guðmundur Kjartansson puts forward his theory on the formation of palagonite during volcanic eruptions beneath a glacier. Final proof was obtained when volcanic material from the suboceanic eruption of Surtsey in 1963–1964 became transformed into hard palagonite

1945   Sveinn Pálsson's treatise on glaciers translated into Icelandic

1945–1946   The US Army Map Service takes vertical aerial photographs of most of Iceland, from which an accurate AMS map is created on a scale of 1:50,000

1946   Motorised vehicles first driven onto Vatnajökull, thus beginning the motor age of glaciers

1947   Sigurjón Rist begins regular hydrological measurements in Iceland. Assisting him are Eberg Elefsen and Davíð B. Guðnason

1950   Meteorologist Jón Eyþórsson establishes Iceland Glaciological Society and is elected its first chairman. Later chairmen include Trausti Einarsson, Sigurður Þórarinsson, Sigurjón Rist, Sveinbjörn Björnsson, Helgi Björnsson and Magnús Tumi Guðmundsson

1951   The journal Jökull is published for the first time. Its first editor was Jón Eyþórsson. Other editors since then have been: Sigurður Þórarinsson, Guðmundur Pálmason, Sveinbjörn Björnsson, Leó Kristjánsson, Helgi Björnsson, Tómas Jóhannesson, Bryndís Brandsdóttir and Áslaug Geirsdóttir

1951   The first cabins of the IGS are set up in the Esjufjöll mountains and at Breiðá on Breiðamerkursandur

1951–1952   British scientists, along with the geologist Sverrir Scheving Thorsteinsson, draw up a map of Breiðamerkurjökull

1951–1954	British students, led by Jack D. Ives, carry out glaciological research in the Öræfi district
1953	Sigurður Þórarinsson presents convincing evidence for his theory that a drop in pressure caused by the lowering of water levels in Grímsvötn leads to volcanic eruptions. Such eruptions have indeed often marked the end of jökulhlaups
1954–1955	Sigurjón Rist is the first hydrologist to measure speed and volume of jökulhlaups in the rivers Skaftá and Skeiðará. Research on outburst floods in the Skaftá begins
1955–1956	Danes, Americans and Icelanders work at a full survey of Iceland using system of triangulation points all over the island
1956	Hydrologist Sigurjón Rist produces a runoff map of Iceland
1960	Extensive measuring of snow levels on Vatnajökull led by Sigurjón Rist
1960–1961	First gravity measurements used to gauge thickness of Vatnajökull glacier
1960–1970	Glaciers cease retreating and some even begin to advance again
1961	Geologist Þorleifur Einarsson presents a systematic overview of advancing glaciers in Iceland during the Older Dryas period 12,200–11,900 years ago. He also publishes a ground-breaking work on vegetation history
1964	Meteorologist Adda Bára Sigfúsdóttir publishes first full rainfall map of Iceland
1972	Satellites and Landsat images first utilised in research into glaciers of Iceland
1972	Under the auspices of the Science Institute of the University of Iceland, the constant monitoring of Katla begins with continuously recorded seismic data
1972	Haukur Tómasson puts forward the conjecture that many canyons in Iceland have been created by catstrophic jökulhlaups
1972–1974	The rivers of Skeiðarársandur plain are bridged
1975	Research into the nature of jökulhlaups intensifies. The Englishman John F. Nye and Helgi Björnsson work on this project
1975–2000	Radio echo soundings are made annually in the spring on all the main glaciers of Iceland, Vatnajökull, Hofsjökull, Langjökull and Mýrdalsjökull, in order to map the thickness of the glaciers and the topology beneath them
1977	Beginning of use of portable huts, which can be hauled by snowmobiles, to enable scientists to remain longer on ice caps
1980	Science Institute of the University of Iceland and the National Power Company of Iceland begin collaborating on glacial research and this cooperation has continued ever since
1982	

	Loran-C data used to locate positions for measuring equipment on a glacier. This speeds up radio echo soundings. Satellite measurements are first used to locate specific measuring points
1983	Volcanic eruption at Grímsvötn
1990s	Off-road jeeps first used in glacial measuring expeditions
1991	Expedition from SIUI maps both the subglacial topography and surface of Mýrdalsjökull. A satellite is employed to gauge the location of measuring devices that are hauled up onto the glacier. (GPS-measurements)
1992	Systematic measurements of mass balances commence on Vatnajökull in a collaborative effort of the SIUI and the NPCI, led by Helgi Björnsson
1992	Freysteinn Sigmundsson and Páll Einarsson begin calculating land rise due to thinning of glaciers
1994	SIUI first sets up automated meteorological observation posts on Vatnajökull and a model developed to calculate connection between weather and mass balance of glaciers
1994	Shallow magma chamber found beneath Mýrdalsjökull using seismic data. Use of helicopters for glacial measurements begins
1995	Ash/tephra layers from volcanic eruptions used to date age of glaciers on snouts of Vatnajökull
1995	Centre of mantle diapir located beneath northwestern Vatnajökull
1995	Interferometric measurements are made of glaciers from satellites every 24 hours and are used to evaluate glacial movements. (Tandem, ERS)
1995	Research into oceanic sediment layers off Iceland advances knowledge of climate changes in the last ice age
1996	Volcanic eruption in Jökulbrjót (Gjálp) north of Grímsvötn, and a jökulhlaup on 5 November sweeps away the bridge over the Gígjukvísl on Skeiðarársandur
1998	Volcanic eruption in Grímsvötn
1998	Exact altitude measurements made of western part of Vatnajökull with radar from an aircraft (SAR); interferometric radar data (InSAR) from satellites and aircraft are also used in producing contour maps of glaciers' surfaces to evaluate their movements
2001	Numerical models created in connection with mass balance and flow of ice on glaciers to evaluate their response to climate change. Forecasts produced which predict that most of Iceland's glaciers might disappear within the next 200 years
2004	Volcanic eruption in Grímsvötn. Measurements of land uplift, led by Freysteinn Sigmundsson, used to forecast eruption
2005	Systematic measurements begin of mass balance of Drangajökull by the Hydrological Service Division of the NEAI
2008–2009	International Polar Year. Glaciologists unite in creating an accurate map of glaciers from laser measurements from aircraft (LiDAR)

2010	Volcanic eruption of Eyjafjallajökull causes massive disruption to air traffic over Atlantic and within Europe
2011	Volcanic eruption in Grímsvötn causes some air traffic disruption from 22 to 25 May in northwestern Europe
2013	Completion of LiDAR mapping of all glaciers of Iceland
2015	Volcanic eruption at Holuhraun, in northern forefield of Dyngjujökull

Appendix B
Surface Areas and Volumes of Several Glaciers Until the Year 2000. See Appendix C for 2014 Figures

GLACIER (year)	Area (km²)	Volume (km³)	Average thickness (m)	Surface: height differential; ave. height. m.a.s.l.	Height differential; ave. height. m. a.s.l.	Max. thickness of ice:
Vatnajökull (~2000)	8100	3100	380	0–2110; 1215	−300–2110; 670	~950
Langjökull (~2000)	900	190	210	430–1440; 1080	410–1330; 865	~650
Hofsjökull (~2000)	890	200	225	620–1790; 1245	470–650; 1020	~760
Mýrdalsjökull (~2000)	590	140	240	120–1510; 995	15–1375; 760	~750
Drangajökull (~1990)	160		~150	60–190; 665		~260
Eyjafjallajökull (~1990)	80			190–1635; 1135		
Tungnafellsjökull (~1990)	43			920–1520; 1335		
Þrándarjökull (~1945)	27			760–1235; 1035		
Eiríksjökull (~1990)	24			550–1665; 1390		
Þórisjökull (~1990)	22			770–1325; 1165		
Tindfjallajökull (~1990)	20			650–1450; 1190		
Snæfellsjökull (~2000)	16	0.5	~45	600–1440; 980	600–1440; 950	~100
Torfajökull (~1990)	13			780–1170; 1015		
Hrútfell (~1990)	8			690–1400; 1205		
Hofsjökull í Lóni (~1990)	6			870–1145; 1045		

(continued)

© Atlantis Press and the author(s) 2017
H. Björnsson, *The Glaciers of Iceland*, Atlantis Advances
in Quaternary Science 2, DOI 10.2991/978-94-6239-207-6

(continued)

GLACIER (year)	Area (km²)	Volume (km³)	Average thickness (m)	Surface: height differential; ave. height. m.a.s.l.	Height differential; ave. height. m. a.s.l.	Max. thickness of ice:
Gljúfurárjökull (~1990)	3			610–1355; 1015		250
Small glaciers in north	190					
Total:	~11,100	~3650				

Appendix C
Surface Areas of Glaciers According to Land Maps, Aerial Photographs and Satellite Images (Until 2014)

	Aerial photos satellite/landsat 8 LM	Satellite IES/NEAI	Satellite DMA map	Landsat aerial photos	Aerial photos AMS map	Map DGS/DGI
	2014 (km^2)	1998–2004 (km^2)	1979–1990 (km^2)	1960/19733 (km^2)	1945–46 (km^2)	1903–1938 (km^2)
Vatnajökull	7764	8192	8130	8300	8276	8538
Langjökull	868	926	932	953	982	1022
Hofsjökull	827	890	907	925	943	996
Mýrdalsjökull	542	598	594	596	642	701
Drangajökull	145	146	159	160	165	199
Eyjafjallajökull	70	80	79	78	84	107
Tungnafellsjökull	33	38	42	48	40	50
Þórisjökull	25	30	30	32	33	33
Eiríksjökull	21	22	24	22	22	23
Þrándarjökull	16	17	–	22	27	27
Tindfjallajökull	11	15	19	19	15	27
Torfajökull	10	11	13	15	16	21
Snæfellsjökull	10	12	–	11	16	22
ICELAND total:	10,462			11,922		~12,000

© Atlantis Press and the author(s) 2017
H. Björnsson, *The Glaciers of Iceland*, Atlantis Advances
in Quaternary Science 2, DOI 10.2991/978-94-6239-207-6

Appendix D
Surface Area (km²) of Individual Outlet Glaciers of Vatnajökull, Langjökull, Hofsjökull and Mýrdalsjökull

Vatnajökull	
Size of ice cap ca. 2000	8190
Síðujökull	401
Skaftárjökull	144
Tungnaárjökull	361
Sylgjujökull	151
Vestari-Skaftárketill	23
Eystri-Skaftárketill	33
Köldukvíslarjökull	314
Bárðarbunga glaciers	91
Kistujökull	89
Dyngjujökull	1053
Gjálp, Jökulbrjótur	71
Grímsvötn	132
Brúarjökull	1601
Eyjabakkajökull	113
Vesturdalsjökull	37
Axarfellsjökull	32
Lambatungnajökull	39
Svínafells- and Hoffelsjökull	212
Viðborðsjökull	36
Fláajökull	176
Heinabergs & Skálafellsjökull	214
Breiðamerkurjökull	1401
Grænalónsjökull	40
Djúparjökull	52
Langjökull	
Size of ice cap 2004	908
Eystri-Hagafellsjökull	111

(continued)

© Atlantis Press and the author(s) 2017
H. Björnsson, *The Glaciers of Iceland*, Atlantis Advances
in Quaternary Science 2, DOI 10.2991/978-94-6239-207-6

(continued)

Langjökull	
Size of ice cap 2004	908
Vestari Hagafelssjökull	137
Lónsjökull	22
Svatárjökull, Kristleifsjökull	39
Flosaskarð glaciers	66
Þrístapajökull	104
Þjófadalajökull	24
Leiðarjökull	86
Kirkjujökull	23
Norðurjökull	62
Suðurjökull	54
Hofsjökull	
Size of ice cap 1983	923
Blágnípujökull	39
Blöndujökull	61
Kvíslajökull	66
Sátujökull	135
Illviðrajökull	59
Háöldujökull	55
Þjórsárjökull	193
Nauthaga- and Múlajökull	94
Blautkvíslarjökull	74
Mýrdalsjökull	
Size of ice cap 1991	599
Sólheimajökull	46
Entujökull	56
Sléttujökull	130
Öldufellsjökull	41
Sandfellsjökull	66
Kötlujökull	148

Appendix E
Computer-Generated Image
of Subglacial Base of Vatnajökull

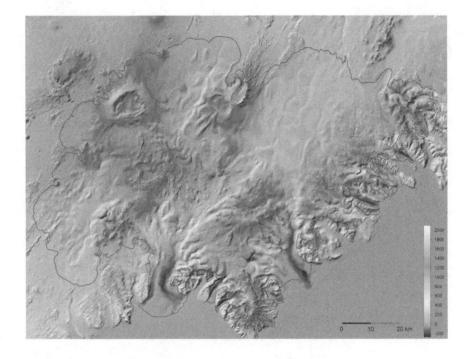

Colour differentials indicate heights measured from sea level (m.)

© Atlantis Press and the author(s) 2017
H. Björnsson, *The Glaciers of Iceland*, Atlantis Advances
in Quaternary Science 2, DOI 10.2991/978-94-6239-207-6

Appendix F
Katla Jökulhlaups since Settlement of Iceland

Year	Eruption/beginning of flood	Duration (days)
1999	17 July	1
1955	25 July	1
1918	12 October	24
1860	8 May	20
1823	26 June	28
1755	17 October	ca. 120
1721	11 May	>100
1660	3 November	>60
1625	2 September	13
1612	12 October	
1580	11 August	
~1500		
15th cent.		
1440		
1416		
~1357		
1262		
1245		
12th cent.		
~1179		
~934		
~920		
900?		
9th cent.		
9th cent.		
8th cent.		

Sources

Larsen, G. (2000) Holocene eruptions within the Katla volcanic system, Iceland: Notes on characteristics and environmental impact. *Jökull*. 49. pp. 1–28.

Larsen, G. (2002) A brief overview of eruptions from ice-covered and ice-capped volcanic systems in Iceland during the past 11 centuries: frequency, periodicity and implications. In Smellie, J. and Chapman M. (eds.) *Ice-Volcano Interaction on Earth and Mars*. London: Geological Society of London Special Publication 202. pp. 81–90.

© Atlantis Press and the author(s) 2017

H. Björnsson, *The Glaciers of Iceland*, Atlantis Advances in Quaternary Science 2, DOI 10.2991/978-94-6239-207-6

Appendix G
Computer-Generated Image
of Subglacial Base of Mýrdalsjökull

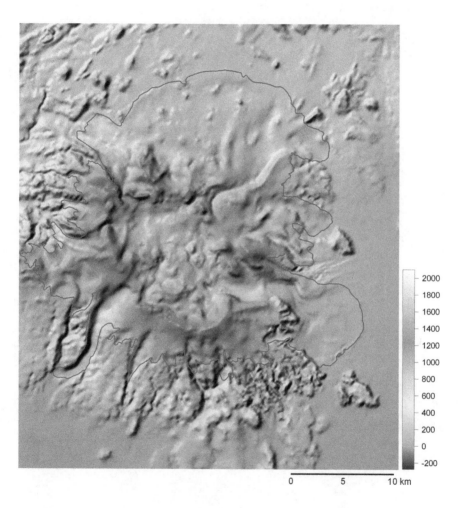

Colour differentials indicate heights measured from sea level (m.)

© Atlantis Press and the author(s) 2017
H. Björnsson, *The Glaciers of Iceland*, Atlantis Advances
in Quaternary Science 2, DOI 10.2991/978-94-6239-207-6

Appendix H
Jökulhlaups from Eystri-Skaftárketill ice cauldron (Vatnajökull)

Year	Month	Max. discharge (m^3/s)
1955	September	755
1957	May	650
1960	September	205
1964	March	885
1966	October	1040
1970	January	1380
1972	July	1090
1974	December	1270
1977	February	1160
1979	September	1220
1982	January	1140
1984	August	1560
1986	November	1308
1989	July	1288
1991	August	1120
1992	September	167
1995	July	1994
1995	October	96
1997	July	921
2000	August	1240
2002	September	689
2003	November	241
2006	April	1370
2008	October	1350
2010	June	1280
2015	September	3000

© Atlantis Press and the author(s) 2017
H. Björnsson, *The Glaciers of Iceland*, Atlantis Advances
in Quaternary Science 2, DOI 10.2991/978-94-6239-207-6

Appendix I
Jökulhlaups from Vestari-Skaftárketill Ice Cauldron (Vatnajökull)

Year	Month	Max. discharge (m^3/s)
1968	September	110
1971	July	658
1973	December	238
1975	September	307
1977	August	545
1980	January	444
1981	August	472
1983	September	222
1986	July	418
1988	August	394
1990	October	129
1994	August	968
1996	August	672
1997	July	335
1998	September	292
2000	August	699
2002	July	720
2003	September	436
2005	August	723
2006	September	194
2008	August	390
2010	June	550

© Atlantis Press and the author(s) 2017
H. Björnsson, *The Glaciers of Iceland*, Atlantis Advances
in Quaternary Science 2, DOI 10.2991/978-94-6239-207-6

Appendix J
Computer-Generated Image
of Subglacial Base of Langjökull

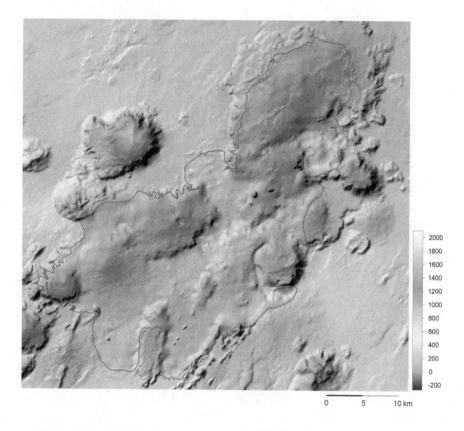

Colour differentials indicate heights measured from sea level (m.)

© Atlantis Press and the author(s) 2017
H. Björnsson, *The Glaciers of Iceland*, Atlantis Advances
in Quaternary Science 2, DOI 10.2991/978-94-6239-207-6

Appendix K
Computer-Generated Image
of Subglacial Base of Hofsjökull

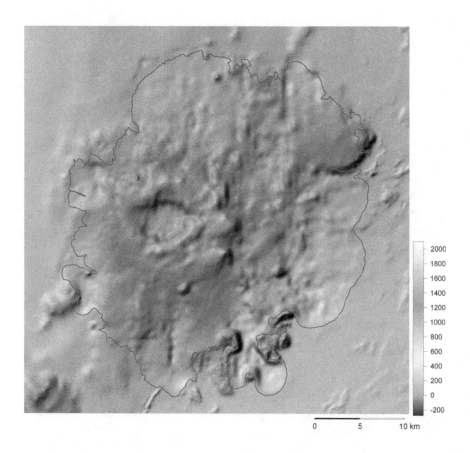

Colour differentials indicate heights measured from sea level (m.)

© Atlantis Press and the author(s) 2017
H. Björnsson, *The Glaciers of Iceland*, Atlantis Advances
in Quaternary Science 2, DOI 10.2991/978-94-6239-207-6

Appendix L
Recorded Volcanic Eruptions Beneath Vatnajökull

Year	Volcano	Jökulhlaup
~900	Bárðarbunga, Veidivotn, fissure swarm[f]	
~905	Grímsvötn system[g]	
~940	Bárðarbunga system	
1000	Uncertain[g]	
1060	Grímsvötn system[g]	
1080	Bárðarbunga system[g]	
1159	Bárðarbunga system[g]	
1332	(Grímsvötn)	(Skeiðará)[a]
1341	(Grímsvötn)[a]	
1354	Grímsvötn system[g]	
1362	Öræfajökull	Several rivers[c]
1410	Bárðarbunga system[g]	
1477	Bárðarbunga, Veiðivötn, fissure swarm	Jökulsá á Fjöllum[f,g]
~1500	Grímsvötn system[g]	
1598	Grímsvötn[a]	
1603	(Grímsvötn system)[a,g]	
1619	Grímsvötn[a, g]	
1629	Grímsvötn	Skeiðará[a]
1638	(Grímsvötn)[a]	
1659	Grímsvötn	Skeiðará[a,q]
1681	Unknown[a]	
1684–85	Grímsvötn, (Dyngjujökull)	Skeiðará, Jökulsá á Fjöllum[a,b]
1697	Bárðarbunga system[q]	
1702?	Unknown[a]	
1706	(Grímsvötn), Bárðarbunga system[a,q]	
1707	Bárðarbunga system[q]	
1711–12	Bárðarbunga system	Jökulsá á Fjöllum[a,b,c]
1716	(Grímsvötn), Bárðarbunga system	Jökulsá á Fjöllum[a,b,q]

(continued)

© Atlantis Press and the author(s) 2017
H. Björnsson, *The Glaciers of Iceland*, Atlantis Advances
in Quaternary Science 2, DOI 10.2991/978-94-6239-207-6

(continued)

Year	Volcano	Jökulhlaup
1717	Bárðarbunga system	Jökulsá á Fjöllum[a,b,g,q]
1720	Bárðarbunga system[q]	
1725	Grímsvötn	Skeiðará[a]
1726	Dyngjujökull (near Grímsvötn)	Jökulsá á Fjöllum[a,b]
1727	Öræfajökull	Several rivers[a]
1729 ?		Jökulsá á Fjöllum[b]
1739	Bárðarbunga system[q]	
1753	Siðujökull	Djupá, Hverfisfljót, Skaftá
1766	Bárðarbunga system	Þjórsá[a,q]
1768	Uncertain[q]	
1769	Grímsvötn or Bárðarbunga system[q]	
1774	Grímsvötn	Skeiðará[a,q]
1783	Grímsvötn, Laki, fissure swarm	Skaftá, Þjórsá[a,r]
1784	Grímsvötn system	Núpsvötn, Skeiðará[a,q,r,t]
1794?	Western Vatnajökull[a]	
1797	NW Vatnajökull (Dyngjuháls)[a,o]	
1807?	(NW Vatnajökull)[a]	
1816	(Grímsvötn)[j]	
1823	Grímsvötn, Þórðarhyrna[a,q]	
1838	Grímsvötn	Skeiðará[a,q]
1854	Grímsvötn system[q]	
1861?	(Grímsvötn)	Skeiðará[a]
1862–64	Tröllagígar, Bárðarbunga system[a]	
1867	Grímsvötn, Háabunga, Þórðarhyrna	Skeiðará[a]
1872 ?	(Dyngjuháls)[a]	
1873	Grímsvötn, (Þórðarhyrna)	Skeiðará, Djúpá[a]
1883	Grímsvötn Skeiðará[a]	
1883	Grímsvötn, (Kverkfjoll)	Skeiðará[a,q]
1887	(Þórðarhyrna)	Súla[a,q]
1892	Grímsvötn	Skeiðará[a,q,s]
1897	(Grímsvötn)[a]	
1902–03	(Dyngjuháls)	Skjálfandafljót, Jökulsá á Fjöllum[a,b]
1903	Þórðarhyrna, (Grímsvötn)	Skeiðará, Sula[a,h]
1903 ?	Unknown	Jökulsá á Brú[a]
1910	Eystri Skaftárketill[a,m]	
1922	Grímsvötn	Skeiðará, Súla[a,q]
1927	(Esjufjöll)	Jökulsá á Breiðamerkursandur[n]
1933	Unknown	Skjálfandafljót?[d,i,h]
1933	North of Grímsvötn[d,i,h]	

(continued)

(continued)

Year	Volcano	Jökulhlaup
1934	Grímsvötn	Skeiðará, Súla[a,q], Skjálfandafljót, Jökulsá á Fjöllum[d,i]
1938	North of Grímsvötn, Jökulbrjótur	Skeiðará, Súla[a]
1939 ?	Grímsvötn	Skeiðará[h,p]
1941 ?	Grímsvötn	Skeiðará[h]
1945 ?	Grímsvötn	Skeiðará[e,h]
1954 ?	Grímsvötn	Skeiðará[e,h]
1983	Grímsvötn[k,l]	
1984 ?	Grímsvötn	
1986 ?	Eystri-Skaftárketill	Skaftá
1996	North of Grímsvötn, Gjálp, Jökulbrjótur	Skeiðará
1998	Grímsvötn	Skeiðará
2004	Grímsvötn	Skeiðará
2011	Grímsvötn	
2014	Bárðarbunga region/Holuhraun[u]	

Sources [a]Þórarinsson (1974b); [b]Þórarinsson (1950); [c]Þórarinsson (1958); [d]Áskelsson (1936a); [e]Áskelsson (1959); [f]Larsen (1984); [g]Larsen (1982); [h]Jóhannesson, Hk. (1983a); [i]Jóhannesson, Hk. (1984); [j]Jóhannesson, Hk. (1987); [k]Grönvold and Jóhannesson (1984); [l]Einarsson and Brandsdóttir (1984); [m]Jónsson, J. (1986); [n]Björnsson, S. (1978); [o]Jónsson, Ó. (1945); [p]Stefánsson (1983); [q]Steinþórsson (1977); [r]Þórðarson (1990); [s]Björnsson, H. (1988); [t]Þórarinsson (1974b); [u]See IES and IMO website. (?) Eruption uncertain. ( ) Location uncertain. Full references at end of Chap. 8.

Appendix M
Grímsvötn: Surface Area and Volume of Grímsvötn and Grímsvötn jökulhlaups; Thickness of Grímsvötn Ice Cover; Water Levels Before and After Outburst Floods; Volume of Water Before and After Floods and Maximum Rates of Discharge (1903–2008)

© Atlantis Press and the author(s) 2017

H. Björnsson, *The Glaciers of Iceland*, Atlantis Advances
in Quaternary Science 2, DOI 10.2991/978-94-6239-207-6

Year	Month	Surface area km²	Ice th. m.	Water lev. bef. flood m.a.s.l.	Water lev. aft. flood m.a.s.l.	Drop in water lev m.	Vol. bef. fld km³	Vol. aft. fld km³	Vol. fr. Grimsv. km³	Vol. of flood. km³	Max. km³ 10³ m³/s
1903	May										
1913	April										
1922	September										
1934	March	37	150	1440	1280	160	4.9	0.6	4.3	4.5	25–30
1938	May	27+15	150	1460	1300	100	5.4	0.9	4.5	4.7	25–30
1939	June	<25	150	1300	1300	40	1.9	0.9	1	1	–
1941	April	<25	150	1360	1305	55	2.3	1	1.3	1.4	–
1945	September	32	150	1425	1325	100	3.7	1.2	2.5	2.6	8–10
1948	February	27	150	1400	1310	90	3	0.9	2.1	2.2	–
1954	July	33	150	1435	1305	130	3.8	0.8	3	3.2	10
1960	January	30	155	1428	1338	90	3.4	1.4	2	2.1	5–6
1965	September	29	175	1432	1317	115	3.1	0.7	2.4	2.5	6
1972	March	27	195	1436	1330	106	2.6	0.6	2	2.1	5
1976	September	26	210	1439	1350	89	2.3	0.7	1.6	1.7	3.5–4
1982	February	22	230	1447	1380	67	1.9	0.75	1.15	1.2	2
1983	December	15	230	1412	1370	42	1.2	0.6	0.6	0.6	0.6
1986	August	17	230	1430	1350	80	1.4	0.3	1.1	1.15	2
1991	November	22	230	1452	1370	82	1.85	0.45	1.4	1.45	2
1996	April	23	230	1454	1379	75	1.5	0.4	1.1	1.15	3
1996	November	40	230	1510	1335	175	3.4	0.2	3.2	3.6	40–50
1998	Feb.–March	13		1407	1348	59	0.051	0.05	0.46		
1999	January	10		1390	1338	52	0.30	0.03	0.27		

(continued)

(continued)

Year	Month	Surface area km²	Ice th. m.	Water lev. bef. flood m.a.s.l.	Water lev. aft. flood m.a.s.l.	Drop in water lev m.	Vol. bef. fld km³	Vol. aft. fld km³	Vol. fr. Grimsv. km³	Vol. of flood. km³	Max. km³ 10³ m³/s
1999	Sept.–October	9.7		1386	1349	37	0.27	0.05	0.22		
2000	July–August	5.5		1369	1350	19	0.12	0.05	0.07		
2001	December	11.6		1397	1391	6	0.38	0.31	0.08		
2002	Feb.-April	12		1399	1361	38	0.41	0.09	0.32		
2004	Oct.–November	16.5		1422	1378	44	0.73	0.19	0.65		~4
2005	March	9.6		1385	1361	24	0.56	0.09	0.47		
2007	October	12		1400	1372	28	0.42	0.15	0.27		
2008	Sept.–October	11		1391	1369	22	0.32	0.13	0.19		

Afterword

What I hope to have achieved with my *The Glaciers of Iceland* is twofold. Firstly, I wanted to present a comprehensive overview, for both laymen and scientists, of the origins, history and current size and condition of all of Iceland's major glaciers (including Vatnajökull, the largest in Europe) at the beginning of the twenty-first century. To this end I have not only used photographs, illustrations, tables, graphs of recent statistics and scientific data, as well references from hundreds of academic books and articles, but also referred to historical writings and drawings from annals, sagas, folk tales, diaries, reports, stories and poems, in an attempt to present a unique approach to the study of glaciers on an island in the North Atlantic. By balancing and comparing the world of man with the world of nature, the perceptions of art and culture with the systematic and pragmatic analyses of science, I have tried to make *The Glaciers of Iceland* present readers with a new and stimulating view of the origins, development and possible future of these massive, mysterious and magical phenomena.

Secondly, I hope that the accumulation of this knowledge on the history and current status of Icelandic glaciers in the early years of a new millennium will provide an essential database for further comparisons dealing with the expansion or retreat of glaciers in a worldwide context. I believe that Iceland can prove to be an invaluable laboratory for the continuing study of temperate glaciers and could provide possible lessons and models for man's future interaction with these colossi of nature in our vital need to understand and predict the current unsettling progress of global warming.

Helgi Björnsson, Professor Emeritus, University of Iceland.

© Atlantis Press and the author(s) 2017
H. Björnsson, *The Glaciers of Iceland*, Atlantis Advances
in Quaternary Science 2, DOI 10.2991/978-94-6239-207-6

CPSIA information can be obtained
at www.ICGtesting.com
Printed in the USA
LVHW081115170319
610947LV00003B/29/P